AIR POLLUTION

Health and Environmental Impacts

AIR POLLUTION

Health and Environmental Impacts

Edited by

BHOLA R. GURJAR
LUISA T. MOLINA
CHANDRA S. P. OJHA

Foreword by Dr. Mario J. Molina

CRC Press
Taylor & Francis Group
Boca Raton   London   New York

CRC Press is an imprint of the
Taylor & Francis Group, an **informa** business

CRC Press
Taylor & Francis Group
6000 Broken Sound Parkway NW, Suite 300
Boca Raton, FL 33487-2742

© 2010 by Taylor and Francis Group, LLC
CRC Press is an imprint of Taylor & Francis Group, an Informa business

No claim to original U.S. Government works

Printed in the United States of America on acid-free paper
10 9 8 7 6 5 4 3 2 1

International Standard Book Number: 978-1-4398-0962-4 (Hardback)

Library of Congress Cataloging-in-Publication Data

Air pollution : health and environmental impacts / editors, Bhola R. Gurjar, Luisa T.
    Molina, and C.S.P. Ojha.
        p. cm.
    Includes bibliographical references and index.
    ISBN 978-1-4398-0962-4 (hard back : alk. paper)
    1. Air--Pollution--Health aspects. 2. Air--Pollution--Environmental aspects. I.
    Gurjar, B. R. II. Molina, Luisa T. III. Ojha, C. Shekhar P. IV. Title.

RA576.A49 2010
363.739'2--dc22   2009053436

Visit the Taylor & Francis Web site at
http://www.taylorandfrancis.com

and the CRC Press Web site at
http://www.crcpress.com

To "Those Committed to Improving Air Quality across the World"

Contents

SECTION I Air Pollution Monitoring and Modeling

SECTION II Air Pollution and Health Effects

SECTION III   Health Risk Assessment and Management

SECTION IV Air Quality Management: Techniques and Policy Aspects

SECTION V Environmental Impacts of Air Pollution

Foreword

Human activities in an increasingly globalized, industrialized, and interconnected world are influencing both air quality and climate change at urban and regional and even continental and global scales. Rapid population growth and increased energy demand are the primary forces causing large quantities of harmful pollutants and greenhouse gases to be emitted into the atmosphere, resulting in serious human health and environmental consequences.

Substantial progress has been made over the past few decades to prevent and control air pollution in many parts of the world through a combination of technology improvements and policy measures. Many countries have clean air laws that set emission and ambient air quality standards to protect public health and the environment. These laws have often been successful in both developed and developing countries. However, increasing human activities are offsetting some of the gains and millions of people are being exposed to harmful levels of air pollutants. Air pollution is especially a problem in many cities of the developing world that are producing goods for the global economy. Also, with rising affluence in these cities, there has been high growth in private car ownership, resulting in increased congestion and pollution. In principle, the problem can be solved by using clean technologies. In practice, however, there are large socioeconomic and political barriers.

This book is focused on the complex problem of controlling air pollution and mitigating its adverse effects on human health and the environment. It is written by leading experts in the field and addresses many aspects of air pollution, including monitoring and source characterization of air pollution, the theory and application of air quality modeling, health effects and risk assessment, air quality management, and relevant policy issues. The book ends with a regional and global perspective on air pollution.

The primary driving force for the design and implementation of emission-control strategies aimed at improving air quality has been the protection of human health within a local or regional area. However, with the growth of multicity "megalopolis" regions in many parts of the world, long-range transport of air pollutants has become a major concern. The regional and global dispersion of pollutants generated locally has been well established in the case of acid deposition and stratospheric ozone depletion. Recently, the long-range transport of tropospheric ozone has increased throughout the northern hemisphere, air pollution and wildfire emissions originating from northern mid-latitudes have given rise to Arctic haze, and atmospheric brown clouds of tiny aerosol particles from anthropogenic emissions have been observed in many regions of the world. The accumulation and dispersion of pollutants, such as tropospheric ozone and airborne particulate matter, not only affect human health and the ecosystem on a local and regional scale, but also influence air quality and the Earth's climate on a global scale. It is clear that air pollution and climate change are intricately interconnected in terms of sources and effects and should be addressed under one common framework.

The means to make rapid progress to improve air quality and mitigate climate change exist, but strong political leadership with effective multistakeholder participation will be essential to achieve this goal. Regional and international cooperation, with suitable mechanisms for facilitating technology transfer, financial resources, and the strengthening of human and institutional capacities, will be necessary to accelerate implementation of emissions reduction strategies around the world.

Air Pollution: Health and Environmental Impacts provides an invaluable and timely contribution to the urgent environmental challenges facing our society today. As illustrated in the book, we need integrated, interdisciplinary approaches to address the effects of human-induced activities.

Mario J. Molina
San Diego, California

Preface

As a consequence of ever-growing anthropogenic activities and interventions, we humans greatly influence different components of the environment (e.g., air, water, and land resources) and get influenced in return. For example, household, workplace, outdoor, and transportation environments pose risks to human health in several different ways (e.g., people involuntarily breathing in poor ambient air). Throughout the world, therefore, reducing adverse effects attributed to environmental exposures is an important public welfare objective aimed at accruing significant societal benefits.

Air pollution is recognized as one of the leading contributors to the global environmental burden of disease. There is extensive scientific evidence of adverse health effects even in countries with relatively low concentrations of air pollution. Air pollution damages terrestrial and aquatic resources, including those of direct economic importance. It is also interwoven with the causes and consequences of global-scale climate change and many other local-scale environmental pressures that confront society, such as poor ambient air quality. In a nutshell, air pollution has strong impacts on both public health and the environment and thus deserves a holistic perspective and integrated policy programs to address the concerned issues. With this perspective, we have produced the present book that has bottom-line information as well as expert knowledge for large audiences, including students, teachers, scientists, policy formulators, executives, engineers, and technocrats dealing with the subject of air pollution and its effects. To meet this objective, we planned and designed the book in such a way that it covers not only the fundamentals of the air pollution problem but also includes field studies and cases from different parts of the world. Thus, the basic premises of the air pollution problem and region-specific uniqueness find a place together in this single book. To help readers appreciate and comprehend the complex problem of air pollution and its adverse effects on human health and the environment in totality, the book's chapters cover almost all aspects of air pollution, for example, monitoring and source characterization of air pollution, modeling, health effects, environmental impacts, risk assessment, air quality management, and relevant policy issues. For an easy grasp of the subject, the book has been divided into five major sections, namely, Air Pollution Monitoring and Modeling; Air Pollution and Health Effects, Health Risk Assessment and Management; Air Quality Management: Techniques and Policy Aspects; and Environmental Impacts of Air Pollution. We have structured the book so that it can act as a primer for students and also as a reference source for researchers and academics working in the field of air pollution. We have made careful efforts to keep the book free from typos and other errors; yet there is a possibility that some gaps may be identified by attentive readers. We welcome feedback so as to incorporate constructive suggestions and comments in future editions of the book.

Bhola R. Gurjar, Luisa T. Molina, and Chandra S.P. Ojha

Acknowledgments

First and foremost, we thank our institutions for providing us conducive working environment to prepare and edit this book. We are thankful to all the contributors who invested their valuable time and intellectual efforts to writing chapters. A number of reviewers helped us greatly improve the manuscript through their comments and suggestions related to individual chapters. We thank all of them, including John C.Y. Chan, Mukesh Khare, Patricia Tord, Juan Felipe Franco Ramírez, Manjola Banja, Sharad Lele, Ravindra Khaiwal, Suman Mor, Charles Kolb, Robert Slott, Linsey Marr, Gerardo Mejia, Benjamin de Foy, and George Hidy. We thank those friends and colleagues, especially Samudra Vijay, who were a constant source of motivation. We thank several researchers, particularly Ajay Singh Nagpure, who willingly helped us format different chapters and put the manuscript in order. Last but not least, we express our deep gratitude to the publishing team (particularly Joseph Clements and Jennifer Ahringer) of CRC Press (Taylor & Francis Group), whose full cooperation and patience kept our spirits high enough to meet the various challenges during the preparation and editing of the book.

Bhola R. Gurjar, Luisa T. Molina, and Chandra S.P. Ojha

Editors

Bhola R. Gurjar is on the faculty of the Civil Engineering Department of India's premier technological institution, the Indian Institute of Technology—IIT Roorkee (URL: http://www.iitr.ac.in/~CE/bholafce). He is also Core Faculty in IIT Roorkee's Centre for Transportation Systems and Centre of Excellence in Disaster Mitigation & Management, and is heading the Max Planck Partner Group for Megacities & Global Change. He holds a PhD in environmental risk analysis from IIT Delhi, and has extensive industrial, teaching, training, and research experience. His present research interests include megacities, air pollution; environmental impact and risk assessment; atmospheric emissions and climate change; and the integrated cross-disciplinary study of science and policy issues of the environment, health, energy, economy, entrepreneurship, technology, infrastructure, and resources—particularly from global change, sustainable development, and risk governance perspectives.

Dr. Gurjar is coauthor/coeditor of six books, and has a number of research papers to his credit. He has received several awards and fellowships, including the prestigious Advanced Postdoctoral Research Fellowship of the Max Planck Society (Germany) to support his research tenure (2002–2005) at the Max Planck Institute for Chemistry (MPIC) in Mainz, Germany. On the basis of his outstanding research work at MPIC—Mainz, the Scientific Steering Committee of the global change SysTem for Analysis, Research and Training (START), Washington, DC, USA, honored him with the 2004 START Young Scientist Award. He is also corecipient of The Nawab Zain Yar Jung Bahadur Memorial Medal (best research paper award) from the Environmental Engineering Division of The Institution of Engineers (India), Kolkata, for the year 1995–1996.

Luisa T. Molina is currently the president of the Molina Center for Strategic Studies in Energy and the Environment (MCE2) in La Jolla, California and principal research scientist at the Department of Earth, Atmospheric, and Planetary Sciences at the Massachusetts Institute of Technology (MIT). After completing her PhD in chemistry from the University of California at Berkeley, she held teaching and research positions at the Santa Cruz and Irvine campuses of the University of California and the Jet Propulsion Laboratory of the California Institute of Technology. She joined MIT in 1990, where she conducted research in atmospheric chemistry and also served as the executive director of the integrated program on urban, regional, and global air pollution.

Dr. Molina's research interests include molecular spectroscopy, chemical kinetics, and atmospheric chemistry. She has been involved in particular with the chemistry of stratospheric ozone depletion and urban air pollution. She demonstrated experimentally a new reaction sequence that explains how chlorofluorocarbons (CFCs) caused the Antarctic ozone hole. Recently, she initiated a multidisciplinary project involving an integrated assessment of air pollution in megacities, aimed at improving the environmental decision-making process through education and the better use of scientific, technical, and socioeconomic understanding. In the spring of 2003, she led a team of researchers from Mexico, the United States, and Europe to investigate air quality in the Mexico City Metropolitan Area (MCMA). In March 2006, she coordinated the MILAGRO campaign in Mexico City, the first international scientific project to study megacity air pollution and its regional and global impacts. She is author/coauthor of over 100 archival publications, editor/lead author of the book entitled *Air Quality in Mexico Megacity: An Integrated Assessment*, and serves as guest editor for several special issues on field studies in *Atmospheric Chemistry and Physics*. Dr. Molina has organized and served on many panels and committees addressing global change issues. She has received several awards for her work, including the Volvo Prize in environment in 2004.

Chandra S.P. Ojha is a professor in the Civil Engineering Department of India's premier technological institution—IIT Roorkee. He holds a PhD in civil engineering from Imperial College of Science, Technology and Medicine, London, UK, where he was a Commonwealth Research Scholar (October 1990 to September 1993). He was also a visiting scholar at Louisiana State University, USA (April–July 2000); an Alexander Von Humboldt fellow at Water Technology Center, Karlsruhe, Germany (December 2001 to July 2002); a guest AvH fellow at the Institute for Hydromechanics, University of Karlsruhe, Germany (January 2002 to July 2002); a visiting professor of civil engineering at AIT Bangkok (August 2004 to November 2004); and a Distinguished visiting fellow of Royal Academy of Engineering at Heriot-Watt University, UK (December 2008 to January 2009).

He specializes in the modeling of environmental and water resources systems. He has published about 100 research papers in peer-reviewed journals and supervised/co-supervised more than 25 PhD theses. He has also coordinated/co-coordinated several international research projects. For example, he was the Indian Project Coordinator for the EU-India River Bank Filtration Network in 2005–2006 and the Principal Indian Investigator for the McGill India Strategic Research Initiative Project in 2007–2008.

In addition to the highly prestigious Alexander Von Humboldt fellowship in 2001–2002 and the distinguished visiting fellowship of Royal Academy of Engineering in 2008–2009, Dr. Ojha received the Young Engineer Award from the Central Board of Irrigation and Power in India in 1996, the Young Teachers Career Award of All India Council of Technical Education in 1997, and the Star Performer for academic excellence by IIT Roorkee in 2006. His research paper awards include the Professor R.C. Singh Gold Medal in 1989, the Nawab Zain Dar Jung Bahadur Memorial Prize in 1992, and the S.S. Sanyal Medal in 2005 from the Institution of Engineers, India; the Shri S.V. Patwardhan Memorial Prize of the Indian Water Works Association and Jal Vigyan Puraskar of Indian Society of Hydraulics in 2007; and the ASCE research paper awards in 2001, 2009, and 2010.

Contributors

Gustavo Acosta
Hospital Juárez de México-SSA
Mexico City, México

Madhoolika Agrawal
Department of Botany
Banaras Hindu University
Varanasi, India

Ernesto Alfaro-Moreno
Investigación Básica
Instituto Nacional de Cancerología
Mexico City, Mexico

and

Lung Toxicology Unit
Pneumology Section
Katholieke Universiteit Leuven
Leuven, Belgium

Lal P. Amgain
Institute of Agriculture and
    Animal Science
Tribhuvan University
Rampur, Nepal

Rajasekhar Balasubramanian
Division of Environmental
    Science and Engineering
National University of Singapore
Singapore

Salvador Blanco
Centro Nacional de Investigación
    y Capacitación Ambiental-Instituto
    Nacional de Ecología
Mexico City, México

Jeffrey R. Brook
Air Quality Research Division
Atmospheric Science and Technology
    Directorate
Environment Canada
Toronto, Ontario, Canada

Patrick Büker
Stockholm Environment Institute
University of York
York, United Kingdom

Eliseo Cantellano
División de Estudios de Posgrado e
    Investigación
FES Zaragoza-Universidad Nacional
    Autónoma de México
Mexico City, México

Ioan Manuel Ciumasu
Centre of Expertise for Sustainable
    Exploitation of Ecosystems (CESEE)
Alexandru Ioan Cuza University
Iasi, Romania

Naela Costica
European Centre for Training the
    Trainers in Ecological Education
    (CEFFEE)
Alexandru Ioan Cuza University
Iasi, Romania

Lorraine Craig
Network for Environmental Risk
    Assessment and Management
University of Waterloo
Waterloo, Ontario, Canada

Dilek Demirbas
Newcastle Business School
Northumbria University
Newcastle Upon Tyne,
    United Kingdom

Andrea De-Vizcaya-Ruiz
Departamento de Toxicología
CINVESTAV-Instituto Politéchnico
    Nacional
Mexico City, México

Guillermo Elizondo
Sección Externa de Toxicología
CINVESTAV-Instituto Politéchnico
    Nacional
Mexico City, México

Lisa D. Emberson
Stockholm Environment Institute
University of York
York, United Kingdom

Magnuz Engardt
Swedish Meteorological and
    Hydrological Institute
Norrköping, Sweden

Elizabeth Estrada-Muñiz
Sección Externa de Toxicología
CINVESTAV-Instituto Politéchnico
    Nacional
Mexico City, México

Claudia García-Cuellar
Investigación Básica
Instituto Nacional de Cancerología
Mexico City, Mexico

Radha Goyal
Department of Civil Engineering
Indian Institute of Technology
New Delhi, India

Sanjeev K. Goyal
National Environmental
    Engineering Research Institute
    (NEERI)
Council of Scientific & Industrial
    Research, CSIR
Nagpur, India

Marco Guarneros
Instituto de Investigaciones
    Biomédicas-Universidad Nacional
    Autónoma de México
Mexico City, México

Bhola R. Gurjar
Department of Civil Engineering
Indian Institute of Technology
Roorkee, India

Sarath Guttikunda
Division of Atmospheric
    Sciences
Desert Research Institute
Reno, Nevada

Olf Herbarth
Department of Human Exposure
    Research and Epidemiology
UFZ Leipzig
Leipzig, Germany

Vicente Hernández
División de Estudios de Posgrado e
    Investigación
FES Zaragoza-Universidad Nacional
    Autónoma de México
Mexico City, México

Kevin Hicks
Stockholm Environment Institute
University of York
York, United Kingdom

Juan J. Hicks
Departamento de Investigación en
  Bioquímica y Medicina Ambiental
Instituto Nacional de Enfermedades
  Respiratorias
Mexico City, México

Robyn Hudson
Instituto de Investigaciones
  Biomédicas-Universidad Nacional
  Autónoma de México
Mexico City, México

Towhid Islam
Department of Environmental
  Science
Bangladesh Agricultural University
Mymensingh, Bangladesh

Sathrugnan Karthikeyan
Division of Environmental Science
  and Engineering
National University of Singapore
Singapore

Mukesh Khare
Department of Civil Engineering
Indian Institute of Technology
New Delhi, India

Gert H.J. Krüger
School of Environmental Science
North-West University
Potchefstroom, South Africa

Twisha Lahiri
Department of Neuroendocrinology
Chittaranjan National Cancer Institute
Kolkata, India

Anita Lakhani
Department of Chemistry
Dayalbagh Educational Institute
Dayalbagh, India

Wenfang Lei
Molina Center for Energy and
  the Environment
La Jolla, California

Rubén Marroquín
División de Estudios de Posgrado
  e Investigación
FES Zaragoza-UNAM
Mexico City, México

Marcelo Mena
Department of Environmental
  Engineering
Universidad Andres Bello
Santiago, Chile

and

Massachusetts Institute of
  Technology
Cambridge, Massachusetts

Manju Mohan
Centre for Atmospheric Sciences
Indian Institute of Technology
New Delhi, India

Luisa T. Molina
Massachusetts Institute of
  Technology
Cambridge, Massachusetts

and

Molina Center for Energy and the
  Environment
La Jolla, California

Michael D. Moran
Air Quality Research Division
Atmospheric Science and
  Technology Directorate
Environment Canada
Toronto, Ontario, Canada

Prabhakar Nema
National Environmental Engineering
 Research Institute(NEERI)
Council of Scientific & Industrial
 Research, CSIR
Nagpur, India

Nguyen T.K. Oanh
Environmental Engineering and
 Management
School of Environment,
 Resources and Development
Asian Institute of Technology
Pathumthani, Thailand

Chandra S.P. Ojha
Department of Civil Engineering
Indian Institute of Technology
Roorkee, India

Ivonne M. Olivares
Departamento de Investigación en
 Bioquímica y Medicina Ambiental
Instituto Nacional de Enfermedades
 Respiratorias
Mexico City, México

Alvaro R. Osornio-Vargas
Investigación Básica
Instituto Nacional de Cancerología
Mexico City, Mexico

William Pennell
NARSTO
Pasco, Washington

G. Anoma D. Perera
Department of Botany
University of Peradeniya
Peradeniya, Sri Lanka

Dora A. Pérez
División de Estudios de Posgrado e
 Investigación
FES Zaragoza-Universidad Nacional
 Autónoma de México
Mexico City, México

Håkan Pleijel
Applied Environmental Science
Göteborg University
Göteborg, Sweden

Nirat Rajput
Department of Chemistry
Dayalbagh Educational Institute
Dayalbagh, India

Manas Ranjan Ray
Department of Experimental Hematology
Chittaranjan National Cancer Institute
Kolkata, India

Ernesto Reyes
División de Estudios de Posgrado e
 Investigación
FES Zaragoza-Universidad Nacional
 Autónoma de México
Mexico City, México

Tracy Rodríguez
División de Estudios de Posgrado
 e Investigación
FES Zaragoza-Universidad Nacional
 Autónoma de México
Mexico City, México

Leonora Rojas-Bracho
Instituto Nacional de Ecología
Secretaría de Medio Ambiente y
 Recursos Naturales
Mexico City, Mexico

Syed R. A. Shamsi
Department of Botany
University of the Punjab, Q.A. Campus
Lahore, Pakistan

Martha P. Sierra-Vargas
Departamento de Investigación
 en Bioquímica y Medicina
 Ambiental
INER-SSA
Mexico City, México

Constantinos Sioutas
Civil and Environmental Engineering
University of Southern California
Los Angeles, California

Pieter R. Smit
School of Environmental Science
North-West University
Potchefstroom, South Africa

Bo Strandberg
Department of Occupational and
Environmental Medicine
Sahlgrenska Academy at
Göteborg University
Göteborg, Sweden

Yessica Torres-Ramos
Departamento de Investigación
en Bioquímica y Medicina
Ambiental
Instituto Nacional de Enfermedades
Respiratorias
Mexico City, Mexico

Horacio Tovalin
División de Estudios de Posgrado e
Investigación
FES Zaragoza-Universidad Nacional
Autónoma de México
Mexico City, Mexico

Frank Ulrich
Department of Human Exposure
Research and Epidemiology
UFZ Leipzig
Leipzig, Germany

Anna M. van Tienhoven
Department of Geography
Environmental Management
and Energy Studies
University of Johannesburg
Johannesburg, South Africa

Libia Vega
Sección Externa de Toxicología
CINVESTAV-Instituto Politéchnico
Nacional
Mexico City, México

See Siao Wei
Department of Chemical and
Biomolecular Engineering
National University of Singapore
Singapore
Instituto Nacional de Enfermedades
Respiratorias
Mexico City, México

Mark Zunckel
uMoya-NILU (Pty) Ltd.
Durban North, South Africa

1 Air Pollution
Health and Environmental Concerns

Bhola R. Gurjar, Luisa T. Molina,
and Chandra S.P. Ojha

CONTENTS

Air pollution emissions are released from both natural and anthropogenic sources. Human-driven activities aimed at providing necessary goods and services to society are responsible for the anthropogenic share of air pollution. Air pollution emissions occur at many stages in the life cycles of products and services, that is, from raw material extraction, energy acquisition, production and manufacturing, use, reuse, recycling, through to ultimate disposal. The resulting emissions undergo several types of physical and chemical transformations and contribute to a wide range of health and environmental impacts, including deterioration of air quality, toxicological stress on human health and ecosystems, photo-oxidant formation (smog), stratospheric ozone (O_3) depletion, climate change, degradation of air resources, and noise, among others (Pennington et al., 2004).

The World Health Organization (WHO) has summarized some of the important facts of air pollution on health (URL: http://www.who.int/mediacentre/factsheets/fs313/en/index.html), which are given below:

Air pollution is a major environmental risk to health and is estimated to cause approximately 2 million premature deaths worldwide per year.

Exposure to air pollutants is largely beyond the control of individuals and requires action by government bodies and public authorities at the national, regional, and even international levels.

More than half of the burden from air pollution on human health is borne by people in developing countries. In many cities, the average annual levels of PM_{10} (particulate matter [PM] with aerodynamic diameter equal to or less than 10 μm—the main source of which is the burning of fossil fuels) exceed 70 μg/m^3. The WHO Air Quality Guidelines (AQGs) say that, in order to prevent ill health, these levels should be lower than 20 μg/m^3.

By reducing air pollution levels, we can help countries reduce the global burden of disease from respiratory infections, heart disease, and lung cancer considerably. For example, by reducing PM_{10} pollution from 70 to 20 $\mu g/m^3$, we can cut air quality-related deaths by around 15%.

It is clear from the above-listed facts that air pollution (both indoors and outdoors) is a major environmental health problem affecting everyone in developed and developing countries alike. Based on various studies, the WHO has developed AQGs (WHO, 2000, 2006; Krzyzanowski, 2008) to represent the most widely agreed and up-to-date assessment of health effects of air pollution and to recommend targets for air quality at which the health risks are significantly reduced. The latest 2005 WHO AQGs offer updated global guidance on reducing the health impacts of air pollution (WHO, 2006; Krzyzanowski, 2008). The new (2005) guidelines (WHO, 2006) apply worldwide and are based on expert evaluation of current scientific evidence. They recommend revised limits for the concentration of selected air pollutants, for example, PM, O_3, nitrogen dioxide (NO_2), and sulfur dioxide (SO_2), applicable across all WHO regions. Key findings in 2005 AQGs are listed below (WHO, 2006):

Even relatively low concentrations of air pollutants are related to a range of adverse health effects.

There are serious risks to health from exposure to PM and O_3 in many cities of developed and developing countries. Poor indoor air quality (IAQ) may pose a risk to the health of over half the world's population. In homes where biomass fuels and coal are used for cooking and heating, PM levels may be 10–50 times higher than the guideline values.

Considerable reduction of exposure to air pollution can be achieved by lowering the concentrations of several of the most common air pollutants emitted during the combustion of fossil fuels. Interestingly, such measures will also reduce greenhouse gases and contribute to the mitigation of global warming.

Among all the air pollutants in ambient air, PM affects more people than any other pollutant (see the review by Pope and Dockery, 2006). The adverse effects of PM on human health occur at levels of exposure currently being experienced by most urban and rural inhabitants in both developed and developing countries. The major constituents of PM are sulfates, nitrates, ammonia, sodium chloride, carbon, mineral dust, and water. PM consists of a complex mixture of solid and liquid particles of organic and inorganic substances suspended in air. Particles are identified according to their aerodynamic diameters as either PM_{10}, mentioned above, or $PM_{2.5}$ (aerodynamic diameters equal to or smaller than 2.5 μm). The latter are considered as more detrimental because, when inhaled, they may reach the peripheral regions of bronchioles and interfere with gas exchange inside the lungs.

Chronic exposure to particles contributes to the risk of developing cardiovascular and respiratory diseases, and also lung cancer. In developing countries, exposure to pollutants from the indoor combustion of solid fuels on open fires or traditional stoves increases the risk of acute lower respiratory infections and associated mortality among young children. Indoor air pollution from solid fuel use is also a major risk factor for chronic obstructive pulmonary disease and lung cancer among adults. The mortality in cities with high levels of pollution exceeds that observed in relatively cleaner cities by 15–20%. Even in the European Union, the average life expectancy is 8.6 months lower due to exposure to $PM_{2.5}$ produced by human activities. According to the 2005 WHO AQGs (WHO, 2006), the PM guideline values are set for $PM_{2.5}$ and PM_{10} separately. The annual mean of $PM_{2.5}$ and PM_{10} should be equal to 10 and 20 $\mu g/m^3$, respectively, whereas the 24-hour mean should not exceed 25 and 50 $\mu g/m^3$, respectively. It is noteworthy that the 2005 AQGs include for the first time a guideline value for PM. The aim is to achieve the lowest concentrations possible. As no threshold for PM has been identified below which no damage to health is observed, the recommended value should represent an acceptable and achievable objective to minimize health effects in the context of local constraints, capabilities, and public health priorities (WHO, 2006).

In the case of O_3, the previously recommended limit, which was fixed at 120 $\mu g/m^3$ 8-hour mean, was reduced to 100 $\mu g/m^3$ based on recent conclusive associations between daily mortality and O_3 levels occurring at O_3 concentrations below 120 $\mu g/m^3$ (WHO, 2006). Ground-level O_3 is one of the major constituents of photochemical smog. It is formed by the reaction with sunlight (photochemical reaction) of pollutants such as nitrogen oxides (NO_x) from vehicle and industry emissions and volatile organic compounds (VOCs) emitted by vehicles, solvents, and industry. The highest levels of O_3 pollution occur during periods of sunny weather.

Tropospheric O_3 levels are sensitive to local concentrations of NO_x and VOCs. Model studies suggest that additional UV-B radiation reduces tropospheric O_3 in clean environments (low NO_x) and increases tropospheric O_3 in polluted areas (high NO_x) (Tang et al., 1998). Excessive O_3 in the air can have a marked effect on human health, material, and vegetation. It can induce breathing problems, trigger asthma, reduce lung function, and cause lung diseases. In Europe, it is currently one of the air pollutants of most concern. Several European studies have reported that the daily mortality rises by 0.3%, and that for heart diseases by 0.4%, per 10 $\mu g/m^3$ increase in O_3 exposure (WHO, 2006).

The major sources of anthropogenic emissions of NO_2 are combustion processes (heating, power generation, and engines in vehicles and ships). Epidemiological studies have shown that symptoms of bronchitis in asthmatic children increase in association with long-term exposure to NO_2. Reduced lung function growth is also linked to NO_2 at concentrations currently measured (or observed) in cities of Europe and North America. The current WHO guideline value of 40 $\mu g/m^3$ (annual mean) set to protect the public from the health effects of gaseous NO_2 remains unchanged from the level recommended in previous AQGs. An additional AQG value of 200 $\mu g/m^3$ for the 1-hour mean is also recommended. As an air pollutant, NO_2 has several correlated activities:

At short-term concentrations exceeding 200 $\mu g/m^3$, it is a toxic gas that causes significant inflammation of the airways.

NO_2 is the main source of nitrate aerosols, which form an important fraction of fine particles and, in the presence of ultraviolet light, of O_3.

SO_2 is produced from the burning of fossil fuels (coal and oil) and the smelting of mineral ores that contain sulfur. The main anthropogenic source of SO_2 is the burning of sulfur-containing fossil fuels (e.g., coal) for domestic heating, power generation, and motor vehicles. Studies indicate that a proportion of people with asthma experience changes in pulmonary function and respiratory symptoms after periods of exposure to SO_2 as short as 10 min. A SO_2 concentration of 500 μg/m^3 should not be exceeded over averaging periods of 10 min, whereas it should not exceed 20 μg/m^3 for the 24-hour mean period. The revision of the 24-hour guideline for SO_2 from 125 to 20 μg/m^3 is based on the consideration that health effects are now known to be associated with much lower levels of SO_2 than previously believed, and thus a greater degree of protection is needed. SO_2 can affect the respiratory system and the functions of the lungs, and causes irritation of the eyes. Inflammation of the respiratory tract causes coughing, mucus secretion, aggravation of asthma, and chronic bronchitis, and makes people more prone to infections of the respiratory tract. Hospital admissions for cardiac disease and mortality increase on days with higher SO_2 levels. When SO_2 combines with water, it forms sulfuric acid; this is the main component of acid rain that affects sensitive ecosystems.

It should be emphasized that the WHO AQGs (e.g., WHO, 2000, 2006) are intended to provide uniform background information and guidance to governments in making health risk management decisions, particularly in setting standards. The guidelines may also be used in planning processes and various kinds of air quality management decisions at community or regional level. The text of the AQG document emphasizes that the guidelines are not standards in themselves. Before transforming them into legally binding standards, the guideline values must be considered in the context of prevailing exposure levels, technical feasibility, source control measures, abatement strategies, and social, economic, and cultural conditions. In certain circumstances there may be valid reasons to pursue policies that may result in air pollutant concentrations above or below the guideline values. Although these guideline levels are considered to protect human health, they are by no means a "green light" for pollution. It should be stressed that attempts need to be made to keep air pollution levels as low as practically achievable, since there is no clear threshold or level below which there are no adverse effects.

The health burden of environmental exposures, including ambient air pollution and climate-change-related health impacts, is not equally distributed between or within regions and countries. These inequalities are currently receiving increased attention in environmental research as well as enhanced appreciation in environmental policy, where calls for environmental equity are more frequently heard. Even the WHO Global Update of the AQGs (WHO, 2006), which makes an attempt to address global-scale inequalities in exposures to air pollution and the burden of diseases due to air pollution, stops short of addressing explicitly the inequalities in exposure and adverse health effects within countries and urban areas due to the differential distribution of sources of air pollution such as motor vehicles and local industry, and differences in susceptibility to the adverse health effects attributed to air pollution. These inequalities may, however, be addressed in local air quality and land use management decisions (O'Neill et al., 2008).

With this perspective, the present book covers not only common knowledge and fundamentals of air pollution problem but also studies and cases from different parts of the world so that the peculiarities of each region are documented. It is hoped that besides understanding the scientific basis of air pollution and its impact on health and the environment, the book will help readers (i) better appreciate social and environmental determinants of public health, and (ii) apply country-based research evidences to reduce health disparities and environmental inequalities. Moreover, the book is expected to stimulate future research and policy action on health and environmental consequences of air pollution from the local to global level.

To help readers appreciate and comprehend the intricate problem of air pollution and its adverse impacts on human health and the environment in totality, the chapters cover almost all the major aspects of air pollution, for example, monitoring and source characterization of air pollution, modeling, health effects, environmental impacts, risk assessment, air quality management, and relevant policy issues. To facilitate an easy grasp of the subject, the book has been divided into five major sections: Air Pollution Monitoring and Modeling, Air Pollution and Health Effects, Health Risk Assessment and Management, Air Quality Management: Techniques and Policy Aspects, and Environmental Impacts of Air Pollution.

To begin with, Chapter 2 presents the fundamentals of air pollution monitoring and source characterization. This primarily covers objectives of air pollution monitoring, sample collection, selection of contaminants for air quality monitoring, selection of monitoring sites, data validation and interpretation, monitoring methods, monitoring of gaseous compounds and PM, direct and indirect mass measurement techniques, and source apportionment and characterization through enrichment factor analysis, chemical mass balance methods, and multivariate receptor models.

The most important activity for management of ambient air quality is collection of data by direct measurements and analysis of these data by fitting into any model. Subsequently, these models can be used to predict air quality under different scenarios and also to quantify the health and environmental risks of emissions from a polluting source. Since air quality modeling is a well-researched and fully developed area and several books are available that exclusively deal with this subject, Chapter 3 presents an overview of the basic theory and application of air pollution modeling. This chapter, therefore, reviews different types of atmospheric dispersion models and commercial software followed by a brief commentary on recently popular approaches of statistical/probabilistic modeling, artificial neural network (ANN) modeling, and fuzzy modeling. It is evident that a great deal of development and advances in air pollution modeling have taken place over the years. Among these, the use of computational fluid dynamic (CFD) models appears to be the most dominant one considering the development of a large number of softwares. The use of specific techniques such as ANN, fuzzy logic, and time series analysis has also been prevalent. Applications of ANN for data generation at different sampling intervals are very promising. In the absence of any rigorous modeling, however, time series can be used locally for useful forecasts of concentrations of air pollutants. Chapter 3 also presents a number of case studies from developing countries such as India and Chile. There are some other chapters in this book that deal with specific aspects of air pollution modeling. For example, Chapter 4 discusses IAQ modeling and Chapter 12 includes the application of appropriate modeling techniques to estimate acute or short-term risk from the accidental release of industrial toxic materials.

Air pollution emissions can pollute both outdoor and indoor air. People spend a large part of their time indoors. Therefore, if the air breathed in indoors is contaminated, it poses a threat to the health of the occupants. According to World Health Reports 2002 (WHO, 2002), indoor air pollution is responsible for 2.7% of the total global burden of diseases. Thus IAQ studies become necessary to evaluate the air quality of indoor spaces in terms of its physical, chemical, and biological properties along with the well-being of the occupants. IAQ studies include first identifying indoor pollutants, their sources, and causes; the various parameters associated with IAQ, that is, building parameters, occupant parameters, meteorological parameters affecting the IAQ, and so on, need extensive survey, monitoring programs, and health investigation studies. Chapters 4 and 5 are dedicated to dealing with these issues in detail. While Chapter 4 presents a general outline of indoor air pollution incorporating briefly indoor air pollutants, their sources and causes, various associated parameters of IAQ, monitoring and modeling of IAQ, health-related aspects and studies and some control measures, Chapter 5 illustrates a comprehensive case study in the Indian context focusing on emissions from biomass fuels and their health effects on women.

The case study presented in Chapter 5 has special significance because indoor air pollution from the use of unprocessed solid biomass fuels such as wood, dung cake, and agricultural wastes for domestic cooking and room heating is a major health concern in developing countries such as India. About 74% of people in the rural areas of India still use biomass as the major source of domestic energy. Biomass fuels are highly polluting and the concentration of respirable suspended PM_{10} in the kitchen during biomass burning is several times higher than vehicular pollution in the cities. Biomass smoke contains a host of toxic substances that are harmful to human health. Women who cook with these fuels and their accompanying children are most vulnerable. Still, the health impact of biomass fuel use in India as well as in many other developing nations is largely unknown. Against this background, this study was undertaken to study the respiratory and systemic toxicity associated with chronic exposures to biomass smoke in the country. In the present case study, a total number of 1260 nonsmoking women (median age 38 years) of rural West Bengal (a state in eastern India) who used to cook exclusively with traditional biomass fuel and 650 age-matched women from the same neighborhood who cooked with cleaner fuel liquid petroleum gas (LPG) were enrolled. However, as illustrated in Chapter 10, it is important to note that not only the cooking fuel but also the cooking method is responsible for ultrafine particles posing health risks to an exposed population.

Compared with LPG-using women, biomass fuel users showed a remarkable increase in the prevalence of respiratory symptoms, lung function reduction, airway inflammation, and covert pulmonary hemorrhage. There was a significant reduction in superoxide dismutase (SOD) enzyme activity in blood plasma, suggesting a decline in the body's antioxidant defense. This was accompanied by a higher frequency of micronucleus formation in buccal and airway epithelial cells and comet formation in lymphocytes, suggesting a higher rate of chromosomal and DNA damage. Moreover, airway epithelial cells of biomass users had greater instances of metaplasia and dysplasia, implying a higher risk of cancer in the airways. In addition, biomass users had prolonged menstrual cycles, a higher risk of spontaneous abortions, still births, and underweight babies. They also suffered more from depression and several other neurobehavioral problems with accompanying alterations in reproductive hormones. The changes were intimately associated with indoor air pollution level after controlling potential confounders such as education, family income, and environmental tobacco smoke. This study found that biomass fuel use for domestic cooking is associated with a high level of indoor air pollution that adversely affects the physical and mental health of women cooking with these fuels. Since millions of poor people in the country still use these fuels, the findings warrant immediate measures for improvement of the situation.

Urban air pollution emissions (e.g., from vehicles) have been known to degrade ambient air quality across the world and pose substantial health risks to city dwellers. Due to high levels of ambient air concentrations of various air pollutants in Indian cities, it is continuing to be a major health concern in India. Although respiratory, cardiovascular, and genotoxic changes are important health effects of air pollution, little is known about their prevalence and risk among urban Indians who are exposed to some of the highest pollution levels in the world. Against this background, the authors of Chapter 6 have conducted epidemiological studies in Kolkata and Delhi, two highly polluted megacities in the country, to investigate the impact of chronic exposures to urban air pollution on respiratory and other organ systems of the body. The study, conducted during 2000–2006, enrolled 6862 nonsmoking residents from Kolkata and Delhi (median age 43 years) and 3715 age- and sex-matched nonsmokers as controls from relatively less polluted rural areas of West Bengal where the PM_{10} level was considerably lower. In addition, 12,688 school-going children (age 8–17 years) of these two cities and 5649 from rural West Bengal were examined.

It was observed that compared with rural controls, urban subjects had a significantly higher prevalence of upper and lower respiratory symptoms, bronchial asthma, and lung function deficits. About 40% of citizens from Delhi had reduced lung function against 20% of matched controls, and a restrictive type of lung function deficit was predominant. Hypertension was more prevalent in urban subjects; they had activated leukocytes and platelets, and a greater number of leukocyte–platelet aggregates in their circulation, which are potential risk factors for cardiovascular diseases. Neurobehavioral symptoms, including depression, were more frequent in these subjects, and their airway epithelial cells had a greater frequency of nuclear anomaly. The adverse health consequences in urban subjects were positively associated with PM_{10} levels in ambient air and personal exposures to benzene after controlling potential confounders such as environmental tobacco smoke and socioeconomic conditions in multivariate logistic regression analysis. The study has revealed that chronic exposures to high levels of urban air pollution in India are adversely affecting the physical and mental health of citizens, especially children and the elderly. Therefore, efforts should be made by all concerned to reduce pollution levels in order to safeguard public health.

Over the years, megacities have emerged as engines of economic growth but also as highly polluted urban air sheds, particularly in developing countries. Mexico City Metropolitan Area (MCMA) is one such most densely populated megacity in the world, with 18 million people according to the 2000 census (INEGI, 2001). In certain city zones of Mexico City, O_3 and PM concentrations have been observed well above the Mexican standard for many days per year. Observation from the intensive

MCMA-2003 campaign shows that MCMA motor vehicles produce high levels of primary PM, particle-bound polycyclic aromatic hydrocarbons (PAHs), and a wide range of air toxics, including formaldehyde, acetaldehyde, benzene, toluene, and xylenes (Molina et al., 2007). In this context, Chapter 7 describes air pollutant exposure and health effects monitored and observed during the MILAGRO–MCMA2006 campaign (Molina et al., 2010). The volunteers investigated were children 9–12 years old, and their parents, recruited from elementary schools in three different downward sites intercepting the Mexico City air pollutant plume during the campaign. The results obtained during the MILAGRO campaign focus attention on the high frequency of many subclinical expressions related to air pollution, even though in the MCMA a significant reduction in air pollution levels has been achieved in recent times through comprehensive air quality management (Molina and Molina, 2002). The authors observe that since children are the most sensitive individuals in a population for many reasons (Chung, 2001), this subgroup may seriously be affected by different chronic diseases if more stringent controls and preventive programs are not established in this metropolitan area.

Among all the air pollutants, PAHs, which is a group of ubiquitous persistent organic pollutants possessing carcinogenic, mutagenic, and immunotoxic properties, have become an issue of increasing concern in recent decades (particularly in areas undergoing rapid urbanization). Chapter 8 describes the sources, distribution, chemical transformation, toxicokinetics, health implications, and control measures of PAHs.

They are formed and released into the atmosphere through natural and anthropogenic sources. Natural sources include volcanoes and forest fires, whereas man-made sources come mainly from the high-temperature combustion of fossil fuels in automobile engines, cooking stoves, power plants, refineries, and other industrial activities. These pollutants have high persistence in the environment, low biodegradability, and high lipophilicity. The association of toxic 4–6-ring PAHs with fine particles in conjunction with their widespread presence in the atmosphere increases the vulnerability of public health upon exposure and inhalation of the aerosols. Consequently, aerosol characterization studies have been carried out in several parts of the world to determine PAHs in airborne particles in order to assess their environmental fate and human exposure. A number of PAHs are mutagenic, having the capacity to bind covalently to DNA in the target tissues and form protein adducts, and activate aryl hydrocarbon receptor (ER)-mediated signaling. Subsequent DNA replication results in mutations leading to carcinogenesis. PAH that can induce mutations can potentially damage the germ line leading to fertility problems and to mutations in future generations. The carcinogenic and mutagenic impact of a single PAH is significantly stronger than the impact of the same PAH occurring at the same concentration in a mixture.

As stated earlier, among all the air pollutants in ambient air, PM affects more people than any other pollutant. However, current evidence is not conclusive on the specific PM characteristics that may be held responsible for adverse health responses and toxicity. In spite of existing epidemiological evidence documenting PM as the air pollutant with the most consistent and strongest associations with diverse adverse health outcomes, we are still far from understanding the mechanisms involved. Chapter 9, therefore, concentrates on cellular mechanisms behind

PM air pollution-related health effects. Cell cultures of target cell types known to be relevant to organs affected by PM, and exposures to particles from various emission sources and of differing nature (from surrogate PM of known composition to real urban settings particles), comprise a range of variables included in experiments described below. Three main biological response patterns have been observed after exposing cells *in vitro* to PM: cytotoxicity (necrosis/apoptosis), cytokine production, and genotoxicity. However, the precise role of these types of cellular responses in the health outcomes observed in humans remains to be better understood. So far, the understanding of the effects of various components of PM has focused on the identification of specific PM components that determine toxicity with a specific cellular effect. There is indication that cellular responses result from interactions among PM components that lead to more complex patterns of cellular responses. Producing and interpreting the results obtained with PM-related mixtures represent an intellectual and methodological challenge. In this respect, recent studies involving concentrated ambient particles (CAPs) (Ghio and Huang, 2004) allow for studying ambient PM with the possibility of investigating PM–component interactions, and interactions between PM and gases. The authors provide experimental evidence attributing respiratory and cardiovascular system toxicity to proinflammatory conditions, oxidative stress (pro-oxidant/antioxidant imbalance), and neural stimulation derived from PM exposure. In spite of the existing uncertainties (precise evaluation of human exposures, sample collection, mechanisms of damage, etc.), the relation of PM air pollution and human health effects has been documented. Future research should focus on using samples from ambient air at relevant doses, relative to open population exposures; better understanding of the potential effects of cumulative exposures and exposures occurring within susceptibility time-windows in the development and growth of children; developing methodologies suitable for addressing multipollutants—multiple effects, as well as the ability of various PM components, and their interactions to convey toxicity and potential to cause adverse health outcomes.

Chapter 10 provides a general framework for assessing human exposure to particulate air pollution and evaluating potential health risks with specific reference to indoor airborne particles. Case studies are presented and discussed in this chapter to illustrate the application of this risk assessment framework. These case studies focus on cooking methodology because it is an important source of particulate air pollution in indoor environments. The objectives of the study presented in this chapter are to use a combination of controlled experiments and real-world measurements to evaluate the potential health risks faced by chefs. To demonstrate the importance of both physical and chemical properties of particles in determining risks, this chapter presents both the physical characteristics (derived from controlled experiments) and the chemical properties (derived from real-world studies), and their relative potential risks. It is observed that the deep-frying method emits the most number of particles and the highest portion of nanoparticles, followed by pan-frying, stir-frying, boiling, and steaming. This trend between particle number concentrations and different cooking methods implies that cooking with oil causes more particulate emissions than cooking with water. A larger proportion of ultrafine and accumulation mode particles was observed during water-based cooking, which is thought to be due to the hygroscopic growth of freshly emitted particles in the presence of high humidity.

The highest concentrations of $PM_{2.5}$ and PAHs were observed during the deep-frying cooking method.

As already seen in previous chapters, there is strong evidence linking urban/rural air pollution to acute and chronic illnesses and premature deaths, and these adverse health impacts in turn result in high economic costs to society. Among the various air pollutants, PM_{10} has been recognized as a major cause of health problems, particularly asthma. Chapter 11 highlights the various sources of PM_{10} along with existing levels of PM_{10} in major urban areas in India. Health impacts in terms of excess mortality and morbidity were estimated for 14 major Indian cities. Analysis of data indicates that in most of the cities, PM_{10} levels exceed 2–3 times the permissible limit of 60 $\mu g/m^3$. Based on the PM_{10} concentration levels of 2001, Delhi is found to be worst affected and Bangalore is found to be least affected in terms of health impacts. Uncertainty associated with the estimation of health impacts highlights that PM_{10} levels vary considerably over time and hence involve large variations in estimates. Accurate measurement of pollution levels and choosing their representative values are important inputs to get to better estimates. Therefore, a major step toward mitigating/minimizing adverse health impacts of air pollution would be to collect systematic and reliable air quality data, which in turn would facilitate delineating appropriate strategies/action plans for air quality management.

Toxic or hazardous air pollutants (HAPs) pose two types of risk in the environment, namely "short-term or acute risk" and "long-term or chronic risk." Short-term risk is associated with one-time acute exposure to potentially hazardous substances accidentally released into the atmospheric environment, whereas long-term risk results from continuous exposure to potentially harmful substances. In Chapter 12, appropriate modeling techniques are applied to estimate acute or short-term risk from the accidental release of industrial toxic materials. Dense gas dispersion and dose–response models have been used. The use of the model is demonstrated with a case study for chlorine storage in the Indian environment. Nomograms are constructed for use during an industrial chemical accident by administrators for evacuation purposes where immediate modeling usages are not required. The nomograms are prepared for eight commonly used toxic chemicals in Indian industries. This chapter also evaluates the potential health risks related to certain carcinogens and noncarcinogens (e.g., cadmium, chromium, and nickel) present in different Indian states (regions). Appropriate dose–response models have been identified and used for this purpose with the assumptions and input data as per the Indian context. Individual and societal risks of extra cancer due to the above toxics have been estimated. The hazard quotients and hazard index representing the noncarcinogenic chronic health effects caused by chromium and cadmium due to their long-term exposure through water and food have also been estimated. The risk results have been compared with the disease surveillance data where a satisfactory validation is observed. Furthermore, an integrated approach for risk estimation is demonstrated with two case studies of chlorine industries in the country. Finally, the current status of quantitative risk assessment (QRA) techniques is elaborated along with limitations of the same.

In addition to epidemiological studies of air pollution, international and national guidelines of emissions and air quality, and approaches and techniques to monitoring and modeling air pollution to eventually control pollution and manage air quality, the development and implementation of suitable policy frameworks are also essential to check the origin of the air pollution problem in the first place. To address this issue from environmental economics and public policy perspective, Chapter 13 deals with different viewpoints, including the Neoclassical Environmental Economic view that is still a major influence for theories of natural capitalism and environmental finances. Even today Neoclassical Economic analysis plays a fundamental role in any economic decision, such as consumption, production, and policy-making. For example, cost–benefit analysis is the major public policy methodology used in reaching environmental decisions and shaping environmental regulations at any national, international, or supranational level. Thus, Neoclassical Environmental Economics has been at the very center of public policies on environmental issues for decades. The question of whether Neoclassical Environmental Economics is the best economic approach for policy decision-making is a very controversial subject, and there are few alternative approaches to Neoclassical Environmental Economics. Chapter 13, therefore, provides insight into air pollution and its impacts on health and the environment, starting from the Neoclassical Economic perspective and reviewing the main alternative approaches (e.g., Austrian Economics, Green Economics, and Ecological Economics) to reach a very balanced global environmental understanding.

The development of air quality policy and methods of assessing the impact of interventions on air quality is a relatively mature part of environmental management. Chapter 14 discusses some of the key scientific elements associated with this process. The foundation of air quality management consists of emission inventories, ambient measurements, and atmospheric models. They provide a quantitative understanding of how atmospheric chemistry, meteorology, and natural emissions influence the fate of human emissions, the buildup of pollutants, and their air concentrations. This information can be used to derive population exposure to quantify risk and assess health impacts, ultimately leading to integrated models capable of estimating the human health benefits associated with specific emission reduction policies. As one of the primary motivations for improving air quality is to reduce human health effects, current evidence on this issue is reviewed in this chapter. The main focus of this chapter is on technical aspects related to successful development and application of air quality models, emission inventories, and measurement programs including a discussion on assessing uncertainty. Therefore, a key goal of this chapter is to discuss how models, emission inventories, and measurements support air quality management and then to address technical issues and uncertainties in the application of these tools. Awareness of these issues helps to ensure the successful application of these tools and to guide effective communication of their results to air quality managers in all levels of government and in the private sector.

As discussed earlier, air pollution not only harms human health but also damages the environment and ecology. Ground-level O_3 is arguably the air pollutant most likely to cause damage to agricultural crops across the globe due to its phytotoxicity and prevalence at high concentrations across agriculturally important regions. Although extensive experimental studies to assess the potential threat posed from O_3 to agriculture have been conducted in Europe and North America, research is far less advanced in South Asia, Southeast Asia, and southern Africa. It is in these regions that research is perhaps most urgently needed, given recent escalations in economic growth and associated pollutant emissions. In this context, Chapter 15 describes one of the first air pollution effect studies to assess impacts on crop productivity in the regions of South and Southeast Asia and southern Africa. The study presented in this chapter has established the current status of knowledge of air pollution impacts on important arable crops across regions and finds evidence of substantial effects (with yield losses of up to 30%) commonly occurring under elevated ambient O_3 concentrations for many important species (e.g., wheat, rice, and beans). Provisional risk assessments have identified high risk across the Indo-Gangetic plain region of South Asia; large parts of Thailand and some areas in Burma; and regions in Zimbabwe and Botswana in southern Africa with photochemical modeling suggesting O_3 concentrations in exceedance of European and WHO AQGs. A coordinated and standardized experimental campaign conducted across South Asia to assess yield losses to mung bean (*V. radiata*) showed that the highest statistically significant yield losses were in the region of 30% and above and tended to occur in the high-risk regions identified by the risk-assessment modeling.

Given the magnitude and extent of yield losses found for key crops across the South Asian region in this and other studies, it would seem that O_3 pollution might well be an additional and significant stress on agro-ecosystems. A comprehensive understanding of the relative importance of all stresses facing current and future agricultural production in the South Asian regions is vitally important given the challenge of the region to provide sustainable increases in productivity to balance reduced per capita area harvested.

Global modeling of ground-level O_3 concentrations for 2030 suggests that O_3 annual mean surface averages for South Asia and southern Africa could vary from -5.9 to $+11.8$ ppb for South Asia and from -2.5 to $+7$ ppb for southern Africa, with the range being dependent on the emission scenario applied (Dentener et al., 2006). The possible increases in mean annual averages would translate into disproportionately higher growing season average concentrations when the climatic conditions (namely temperature and solar radiation) enhance the chemical formation of O_3 in the atmosphere. As such, it seems more than likely that O_3 concentrations, which are already at concentrations capable of causing yield and productivity losses across many parts of South Asia and southern Africa, will continue to worsen over the next 20–50 years. It would therefore seem prudent to consider this pollutant in future research to assess the effect of multiple stresses on sustainable crop production across Asia and southern Africa.

Future assessments related to crop impacts from O_3 would ideally incorporate the effects of climate change, and seek to involve specialists on adaptation options. Ideally, key decision-makers from governments would come together to discuss likely combined impacts, measures to reduce the vulnerability of end users, national risk assessments, and policy options to reduce the threat from this environmental problem. The opportunity for cobenefits for air pollution and climate change in emission reduction policy (e.g., Ramanathan and Feng, 2008) is of particular importance in many developing regions where there is a suspicion of climate change policies, but where the importance of abating air pollution is recognized. Most work on cobenefits has however been undertaken with developed world perspectives and priorities. This now needs to be widened to reflect the perspectives and priorities of developing countries.

Chapter 16 presents the impacts of air pollution on ecosystem and human health with a sustainability perspective. The authors propose a short journey through the scientific fundamentals of sustainable development and the relationship between biosphere, society, and economy. The journey begins with a brief conceptual description of the relationship between natural, social, and economic systems, including the ecosystems benefits for humans and the threats to their maintenance. A healthy environment is a precondition for long-term development of the world economy. A sustainable development starts with the acknowledgment that economy, society, and environment must be considered simultaneously in any plan of development. Air pollution, therefore, has been discussed in this chapter in the context of human search for sustainable development.

The bottom line of Chapter 16 is the environmentally conscious development, following several main aspects: for example, measurement and assessment of the causes and effects of air pollution, the natural capital and its irreplaceable character, the carrying capacity of ecosystems and the ecological footprint of human activities, and the importance of several fields of research (e.g., environmental science and technology, ecological economics, and ecological education). Environmental education includes the dimension of ecological sustainability, and the ability to develop and operate with sustainability indices: for example, the Environmental Sustainability Index—ESI, the Ecological Footprint Index—EFI, the Environmental Vulnerability Index—EVI, the Millennium Development Goal 7 Index—MDG7 (e.g., Esty et al., 2005), and the Living Planet Index—LPI (WWF, 2004). In a wider perspective, many specific environmental indices exist for all major environmental issues, such as those used by Organization for Economic Cooperation and Development (OECD). These also include measurements of greenhouse gas emissions, apparent consumption of O_3-depleting substances (ODS), SO_x and NO_x emission intensities, municipal waste generation intensities, wastewater treatment rates, intensities of use of energy and water, forest, and fish resources, and threatened species (OECD, 2004). But they do not offer an integrated picture of the state of the environment. Therefore, there is a strong need to use composite indices, which relate a global environmental degradation to economic development. Nevertheless, no matter how efficient they are in encapsulating complex realities in simple terms, aggregated indices must be used critically, since the more complex the reality they concentrate, the higher the danger that they do not apply to the intended conditions.

Education for sustainable development is our best chance in what Edward O. Wilson called the "Century of the Environment" and the "bottleneck" of the immediate future. We subscribe to his view that combining science and technology with foresight and moral courage is the way to secure a long-term perspective of humanity (Wilson, 2002). Above all, this education should aim at developing a reflex of judging immediate economic gains in the context of long-term effects on our ecological–social–economic systems.

The book ends with Chapter 17, which comprehensively illustrates the regional and global environmental issues related to air pollution. Air pollution is generally considered to be a local concern rather than a long-term global change issue. However, there is growing evidence that human activities in an increasing globalized, industrialized, and interconnected world are influencing both air quality and climate change ranging from urban and regional up to continental and global scales. Rapid population growth and increased energy demand are the primary forces driving unprecedented environmental changes. Emissions from urban and industrial centers of the developed world, and increasingly from large cities of the developing world, are changing the chemical content of the downwind troposphere in a number of fundamental ways. As mentioned above, emissions of NO_x and VOCs drive the formation of photochemical smog and its associated oxidants, degrading air quality and threatening both human and ecosystem health. On a larger scale, these same emissions drive the production of O_3 (a powerful greenhouse gas) in the free troposphere, contributing significantly to global warming. Urban and industrial areas are also major sources of the major directly forcing greenhouse gases, including carbon dioxide, methane, NO_x, and halocarbons. NO_x and SO_2 emissions are also processed to strong acids by atmospheric photochemistry on regional to continental scales, driving acid deposition to sensitive ecosystems and damage to materials, including historic buildings and monuments. Direct urban/industrial emissions of carbonaceous aerosol particles are compounded by the emission of high levels of secondary aerosol precursors, including NO_x, VOCs, SO_2, and NH_3. The resulting mix of primary (directly emitted) and secondary (formed in subsequent photochemical and chemical reactions) aerosols is now recognized to play an important role in the Earth's climate (IPCC, 2007).

This chapter presents some of the regional and global consequences of air pollution and the challenges and opportunities in addressing these complex issues. One of the key steps in any rational approach to addressing regional and global environmental issues is to promote internationalism—a widespread understanding that all of our human problems are interconnected. Regional and international cooperation will be essential to the solution of environmental problems. Strong involvement of stakeholders at all levels, changes in human behavior, coupled with suitable mechanisms for facilitating technological and financial flows, and the strengthening of human and institutional capacities will be crucial to the future success of efforts to control atmospheric pollutions.

REFERENCES

Chung, F. 2001. Anti-inflammatory cytokines in asthma and allergy: Interleukin-10, interleukin-12, interferon-gamma. *Mediators of Inflammation* 10 (2), 51–59.

Dentener, F., Stevenson, D., Ellingsen, K., et al. 2006. The global atmospheric environment for the next generation. *Environmental Science & Technology* 40, 3586–3594.

Esty, D.C., Levy, M., Srebotnjak, T., and De Sherbinin, A. 2005. *Environmental Sustainability Index: Benchmarking National Environmental Stewardship*. New Haven: Yale Center for Environmental Law & Policy.

Ghio, A.J. and Huang, Y.C. 2004. Exposure to concentrated ambient particles (CAPs): A review. *Inhalation Toxicology* 16, 153–159.

INEGI. 2001. *Censo de Población 2000*. México: INEGI.

IPCC. 2007. *Climate Change 2007: The Physical Science Basis, Contribution of Working Group I to the Fourth Assessment Report of the Intergovernmental Panel on Climate Change* (S. Solomon, D. Qin, M. Manning, Z. Chen, M. Marquis, K.B. Averyt, M. Tignor, and H.L. Miller [Eds]). Cambridge, UK and New York, NY: Cambridge University Press, 996pp.

Krzyzanowski, M. 2008. WHO Air Quality Guidelines for Europe. *Journal of Toxicology and Environmental Health, Part A* 71, 47–50.

Molina, L.T., Kolb, C.E., de Foy, B., et al. 2007. Air quality in North America's most populous city—overview of MCMA-2003 campaign. *Atmospheric Chemistry and Physics* 7, 2447–2473.

Molina, L.T., et al., 2010. An overview of the MILAGRO 2006 campaign: Mexico City emissions and their transport and transformation, *Atmospheric Chemistry and Physics Discussion*, 7819–7983.

Molina, M.J. and Molina, L.T. (Eds). 2002. *Air Quality in the Mexico Megacity: An Integrated Assessment*. New York: Springer.

Pennington, D.W., Pottingb, J., and Finnveden, G. 2004. Life cycle assessment Part 2: Current impact assessment practice. *Environment International* 30 (5), 721–739.

Pope, C.A. and Dockery, D.W. 2006. Health effects of fine particulate air pollution: Lines that connect. *Journal of the Air and Waste Management Association* 54, 709–742.

OECD. 2004. *OECD Key Environmental Indicators*. Paris: OECD.

O'Neill, M.S., Kinney, P.L., and Cohen, A.J. 2008. Environmental equity in air quality management: Local and international implications for human health and climate change. *Journal of Toxicology and Environmental Health, Part A* 71(9), 570–577.

Ramanathan, V. and Feng, Y. 2008. On avoiding dangerous anthropogenic interference with the climate system: Formidable challenges ahead. *PNAS* 105, 14245–14250.

Tang, X., Madronich, S., Wallington, T., and Calamarid, D. 1998. Changes in tropospheric composition and air quality. *Journal of Photochemistry and Photobiology B: Biology* 46 (1–3), 83–95.

WHO. 2000. Air Quality Guidelines for Europe, second edition (WHO regional publications, European series, No. 91). World Health Organization, Regional Office for Europe, Copenhagen. European commission air quality website (2001). URL: http://www.euro.who.int/document/e71922.pdf

WHO. 2002. *The World Health Report 2002: Reducing Risks*.Geneva: Promoting Healthy Life.

WHO. 2006. Air Quality Guidelines Global Update 2005. WHO Regional Office for Europe Scherfigsvej 8, DK-2100 Copenhagen Ø, Denmark, ISBN 92 890 2192 6.

Wilson, E.O. 2002. The future of life. Little, Brown by arrangement with Alfred A. Knopf, Inc. Published again in 2003 by Abacus, 230pp., ISBN 0-349-115795. Chapter two "The Bottleneck," published in Scientific American, February 24, 2002.

WWF. 2004. The Living Planet Report 2004. World Wildlife Fund in the US and Canada, United Nations Environmental Program—World Conservation Monitoring Program, Global Footprint Network. URL: http://www.panda.org/downloads/lpr2004.pdf

Section I

Air Pollution Monitoring and Modeling

2 Air Pollution Monitoring and Source Characterization

Anita Lakhani, Rajasekhar Balasubramanian, and Bhola R. Gurjar

CONTENTS

2.1 INTRODUCTION

Air pollution refers to the introduction of physico-chemical or biological materials into the atmosphere that may cause harm or discomfort to humans or other living organisms, or deterioration of the natural environment. The ambient and indoor air composition has a considerable impact on our health and quality of life. Air pollution and greenhouse gas emissions can have a considerable impact on the environment, including broader global environmental issues such as stratospheric ozone depletion and climate change. Air pollutants can be broadly classified as either primary or secondary. Usually, primary air pollutants are substances that are directly emitted from a natural or anthropogenic process, such as ash from a volcanic eruption, carbon monoxide (CO) gas from motor vehicle exhaust, or sulfur dioxide (SO_2) released from factories. However, primary pollutants do not, by themselves, produce all of the adverse effects of air pollution. Chemical reactions may occur among primary pollutants and constituents of the atmosphere, as illustrated in Figure 2.1. Subsequently, gaseous pollutants reacting with each other and with particles in the air produce a complex array of new chemical compounds. Air pollutants not directly emitted as such, but formed in the air, are called secondary air pollutants, which are responsible for several ill effects of air pollution such as smog, haze, eye irritation, and damage to vegetation and material.

The first step in instituting measures against air pollution is to have a clear picture of the current pollution situation through air pollution monitoring. Measuring air quality and understanding its impacts provide a sound scientific basis for its management and for control of air pollution sources. Thus, air quality or pollution monitoring plays a vital role in developing policies and strategies, for measuring compliance with guideline values and tracking progress toward environmental goals or targets. Considerable effort should, therefore, be devoted to the systematic measurement of levels of air pollution in different scales ranging from local to global. Air pollution monitoring, together with the information derived from it, is not an end in itself. Instead, it offers us the best way of understanding air pollution problems, assessing and reviewing environmental measures so that pollution problems can be tackled effectively at local, national, and international levels (Bower and Mucke, 2004).

The ultimate objective of air pollution monitoring is to collect reliable data that can be used by scientists, policy makers, and planners so as to enable them to make informed decisions on managing and improving the overall quality of the environment (Bower and Mucke, 2004). Along with emission inventories and clearly defined targets, monitoring data can be used to develop policies that are tailored to address the most important issues of concern. Well-presented monitoring data also help in communicating air quality issues to local communities and ensuring that the impacts of management decisions can be properly assessed. Good quality monitoring data that indicate potential health or environmental effects, and track changes in pollution levels are a critical part of any effective air quality management strategy (Mucke et al., 1995). However, there are limitations with any monitoring program and the use of analytical data from one technique; hence, it should be used in conjunction with the results of other assessment techniques, including modeling, emission inventories, interpolation, and mapping.

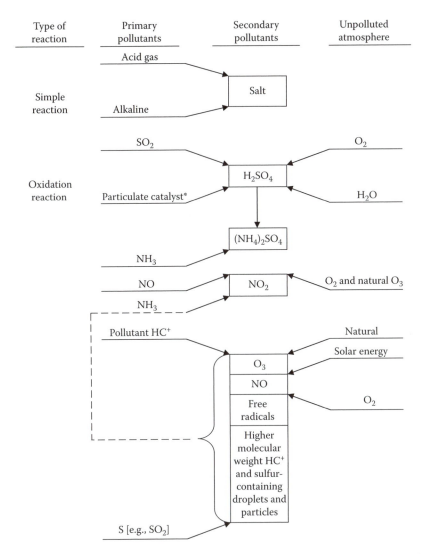

FIGURE 2.1 Primary and secondary pollutants. *Reaction can occur without catalyst (HC+, hydrocarbons). (Adapted from Boubel R.W., et al. 1994. *Fundamentals of Air Pollution* (3rd ed.), Academic Press, San Diego, CA, pp. 195–203. Copyright Elsevier 1994. With permission.)

2.2 OBJECTIVES OF AIR POLLUTION MONITORING

Any air pollution monitoring program generally has the following broad objectives (Ahmet and Dijk, 1994; Fisher et al., 1995):

- To provide a sound scientific basis for the development of cost-effective control policies and solutions to abate air pollution

- To assess how far air quality standards, limit values, and objectives are being met
- To evaluate potential impacts of air pollution on population health and welfare
- To determine the impact of air pollution on ecosystems and our natural environment
- To provide the public with reliable and up-to-date information on air pollution
- To fulfill statutory air quality reporting requirements

An important step in developing an effective monitoring program is to assess where air quality in the region is likely to be deteriorated or at risk and therefore where monitoring should be carried out. The areas at risk can be determined by considering the following (Harrison and Perry, 1986):

- Emissions sources (e.g., domestic, industrial, transport, agricultural, and natural) and contaminants emitted, and also location and magnitude
- Meteorology (areas prone to temperature inversions, etc.)
- Topography
- Geography
- Population centers (especially where domestic fires and traffic emissions occur)
- Historical monitoring data (if available)
- Areas with high natural environmental value (e.g., in and around natural parks, forests, wilderness, and wetlands)
- Areas vulnerable to air pollution plumes from other areas
- Areas planned for development (e.g., to get a picture of background concentrations)
- Any public complaints or issues of concern relating to air pollution
- Any epidemiological studies on air quality effects on health already carried out

2.3 SAMPLE COLLECTION

2.3.1 Selection of Contaminants for Air Quality Monitoring

It is impossible to monitor all contaminants in the air; hence, monitoring programs usually measure only a small group of contaminants that act as indicators of air quality. Particular contaminants are selected on the basis that

- They cause significant adverse health or environmental effects
- They are commonly discharged from known or suspected sources within the area
- They provide a good indication of the overall quality of air

The nature of sources in an area can provide a good indication of which contaminants to monitor (WHO, 1998; ISO/TR 4266, 1989). For example, if vehicles are the primary source of pollutants, nitrogen dioxide (NO_2), CO and possibly benzene and fine particles ($PM_{2.5}$) should be monitored. If the area is affected by domestic fire emissions from wood burning, particles and possibly CO and polycyclic aromatic hydrocarbons (PAHs) should be examined. Where coal is commonly used as domestic fuel or by industry, SO_2 should also be monitored.

When selecting contaminants for monitoring, it is also important to consider whether there is a guideline value or standard against which the results can be compared. Comparison of monitoring data with guideline values can indicate whether the contaminant is likely to cause adverse health (or environmental) effects (Boubel et al., 1994). Regional, national, and sometimes international guideline values are typically referred for this purpose. Guidelines can be quantitative or highly qualitative.

2.3.2 SELECTION OF MONITORING SITES

Site selection is a very important part of any air pollution monitoring program. The location of a monitoring station is directly related to the area that the data would represent. The selection of the sampling site depends on the objectives of the monitoring program. For monitoring air pollution within an urban airshed, it is usually inappropriate to locate a monitoring site close to a strong localized source, such as a vent from a factory or garage (Harrison and Perry, 1986; Boubel et al., 1994). At these sites, high contaminant values could be recorded that do not generally cause problems and do not reflect overall air quality in an area. However, for specific projects such as for determining the relationship between CO concentrations at a roadside with vehicle emissions, or between an industrial discharge and its impacts on air quality, it is appropriate to place the monitoring equipment close to the source. Care is also needed to select a site that is not strongly affected by local airflows, with either shielding or funneling. The monitoring location should reflect the environment that is being affected or that may be affected. In other words, if adverse effects on human health are a concern, then the monitor should be located where humans have the potential to be affected, either now or in the future. Likewise, if impacts on ecosystems are the issue, then the monitor should be located near a sensitive ecosystem that may be affected.

When selecting a site, it is also important to take into account the relevant averaging period for the guideline of the contaminant of concern. For example, CO causes health problems when people are exposed to levels above the guideline for a 1-hour period or an 8-hour period. Therefore, for CO it is appropriate to place the equipment where people may be exposed over 1- or 8-hour periods. Such sites are located near busy traffic intersections and along traffic corridors, or within wider airsheds, if domestic emissions comprise significant amounts of CO. On the other hand, the health effects of benzene are usually judged on an annual exposure period. Therefore, to ascertain the overall annual exposure of most urban residents, it is appropriate to locate a benzene monitor in an urban residential area, away from the major influence of a busy road. This is because on an annual basis, people do not spend much of their time right next to a busy road. However, for carrying out a more detailed personal exposure assessment, it may be appropriate to site the monitor near the road. Moreover, to collect data for epidemiological studies, it is important to collect information from all the main areas where a person may be exposed. This could include indoor air or in-vehicle air.

Site selection should also take into account atmospheric reactions affecting the formation and destruction of the contaminant. For example, ozone is formed over time, as pollutants in the atmosphere react with each other in the presence of sunlight. It is therefore appropriate to monitor ozone (O_3) where it is most likely to be present, namely downwind from the major sources of its precursors (e.g., NO_x, VOCs, etc.). Similarly, NO_2 is another contaminant that is not discharged into air in significant concentrations, but rather formed over time from the NO released from combustion sources. Therefore a monitor close to a busy road, where NO has not had time to convert to NO_2, may not be the best location for determining NO_2 concentrations.

2.3.3 EQUIPMENT MAINTENANCE AND INSPECTION

Proper equipment maintenance and inspection are crucial for obtaining accurate and reliable monitoring results. Most air pollution monitoring equipment (including data loggers) must be calibrated at certain intervals. Instrumental bias and drift are very common features. Data should be corrected on the basis of calibration results. A record of the calibration procedures and history should be maintained, and must be available when requested. Calibrations can range from simple visual checks on equipment operation through to a detailed examination of individual components and multipoint measurements on complex mixtures of contaminants. Calibration often involves checking whether the monitoring equipment accurately measures the concentration of a contaminant in a sample containing a known concentration. Wherever possible, air quality data should be recorded automatically, with as high a time resolution as practicable, with the use of data loggers (WHO, 1994c).

The accuracy and variability between different methods/equipment should be checked by operating them at the same location simultaneously or using them at different labs to analyze duplicate samples. This is particularly useful for identifying the difference between particulate sampling methods (Larsen et al., 1999). Such comparisons should be made when there is doubt regarding the reliability or accuracy of a particular method or where other, less expensive methods are being evaluated.

2.3.4 DATA VALIDATION AND INTERPRETATION

The purpose of monitoring is not merely to collect data, but to produce information that is useful for technical, policy, and public end-users. Raw data in themselves are of very limited utility. These data first need to be screened by validation and ratification and then collated to produce a reliable and credible dataset (WHO, 1994a,b,c; Kotlik and Vondruskova, 1996; CRC, 1997).

Interpretation of data from any air pollution monitoring site should take into account the site situation and its implications. A general description of the site characteristics and any local sources of air contamination should be included. A record of meteorological variables such as wind speed, wind direction, temperature, and so on should be recorded, and a brief description of meteorological conditions that are likely to affect air quality at the site, for example, inversions, prevailing wind direction, and height above sea level, should also be incorporated (USEPA, 1987; WMO, 1992; Larsen and Schaug, 1995).

2.4   SELECTION OF MONITORING METHODS

A wide variety of methods are available for measuring contaminants in ambient air, with an equally wide variation in cost and precision. Specific monitoring methods should be chosen taking into consideration the purpose and objectives of the monitoring program, the available budget, and whether there is a need to comply with any standard methods and/or national recommendations (Larsen et al., 1999). Air monitoring methods can be categorized as follows.

2.4.1   HIGH-PRECISION INSTRUMENTAL METHODS

These methods provide continuous records of pollutant levels over extended periods of time (weeks or months) with minimal operator intervention, and have a high degree of measurement precision. The detection levels of these systems are often one order of magnitude or more below typical background levels in urban areas. These are also the most expensive monitoring methods, and appropriate calibration and operation are required to ensure that high precision is achieved. High-precision instrumental methods such as differential optical absorption spectroscopy (DOAS) are generally used for research studies or other specific investigations, where there is a need to understand the ways in which pollutant levels fluctuate over short time periods (hours or days).

2.4.2   LOWER-LEVEL INSTRUMENTAL METHODS

Instrumental methods such as automated monitors used for routine air quality monitoring are cheaper than the high-precision methods. However, they usually require more frequent operating checks and adjustments, and the measurement precision is only marginally below typical background levels.

2.4.3   MANUAL PARTICULATE METHODS

Manual filter-based particulate collection methods give time-averaged results of PM, typically over 24 h, and usually require manual changes of the sampling filters in between each sample; however, a number of semiautomated systems are also available. The precision of most manual particulate methods is about 10–20% of typical background levels.

2.4.4   MANUAL WET-CHEMICAL METHODS

Manual wet-chemical methods are usually used for a wide variety of gaseous pollutants such as SO_2 and NO_2; they require regular operator intervention and also produce time-averaged results (typically over 24 h). The precision of most wet-chemical methods is only marginally below typical background levels, and the methods are sometimes subject to interference from other pollutants. The operational costs are very similar to those of the lower-level instrumental monitors (Larsen et al., 1999).

2.4.5 Passive Monitoring Methods

Passive samplers are normally adopted for survey or to classify areas as having low, medium, or high pollution levels, based on long-term (usually monthly) sampling. They do not usually provide anything about short-term pollution levels nor do they allow precise measurement of the actual pollutant concentrations without calibration under field conditions through concurrent real-time measurements using active analyzers. This is the cheapest monitoring option (WHO, 1994a) because it does not require electricity and routine maintenance. Passive samplers are widely used for measurements of trace gases such as O_3, SO_2, and NO_2 in rural or remote areas.

2.5 COMMON MONITORING METHODS FOR SO_x, NO_x, AND PM

2.5.1 Monitoring of Gaseous Sulfur and Nitrogen Compounds

The gaseous compounds of sulfur (S) and nitrogen (N) that are of concern in atmospheric pollution are their corresponding oxides, hydrides, and organic compounds. Of the oxides of sulfur, SO_2 and sulfur trioxide (SO_3) are important pollutants. The main source of SO_2 is the combustion of fossil fuels containing sulfur, in which S is oxidized to SO_2 and to a lesser extent to SO_3. The oxides of nitrogen that are of major concern are nitric oxide (NO) and NO_2. Both chemical and physical methods can be adopted for quantification of these pollutants. Chemical methods for the analysis of gaseous sulfur and nitrogen pollutants can be classified into acidimetric, colorimetric, chromatographic, and coulometric methods, details of which can be seen in the relevant literature (Harrison and Perry, 1986; Lodge, 1989a,b; Vadjic et al., 1992; Quirit et al., 2007). Further, gaseous S and N compounds are measured by physical methods that employ optical techniques such as chemiluminescence, fluorescence, and absorption spectroscopy.

2.5.2 Measurement of PM

Ambient aerosol is a complex mixture of organic and inorganic, volatile and semivolatile, water-soluble and -insoluble matter (Wittmaack and Klek, 2004), possessing a range of morphological, chemical, physical, and thermodynamic properties (Wilson et al., 2002). Atmospheric PM includes combustion-generated particles such as diesel, soot, and fly ash; photochemically produced particles such as those found in urban haze; salt particles produced by sea spray; and crustal particles from resuspended dust (Wilson et al., 2002). Some particles may be hygroscopic and contain particle-bound moisture. The amount of moisture is dependent on the particles' composition and ambient relative humidity. At a relative humidity exceeding 80%, water typically constitutes more than half of the mass of fine atmospheric particles (McMurry, 2000). Measurements of airborne PM are often used for a variety of reasons, including source apportionment, reviewing the effectiveness of control strategies, and investigating the relationship between air quality and health. The highly variable nature of airborne PM makes measurements of both PM_{10} (PM with aerodynamic diameter ≤ 10 μm) and $PM_{2.5}$ (PM with aerodynamic diameter ≤ 2.5 μm) problematic.

Measurement of the mass of PM suspended in air, whether it is ambient air PM or coal dust in mines, requires the accurate quantification of minute amounts of mass. Broadly speaking, there are two ways of approaching this measurement: direct microweighing techniques that fundamentally determine the mass and indirect methods using other properties of particles to estimate the mass.

2.5.2.1 Direct Mass Measurement Techniques

Direct mass measurement techniques include the gravimetric technique. The most simple and direct sampling technique is to collect particles on filters. Through the use of a laboratory gravimetric balance, the difference in pre- and postsample weights yields the PM mass collected. Knowing the volume of air passed through the filter allows the determination of the PM mass concentration. Frequently, this classic method of weighing the deposited mass on a filter is considered as an absolute standard (Federal Register, 1997; Noble et al., 2001).

Soluble portions of aerosol particles can be extracted with water and determined with an ion chromatograph for cation and anion concentrations. Insoluble particles are typically analyzed with instrumental neutron activation analysis (INAA), proton-induced x-ray emission (PIXE), or inductively coupled plasma mass spectrometry (ICP-MS) for elemental composition. Semivolatile components are typically analyzed by gas chromatography by using an electron capture detector, or one coupled with a mass spectrometer (Parungo et al., 1992).

The bulk composition of individual particles collected by particulate monitors varies significantly, reflecting the particles' diverse origin and atmospheric processing (McMurry, 2000). It has been noted that there are numerous uncertainties regarding the relationship between the mass and composition of particles collected on a filter, and the actual composition of PM in the atmosphere (Wilson et al., 2002). Sampling artifacts, such as the loss of semivolatile material, can be found for a range of filter-based measurement systems as reported by Chuen-Jinn and Hsin-Ying (1995). It is observed that due to their high alkalinities, glass fiber filters often adsorb SO_2 and HNO_3 gases to form extraneous sulfate and nitrate in collected particle samples, causing an overestimation of both and possibly affecting the recorded mass of PM. This may occur within both the filter material and the deposited PM. Negative artifacts can occur due to evaporation and/or chemical reactions during sampling. One example is the loss of nitrate due to the reactions of acidic aerosol, such as particulate H_2SO_4, with collected ammonium nitrate on the filter media. This results in the evaporation of nitric acid gas and the loss of ammonium nitrate. In addition, the pressure drop that exists in aerosol sampling systems leads to a decrease in the gas phase concentration of certain species as air flows through the system. This creates a concentration gradient between the air stream and the particulate mass, again leading to the evaporative loss of semivolatile material from the mass collected on the filter (Chuen-Jin and Hsin-Ying, 1995). The extent of these processes varies with location, depending on aerosol mass concentration and composition together with meteorological conditions. Artifacts associated with sampling, transport, and storage have also been reported (McMurry, 2000). One example is the loss of volatile compounds, which can leave the filter after sampling, but prior to weighing or chemical analysis (Chow, 1995). Furthermore, the temperature and humidity history of the dust-loaded filter greatly influences the actual mass on the filter (Allen et al., 1997; Eatough et al., 2003).

2.5.2.2 Indirect Mass Measurement Techniques

The tapered element oscillating microbalance (TEOM) and beta gauge are examples of indirect (equivalent) methods that are able to provide near-continuous measurements of ambient particulate concentrations. Other examples of indirect techniques include optical and filter pressure drop methods. Indirect methods are inherently problematic as there is no consistent physical connection between other particle properties and particle mass.

2.5.2.2.1 Tapered Element Oscillating Microbalances

This is a filter-based, true mass measurement technique offering high-time-resolution capability that represents a major advance in PM monitoring. Over the past several decades, inertial mass detectors called TEOMs have been designed to provide this capability for a number of demanding applications (Meyer et al., 2000). Initially designed for space-related applications, TEOM mass detectors are used in near-real-time PM monitors for ambient air, diesel exhaust, stack emissions, fly ash carbon concentration, and in instruments studying the chemical kinetics of catalysts (Allen et al., 1997; Ayers et al., 1999; Hering et al., 2004).

The TEOM monitor provides a filter-based, direct mass measurement of PM in ambient air by drawing a sample through a sampling inlet, followed by the sample filter, and by continuously weighing the sample filter. TEOM technology removes the mass calibration uncertainty that exists in systems that do not measure mass directly. In addition, possible filter handling errors (at both the sampling site and laboratory) using manual, gravimetric methods are eliminated while having the advantage of providing filter-based mass measurements in near real time. The engineering design of all TEOM-based monitoring instrumentation provides automatic, real-time flow control, sample conditioning, and quantification of PM mass. The TEOM can be used to measure total suspended particles (TSPs) or size-fractionated PM (e.g., PM_{10}, $PM_{2.5}$, and PM_1) mass concentration when used in conjunction with a size-selective inlet. In addition, the TEOM is used as the central mass measurement system in a number of prominent research programs involving studies of semivolatile particles, fine and coarse particles, and particle behavior. Optional add-on components may be used to optimize the TEOM to eliminate particle-bound water, retain specific semivolatile particle constituents, or measure total nonvolatile and volatile particle mass (Meyer et al., 2000; Eatough et al., 2003). Nevertheless, a number of concerns have been raised about the ability of real-time monitors to accurately measure atmospheric particles in that those operated at elevated temperatures, including the TEOM, are reported to lose semivolatile material (Allen et al., 1997; Ayers et al., 1999; Meyer et al., 2000; Hering et al., 2004) with the mass recorded by the TEOM being on average 18.3% lower than the reference method across Europe (Muller et al., 2004).

2.5.2.2.2   Beta Gauge

The beta gauge is a widely used continuous monitoring method used for PM_{10} concentration measurement (Chueinta and Hopke, 2001). It stands on beta-ray absorption in a sample captured on filtering material, which works by measuring the attenuation of ionizing radiation through particulate mass deposited on a filter. It allows for unattended operation over extended periods of time, with a time resolution of about 0.5–2 h. The response of the instrument depends on the beta absorption coefficient of the particulate, and this can vary with chemical composition. However, the variation is not especially great, and this is not a significant limitation in most monitoring applications. The difference between the beta-ray absorption of the exposed and nonexposed filtering material, which is proportional to the mass of the captured suspended particle matter, gives information on its concentration.

2.5.2.2.3   Light-Scattering Instruments

Light-scattering instruments have been available for many years, but were mainly used for monitoring workplace dust exposures. Over the last few years, some of these instruments have been adapted for ambient monitoring, with variable degrees of success. The "workplace" units are relatively cheap and portable, and give a direct readout of particle concentrations. However, their measurement precision and sensitivity are usually quite poor. As such, they are only suitable for low-level survey work. On the other hand, there are some instruments now available that give a much higher level of performance, to the extent that they are acceptable for use in regional networks and for monitoring against ambient guidelines. One such unit is the GRIMM aerosol spectrometer (Massey et al., 2009), which the manufacturers expect to be granted equivalency status by the USEPA in the near future.

The main limitation with light-scattering instruments is that the instrument response depends on the size distribution and the numbers of particles rather than on the total mass of airborne particulate. This can be overcome to some extent by carrying out periodic calibrations using manual filter sampling. However, such calibration "factors" are likely to vary with different monitoring locations and at different times of the year, because of the changes in composition and nature of the airborne particles. Some of the light-scattering instruments can also give an indication of particle size distributions. This may be of value in specific investigations.

2.6   SOURCE APPORTIONMENT AND CHARACTERIZATION

Identifying important pollution sources that contribute to ambient concentrations of pollutants is essential for developing an effective air quality management plan. Air quality models (see details in the next chapter) use mathematical and numerical techniques to simulate the physical and chemical processes that affect air pollutants as they disperse and react in the atmosphere. Based on inputs of meteorological data and source information such as emission rates and stack height, these models are designed to characterize primary pollutants that are emitted directly into the atmosphere and, in some cases, secondary pollutants that are formed as a result of complex chemical reactions within the atmosphere. These models are important to the air quality management system because they are widely used by agencies tasked with controlling air pollution to both identify source contributions to air quality problems and assist in the design of effective strategies to reduce harmful air pollutants. For example, air quality models can be used during the permitting process to verify that a new source will not exceed ambient air quality standards or, if necessary, determine appropriate additional control requirements. Air quality models can also be used to predict future pollutant concentrations from multiple sources after the implementation of a new regulatory program, in order to estimate the effectiveness of the program in reducing harmful exposures to humans and the environment.

Receptor modeling includes observational techniques that use the chemical and physical characteristics of gases and particles measured at source and receptor to both identify the presence of and quantify source contributions to receptor concentrations. They start with the measurement of a specific feature of the air pollutant; for example, in the case of aerosol, it includes particle sizes, size distribution, component identification, chemical state and concentration, and time and spatial variation at the receptor, and calculates the contribution of a specific source type. Receptor modeling focuses on the behavior of the ambient environment at the point of impact (Hopke, 1985). The fundamental principle of receptor models is that mass conservation can be assumed and a mass balance analysis can be used to identify and apportion sources of airborne PM in the atmosphere (Hopke, 1985; Hopke et al., 1991).

Receptor models are grouped into two basic categories: microscopic and chemical methods. Microscopic methods resolve the sources on the basis of characteristic morphological features such as wood fiber, tire rubber, pollen, and so on (Cooper and Watson, 1980). For quantitative predictions, they estimate the number of particles, their density, and their volume. Chemical methods require knowledge of the chemical composition of both ambient and source particulates and are based on an assumed conservation of mass. The degree of validity in this assumption depends on the chemical and physical properties of the species and its potential for atmospheric modification such as condensation, volatilization, chemical reactions, and sedimentation. Chemical methods can be divided into subgroups such as enrichment factor analysis, time and spatial series analysis that primarily provide quantitative information about possible sources, while the chemical mass balance (CMB) and advanced multivariate methods provide quantitative information about sources and are used largely for source impact assessment studies.

2.6.1 ENRICHMENT FACTOR (EF) ANALYSIS

Atmospheric aerosols are expected to have the crust and the sea as their major natural sources in continental and marine areas, respectively. The comparison between the concentration of atmospheric aerosol elements, when they are found to be higher than expected in their natural forms according to their proportions in the background aerosol (earth's crust and sea), is of utmost importance (Zoller et al., 1974). By this comparison, it can be decided whether these elements are anthropogenic or natural in origin. This comparison is made by calculating EFs for the various elements in the aerosol relative to the background usually normalized to an element considered as being the most unambiguous indicator of the source material (Rahn, 1976), which is expressed by Equation 2.1:

$$EF = \frac{[C]_{sample}/[C_{ref}]_{sample}}{[C]_{crust}/[C_{ref}]_{crust}}, \qquad (2.1)$$

where $[C]_{sample}$ and $[C_{ref}]_{sample}$ are the concentrations of C (analyte element) and the normalization reference element in the sample and $[C]_{crust}$ and $[C_{ref}]_{crust}$ are their concentrations in the continental crust. Al, Si, Fe, and Sc are generally used as the normalization reference element for the crust because they are chemically inactive or stable elements while Na is the tracer for sea salt. Elements with EFs close to unity likely have the crust or sea as the main source; they are called "nonenriched" elements. Elements with EFs significantly greater than unity are called "enriched" elements and probably have another major source besides the crust/sea (Rahn, 1976; Gao et al., 2002).

CASE 2.1

Here is an example where Al was used as a reference element to study the particles from biomass burning. See et al. (2007) conducted an intensive field study in Sumatra, Indonesia, during a peat episode to investigate the physical and chemical characteristics of particulate emissions in peat smoke and to provide necessary data for source–receptor analyses. Ambient air sampling was carried out at three different sites at varying distances from the peat fire: the first location was Sungai Sembilan in Dumai, a rural site where peat fires occurred; the second location was in the nearby village of Belakang Rumah in Dumai, representing a semirural site; the third location, Pekanbaru, was the nearest provincial capital in Riau, representing an urban site in this study. To gain insights into the possible sources of trace metals from peat fires, EFs were calculated according to $EF = (X/Al)_{air}/(X/Al)_{crust}$, where $(X/Al)_{air}$ and $(X/Al)_{crust}$ refer to the ratios of the concentration of metal X to that of reference metal, Al, in the air and crust, respectively. $EF \approx 1$ suggests low-temperature crustal weathering and soil remobilization, whereas $EF > 1$ suggests high-temperature man-made processes. However, because the types of crustal materials are unique in different areas and not much is known about uncertainties concerning fractionation during weathering, EFs are not resolvable within an order of magnitude ($EF < 10$) for crustal materials (Kaya and Tuncel, 1997). This means that $EF < 10$ implies that the metal is most likely derived from crustal sources, whereas $EF > 10$ implies that the metal primarily originates from anthropogenic sources.

EFs, based on the average values of total metals, are shown in Figure 2.2 for all sampling sites. The x-axis is intentionally set to cross at 10, so those metals

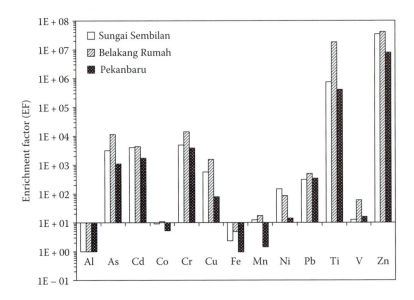

FIGURE 2.2   EF analysis of trace metals with the use of Al as an indicator of crustal source. (Reprinted from See S.W., et al. 2007. *Environmental Science Technology*, 41, 3488–3494. Copyright 2007 American Chemical Society. With permission.)

with EFs < 10 are crustal materials and those with EFs > 10 are anthropogenic or combustion-derived materials. Since the trace metal composition of local peat soils in Indonesia is not available, continental crustal averages (Mason, 1966) were used in this study. Most metals exhibited EF > 10, suggesting that the combustion of peat contributed significantly to the metal loadings in the airborne PM.

The EF approach has been most useful when there has been a limited amount of information available. It has certain limitations, for example, it cannot quantify a source's contribution, relies on the assumed background composition, and is not applicable to complex source mixtures where multiple sources contribute to the same element. The use of aerosol–crust EFs has spread rapidly; however, there is no standardization of reference material or reference element. Both rock and soil vary in composition from place to place, and hence the aerosol originating from them should vary correspondingly. EFs calculated relative to local crust may therefore be more realistic than those calculated relative to a global average. Calculation of EFs using globally averaged crust/soil, however, has the advantage that it can be standardized for all datasets, but local information could profitably be reported on the basis of local EFs.

2.6.2   CMB METHODS

The CMB receptor method tends to extract information about a source's contribution on the basis of the variability of elements measured in a large number of samples. If two or more compounds originate from the same source, their variability as a function of time measured at a receptor will be similar. The objective of this method is to detect this common variability and imply source identity by comparing elements with common variability with elements associated with specific sources (Henry et al., 1984).

CMB methods identify aerosol sources by comparing ambient chemical patterns, or fingerprints, with source chemical patterns (Friedlander, 1973; Kowalczyk et al., 1978). From the viewpoint of the receptor model, conservation of mass is assumed, and the total mass of a given element is the linear sum of the masses of the individual species that arrive at the receptor from each source (Hopke, 1985, 1991; Watson et al., 1991). Thus, CMB is applicable whenever a quantity can be expressed as a linear combination of other species and a mass balance equation can be written to account for all m chemical species in the n samples as contributions from p independent sources:

$$X_{ij} = \sum_{k=1}^{p} C_{ik} S_{kj} + E_j, \tag{2.2}$$

where x_{ij} is the ith elemental concentration measured in the jth sample, C_{ik} is the gravimetric concentration of the ith element in material from the kth source, and S_{kj} is the airborne mass concentration of material from the kth source contributing to the jth sample. E_j represents random errors in the measurement of C_{ik} and S_{ki}, or unaccounted for sources.

There are many ways of solving Equation 2.2, depending on what information is available. Five methods of performing this calculation have been applied: the tracer property (Miller et al., 1972), linear programming, ordinary linear least-squares fitting (Friedlander, 1973; Kowalczyk et al., 1978), effective variance least-squares fitting (Cooper et al., 1984; Watson et al., 1990), and ridge regression.

An implicit assumption in Equation 2.2 is that only inert species can be accepted as tracers in the CMB model. This makes it difficult to use species such as organic compounds as tracers because they may react or degrade during atmospheric transport. In the case where a specific reaction rate constant or decay factor is known or can be determined for a possible tracer, the mass balance equation can be modified to include concentration degradation α. The revised mass balance equation is given by Equation 2.3:

$$X_{ij} = \sum_{k=1}^{p} C_{ik} S_{kj} \alpha_{ij} + E_j, \tag{2.3}$$

where α_{ij} is the decay factor of compound i for source j (Wang and Larson, 1993).

CASE 2.2

The CMB air quality receptor model EPA-CMB8.2 (USEPA, 2004) was used to estimate the relative contributions of peat smoke and other sources to ambient $PM_{2.5}$ concentrations measured at downwind locations by See et al. (2007).

This CMB model has been successfully used by a number of researchers to quantify the relative contribution of various particulate sources in both urban and rural areas (Chan et al., 1999; Chow et al., 1996). The source profiles of inorganic ions and trace elements used in the model could be obtained from SPECIATE 3.2 (USEPA, 2002). In the study on peat smoke by See et al. (2007), source apportionment was attempted to gain further insights into the effect of peat smoke on ambient air quality at Belakang Rumah (Figure 2.3). The contribution of peat fires to $PM_{2.5}$ at Sungai Sembilan is assumed to be 100% and is thus not presented. CMB performance indices are all within the target ranges recommended by the USEPA: $r^2 > 0.8$, $\chi^2 < 4$, and % mass = 80–120% (normalized to 100% in Figure 2.3). The major sources of $PM_{2.5}$ at the two sites include peat fires, motor vehicles, refineries, crustal matter, and secondary aerosols. As deduced from the results of chemical speciation, $PM_{2.5}$ source contribution estimates demonstrated that the influence of peat smoke tended to decrease in significance as the distance from the fires increased, falling to 51.0% at Belakang Rumah and just 18.1% at Pekanbaru.

The CMB approach has shown considerable promise as a receptor model and has been extensively used for source apportionment studies. However, it also suffers from certain limitations; for example, the composition of all components of all possible sources is not known and may also differ from one area to another. For example, oil and oil-combustion components depend on the composition of the fuel, the nature of the plant, and the pollution control technology adopted. Similarly, these models do not account for secondary aerosol particles in the atmosphere, are not definitive enough to identify contributions of individual sources within a class, and do not incorporate the time variation of ambient concentrations and source emissions.

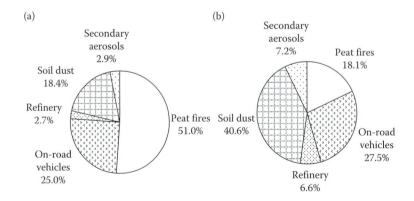

FIGURE 2.3   Source contributions to $PM_{2.5}$ at (a) Sungai Sembilan and (b) Belakang Rumah. (Reprinted from See S.W., et al. 2007. *Environmental Science Technology*, 41, 3488–3494. Copyright 2007 American Chemical Society. With permission.)

2.6.3 MULTIVARIATE RECEPTOR MODELS

Multivariate receptor models utilize statistical techniques to reduce the data to meaningful terms for identifying sources of air pollutants and to estimate the source contributions. Among the tools that have been used for this purpose are cluster analysis, factor analysis, principal component analysis (PCA), target transformation factor analysis (TTFA), positive matrix factorization (PMF), and Q-mode and R-mode factor analysis. These analyses have typically focused on particle composition (Artaxo et al., 1999; Hien et al., 2001; Song et al., 2001; Manoli et al., 2002; Hopke, 2003; Qin and Oduyemi, 2003) and gas concentration (Guo et al., 2004a,b) measurements since their basic assumptions would appear to be the most valid for these types of measurements.

As shown in Equation 2.4, multivariate models (except the Q-mode method) deal with a series of m measurements of aerosol component i during the sampling period or at the sampling site k.

$$X_{ij} = \sum_{k=1}^{p} C_{ik} S_{kj}, \quad k = 1,\ldots,m. \tag{2.4}$$

These models use X_{ij} with the objective of predicting the number of sources, p, and predicting which C_{ik} is associated with which S_{kj}.

2.6.3.1 Cluster Analysis

Cluster analysis is an exploratory data analysis technique that has been widely used for solving classification problems (Facchinelli et al., 2001; Cheng et al., 2007). This technique comprises an unsupervised classification procedure that involves measuring either the distance or the similarity between objects to be clustered. The information obtained from the measured variables is used to reveal the natural clusters existing between the studied samples. Objects are grouped in clusters in terms of their similarity, so that the degree of association is strong between members of the same cluster and weak between members of different clusters. The initial assumption is that the nearness of objects in the space defined by the variables reflects the similarity of their properties.

2.6.3.2 Factor Analysis, PCA, and PMF

In factor analysis, the data are transformed into a standardized form by normalizing the concentration of each element in each sample with respect to the mean value and standard deviation for the element (Equation 2.5) so that each standardized variable has a mean value of 0 and a standard deviation of 1. Factor analysis begins with an eigenvector analysis of a cross-product matrix of the data, frequently the matrix of correlation coefficients. The correlation coefficient is a measure of the extent to which ambient concentrations vary in a similar way:

$$Z_{ij} = X_{ij} - \frac{X}{\sigma_j}. \tag{2.5}$$

The factor model assumes that each variable is linearly related to some number of underlying factors so that the values of variables may be expressed as a set of n linear equations and transforms Equation 2.5 to

$$X_{ij} = \sum_{k=1}^{p} C_{ik} S_{kj} + d_j U_{ij}, \quad k=1,\ldots,m,$$ (2.6)

where

$$X_{ij} = \frac{X_{ij} - \overline{X}_i}{\left(\overline{X}_i^2 - \overline{X}_i^2\right)^{1/2}},$$ (2.7)

$$S_{ik} = \frac{S_{kj} - \overline{S}_k}{\left(\overline{S}_k^2 - \overline{S}_k^2\right)^{1/2}},$$ (2.8)

$$C_{ik} = C_{ik} \left(\frac{\overline{S}_k^2 - \overline{S}_k^2}{\overline{C}_i^2 - \overline{C}_i^2}\right),$$ (2.9)

where C_{ik}, called the factor loading, is related to the source compositions through Equation 2.9 and S_{kj}, the factor score, is related to the source contribution through Equation 2.8. d_j and U_{ij} are unique factor loadings and scores, respectively. The purpose of a factor analysis is to determine a number of such common factors to account for an acceptable amount of variance in the observed data. Common factors can account for varying amounts of the total variance, which is expressed by communality of the variables. The communality of a variable is equal to the sum of squares of the factor loadings given as Equation 2.10:

$$h_i^2 \sum_{k=1}^{m} C_{ki}^2.$$ (2.10)

The communality is 1 if the total variance of the variable in the system is accounted for by common factors, but in most systems it is less than 1.

Factor analysis differs from PCA in which the unique factor is ignored. However, both do not have any constraints on the values of either component loadings or scores, but require that the resulting components be orthogonal. PMF differs from the two in that it requires component loadings and scores to be non-negative, but has no orthogonality requirement. The lack of a non-negativity requirement in PCA has the potential of giving physically unreasonable results in the form of negative values for quantities that must be non-negative. However, in practice this is not usually a problem since, after Varimax rotation, it is typical that for each component all scores (the amounts of the component present) that are not near zero have the same sign;

these can be chosen to be positive. Thus, in practice it is possible to implement an effective non-negativity constraint for absolute PCA scores. The same cannot be said for loadings (the relative amounts of each measured species in the component); it is not clear whether a non-negativity constraint is always appropriate for loadings. For example, negative loadings might represent an anticorrelation between species. For these reasons, it is believed that the non-negativity constraint of PMF is not a large advantage unless physically reasonable results are not obtained with PCA. However, both PCA and PMF are capable of identifying different sources and their composition features without any prior knowledge about the sources (Henry et al., 1984).

Factor analysis of a data matrix can involve correlation between the rows or columns of the data matrix. The method involving correlations between rows is called "R-mode factor analysis" (Oprea and Mihul, 2003), whereas that involving correlations between columns is called "Q-mode factor analysis." An extended Q-mode factor analysis method has also been adopted by Johnson et al. (1984) and differs from these multivariate methods in that it is applied to a single sample and does not depend on time correlations of concentrations.

One of the major problems of factor analysis and PCA is the decision regarding the number of factors m to be retained. As a general rule, the factors should include the maximum variance that is possible and provide a simple interpretation of the factors to sources. Both factor analysis and PCA have been extensively employed to apportion sources of PM along with gaseous species at various locations of the world (Paterson et al., 1999; Vallius et al., 2003; Chan and Mozurkewich, 2007a,b). Recently, a modified version of PCA called supervised PCA (SPCA) for analyzing the adverse health effects of multiple pollutants has been proposed and applied.

CASE 2.3

Tan et al. (2007a,b) used PCA to investigate the potential source of polychlorinated biphenyls (PCBs) in house dust samples collected in Singapore. In this study, PCA was applied to compare the composition of samples relative to commercial mixtures (Aroclor 1242, 1254, 1260; Delor 103; Clophen A30, A40) and the results are shown in Figure 2.4. Most samples matched Aroclor 1254, which is characterized by penta- and hexa-PCBs. Aroclor 1254 is one of the main components of Askarel, a generic PCB-containing product previously used in transformers and capacitors. A similar match has also been observed in marine sediments and green mussels (Bayen et al., 2004; Wurl and Obbard, 2005) in Singapore. A few samples (i.e., sample nos 16 and 25) showed a similar profile to Aroclor 1260 with a predominance of higher chlorinated PCBs, whereas other samples (i.e., sample nos 3, 6, and 7) were close to lower chlorinated PCB-containing commercial products such as Aroclor 1242. However, congener patterns in dust were merely close to, but did not perfectly match, those of commercial mixtures, possibly due to the occurrence of PCBs in homes as a result of contamination of various commercial mixtures.

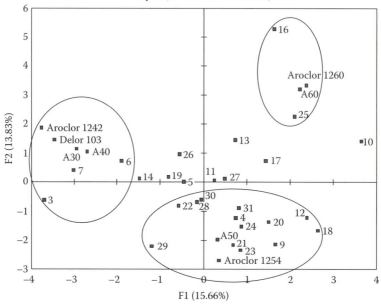

FIGURE 2.4 PCA of dust samples from singapore and commercial PCB mixtures. (Reprinted from Tan J., et al. 2007b. *Chemosphere*, 68 (9), 1675–1682. With permission.)

2.6.3.3 Target Transformation Factor Analysis

The TTFA model tends to extract a maximum of information on the number and nature of sources with no or very limited *a priori* information other than elemental compositional data. This approach has been detailed by Malinowski et al. (1980) and has been applied widely in the apportionment of aerosol mass to sources (Hopke, 1980; Chang et al., 1988). The objectives of TTFA are to determine the number of independent sources that contribute to the system, identify the elemental source profiles, and calculate the contribution of each source to each sample. The number of sources is determined by performing an eigenvalue analysis on the matrix of correlations between samples (Hopke, 1985).

A target transformation determines the degree of overlap between an input source and one of the calculated factor axes. Input source profiles, called test vectors, are developed from existing knowledge of emission profiles of various sources. Unique test vectors, with the value for one element set to 1.0 and those of all other elements set to 0.0, are also analyzed. The uniqueness test can resolve sources that are identified primarily with only one element. To give each element equal importance in target transformation, a weighting scheme has been added to the analysis. Weighting factors for each element are the calculated variance of the element weighted with the experimental error for each datum point. The weighted rotation greatly enhances the ability of target transformation to resolve sources with similar concentration profiles (Liu et al., 1982; Severin et al., 1983).

2.6.3.4 Multiple Linear Regression

Multiple linear regression (MLR) is midway between CMB models and PCA/factor analysis and is based on a linear least-squares fitting process such as CMB. It requires a tracer element or property to be determined for each source category and it is assumed that the amount of tracer in the sample is proportional to the source strength at the receptor. It is performed on the observed total mass using tracers as independent variables. The choice of tracer elements or properties is often based on the knowledge of emission sources and their compositions and can be supplemented by factor analysis.

2.7 SUMMARY

Due to increasing emissions from various sources of modern anthropogenic activities, the problem of air pollution has acquired a level of intense seriousness and urgency. The importance of the issue can be gauged by an initiative taken by the World Meteorological Organization (WMO) in the form of a Background Air Pollution Monitoring Network (BAPMON) program in July 1970, which was consolidated under Global Atmosphere Watch (GAW) in 1989. The objective of the WMO GAW program is to

- Make reliable, comprehensive observations of the chemical composition and selected physical characteristics of the atmosphere on global and regional scales
- Provide the scientific community with the means to predict future atmospheric states
- Organize assessments in support of formulating environmental policy

Various types of instruments and techniques that can be used to carry out direct or indirect air pollution monitoring are in practice. Also, many receptor models are available for identifying sources of air pollutants and for estimating source contributions. Among the tools that have been used for this are factor analysis, PCA, PMF, and TTFA; also, multiple regression in conjunction with these techniques has been employed. These analyses have typically focused on particle composition and gas concentration measurements since their basic assumptions would appear to be the most valid for these types of measurements. In some cases, the composition data are divided into coarse and fine particle fractions. To obtain more insights into particle origins, there have been several attempts to include some particle size information in these analyses. A few studies have applied factor analysis techniques to datasets that included detailed size distribution data. Several studies have used either PCA or PMF. It is observed that in order to obtain meaningful results, the inclusion of appropriate input data and appropriate usage of the method is more important than the specific method used.

In this context, brief information on measurement related issues of other important air pollutants such as ammonia (NH_3), methane (CH_4), volatile organic compounds (VOCs), pesticides (persistent organic pollutants, POPs), and air toxics like benzene, tolulene, and so on are given in Chapter 14.

REFERENCES

Ahmet S. and Van Dijk M. 1994. Monitoring of air quality in the Port Phillip Control Region, 1979–1991, Environment Protection Authority, Government of Victoria (EPAV), Publication 421.

Allen G., Sioutas C., Koutrakis P., Reiss R., Lurman F.W., and Roberts P.T. 1997. Evaluation of the TEOM method for measurement of ambient particulate mass in urban areas, *Journal of the Air and Waste Management Association*, 47, 682–689.

Artaxo P., Oyola P., and Martinez R. 1999. Aerosol composition and source apportionment in Santiago de Chile, *Nuclear Instruments and Methods in Physics Research, Section B*, 150, 409–416.

Ayers G.P., Keywood M.D., and Gras J.L. 1999. TEOM vs. manual gravimetric methods for determination of $PM_{2.5}$ aerosol mass concentrations, *Atmospheric Environment*, 33, 3717–3721.

Bayen S., Gong Y.H., Chin H.S., Lee H.K., Leong E.Y., and Obbard J.P. 2004. Determination of polybrominated diphenyl ethers (PBDEs) in marine biological tissues using microwave assisted extraction (MAE), *Journal of Chromatography A*, 1035, 291–294.

Boubel R.W., Fox D.L., Turner D.B., and Stern A.C. (Eds). 1994. *Fundamentals of Air Pollution* (3rd ed.), Academic Press, San Diego, CA, pp. 195–203.

Bower J.S. and Mucke H.G. 2004. Design, operation and quality assurance and control in a monitoring system. Report on Air pollution in the U.K.

CRC. 1997. Quality assurance procedures manual for ambient air quality monitoring. Canterbury Regional Council (CRC) Internal Publication.

Chan T.W. and Mozurkewich M. 2007a. Simplified representation of atmospheric aerosol size distributions using absolute principal component analysis, *Atmospheric Chemistry and Physics*, 7, 875–886.

Chan T.W. and Mozurkewich M. 2007b. Application of absolute principal component analysis to size distribution data: Identification of particle origins, *Atmospheric Chemistry and Physics*, 7, 887–897.

Chan Y.C., Simpson R.W., Mctainsh G.H., Vowles P.D., Cohen D.D., and Bailey G.M. 1999. Source apportionment of $PM_{2.5}$ and PM_{10} aerosols in Brisbane (Australia) by receptor modeling, *Atmospheric Environment*, 33, 3251–3268.

Chang S.N., Hopke P.K., Gordon G.E., and Rheingrover S.W. 1988. Target transformation factor analysis of airborne particulate samples selected by wind-trajectory analysis, *Aerosol Science and Technology*, 8, 63–80.

Cheng F.L., Hsiao J.H., Cheng S.J., and Hsieh H.H. 2007. Identification of regional air pollution characteristic and the correlation with public health in Taiwan, *International Journal of Environmental Research and Public Health*, 4, 106–110.

Chow J.C. 1995. Measurement methods to determine compliance with ambient air quality standards for suspended particles, *Journal of the Air and Waste Management Association*, 45, 320–382.

Chow J.C., Watson J.G., Lowenthal D.H., and Countess R.J. 1996. Sources and chemistry of PM10 aerosol in Santa Barbara County, CA, *Atmospheric Environment*, 30, 1489–1499.

Chueinta W. and Hopke P.K. 2001. Beta gauge for aerosol mass measurement, *Aerosol Science and Technology*, 35 (4), 840–843.

Chuen-Jinn T. and Hsin-Ying H. 1995. Atmospheric aerosol sampling by an annular denuder system and a high volume PM10 sampler, *Environment International*, 21, 283–291.

Cooper J.A. and Watson Jr. J.G. 1980. Receptor oriented methods of air particulate source apportionment, *Journal of Air Pollution Control Association*, 30, 1116–1125.

Cooper J.A., Watson J.G., and Huntzicker J.J. 1984. The effective variance weighting for least squares calculations applied to the mass balance receptor model, *Atmospheric Environment*, 18, 1347–1355.

Eatough D.J., Long R.W., Modey W.K., and Eatough N.L. 2003. Semi-volatile secondary organic aerosol in urban atmospheres: Meeting a measurement challenge, *Atmospheric Environment*, 37, 1277–1292.

Facchinelli A., Sacchi E., and Mallen L. 2001. Multivariate statistical and GIS-based approach to identify heavy metals sources in soils, *Environmental Pollution*, 114, 313–324.

Federal Register. 1997. Revised requirements for designation of reference and equivalent methods for $PM_{2.5}$ and ambient air quality surveillance for particulate matter; Final Rule: 40 CFR Parts 53 and 58, *Federal Register*, 62, 38764.

Fisher G.W., Graham B.W., and Bell M.J. 1995. Design of a national ambient air quality monitoring network for New Zealand. NIWA Report AK95001 to the Ministry for the Environment, 107pp.

Friedlander S.K. 1973. Chemical element balances and identification of air pollution sources, *Environmental Science and Technology*, 7, 235–240.

Gao Y., Nelson E.D., Field M.C., et al. 2002. Characterization of atmospheric trace elements on $PM_{2.5}$ particulate matter over the New York–New Jersey harbor estuary, *Atmospheric Environment*, 36, 1077.

Guo H., Wang T., and Louie P.K.K. 2004a. Source apportionment of ambient non-methane hydrocarbons in Hong Kong: Application of a principal component analysis/absolute principal component scores (PCA/APCS) receptor model, *Environmental Pollution*, 129, 489–498.

Guo H., Wang T., Simpson I.J., et al. 2004b. Source contributions to ambient VOCs and CO at a rural site in eastern China, *Atmospheric Environment*, 38, 4551–4560.

Harrison R.M. and Perry R. (Eds). 1986. *Handbook of Air Pollution Analysis*, Chapman and Hall, London.

Henry R.C., Lewis C.W., Hopke P.K., and Williamson H.J. 1984. Review of receptor model fundamentals. *Atmospheric Environment*, 18 (8), 1507–1515.

Hering S., Fine P.M., Constantinos S., et al. 2004. Field assessment of the dynamics of particulate nitrate vaporization using differential TEOM and automated nitrate monitors, *Atmospheric Environment*, 38, 637–651.

Hien P.D., Binh N.T., Truong Y., Ngo N.T., and Sieu L.N. 2001. Comparative receptor modeling study of TSP, PM_2 and PM_{2-10} in Ho Chi Minh City, *Atmospheric Environment*, 35, 2669–2678.

Hopke P.K. 1980. Source identification and resolution through application of factor and cluster analysis, *Annals of New York Academy of Sciences*, 338. 103–115.

Hopke P.K. 1985. *Receptor Modeling in Environmental Chemistry*, Wiley, New York.

Hopke P.K. (Ed.). 1991. *Receptor Modeling for Air Quality Management*, Elsevier, Amsterdam.

Hopke P.K. 2003. Recent developments in receptor modeling, *Journal of Chemometrics*, 17, 255–265.

ISO/TR 4266. 1989. Planning of ambient air quality monitoring. International Organisation for Standardisation Technical Report, 15pp.

Johnson D.L., Davis B.L., Dzubay T.J., et al. 1984. Chemical and physical analyses of Houston aerosol from interlaboratory comparison of source apportionment procedures, *Atmospheric Environment*, 18, 1539–1553.

Kaya G. and Tuncel G. 1997. Trace element and major ion composition of wet and dry deposition in Ankara, Turkey, *Atmospheric Environment*, 31, 3985–3998.

Kotlik B. and Vondruskova I. 1996. Workshop of the staff of the Mobile Measuring Systems for Outdoor Air Quality Monitoring—Internal Report. National Institute for Public Health, Prague, Czech Republic.

Kowalczyk G.S., Choquette C.E., and Gordon G.E. 1978. Chemical element balances and identification of air pollution sources in Washington, DC, *Atmospheric Environment*, 12, 1143–1153.

Larsen G. and Schaug J. (Eds). 1995. EMEP Workshop on Quality Assurance of Measurements, Berlin, November 20–23, 1995.

Larsen S., Sluyter R., and Helmis C. 1999. Criteria for EUROAIRNET. The EEA Air Quality Monitoring and Information Network (http://www.eea.eu.int/ Document/Entecrep/default. htm#top). Technical Report No. 12, European Environment Agency, Copenhagen (accessed August 12, 1999).

Liu C.K., Roscoe B.A., Severin K.G., and Hopke P.K. 1982. Source identifications and spatial distributions of aerosol measured at multiple sites across St. Louis, *American Industrial Hygiene Association Journal*, 43, 314–318.

Lodge J.P. (Ed.). 1989a. Method 406: Determination of the NO_2 content of the atmosphere (Griess–Saltzman reaction), In *Methods of Air Sampling and Analysis* (3rd ed.), InterSociety Committee (APCA, ACS, AIChE, APCSA, ASME, AOAC, HPS, ISA), Michigan, USA, Lewis Publishers Inc., pp. 389–393.

Lodge J.P. (Ed.) 1989b. Method 878: Determination of the NO_2 content of the atmosphere (Griess–Saltzman reaction), In *Methods of Air Sampling and Analysis* (3rd ed.), InterSociety Committee (APCA, ACS, AIChE, APCSA, ASME, AOAC, HPS, ISA), Michigan, USA, Lewis Publishers Inc., pp. 389–393.

Malinowski's E.R., Howery D.J., Weiner P.H., et al. 1980. FACTANL-target Transformation Factor Analysis. Quantum Chemistry Program Exchange, Indiana University Bloomington, Inc., 1979, Program 320.

Manoli E., Voutsa D., and Samara C. 2002. Chemical characterization and source identification/apportionment of fine and coarse air particles in Thessaloniki, Greece, *Atmospheric Environment*, 36, 949–961.

Mason B. 1966. The structure and composition of the Earth. In *Principles of Geochemistry* (3rd ed.), Wiley, New York, pp. 45–46.

Massey D., Masih J., Kulshrestha A., Habil M., and Taneja A. 2009. Indoor/outdoor relationship of fine particles less than 2.5 μm ($PM_{2.5}$) in residential homes locations in central Indian region, *Building and Environment*, 44 (10), 2037–2045.

McMurry P.H. 2000. A review of atmospheric aerosol measurements, *Atmospheric Environment*, 34, 1959–1999.

Meyer M.B., Patashnick H., Ambs J.L., and Rupprecht E. 2000. Development of a sample equilibration system for the TEOM continuous PM monitor, *Journal of the Air and Waste Management Association*, 50, 1345–1349.

Miller M.S., Friedlander S.K., and Hidy G.M. 1972. A chemical element balance for the Pasadena aerosol, *Journal of Colloid and Interface Science*, 39, 65–176.

Mucke H.G., Manns H., Turowski E., et al. 1995. European Intercomparison Workshops on Air Quality Monitoring, Vol. 1. Measuring of SO_2, NO and NO_2. Air Hygiene Report 7, WHO Collaborating Centre for Air Quality Management and Air Pollution Control, Berlin.

Muller K., Spindler G., Maenhut W., et al. 2004. INTERCOMP 2000, a campaign to assess the comparability of methods in use in Europe for measuring and composition, *Atmospheric Environment*, 38, 6459–6466.

Noble C.A., Vanderpool R.W., Peters T.M., McElroy F.F., Gemmill D.B., and Wiener, R.W. 2001. Federal reference and equivalent methods for measuring fine particulate matter, *Aerosol Science and Technology*, 34, 457–464.

Oprea C.D. and Mihul A.L. 2003. R-mode factor analysis applied to the exploration of air pollution in the south of Romania, *Romanian Reports in Physics*, 55 (32), 91–110.

Parungo F., Kopcewicz B., Nagamoto C., et al. 1992. Aerosol particles in the Kuwait oil fire plumes: Their morphology, size distribution, chemical composition, transport, and potential effect on climate, *Journal of Geophysical Research*, 97 (D14), 15867–15882.

Paterson K.G., Sagady J.L., Hooper D.L., Bertman S.B., Carrol M.A., and Shepson P.B. 1999. Analysis of air quality data using positive matrix factorization, *Environmental Science Technology*, 33, 635–641.

Qin Y. and Oduyemi K. 2003. Atmospheric aerosol source identification and estimates of source contributions to air pollution in Dundee, UK, *Atmospheric Environment*, 37, 1799–1809.

Quirit L.L., Hernandez K.N., and Lee B.J. 2007. Comparison of several methods for nitrogen dioxide and sulphur dioxide in Metro Manila air, *Science Dilman*, 19 (2), 7–11.

Rahn K.A. 1976. The chemical composition of the atmospheric aerosol. Technical Report, Graduate School of Oceanography, University of Rhode Island, Kingston.

See S.W., Balasubramanian R., Rianawati E., Karthikeyan S., and Streets D.G. 2007. Characterization and source apportionment of particulate matter ≤2.5 μm in Sumatra, Indonesia, during a recent peat fire episode, *Environmental Science Technology*, 41, 3488–3494.

Severin K.G., Roscoe B.A, and Hopke P.K. 1983. The use of factor analysis in source determination of particulate emissions, *Particle Science Technology*, 1, 183–192.

Song X.H., Polissar A.V., and Hopke P.K. 2001. Sources of fine particle composition in the northeastern U.S., *Atmospheric Environment*, 35, 5277–5286.

Tan J., Cheng S.M., Loganath A., Chong Y.S., and Obbard J.P. 2007a. Polybrominated diphenyl ethers in house dust in Singapore, *Chemosphere*, 66 (6), 985–992.

Tan J., Cheng S.M., Loganath A., Chong Y.S. and Obbard J.P. 2007b. Selected organochlorine pesticide and polychlorinated biphenyl residues in house dust in Singapore, *Chemosphere*, 68 (9), 1675–1682.

USEPA. 1987. On-site meteorological program guidance for regulatory modeling applications, EPA 450/4-87-013, Office of Air and Radiation, June 1987, 210pp.

USEPA. 2002. SPECIATE: VOC/PM Speciation Database Management System, Version 3.2, Research Triangle Park, NC.

USEPA. 2004. EPA-CMB8.2 Users Manual. EPA-452/r-04-011, Office of Air Quality Planning & Standards, Research Triangle Park, NC.

Vadjic V., Gentilliza M., Cackovi M., and Eskina I. 1992. Comparison of mass concentrations of SO_2 determined in air by two different methods, *Environmental Monitoring and Assessment*, 21, 19–26.

Vallius M., Lanki T., Tiittanen P., Koistinen K., Ruuskanen J., and Pekkanen J. 2003. Source apportionment of urban ambient $PM_{2.5}$ in two successive measurement campaigns in Helsinki, Finland, *Atmospheric Environment*, 37, 615–623.

Wang S. and Larson T. 1993. Ambient error weighted partial least-squares regression: A new receptor model, *Analytica Chimica Acta*, 272, 333–337.

Watson J.G., Chow J.C., and Pace T.G. 1991. Chemical mass balance, In *Receptor Modeling for Air Quality Management*, P.K. Hopke (Ed.), Elsevier, Amsterdam.

Watson J.G., Robinson N.F., Chow J.C., et al. 1990. The USEPA/DRI chemical mass balance receptor model CMB 7.0, *Environmental Software*, 5, 38–49.

WHO. 1994a. *Passive and Active Sampling Methodologies for Measurement of Air Quality*, World Health Organization, Geneva (GEMS/AIR Methodology Review Handbook Series, Vol. 4, document WHO/EOS/94.3).

WHO. 1994b. *Primary Standard Calibration Methods and Network Intercalibrations for Air Quality Monitoring*, World Health Organization, Geneva (GEMS/AIR Methodology Review Handbook Series, Vol. 2, document WHO/EOS/94.2).

WHO.1994c. *Quality Assurance in Urban Air Quality Monitoring*, World Health Organization, Geneva (GEMS/AIR Methodology Review Handbook Series, Vol. 1, document WHO/EOS/94.1).

WHO. 1998. Air Quality Guidelines for Europe. World Health Organization, Regional Office for Europe, Copenhagen, WHO Regional Publications, European Series-I.

WMO. 1992. Report on the WMO Meeting of Experts on the Quality Assurance Plan for the Global Atmosphere Watch, Garmisch-Partenkirchen, March 26–30, World Meteorological Organization (WMO), Geneva (No. 80, WMO/TD-NO 531).

Wilson W.E., Chow J.C., Claiborn C., Fusheng W., Engelbrecht J., and Watson J.G. 2002. Monitoring of particulate matter outdoors, *Chemosphere*, 49, 1009–1043.

Wittmaack K. and Klek L. 2004. Thermodesorption of aerosol matter on multiple filters of different materials for a more detailed evaluation of sampling artifacts, *Atmospheric Environment*, 38, 5205–5215.

Wurl O. and Obbard J.P. 2005. Organochlorine pesticides, polychlorinated biphenyls and polybromianted diphenyl ethers in Singapore's coastal marine sediments, *Chemosphere*, 58, 925–933.

Zoller W.H., Gladney E.S., and Duce R.A. 1974. Atmospheric concentrations and sources of trace metals at the south, *Science*, 183, 198–200.

3 Air Pollution Modeling
Theory and Application

Chandra S.P. Ojha, Marcelo Mena, Sarath
Guttikunda, Bhola R. Gurjar, and Wenfang Lei

CONTENTS

3.1   INTRODUCTION

Air pollution models are powerful tools for policy makers as they can be used to relate emissions and concentrations. Also, considering that observations are often sparse, models can be used to make inferences on concentrations where there is no information. Combined with the information on population, models can readily be used to estimate exposure and, ultimately, health effects. In the correct framework, the estimation of costs of emission reductions and the economic benefits of reduced health effects and environmental effects can be used as a basis for cost–benefit analysis of emission standards, pollution prevention plans, or day-to-day mitigation strategies. Air pollution modeling has been used for a long time in support of long-term and short-term policies. In the short term, forecast models, for instance, are used to inform the public about possible bad air quality in the coming days (Zeldin and Cassmassi, 1979) and have even been used to decide whether real-time pollution abatement strategies are being put into place, in the case of Santiago and Temuco in Chile, for example (Ulriksen and Merino, 2003; Diaz-Robles et al., 2008). In the long term, air quality models are used to evaluate the effect of pollution abatement strategies in support of studies on emission standards (Mediavilla-Sahagun and ApSimon, 2006) and on attainment of air quality standards in both developed (Chock, 1999; Seigneur, 1999; USEPA, 2007) and developing countries (Hao et al., 2007). In Chile, when nonattainment of particulate matter (PM_{10}) is detected through monitoring stations, dispersion models are used to delimit the area under which specific emission standards will be put into place (Jorquera, 2008). However, it is important to note that for models to be used in support of policy decisions, it is necessary for them to correctly represent "reality."

This can be assessed by using statistical parameters such as root mean square error, correlation coefficients, or mean bias. Indeed, an important part of the air pollution modeling process is the reduction of model error, albeit through systematic (inverse modeling) or unsystematic (brute force sensitivity analysis) evaluation. In this respect, recent advances in inverse modeling, in the form of data assimilation (Sandu et al., 2005; Carmichael et al., 2008), have shown great promise in improving model performance through optimal calculation of initial conditions (Chai et al., 2007; Zhang et al., 2008) or emission inputs (Pan et al., 2007) by a systematic comparison of observations and modeled concentrations. In addition, inverse methods have been used to analyze which geographical areas have the most impacts in nonattainment of pollution standards (Hakami et al., 2006). Furthermore, inverse modeling has been used to identify the location of unknown sources of air pollution again by combining observations and model values (Bady et al., 2009).

In this chapter, we review atmospheric modeling from a deterministic approach (atmospheric dispersion models) alongside more data-intensive statistical models, such as artificial neural network modeling, fuzzy modeling, and certain aspects of statistical/stochastic modeling. However, we will begin with the most basic air pollution model and describe how more complex models evolve around basic principles of simple air pollution models.

3.2 ATMOSPHERIC DISPERSION BASICS

In its simplest representation, an emission source reaching the atmosphere can spread horizontally and vertically through a phenomenon called dispersion, and the shape traced out is approximately conical and called a plume. How the plume spreads will depend on the meteorological conditions (wind velocities, e.g.) and process conditions (temperature of emission, plume rise velocity, etc.) prevailing in the atmospheric boundary layer. The variability of temperature with altitude influences the turbulence characteristics and thus the dispersion of pollutants. The temperature in the atmosphere is governed by incident solar radiation, prevailing wind velocity, and percentage of cloud cover. Depending on the magnitude of these parameters, atmospheric conditions can be classified as follows: class A, extremely unstable; class B, moderately unstable; class C, slightly unstable; class D, neutral conditions; class E, slightly stable conditions; and class F, moderately stable conditions. For example, very low wind speeds (i.e., <2 m/s) and very strong incident solar radiation will lead to extremely unstable conditions (class A). There has been extensive work on providing further details on the classification of different stability levels in the atmosphere and the impact of these on dispersion (Cheremisinoff, 2002). For unstable conditions (L, the Monon–Obukhov length) and stable conditions (L greater than zero), different regimes of the atmospheric boundary layer are described by Holtslag and Nieuwstadt (1986) along with the relevant scaling parameters. Assuming a dominant wind direction along the x axis with a mean wind speed U, vertical and transverse turbulent diffusion (along the y and z axes, respectively), and an R_i term representing the rate of reaction of a specific compound i in sources and sinks, the conservation of species mass can be expressed by the widely used Gaussian plume model as

Equation 3.1: Gaussian plume model mass balance

$$\frac{\partial C_i}{\partial t} + U\frac{\partial C_i}{\partial x} = \frac{\partial_i}{\partial y}\left(K_y\frac{\partial C_i}{\partial y}\right) + \frac{\partial_i}{\partial z}\left(K_z\frac{\partial C_i}{\partial z}\right) + R_i, \tag{3.1}$$

where t is the time, and K_y and K_z are diffusion terms as a function of distance, stability conditions, and wind speed. The steady-state solution of Equation 3.1 for nonreactive species (the change in concentration of i with respect to time and the R_i term are zero), for a continuous emission of mass per unit time of inactive species at a constant rate Q in a stack of height H, is given by Equation 3.2.

Equation 3.2: Gaussian plume steady-state concentration of pollution for a constant source and no reactivity

$$C(x,y,z,H) = \frac{Q}{2\pi\sigma_y\sigma_z U}\exp\left(-\frac{1}{2}\frac{y^2}{\sigma_y^2}\right)\left\{\exp\left[-\frac{1}{2}\frac{(z-H)^2}{\sigma_z^2}\right] + \exp\left[-\frac{1}{2}\frac{(z+H)^2}{\sigma_z^2}\right]\right\},$$

$$\tag{3.2}$$

where σ_z and σ_y are diffusion coefficients in the vertical and transverse directions, respectively. If R_i were not zero but a first-order decay expression, then the solution would be of the form described in Equation 3.3, where k represents the first-order reaction constant.

Equation 3.3: Gaussian plume steady-state concentration of pollution using first-order decay

$$C(x,y,z,H) = \frac{Q}{2\pi\sigma_y\sigma_z U}\exp\left(-\frac{1}{2}\frac{y^2}{\sigma_y^2}\right)\left\{\exp\left[-\frac{1}{2}\frac{(z-H)^2}{\sigma_z^2}\right] + \exp\left[-\frac{1}{2}\frac{(z+H)^2}{\sigma_z^2}\right]\right\}$$

$$\times \exp\left(-\frac{kx}{U}\right).$$

$$\tag{3.3}$$

These diffusion coefficients can be estimated by Equation 3.4.

Equation 3.4: Calculation of diffusion coefficients through atmospheric stability class

$$\sigma_y = \frac{k_1 x}{(1 + x/k_2)^{k_3}}, \tag{3.4a}$$

$$\sigma_z = \frac{k_4 x}{(1 + x/k_2)^{k_5}},$$ (3.4b)

where k_1 through k_5 are constants depending on the atmospheric stability class (Zanetti, 1990).

When wind is very low, U tends to be zero and C will tend to increase to infinity. To deal with such situations, the concept of puff modeling has evolved. Puffs are very similar to pulses, and plumes can be reasonably approximated by a series of equivalent or fictitious puffs. The concentration contribution of a single puff at a point (receptor) can be written in terms of the prevailing values of the puff's variables, that is, M mass of pollutant stored in the puff) and σ_y and σ_z (concentration standard deviations in the y and z directions, respectively), as shown in Equation 3.5 (Nema and Tare, 1989).

Equation 3.5: Gaussian puff model concentration for a receptor

$$C = \frac{M}{(2\pi)^{1.5} \sigma_y^2 \sigma_z} \exp\left[-\frac{(CPX - XR)^2}{2\sigma_y^2}\right]$$

$$\times \exp\left[-\frac{(CPY - YR)^2}{2\sigma_y^2}\right] \exp\left[-\frac{(CPZ - ZR)^2}{2\sigma_z^2}\right],$$ (3.5)

where CPX, CPY, and CPR are the coordinates of the center point of a puff, and XR, YR, and ZR are the coordinates of the receptor.

Details about puffs and pollution dispersion analysis using the puff model are given by Jung et al. (2003). It is interesting to note that in the work of Jung et al. (2003), under the assumption of stationarity, spatially homogeneous flow over the flat terrain represents the Gaussian plume model, as shown in Equation 3.6.

Equation 3.6: Steady-state concentration of pollutant in the Gaussian puff model

$$C(x, y, z, H) = \frac{Q}{2\pi \sigma_y \sigma_z U} \exp\left(-\frac{1}{2}\frac{y^2}{\sigma_y^2}\right)\left\{\exp\left[-\frac{1}{2}\frac{(z - H)^2}{\sigma_z^2}\right]\right\}.$$ (3.6)

The plume model has been commonly used to analyze steady-state continuous gas diffusion, whereas the puff model has been considered for calm wind conditions. These approaches are useful in understanding the nature of dispersion. However, when these simplifications cannot be used in the presence of variable meteorological conditions, transient emission patterns, or reactive species, analytical solutions will not be usually possible and numerical solutions are the only recourse. Similarly, when changes in the air stream are taken as arbitrary in three directions, there is no way of obtaining the concentration other than by numerical calculations (Mori, 2000).

3.3   DISPERSION MODELING AND TYPES
OF DISPERSION MODELS

It is apparent that simple Gaussian models cannot be used to describe more complex conditions. Over the years, other approaches have been introduced to incorporate more processes, and time varying conditions. Initially, models were used to predict the downwind concentration of pollutants, which is useful for the environmental impact assessment of existing or proposed new sources in compliance with local air quality standards. These models work as important tools for developing effective control strategies that reduce the emissions of harmful air pollutants. The dispersion models require the input of data, which may include the following: meteorological conditions such as wind speed, wind direction, temperature (which affects reaction kinetics), sunlight and cloud cover (which affects photochemistry), and rainfall (which affects wet removal processes); emission parameters such as source location and height, stack diameter, exit velocity, exit temperature, and mass flow rate; terrain elevation at the source location (location, height, and width of any obstruction in the path of an emitted gaseous plume); and land use of the study area.

In addition to the above, nonhomogeneous and unsteady conditions also influence the atmospheric dispersion. Nema and Tare (1989) emphasized the need to resort to computational simulation procedures to deal with such situations. Based on the algorithms of Zannetti (1986), these authors developed a set of computer programs to deal with (i) a single steady and continuous source, (ii) multiple steady and continuous sources, (iii) calm wind conditions, (iv) light wind conditions, and (v) transport conditions. While classification of dispersion models into analytical as well as numerical can be one possible approach, this does not appear to be a widely accepted classification. Therefore, it is relevant to look into the types of dispersion models that are prevalent in the literature. Holmes and Morawska (2006) presented a detailed review of dispersion modeling, and much of the material contained in Tables 3.1 through 3.3 is reproduced here with permission from the publisher.

A broad classification of air pollution models includes

- Box model
- Gaussian plume model
- Gaussian puff model
- Lagrangian model
- Eulerian model
- Computational fluid dynamics (CFD) model

These models can be used to model dispersion on a local scale (model street canyons, local influence of stacks, etc.) or to model urban- and regional-scale processes (interactions of multiple source categories and even the interaction between emissions from different cities). The representation of chemistry also varies. Gas-phase chemistry is well established, but modeling particle concentrations and number incorporates more complex processes involved in aerosol chemistry and aerosol dynamics.

TABLE 3.1
Basic Parameters for Models not Containing Aerosol Dynamics Modules

Name Developer	Model Type[a]	Scale[b]	Grid Size	Resolution	Source Type[c]	Pollutants[d]	Output Frequency	Atmospheric Stability[e]	Turbulence[f]
AURORA VITO	B	L	1 × 1 km	NA	L	CO, NO$_2$, SO$_2$, PM$_{10}$	1 h, 24 h, 1 year	NA	Limited AMB
CPB GEOMET	B	L		NA	L	NO$_2$, inert gases		NA	NA
CALINE 4 California Department of Transportation	GP	L	H: 100–500 m	1 m	L	CO, NO$_2$, TSP	1 h, 8 h Worst case	P	VIT, AMB
HIWAY2 USEPA	GP	L	10–100 m but up to 10 km depending	1 m	L	Nonreactive gases	1 h	P	VIT, AMB
CAR-FMI Finnish Meteorological Institute	GP	L	Up to 10 km	H: adjustable V: not defined	L	CO, NO, NO$_2$, NO$_x$, PM$_{2.5}$	1 h, 8 h, 24 h, 1 year	BL	VIT, AMB
AEROPOL Bulgaria	GP	L	H: up to 100 km V: up to 2 km	H: 10–1000 m	P, V	G, P	1 h	P	AMB
ADMS	3D quasi-GP	L, R	3000 grid cells up to 50 km	H: no limits V: no limits	P, A, L	G, P	10 min to 1 year	BL	VIT
CERC									AMB

continued

TABLE 3.1 (continued)

Basic Parameters for Models not Containing Aerosol Dynamics Modules

Name Developer	Model Type[a]	Scale[b]	Grid Size	Resolution	Source Type[c]	Pollutants[d]	Output Frequency	Atmospheric Stability[e]	Turbulence[f]
GRAL	L	L	100 m to 20 km	H: no limits	P, L	G, P	10 min–1 h	BL	Local (k–L model) vertical inhomogeneous turbulence and inhomogeneous 3D wind fields
GATOR	E	L, R, G	Up to global	Depends on scale of area	P, L, A, V	G, P	1 h to 1 year	BL	AMB
OSPM National Environmental, Denmark	GP/Box	L	NA	NA	L	NO_x, NO_2, O_3	1 h	NA	VIT, empirical wind turbulence
STAR-CD	CFD	L	<1 km	H: <1 m; +V: <1 m +	P, L, A, V	G, P	1 min	BL	VIT
ARIA Local ARIA Technologies	CFD	L	Depends on the scaling factor	H: <1 m; +V: <1 m +	P, L, A, V	G, P	Real time	P	VIT, Local (k–L model) vertical inhomogeneous turbulence and inhomogeneous 3D wind fields
PBM	Box	R	H: <50 km; V: variable <2 km	NA	P, L, A	G		NA	NA
CALPUFF California Department of Transportation	Multilayer Non-steady-state GPuff	R	<200 km	H: no limits	P, L, A, V	G, P	>1 h	BL	AMB

Model	Type	Scale	Spatial extent	Dimensions	Source types	Pollutants	Averaging time	TY	Turbulence
SCREEN3	G, P	R	<50 km	H: no limits; V: no limits	P, A, V	G, P	1 h in simple terrain and >24 h in complex terrain	worst-case scenario meteorology	
TAPM CSIRO, Australia	E/L	R	<1000 × 1000 km	H: 0.3 to 30 km; V: >10 m	P, A, V	G, P	1 h, 8 h, 1 year	B, L	$k-\varepsilon$
AERMOD American	Bi Gaussian	L, R	<50 km	H: no limits	P, A, V (L treated as series of V)	G, P	1 h, 24 h, 1 year	BL	AMB
SPRAY ARIA Technologies	L	L, R	<1–100 km	H: 1 m to 4 km; V: 1 m to 4 km	P, L, V	G, P	1 min +	BL	
MISKAM	CFD	L	<300 m	H: 1 m (60 cells In each direction); V: 1 m (20 cells)	P, L, V	G, P	1 min +	BL	AMB
MICRO-CALGRID	CFD	L	<10 km	H: 1 m; V: 1 m	P, L, V	G, P	1 min +	BL	VIT, AMB

Source: Reproduced by Holmes N.S. and Morawska L. 2006. *Atmospheric Environment,* 40, 5902–5928. With permission from Elsevier.

NA, not applicable.

a Model types: B, Box; GP, Gaussian plume; L, Lagrangian; E, Eulerian; CFD, computational fluid dynamics; and Gpuff, Gaussian puff.

b Scale: L, local; R, regional.

c Source types: L, line; P, point; A, area; V, volume.

d Pollutants: G, gases; P, particles.

e Atmospheric stability: P, Pasquill; BL, boundary layer scaling; T, Turner.

f Turbulence: VIT, vehicle-induced turbulence; AMB, turbulence of ambient air.

TABLE 3.2

Processes Included in the Dispersion Models not Containing an Aerosol Dynamics Package

Name Developer	Street Canyon	Building Wake Effects[a]	Topography	Intersections	Plume Rise	Chemistry	Aerosol Dynamics
AURORA VITO	Y	Y	Simple	X	X	X	X
CPB GEOMET	Y	X	Simple	X	X	X	X
CALINE 4 California Department of Transportation	X	Y	Simple	Y	X	DPM	X
HIWAY2 USEPA	X	X	Simple	X	X	X	X
USEPA	X	X			X		
CAR-FMI Finnish Meteorological Institute	X	X	Simple	X	X	DPM	X
AEROPOL Bulgaria	X	X	Simple	X	Y	Y	Deposition
ADMS	Y	Y	Complex	Y	Y	Y	X
CERC				X			
GRAL		X	Complex	X	Y	X	X
GATOR	X	X	Simple		X	Y	Y
OSPM National Environmental Research Institute, Denmark		Y				Y	
STAR-CD	Y	Y	Complex			$NO-NO_2-O_3$ chemistry	
ARIA Local ARIA Technologies	Y	S–S H–S	Complex	Y	Y	Y	X

Model							
PBM	X	Y	X	X	X	S–S, H–S	X
CALPUFF California Department of Transportation	X	X	X	X	Complex	S–S, H–S	X
SCREEN3	X	X	Y	X	Simple and complex	S–S, H–S	Y
TAPM CSIRO, Australia	X	Y GRS	Simplified Glendinning et al. (1984)	X	Complex	S–S, H–S	X
AERMOD American Meteorological Society	X	Y	X	X	Simple and complex	Evaluation version	X
SPRAY ARIA Technologies	X	Simple SO_2 decay	X	X	Simple	Y	Y
MISKAM		Simple (NO–NO_2–O_2 conversion model)		X			
MICRO-CALGRID	Y	Y	Y	X	Simple and complex	Y	Y

Source: Reproduced from Holmes N.S. and Morawska L. 2006. *Atmospheric Environment*, 40, 5902–5928. With permission from Elsevier.

X, not included; Y, included.

[a] Building wake effects: S–S = Schulman–Scire, H–S = Huber–Snyder.

TABLE 3.3
Aerosol Dynamics Models

Name Developer	Dispersion Model	Nucleation	Coagulation	Condensation/ Evaporation	Deposition	Particle Size Method	Particle Composition
UHMA University Helsinki		B + T	Y	Y	D:Y W:X	Hybrid/moving center of retracking methods, 0.7 nm–2 μm	H_2SO_4 inorganics, organics
MONO32	Coupled to OSPM	B + T	Y	Y	D:Y W:X	Four size modes. Monodisperse Approach 7–450 nm	None
AERO	Coupled to UAM IV	Y		Y	D:Y W:X	0.01–10 mm	Inorganic, organic, and elemental carbon. Internally mixed
GATOR	Eulerian	B	Y	Y	D:Y W:X	Moving size or stationary size multiple size sectional 200 groupings	None
MADRID	Coupled to CMAQ	SOA, B		Y	D:Y W:X		
AEROFOR	Sectional Box	B, T	Y	Y	D:Y W:X		Externally or internally mixed varying within each size group

URM	Eulerian	B	X	Y	D:Y W:X	Four groups <10 m	Internally mixed
RPM	Incorporated into RADMII	B	Y	Y	D:Y W:X	0.01–0.07 m	Ammonium sulfate, ammonium nitrate
CIT, California Institute of Technology	Eulerian	B	X	Y	D:Y W:X	0.5–10 m	Organic inorganics
STEM	Eulerian	B	Y	Y	D:Y W:Y		Inorganics
CAMx/PMCAMx	Eulerian	B	Y	Y	D:Y W:Y		Inorganics and organics
CMAQ	Eulerian	B	Y	Y	D:Y W:Y		Inorganics and organics
WRF-Chem	Eulerian	B	Y	Y	D:Y W:N		Inorganics and organics

Y, process included; X, process not included.

[a] Nucleation: B, binary; T, ternary; SOA, secondary organic aerosol formation.

[b] Deposition: D, dry deposition; W, wet deposition.

3.3.1   Box Model

A box model is the most basic model, largely based on mass balance. While its treatment of transport is simplified, the chemistry involved can be very complex. Here, the site is idealized as a box. Pollutants emitted into the box are subjected to physical and chemical changes. The box model also requires the input of simple meteorology and emissions. The movement of pollutants in and out of the box is allowed. The air mass inside the box is assumed to be well mixed. It can include more detailed and complex reaction schemes for gases and particles, which can be tested in a box model before being incorporated into more complex models. However, well-mixed and homogeneous conditions are sometimes unrealistic. These models should be used with the caveat that they do not represent large areas, or conditions other than those incorporated in the model.

3.3.2   Gaussian Plume Model

Gaussian-type models are widely used in atmospheric dispersion modeling for regulatory purposes. This type of model is based on Gaussian distribution of the plume in the vertical and horizontal directions under steady-state conditions, as described in detail in the previous section.

Most Gaussian models consider only the diffusion and advection of pollutants. Recent developments in Gaussian models now include physical processes such as deposition and fast chemical reactions. These models also assume a homogeneous wind field. The simplicity of Gaussian plume models makes their use widespread, despite several limitations:

1. The steady-state assumption: Here, models do not take into account the time required for the pollutant to travel to the receptor.
2. Gaussian models are not well suited for regional modeling of particulates, mostly due to their simplified treatment of secondary aerosol formation.
3. The Gaussian equation is not able to calculate recirculation effects caused by multiples of buildings or at intersections.

Gaussian models are not designed to model dispersion under low wind conditions or at sites close to the source, that is, at distances less than 100 m. These models often overpredict concentrations in low wind conditions.

3.3.3   Gaussian Puff Model

Puff models can handle temporal and spatial variations in meteorological fields. This is done by approximating the continuous emissions from sources with a series of discrete puffs that are advected according to the prevailing wind vector and that grow at a rate appropriate to the prevailing state of atmospheric turbulence (Ludwig et al., 1977). There are two approaches: puff superposition and the segmented plume approach (Ludwig et al., 1977; Zannetti, 1986). The characteristic feature of these models is that the calculation of pollutant diffusion, transportation, and removal is performed in a moving frame of reference attached to a number of parcels as they are transported around the geographical region of interest, in accordance with an observed and a calculated wind field. The growth of the plume is proportional to the square root of downwind distance whereas the growth of the puff is a function of 3/2 the power of downwind distance (Pasquill and Smith, 1983). Puff algorithms consist of the following steps (Zannetti, 1981):

Emission: One puff is generated at each emission exit point.

Advection: The center of the existing puff is moved according to local winds.

Diffusion: All puff diffusion coefficients are calculated according to the local turbulence state.

Deposition, precipitation, scavenging, and chemical decay: The mass of each puff is reduced to take into account these effects.

Contribution to receptors: Computed by summing all single contributions of the existing puffs.

3.3.4 LAGRANGIAN MODEL

The Lagrangian approach is based on studying the property of a particular fluid particle by following its trajectory. Lagrangian models are very similar to box models. The region of air containing an initial concentration of pollutants is considered as a box. The box is considered to be moving and the model follows the trajectory of the box. The concentration is the product of the source term and the probability density function as pollutants move from one place to another. This model incorporates changes in concentration due to mean fluid velocity, turbulence of the wind component, and molecular diffusion. Lagrangian models work well both for homogeneous and stationary conditions over the flat terrain and for inhomogeneous and unstable media conditions for the complex terrain. Lagrangian models are particularly important to study the forward and backward trajectories of emissions. Perhaps the most well-known Lagrangian models are FLEXPART (http://transport.nilu.no/flexpart/model-information) and HYSPLIT (http://www.arl.noaa.gov/HYSPLIT.php), which are available from the Norwegian Institute for Air Research (NILU) and the National Oceanic and Atmospheric Administration (NOAA), respectively. This is mainly because they are easy to use and can be run on web-based tools.

3.3.5 EULERIAN MODEL

The Eulerian approach is based on studying fluid property in a control volume at a fixed point in space, that is, the control volume is stationary and fluid moves through the control volume. Eulerian models use two- or three-dimensional grid cells for solving the differential equations. A mathematical expression containing advection, diffusion, transport, and removal of pollutant emissions is simulated in each cell. Equations 3.7 and 3.8 represent the governing equations.

Equation 3.7: Mass balance for the Eulerian model in gaseous phase

$$\frac{\partial C_i}{\partial t} + \nabla(\mathbf{v}C_i) = \nabla \mathbf{K} \nabla C_i + R_i + S_i + G_i. \tag{3.7}$$

Equation 3.8: Mass balance for the Eulerian model in cloud, rain, or snow phase

$$\frac{\partial s_p C_{ip}}{\partial t} + \nabla(\mathbf{v} - \mathbf{v}_{sp})s_p C_i = \nabla C_i \mathbf{K}_p \nabla s_p + R_{ip} + G_{ip}, \tag{3.8}$$

where C_i is the concentration of species i, C_{ip} is the concentration of species i in phase p (cloud, rain, or snow), $\mathbf{v}$ is the time-averaged wind field, $\mathbf{v}_{sp}$ is the settling velocity for the hydrometeor, sp is the liquid water content of phase p, $\mathbf{K}$ is the turbulent eddy diffusivity tensor, R_i is the rate of reaction of species i, S_i is the emission rate of species i, and G_i and G_{ip} are the rates of interphase mass transfer of species i. The processes of dynamics (transport and dispersion), physics (dry and wet deposition), emissions, and chemistry/photochemistry are taken into account in the governing equation. The equations are discretized and are usually solved numerically using the operator splitting method.

3.3.6 CFD MODEL

CFD models deal with fluid flow in a complex geometry by solving the Navier–Stokes equation (Equation 3.9) and the continuity equation (Equation 3.10) when the flow field is idealized as laminar flow.

Equation 3.9: Navier–Stokes equation for CFD

$$\frac{\partial U}{\partial t} + U \cdot \nabla U = -\frac{1}{p}\nabla p - gk - 2\Omega \times U + v\nabla^2 U, \tag{3.9}$$

where $\partial U/\partial t + U \cdot \nabla U$ represents acceleration, $-(1/p)\nabla p$ represents pressure gradient force, gk represents gravity force, $2\Omega \times U$ is Coriolis force, and $v\nabla^2 U$ is turbulent diffusion.

Equation 3.10: Continuity equation

$$\frac{\partial \rho}{\partial t} + \nabla \cdot \rho U = 0. \tag{3.10}$$

The Navier–Stokes equation and the continuity equation can be solved simultaneously using finite difference or finite volume methods. If the flow is turbulent, then the suitable turbulence model is also used to solve the Navier–Stokes equation. The Reynolds Navier–Stokes equation along with the continuity and turbulence closure models is used for this case. There is enough literature on turbulence closure models (Versteeg and Malalasekera, 2008). The use of CFD in the simulation of air quality is gaining popularity, and many CFD packages are able to simulate air quality variations in complex flow domains. Simulation results are often sensitive to the use of particular turbulence closure models, and therefore there is often a need to tune the model parameters before using them for predictive purposes.

3.4 COMMERCIAL AND OPEN SOURCE SOFTWARE

The previous section described basic components and approaches to dispersion modeling. While there are many versions of models in an open source format, they are not always user friendly. A fairly large number of commercial software have been developed using the different versions of models, as mentioned above. Sometimes specific models are also available for free, but easy-to-use packages are available commercially. The US Environmental Protection Agency (USEPA), in its Technology Transfer Network Support Center for Regulatory Modeling, has compiled a list of recommended and preferred dispersion, photochemical, and receptor models (USEPA, 2009).

It is essential to know the capabilities as well as limitations of existing software. For example, a short-range model that does not incorporate photochemistry cannot be used for ozone formation studies. To present the development of a large number of software, we have developed the presentation into two major categories: software that excludes the specific treatment of aerosol dynamics and software that includes aerosol dynamics. Under the first category, we intend to describe briefly the available software under the subheads of Box, Gaussian, Lagrangian, and CFD models.

3.4.1 Software Excluding Aerosol Dynamics

3.4.1.1 Box Models

Air Quality Modeling in Urban Regions Using an Optimal Resolution Approach (AURORA): This model has been used to represent the concentration of inert and reactive gases and particles (Mensink and Colles, 2003). The model calculates the concentration of pollutants within a street canyon. It assumes uniform concentration over the street. Convections in both the x and z directions are considered inside the box.

Canyon Plume Box (CPB): This model is appropriate for street canyons with height-to-width ratios between 0.5 and 2 (Yamartino and Wiegand, 1986), and was developed by the United States Federal Highway Administration to evaluate concentrations within streets. The model calculates the average concentration of inert gases and NO_2.

Photochemical Box Model (PBM): This is an extension of simpler box models that simulate photochemical smog at an urban scale (Jin et al., 1993). The model is ideal for low and variable wind conditions in the presence of sunlight. It assumes that emissions from point, line, or area sources are homogeneously distributed across the surface of the box and that the volume within the box is well mixed.

3.4.1.2 Gaussian Models

California Line Source Dispersion Model (CALINE4): This model was developed for regulatory purposes (Benson, 1992). It is used for treating traffic emissions by distributing them as an infinite line source divided into a series of elements located perpendicular to the wind direction. Vertical dispersion parameters take into account both thermal and mechanical turbulence caused by vehicles. But since it is a Gaussian model in nature, it is not recommended for modeling at low wind speeds, cannot be used for short distances, and lacks the capacity required for modeling in street canyons.

HIWAY2: This USEPA-developed model (Petersen et al., 1980) is similar to the CALINE4 model. The only difference is that the vertical dispersion parameter considers the effect of vehicles and ignores the effect of the thermal turbulence caused by vehicles.

Contaminants in the Air from a Road (CAR-FMI): This model is based on the equations of Luhar and Patil (1989). It is designed to calculate the hourly concentrations of CO, NO, NO_2, NO_x, and $PM_{2.5}$ from vehicles. The horizontal and vertical dispersion parameters included turbulence terms from ambient wind speed, exhaust velocities, and vehicles.

Operational Street Pollution Model (OSPM): This is a semiempirical model that uses the Gaussian plume equation to derive the direct contribution from the source and a box model to calculate the effect of turbulence on the concentration (Vignati and Berkowicz, 1999). Crosswind diffusion within the plume is disregarded, and the sources are treated as infinite line sources. The model assumes that traffic emissions are uniformly distributed across the canyon. It is unable to represent intermittent fluctuations of wind flow and is therefore not recommended for calculating concentrations on timescales shorter than 1 h; it also does not consider cooling of the exhaust plume after emission.

California Puff Model (CALPUFF): This model has been used to investigate gas and particulate dispersion (USEPA, 1998; Elbir, 2003). It has the ability to model four types of sources, that is, point, line, area, and volume, using an integrated puff formulation incorporating the effects of plume rise, partial penetration, and buoyant and momentum plume rise. It is not recommended for the calculation of timescales shorter than 1 h or where dispersion is heavily influenced by turbulence.

AEROPOL: This is a steady-state dispersion model for inert gases and particles up to 100 km from the source. It can be used for local-scale dispersion, and treatment of building effects is also included. It is limited to flat terrain applications. The model includes an algorithm for plume rise based on the equations developed by Briggs (1975). The model calculates dry and wet deposition as a function of deposition velocity and precipitation, respectively.

AERMOD (AMS/USEPA Regulatory Model Improvement): This is a steady-state model that includes treatment of both surface and elevated sources over simple and complex terrain (Cimorelli et al., 2005). It is able to model multiple sources of different types, including point, area, and volume sources. In the stable boundary layer, distribution is assumed to be Gaussian in both the horizontal and vertical directions. In the convective boundary layer, vertical distribution is described using a bi-Gaussian probability density function, developed by Willis and Deardorff (1981), while horizontal distribution is again considered to be Gaussian in nature. The model is also able to represent buoyant plumes and incorporates a treatment of lofting, whereby the plume remains near the top of the boundary layer before mixing with the convective boundary layer.

UK Atmospheric Dispersion Modeling System (UK-ADMS): This model is developed for the dispersion of buoyant or neutrally buoyant particles and gases (Carruthers and Holroy, 1994). It predicts the boundary layer structure using the similarity scaling approach in a method similar to that by Berkowiicz et al. (1986).

Screening Version of the ISC3 Model (SCREEN3): This very well-known and used model was developed by the USEPA (1995). It is used for regulatory purposes to calculate concentrations up to 50 km from an industrial emission. It can calculate the effect of simple elevated terrain and also the 24 h concentration due to plume impaction in complex terrain.

3.4.1.3 Lagrangian/Eulerian Models

Graz Lagrangian Model (GRAL): This model is designed to represent the dispersion of inert compounds within heterogeneous wind fields. The model calculates concentrations from 10 min up to 1 h for line sources, point sources, and tunnel portals within flat (Oettl and Sturm, 2005) and complex terrain (Oettl and Sturm, 2003). It assumes constant plume rise in the vicinity of the tunnel portal as a function of the temperature difference between ambient air and tunnel flow. The limitation of this model is that it does not consider secondary aerosol formation.

The Air Pollution Model (TAPM): This model was developed by the Australian Commonwealth Scientific and Industrial Research Organization (CSIR) (Hurley et al., 2005). It is an Eulerian grid-based regional dispersion model that includes a Lagrangian particle model for near-source concentrations. The atmosphere is treated as an incompressible nonhydrostatic fluid with the horizontal wind components determined from the momentum equations. This model is characterized as a user-friendly graphic interface, as it incorporates its own meteorological model.

ARIA Flexible Air Quality Regional Model (FARM): ARIA FARM has been developed in order to analyze the dispersion of gases and particles (Silibello et al., 2008), from industrial, transportation, and area sources, up to 1000 km distance with a resolution of between 1 and 10 km. ARIA can process multi- and single-constituent isothermal and nonisothermal gas flows as a function of the thermodynamic properties of gases. The ARIA regional model uses two different theoretical approaches that allow the user to choose the most suitable dispersion model for the application: FARM, which is based on the Eulerian approach, and SPRAY, which is based on the Lagrangian approach. FARM is used to calculate the concentration and deposition of reactive emissions, including photochemistry gases and particles between 50 and 1000 km. SPRAY is used to determine the concentration and deposition of nonreactive emissions over complex terrain and focuses on particle emission. This model has been used for policy-oriented applications as well as for operational air quality forecasts in Europe (Calori et al., 2008).

3.4.1.4 CFD Models

The following models are based on models for dispersion within a street environment and models that exclude specific treatment of aerosol dynamics.

ARIA local: ARIA local is a CFD model that has been used to calculate the real-time dispersion of gases and particles from buses and trains within urban environments (Moon and Albergel, 1997; Albergel and Jasmin, 1998). Different sizes of a grid can be used. The smallest grid size can be below 1 m. If the flow is turbulent, then the k–ε turbulence model can also be used. Pollution sources include point, line, area, and volume releases with the emission generated as either continuous or intermittent release.

Microscale Flow and Dispersion Model (MISKAM): This model can be used for specific applications in urban-scale modeling, incorporating buildings and even landscaping (Balczo et al., 2009). The model does not allow steep topography. Also, it does not include thermal effects, buoyant release, or chemical reactions.

Microscale California Photochemical Grid Model (MICRO-CALGRID): This is an urban canopy-scale photochemical model (Stern and Yamartino, 2001) that uses the flow fields and turbulence generated by the MISKAM model. It incorporates a traffic-induced emission model, MOBILEV (Fige, 1997), and horizontal and vertical advection and diffusion schemes.

Atmospheric Transport Modeling System (ATMoS) Dispersion Model: ATMoS is a mesoscale three-layer forward trajectory Lagrangian puff-transport model. The model was developed for sulfur pollution modeling as part of the Regional Air Pollution Information System for Asia (RAINS-Asia). It was extensively applied for sulfur and particulate modeling studies in Asia at the regional, national, and urban scale (Arndt et al., 1998; Calori et al., 1999; Guttikunda, et al., 2002; Li et al., 2004). The model has flexible temporal and spatial resolution. The horizontal spatial resolution can be varied from 1000 m for an urban-scale study to $1° \times 1°$ (~90 km) for a regional-scale study. However, the use of this model for episodic analysis is not recommended owing to the applied advection schemes, which tend to simplify the interaction between horizontal and vertical layers. The model is applicable to regional- and urban-scale studies, analyzing seasonal and annual air quality for long-term trends and evaluating "what-if" scenarios. The model is run separately for sulfur, nitrogen, and particulate pollution. For particulates, owing to physical and chemical differences in PM_{10} and $PM_{2.5}$, two separate bins are distinguished with varying dry and wet deposition functions. In addition, sulfate and nitrate concentrations are added to the PM_{10} and $PM_{2.5}$ fractions to reflect the contribution of secondary particles.

3.4.2 Models/Software Including Aerosol Dynamics

The following section describes models that incorporate some sort of aerosol dynamics, ranging from local (street and building level) to regional scales. It is not a comprehensive list, but does consider models that have been used in policy applications.

3.4.2.1 Eulerian and Lagrangian Models

Gas Aerosol Transport Radiation Model (GATOR): GATOR is an Eulerian dispersion model that calculates the dispersion of gases and aerosols in urban-scale (Jacobson, 1996, 1997) and mesoscale environments (Lu and Turco, 1997; Jacobson, 2001). It has the option of using either a moving size or a stationary size particle dynamic module and it couples to the Eulerian dispersion approach. The model also calculates the solar irradiance that is vital for calculating photo dissociation from the scattering and absorption curves of particles and gases.

Multimono Operational Street Pollution Model (MONO32OSPM): The particle dynamic model MONO32 is coupled with the simple plume model OSPM to calculate the concentration, size distribution, and chemical composition of particles. For the study of road tunnels, the MONO32 model coupled with STARCD (Gidhagen and Johansson, 2004) simulates the particle number concentration from vehicle traffic very well. MONO32, in conjunction with the CFD model, can be used to predict the aerosol dynamics of particles emitted from vehicles accurately.

University of Helsinki Multicomponent Aerosol Model (UHMA): Korhonen and Lehtinen (2004) developed this size-segregated aerosol dynamics model. The model is designed to include treatment of aerosol dynamics with a focus on new particle formation and growth. The growth of particles depends on both coagulation and condensation onto the particles. The treatment of coagulation in the model is based on Brownian motion, which is the major factor responsible for coagulation of submicrometer particles and is recalculated as a function of particle size at each time step.

California/Carnegie-mellon Institute of Technology (CIT): This model is designed to model dispersion and chemistry within an air shed and incorporates the aerosol model of Pilinis and Seinfeld (1988). It uses the sectional approach to particle size distribution (with three size sections between 0.05 and 10 μm) and aerosols composed of a mixture of organic and inorganic compounds. It assumes that aerosols are in thermodynamic equilibrium. Gas-phase chemistry was modeled using the mechanisms of Russell (1988). Secondary organic aerosol formation is assumed to be from three sources: aromatics, diolefins, and the cyclic ethenes cyclopentene and cyclohexene.

The Urban-to-Regional Multiscale—One Atmosphere Model (URM-1ATM): This model is an updated version of the CIT model (Boylan et al., 2002). It calculates the dispersion and chemistry of pollutants by solving the Eulerian equation for conservation of mass using the finite element variable transport scheme coupled to the updated SAPRC (Statewide Air Pollution Research Center) chemical mechanism (Carter, 2000; Carter, 2003). Aerosol dynamics are modeled using a sectional approach, with four size groups up to 10 μm made up of internally mixed atmospherically relevant particles.

AERO: Lurmann et al. (1997) coupled the AERO model to the UAM IV dispersion model. Size distribution was represented by eight sections between 0.04 and 10 μm and assumed a uniform composition of aerosols composed of organic and inorganic compounds and elemental carbon.

Regional Particulate Model (RPM): This model includes the treatment of particle dynamics incorporated into the Regional Acid Decomposition Mechanism II (RADM II) dispersion model (Binkowski and Roselle, 2003).

AEROFOR2 (Model for Aerosol Formation and Dynamics): The AEROFOR2 model (Pirjola and Kulmala, 2001) uses a sectional modeling approach of over 200 evenly distributed size sections, with logarithmic distribution within each section. The composition of aerosols within each section can be varied for soluble, weakly soluble, and insoluble particles, and through temporal treatment of the dynamics it is possible to follow particle number concentration as well as composition with time.

UNI-AERO (EMEP Aerosol Dynamics Model): This is an aerosol model that incorporates the EMEP dispersion model within equidistant 50 × 50 km horizontal grid cells and 20 size varying vertical layers. Horizontal and vertical advections are determined according to schemes designed by Bott (1989). This model includes the treatment of both primary and secondary particles.

California Photochemical Grid (CALGRID) Model: This model is an Eulerian dispersion model based on the UAM-IV (Urban Airshed Model with Aerosols version 4) model with improvements to horizontal advection (Yamartino and Scire, 1989), vertical transport, deposition, and chemical transformation (Scire and Yamartino, 1989). The model uses horizontal grid sizes between 500 m and 20 km and vertical grid sizes between 20 m and 2 km to create domain sizes between 20 and 1000 km horizontally and up to 10 km vertically to calculate the hourly concentration of both reactive and inert gases and particles within a complex terrain. Atmospheric stability and boundary layer height are calculated using stability categories (Briggs, 1973). Plume rise of buoyant sources within a stable, neutral, or unstable atmosphere is calculated using the treatment of Briggs (1975). This model uses the photochemical mechanism SAPRC to predict the formation of secondary gases (Carter, 2000; Carter, 2003).

The University of Iowa Sulfur Transport Eulerian Model (STEM): This model was developed originally to study the regional transport of sulfur for acid rain regulatory purposes (Carmichael et al., 1991), but has since been used extensively for chemical weather modeling and forecasting because it incorporates meteorologically driven size-resolved primary emissions of sea and dust aerosol and secondary formation of inorganic aerosol (Tang et al., 2004). It uses the SAPRC-99 lumped species chemical mechanism. It has also been used to evaluate the interactions between aerosol species and photochemistry (Tang et al., 2003; Mena-Carrasco et al., 2009a). Finally, it has been used extensively to evaluate the regional export of secondary pollutants, such as ozone or sulfur, from megacities (Carmichael et al., 1998). This model has been used to evaluate and improve estimations of emissions inventories in the United States (Mena-Carrasco et al., 2007). Recent developments of an adjoint version of the model have made it one of the first full chemistry inverse models. This version uses a four-dimensional variational approach to improve modeling performance by optimal estimation of emissions or by assimilating initial conditions (Chai et al., 2007). Practical applications have been improving the estimation of anthropogenic emissions of CO and NO_x (Chai et al., 2007) and the estimation of continent-wide respiration fluxes based on carbonyl sulfide and carbon dioxide aircraft observations (Campbell et al., 2008).

Weather Research Forecast Chemistry (WRF-Chem) Model: The *WRF-Chem* model is a next-generation mesoscale numerical weather prediction system (WRF) fully coupled with chemistry. WRF is a flexible, state-of-the-art atmospheric simulation system developed collaboratively among many institutions (http://www.wrf-model.org), designed to serve both operational forecasting and atmospheric research needs across scales ranging from meters to thousands of kilometers. It features multiple dynamical cores, data assimilation system and a software architecture allowing for computational parallelism and system extensibility. The *WRF-Chem* is developed by NOAA in collaboration with multiple institutions (Grell et al., 2005; Fast et al., 2006). The model simultaneously calculates the transport, mixing, deposition, chemical transformation, and surface emissions (including on-line calculation of biogenic emissions) of tracer gases and aerosols, and the radiation and photolysis rates. It offers a variety of chemistry and aerosol module options, including the SOGRAM (Schell et al., 2001) and MOSAIC (Zaveri et al., 2008) modules that incorporate inorganic and secondary organic aerosol formation.

The model is currently being used for an operational forecast of air quality in the United States (http://www-frd.fsl.noaa.gov/aq/wrf/). It has been used extensively to study the photochemistry in polluted atmosphere in North America, such as examining the impact of power plant NO_x emission reduction on ozone in the eastern United States (Frost et al., 2006), characterizing ozone formation (Tie et al., 2007, 2009; Li et al., 2010) as well as assessing aerosol emissions (Fast et al., 2009) in Mexico City. WRF has also been applied to investigate the circulation pattern in Mexico City (de Foy et al., 2008).

Community Multiscale Air Quality (CMAQ) Model: The CMAQ modeling system (Byun and Schere, 2006) was designed for local and regional modeling of photochemical smog, primary and secondary aerosol, air toxics, and visibility degradation. The approach was to have one model that was able to work at multiple scales. The model was developed by the USEPA for both policy- and science-related goals. For this reason, there are a number of policy-oriented programs associated with CMAQ, such as BENMAP (McCubbin et al., 2004) for the estimation of exposure and benefits of air quality control strategies (Davidson et al., 2007).

Model of Aerosol Dynamics, Reaction, Ionization, and Dissolution (MADRID): The MADRID model (Zhang and Pun, 2004) was coupled to the CMAQ dispersion model in order to simulate dispersion. It uses a multiple size sectional approach with internally mixed particles to describe size distribution. The model includes explicit treatment of all processes except for coagulation.

CAMx and PMCAMx: This Eulerian photochemical dispersion model (www.camx.com) is designed to model gaseous and particulate pollution. The model can use a variety of meteorological models and emission preprocessors. It can use multiple chemical mechanisms, such as CB4 (Gery et al., 1989) or SAPRC-99 (Carter, 2000). It includes complete wet and dry deposition mechanisms and simplified secondary particulate formation. It has been used extensively for policy applications for government agencies. The model has been used to evaluate the effectiveness of state implementation plans for ozone for the USEPA (Morris et al., 2002) and the sensitivity of ozone production to precursor gases in Mexico City (Lei et al., 2007, 2008; Molina et al., 2010). Its most recent inception, PMCAM*x* (Gaydos et al., 2007), was developed to include more complete treatment of secondary aerosol formation, including inorganic aerosol growth, secondary organic aerosol formation, nucleation, and coagulation. Initial results show that the model has shown great promise in predicting the mass and composition of $PM_{2.5}$. Tsimpidi et al. (2009) have added new primary and secondary organic aerosol modules to PMCAM*x* based on the volatility basis-set approach. The resulting PMCAM*x*-2008 was applied to the simulation of secondary organic aerosol formation in Mexico City Metropolitan Area.

3.5 STATISTICAL/PROBABILISTIC AIR QUALITY MODELS

While basic theory on air quality models has been known for quite some time, deterministic models have recently been used in support of emission permits and policy applications. Recent advances in computational capacity have made them useful for air quality forecasting. However, their performance is limited to the quality of inputs: meteorological modeling, emission inventory estimations, boundary conditions, and their representation of complex phenomena (such as wind-driven emissions of sea salt and dust). A study compared deterministic models with statistical models in their performance on ozone modeling and showed that statistical models still outperform deterministic models (Comrie, 2006). Statistical models have been developed with the intention of predicting concentrations based strictly on observations. The main types of statistical models are regression-type models and neural network-type models. The first are models that use correlations to infer information for future concentrations. For example, tomorrow's PM concentration can be correlated to today's air quality and some meteorological factors. These correlations are estimated through multiple variable regressions. On the other hand, neural network models "learn" from previous experience, correlating tomorrow's forecast with the outcome of similar patterns and trends. In comparison to deterministic models, statistical models are simpler and have fewer sources of model error, but they do not provide information on the processes that cause air pollution. Their model outputs are usually discontinuous both temporally (only predict maximum concentrations, no time series) and geographically (only provide information on locations where observations are taken). Statistical models do not explicitly incorporate model inputs such as emissions inventories or episodic events (such as wind storms, biomass burning, etc.).

3.5.1 ARTIFICIAL NEURAL NETWORK MODELS

As indicated in the preceding sections, the ultimate goal of any modeling exercise is to calculate the concentrations of a given pollutant for a known set of input data. Obviously, in deterministic models, the equations of continuity, momentum, energy, turbulence, chemical reactions, etc. need to be solved. Computational efforts will increase, depending on the size of grids and the scale of models used. Suppose that one is interested in computing air quality at a distance of a few kilometers from the polluting source. To know this, computations have to be done for the entire flow domain, including the point of interest where air quality is desired. On the other hand, an artificial neural network model (ANN) can accomplish this job much more easily if it is trained using some known results. To understand the capabilities of ANN, we first begin with a brief introduction. Later, we consider an application that shows why its use is relevant to the subject of air quality modeling.

ANN is an advanced soft computing technology that is widely used in the field of pattern identification, dynamic system prediction, control, and optimization. Artificial neurons are the processing nodes through which information processing occurs. Signals are passed between nodes through connection links. Each connection link has an associated weight that represents its strength and each node typically applies a nonlinear transformation, called an activation function, to its net input to determine its output signal. The architecture of the neural network represents the pattern of connection between nodes, its method of determining the connection weights, and the activation function. The nodes are arranged in layers, namely input layer, hidden layer, and output layer. The input layer receives the input variables and provides information to the network. The output layer consists of the values predicted by the network. The nodes of one layer are connected to the nodes of neighboring layers but are not connected to the nodes of the same layer. The number of hidden layers and the number of nodes in a hidden layer are determined by a trial-and-error procedure. The output of a node in a layer depends on the input it receives from nodes of the previous layer to which it links and the corresponding weight.

The inputs form an input vector $X = (x_1, x_2, x_3, \ldots, x_N)$ and the corresponding weights can be represented as w_{ij}. The output vector $Yj = f(X \cdot W_j - b_j)$, where b_j is the threshold value and called bias, X is the input vector, and W_j is the weight vector. The function f is called an activation function. Its functional form determines the response of a node to the total input signal it receives. The most commonly used form of the function is the sigmoid function.

The ANN model needs to be trained and validated. Hence the datasets are divided into training datasets and testing datasets. After the ANN model is trained with the training dataset, it should be validated with the testing dataset. The ANN model can be considered to be appropriate if it reproduces the testing dataset with reasonable accuracy.

Neural networks are best suited where there is no mathematical relationship between the variables. The ANN model can be applied to the area of temporal interpolation of air quality because the relationship between the variables involved in the system is not exactly known. The number of input and output nodes is problem dependent. The number of hidden layers influences the performance of the network significantly. Also, the error during training can be dependent on the scaling of outputs (Ojha and Singh, 2002). The network will approximate poorly with too few nodes and it will overfit the training data with too many nodes. Hence a trial-and-error procedure is generally applied to decide on the optimal architecture.

3.5.2 Fuzzy Logic-Based Modeling

Recently, fuzzy theory has been used for the ranking of air quality models. For this reason it is considered appropriate to present a brief discussion of fuzzy logic (FL), which is relevant to the work reported here.

Fuzzy set theory, which was pioneered by Zadeh, has been used to represent uncertain or noisy information in mathematical form (Zadeh, 1965). It has been applied for different purposes in engineering, business, psychology, and many other areas. It is a superset of conventional (Boolean) logic that has been extended to handle the concept of partial truth–truth values between "completely true" and "completely false." It is a convenient way of mapping input space to output space. A fuzzy set is a set without a crisp clear defined boundary. The central concept of fuzzy set theory is the membership function that represents numerically the degree to which an element belongs to a set. For example, if an element is a member of a fuzzy set to some degree, the value of its membership function can be between 0 and 1, as determined by eliminating the sharp boundary dividing members of the set from nonmembers (Klir and Foger, 1988). Several fuzzy sets representing linguistic concepts, such as low, medium, high, and so on, are often employed to define the state of variables and are called fuzzy variables. FL has to be a useful and practical technique for modeling a complex phenomenon that may not yet be fully understood owing to its ability to deal with imprecise, uncertain, data or ambiguous relationships among datasets (Metternicht, 2001). FL theory and fuzzy set theory provide an excellent means of representing imprecision and uncertainty in the decision-making process, and for reasoning in such a process (Zadeh, 1983).

FL is basically a multivalued logic that allows intermediate values to be defined between conventional evaluations such as yes/no, true/false, black/white, and so on; it provides a remarkably simple way of drawing definite conclusions from vague, ambiguous, or imprecise information (Klir and Foger, 1988).

3.5.3 RANKING OF MODELS

The ranking of atmospheric dispersion models can be done using statistical methods. But it is difficult to decide about the superiority of a model over all the other models because each model has its own functional characteristics. Hence it is desirable to develop a methodology for selecting an appropriate model. Here, for a better understanding, a model to predict plume dispersion in coastal areas is explained by Park and Seok (2007). To evaluate various modeling strategies, input data are selected from a coastal area near a power plant in Boryeung, Korea. Meteorological data from 1.5 to 60 m height, including upper air data of pressure and temperature with altitude and 1-h SO_2 concentration data, constitute the input data.

In the first step of this method, eight dispersion modeling schemes applicable to complex coastal areas are identified by considering the presence of fumigation phenomena and thermal internal boundary layers and then the performances of each modeling scheme in the prediction of 1-h SO_2 complex coastal terrain are roughly reviewed using as many as eight statistical measures.

In the second step, the statistical score as a single index is calculated by applying fuzzy inference and the best model is determined by comparing the indices. In order to provide a single index based on various statistical measures for selecting an appropriate scheme among eight fumigation modeling schemes, fuzzy inference is used. Here variables are expressed in a triangle or trapezoid membership function. Inference results are quantified into several scores by defuzzification of the output.

3.5.4 ESTABLISHMENT OF MEMBERSHIP FUNCTIONS FOR PREMISE VARIABLE

Different statistical measures are taken as variables of the premise part. They are fractional bias (FB), normalized mean square error (NMSE), geometric bias mean (MG), geometric bias variance (VG), within a factor of two (FAC2), index of agreement (IOA), unpaired accuracy of peak concentration (UAPC), and mean relative error (MRE). Membership functions are classified as good, fair, and poor according to model performance level by considering judging criteria on the measures proposed by many investigators, including Chang and Hanna (2004), Ziomas et al. (1998), and Zawar-Reza et al. (2005).

3.5.5 FUZZY INFERENCE

A fuzzy model with eight inputs and one output structure is used. In this model, eight statistical measures are taken as premise part variables and model performance is a consequent part variable. A set of rules according to the respective number of good, fair, or poor memberships is evaluated. To determine the order of rules, weight is given to each membership. The weights arrived are 7–10 for good (average 8.5), 4–7 for fair (average 5.5), 6 for fair (over), 5 for fair (under), and 1–4 for poor (average 2.5) membership. It is noted that such weights will be problem specific (Ojha et al., 2007).

3.5.6 Selection of an Appropriate Modeling Scheme

Model performance can be evaluated by quantifying the result of fuzzy inference into statistical scores. The results of fuzzy inference based on statistical measures for 1-h SO_2 concentration calculated using eight schemes of fumigation modeling by Park and Seok (2007) indicated that expressing the lateral dispersion of fumigating plumes $\sigma_{yf} = \sigma_y$ (before deposition) +12.5% of the effective stack height appears to be a superior one.

3.5.7 Time Series Analysis

For areas that are susceptible to high pollution concentrations, accurate forecasts of pollutant concentrations as a function of time are necessary so that a person affected with pollutants can plan his activities in advance. Further, prior knowledge of high pollution levels can be used for the timely reduction of emissions by traffic diversion and the shutdown or control of specific industrial units. Using the amounts of data available, one can forecast air pollution. Statistical approaches may be effectively used for developing and testing such a system that can forecast accurately.

3.5.8 Methodology of Time Series Modeling

The initial steps involved in the time series analysis involve plotting the data, checking for stationarity, and examining autocorrelations and partial autocorrelations. The next step includes the use of an iterative procedure for model identification, estimation of parameters, and diagnostic checking (Box and Jenkins, 1976).

3.5.9 Preparation and Preliminary Investigation of Data

The first step in the time series analysis is usually to plot the data and to obtain simple descriptive measures of main properties such as outliers, trend, etc. Another general component is seasonal dependency, which can be visually identified in the series as a pattern that repeats in a time series data plot. The time series to be analyzed should not contain any missing values. The data should be in chronological order with all values lagged at the same time interval. Missing values should be replaced by some reasonable values such as the mean of nearby observations.

3.5.10 Correlogram Analysis

The primary aim of any time series modeling is future prediction about system behavior from existing information and knowledge contained in the system. Hence one needs to find out the dependencies between ordered observations. This can be achieved by correlation analysis. Autocorrelation is the correlation of a series with itself, shifted by a particular lag of k observations. The autocorrelation function is a plot of serial correlation coefficients for consecutive lags.

3.5.11   MODEL IDENTIFICATION PROCESS

Selection of the most appropriate model is a step-by-step procedure. Since most of the probability theory of a time series is concerned with a stationary time series, time series analysis requires turning a nonstationary series into a stationary one. The transformations usually done for stabilizing the variance in the series are logarithmic, square root, or power transformations. If the estimated autocorrelation coefficients decline slowly at longer lags, first-order differencing is usually needed. Once the series is made stationary by proper transformation, iterative model building starts. A tentative model may be specified based on the shape of the autocorrelation function and the partial autocorrelation function. At this stage, one needs to decide how many autoregressive and moving average parameters are necessary to yield an effective but parsimonious model of the process. Parsimonious means that it has the fewest parameters and the greatest number of degrees of freedom among all models that fit the data.

3.5.12   ESTIMATION OF PARAMETERS

The parameters are estimated using iterative methods such as Newton's method, the steepest descent method, or the Levenberg–Marquardt method. In these methods, the sum of squared residuals is minimized. Statistical adequacy of the model is checked by performing diagnostic tests. This is done by examination of the residual series for interdependence. If residuals series do not satisfy the diagnostic requirements, the model should be refined and re-estimation of the parameters should be carried out. This process of checking the residuals and adjusting the values of parameters continues until the resulting residuals contain no additional structure. The statistics used for diagnostic purposes include an autocorrelation function plot of residuals: If all the autocorrelations and partial correlations are small, it can be assumed that the model is appropriate for forecasting. Also, the model should minimize the sum of squared residuals.

3.5.13   APPLICATION OF TIME SERIES ANALYSIS

The time series analysis method can be applied to the spatial and temporal interpolation of air quality data. Romanowicz et al. (2006) applied time series analysis methodology to the daily nitric oxide (NO) concentrations measured at 23 stations around Paris. The analysis is divided into two parts: (1) time series analysis of the data and (2) development of combined spatial and temporal analysis techniques using NO observations from 19 stations. Log-transformed daily data from 19 stations were decomposed separately into log-medium and harmonic components using dynamic harmonic regression (DHR) analysis (Young et al., 1999). The DHR model can be represented as

$$y_t = T_t + S_t + e_t, \quad e_t : N\{0, \sigma^2\}, \tag{3.11}$$

where y_t is the observed time series, St is a seasonal/cyclical component, e_t is a noise component, modeled as mutually independent Gaussian random variables with zero mean and variance σ^2, and T_t is a smoother, long- to medium-term trend component, without any periodicity, that reflects the part of the series not accounted for by the seasonal/cyclical and irregular components. The seasonal/cyclical S_t component was modeled in time variable parameter trigonometric form:

$$S_t = \sum_{i=1}^{R} \{a_{it} \cos(\omega_i t) + b_{it} \sin(\omega_i t)\}, \tag{3.12}$$

where ω_i, $i = 1 - R$, are the fundamental and harmonic frequencies associated with the seasonality/cycles in the series and a_{it}, b_{it} are the parameters that are allowed to vary in time over the observation interval if this is indicated by the analysis and optimization of the model. The results of this analysis of 19 stations are used for further spatiotemporal analysis of the data. For spatiotemporal analysis, the DHR model was represented as

$$y_{it} = T_{it} + \sum_{j=1}^{2} \{a_{ijt} \cos(2j\pi t/7) + b_{ijt} \sin(2j\pi t/7)\} + \varepsilon_{it}, \tag{3.13}$$

where $i = 1, 2, \ldots, 19$, $t = 1, 2, \ldots, N$ at the ith location, y_{it} is the logarithm of NO concentration measurement, N is the number of time periods, and ε_{it} is the observation noise, which is assumed to be zero mean, normally distributed, white noise. Thus, it is possible to satisfactorily describe the spatiotemporal characteristics of data using time series analysis.

3.6   CASE STUDIES

The following is a selection of examples that use statistical and deterministic models in support for policy decisions. Many of these examples are taken from the Chilean experience as the Chilean Environmental Commission (CONAMA) has historically used deterministic models in support of long-range air pollution control strategies and statistical models in support of real-time pollution abatement strategies (i.e., condition some emission restrictions based on model predictions). Also, there are some selected case studies in Asia, with the intention of showing that great progress can be achieved in air quality modeling with very simple tools. The selections from Asia have the intention of showing a broad range of applications in estimating specific benefits on health due to restrictions on emissions on specific sectors.

3.6.1   Deterministic Modeling

3.6.1.1   Using CALMET + CALPUFF to Define Geographical Extension of Nonattainment Areas

A very common activity for regulators is to estimate the geographical extension of nonattainment areas. This type of study originates when a particular region shows exceedance in its air quality standard. Usually nonattainment is inferred if a particular monitoring station's three-year mean exceeds 50 µg/m³ or if the 98% percentile of 24-h PM_{10} exceeds 150 µg/m³. To evaluate the geographical area that is in nonattainment (under which emissions are regulated according to a pollution prevention plan) in Chile, dispersion models are routinely used. Figure 3.1 shows how CALMET and CAM_x are used to estimate which geographical area is under nonattainment (considering that observations are only available in a limited area). This is done by modeling a full year of PM_{10} concentrations using the currently available emissions inventories developed by CONAMA for the region, including mobile sources, residential sources, and regional industrial sources. This work was carried out by Universidad Católica by request of CONAMA (Jorquera, 2006).

Similarly, CONAMA assessed the impact of particular large industrial sources (smelters and power plants) in nonattainment of Chile's SO_2 96 ppbv 24-h mean air quality standard. For this specific air quality, episodes were modeled under CALMET CAM_x (Figure 3.2). Source contributions were also estimated by the use of tracers, which allowed estimating the specific contribution of a large point source to the total observed values. Figure 3.3 shows that the high observed values of SO_2 at different

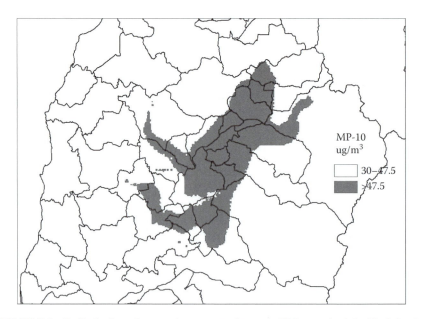

FIGURE 3.1   Delimitation of nonattainment area in yearly PM_{10} standard for Sixth Region of Chile using CALMET-CAMx during Rancagua Region State Implementation Plan.

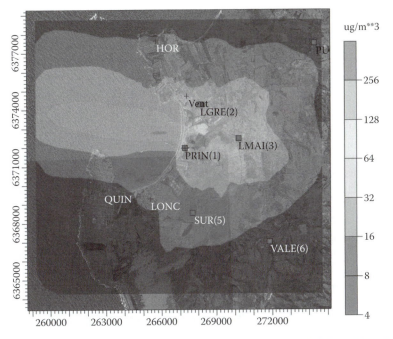

FIGURE 3.2 Map of mean surface SO_2 (g/m³) modeled under CALMET/CALPUFF for Ventanas Pollution Prevention Plan

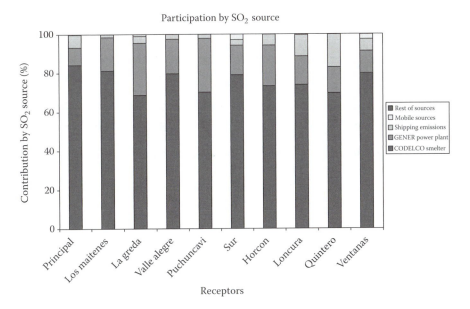

FIGURE 3.3 Percent contribution of SO_2 sources to total observed concentrations, based on CALMET/CALPUFF receptor modeling.

monitoring stations are largely due to contributions from the CODELCO (Chilean National Copper Company)-owned Ventanas smelter.

3.6.1.2 Using WRF-Chem for Regional Modeling of Ozone in Chile

WRF-Chem has been used in Chile for diagnostic work in terms of contributions of specific sources toward air quality violations, specifically for ozone. Also, the work analyzed the interactions between regional sources of pollution (industrial, urban, etc.). The model was run by Universidad de Chile for the Chilean National Environmental Commission, CONAMA (Schmitz et al., 2008). Full chemistry analysis was carried out during January 1–15, 2008 (high-ozone season), with a 4-km resolution, using a 115 × 127 grid, with 36 vertical levels, using a 15-s integration time step and hourly outputs. Figure 3.4 shows the mean maximum concentrations for Central Chile. These results were valuable in showing that maximum ozone concentrations in Santiago were located northeast of the city, beyond the location of ozone monitors. This study was also important in suggesting that Santiago tropospheric ozone formation was largely volatile organic compounds (VOC) limited, which is important to understand the design of ozone pollution control strategies. Figure 3.5 shows a sensitivity run with the objective of evaluating the impact of a specific power plant on ozone formation. This is calculated as the difference in ozone concentration with and without the point source. In this case, the large point source locally decreases ozone formation (possibly due to titration) and, further downwind from the source, ozone formation increases.

3.6.1.3 Deterministic Air Quality Forecasting in Chile

The Chilean Meteorological Office (Dirección Meteorológica de Chile) has been developing a chemical weather forecast using the POLYPHEMUS model (Mallet et al., 2007) coupled to the MM5 meteorological model. The model was run using 3-km resolution using the CONAMA 2005 emissions inventory for Santiago. This model is available online (http://www.meteochile.cl/modelos.html) and is also projected in Google Earth. Indeed, ozone modeling has shown a lot of progress, as gaseous-phase chemistry is better understood and represented in models than PM modeling, which involves heterogeneous chemistry, and a broader scope of emission sources and precursors. Model outputs and comparison to observations are shown in Figure 3.6.

3.6.1.4 Regional Air Quality Forecasting in Support of VOCALS Campaign

During October and November 2008, the VOCALS (Vamos Ocean Cloud Atmosphere Land Study) was carried out off the coasts of northern Chile. The campaign consisted in evaluating the effect of anthropogenic aerosol on stratocumulus in the Southeast Pacific. The Universidad Andres Bello, University of Iowa, and MIT collaborated in implementing Chile's first regional chemical weather forecast using the STEM model for chemical transport and WRF for meteorological forecasting. One of the first challenges was to develop a South American Emissions Inventory, which consisted of compiling the best available information on anthropogenic emissions. For this area, some emissions inventories were developed by the Chilean Environmental Commission (CONAMA) and SECTRA (Chilean Secretariat for Transportation),

FIGURE 3.4 Mean simulated daily maximum ozone concentrations in Central Chile, January 1–15, 2008, using WRF-Chem at 3 km resolution. (From Park O.H. and Seok M.G., *Atmospheric Environment*, 41, 2007. 6095–6101. With permission.)

which included primary emissions of $PM_{2.5}$, PM_{10}, total NO_x, CO, SO_2, and unspeciated total VOCs, for the industrial, transport, and residential sectors. For regions in which no emissions inventories were available, the EDGARV3 FT2000 (Olivier et al., 2005) global emissions inventory was used for gaseous species. Emissions of organic and black carbon for all anthropogenic categories were estimated based on the inventory developed by Bond et al. (2004). These regional and national totals were distributed based on population densities obtained from the LANDSCAN 2006 ambient population model, developed by the Oakridge National Laboratory (Dobson et al., 2000). Figure 3.7 shows how the model was able to correctly predict regional

FIGURE 3.5 Influence of a large power plant on ozone formation, calculated as the difference between modeled ozone with and without emissions from the source. Light shading denotes an increase in ozone concentration greater than 1 μg/m³; dark shading denotes a decrease in ozone concentrations. The dot denotes the location of the point source. WRF-Chem at 3 km resolution was used.

transport of anthropogenic pollution (represented as CO) from Central Chile to the Pacific Ocean, as compared to the observed cloud brightness temperature difference as detected by the GOES satellite. Air masses impacted by anthropogenic sources tended to show brighter clouds, with smaller cloud droplet sizes (Mena-Carrasco et al., 2009b). More information on model configuration is available at http://www. cgrer.uiowa.edu/VOCALS-BA/.

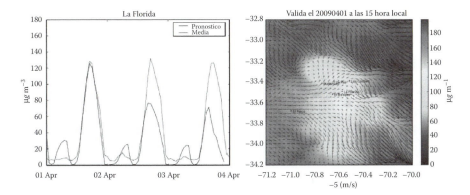

FIGURE 3.6 Left panel: Observed and modeled ozone in La Florida Station in Santiago using Polyphemus. Right panel: Low layer modeled ozone during April 1, 2009 using Polyphemus at 3 km resolution.

3.6.1.5   Inverse Modeling of Emissions Inventories in Asia During Trace-P Campaign

Improvement in model performance will be achieved by error covariance analysis (initial approach; Mena-Carrasco et al., 2007) and further improvements will be achieved by data assimilation using the STEM4DVar (Chai et al., 2006, 2007; Mena Carrasco, 2008) or WRF-Chem 4DVar (under development). Previous experience in data assimilation in Asia during Trace-P showed that assimilation of CO observations led to increased model performance (Mena-Carrasco, 2008). This strategy will be useful to scale related emissions (VOC and NO_x) accordingly, as many times these species are coemitted. The 4DVar approach minimizes the cost function, J, by reducing the difference between observations (y) and modeled values ($h(c)$) by modifying emission scaling factors (E_f), taking the uncertainty of emissions (B)

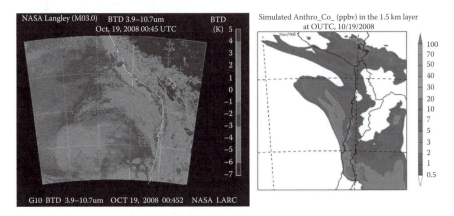

FIGURE 3.7 Left: Observed GOES satellite cloud brightness temperature difference. Right: Modeled anthropogenic CO using WRF-STEM at 12 km resolution and South American Integrated Emissions Inventory.

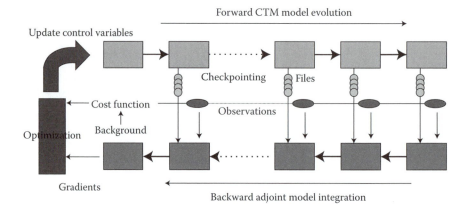

FIGURE 3.8 Modeling framework for STEM 4DVar chemical data assimilation [Adapted from Chai T.F., Carmichael G.R. et al. 2007. *Journal of Geophysical Research—Atmospheres*, 112 (D12): 1–18.]

and observations (*O*) into account, and by balancing the *a priori* (left term) with the *a posteriori* using the μ term. The linearization of the cost function is possible through the use of Lagrangian operators. A schematic that shows data assimilation is shown in Figure 3.8. Usually 4DVar is CPU intensive, but this simplified approach allows multiple optimizations at a lower cost. Figure 3.9 results from a previous project in which correlation coefficients improved from $R^2 = 0.59$ to $R^2 = 0.73$, and

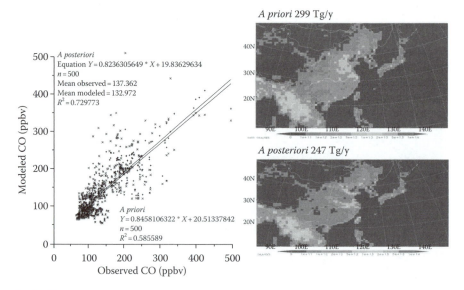

FIGURE 3.9 Left: Comparison of *a priori* and *a posteriori* modeled values (90 km tracer STEM model) to observed CO values (Glen Sachse) for the Trace P (March 2001) period. Right: Prior and posterior (assimilated) anthropogenic and biomass burning emissions of CO (molecules/cm²/s) for the Trace P period.

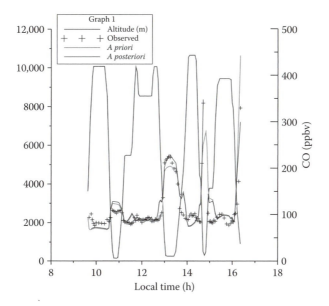

FIGURE 3.10 Observed versus modeled CO during Trace P for *a priori* and *a posteriori* STEM modeling of DC-8 observations.

suggested underestimation of emissions in large megacities, and overestimation of emissions in remote areas. Ultimately, the methodology constrained total emissions (from 299 and 249 Tg of CO). This approach allowed rapid improvement of model performance with a relatively small CPU time. Figure 3.10 shows how improved recovery of emissions allows improved representation of observations through the model.

$$J = \frac{1}{2\mu}[E_f - 1]^T B^{-1}[E_f - 1] + \frac{1}{2}[y - h(c)]^T O^{-1}[y - h(c)]. \qquad (3.14)$$

3.6.2 Methodology for Urban Health Impacts Analysis

Previous studies showed specific applications of models to predict the concentrations of pollutants, to delimit air quality nonattainment areas, to assess the effect of specific sources, or to forecast air quality locally or regionally. Air pollution has long been associated with large health impacts that, if unaddressed, become a load on emerging economies, as many times the government must fund treatment for disease associated with pollution. Table 3.4 shows a summary of recent news reports regarding health impacts. To quantify the cost effectiveness of implementing pollution, there is a need to estimate effects based on environmental exposure. This can be done once the concentrations of pollutants are estimated for a base case and a regulated case. For the case studies presented below, following ambient air pollution analysis, the simulated baseline and controlled scenarios were modeled using

TABLE 3.4

Reports of Health Impacts Due to Air Pollution in the Public Information Domain

Country	City	Mortality Reported	Reference
Worldwide		Climate change causes 315,000 deaths a year	*Reuters*, May 29, 2009
Pakistan		Air pollution kills ~23,000 annually	*The News*, May 15, 2009
Worldwide		100 million premature deaths could be prevented by cutting global emissions by 50% by 2050	*Guardian*, May 12, 2009
India		Over 20 million people have asthma and susceptible to higher risks	*Hindustan Times*, April 25, 2009
Pakistan		Pollution kills 55,000 every year	*The Nation*, April 24, 2009
United States		Air pollution shortens life	*BBC*, April 12, 2009
UAE		Air pollution is a major public health issue	*Gulf News*, April 12, 2009
Worldwide		Up to 60,000 premature deaths a year worldwide are due to PM emissions from ocean-going ship engines	*Guardian*, April 9, 2009
Bangladesh	Dhaka	Air pollution is causing ~15,000 premature deaths a year	*IRIN*, April 3, 2009
Worldwide	By WMO	Lethal air pollution booms in emerging nations	*AFP*, March 22, 2009
India	Delhi	Country's asthma capital	*Mail Today*, March 1, 2009
Indonesia	Jakarta	50% of the professionals may be literally sick of work due to air quality	*Jakarta Globe*, February 19, 2009
Russia		Air pollution is responsible for 17% of diseases in children and 10% in adults, and affects 44% of the population	*Russian News & Information Agency*, February 13, 2009

continued

TABLE 3.4   (continued)
Reports of Health Impacts Due to Air Pollution in the Public
Information Domain

Country	City	Mortality Reported	Reference
Uganda		Do we have to live with pollution?	*The New Vision*, February 10, 2009
Afghanistan	Kabul	Air pollution is hastening the deaths of 3000 every year	*IRIN*, January 29, 2009
Bahrain		More than 10% of the population suffers from asthma and the number is set to rise as air pollution increases	*Gulf News*, February 8, 2009
Hong Kong	Hong Kong	Air pollution caused more than 6600 premature deaths	*AFP*, January 22, 2009
Philippines	Manila	Air pollution kills 5000 annually	*Manila Times*, November 7, 2009
China	Beijing	Pollution sparks health worries	*Telegraph*, August 7, 2007
India	Hyderabad	Air pollution causes ~2500 premature deaths annually	*IES*, March 15, 2008
China		The combined health and nonhealth costs of outdoor air and water pollution for China's economy comes to around US$100 billion a year	*World Bank*, July 7, 2007
Worldwide	By WHO	Air pollution in the world cities is causing some 2 million premature deaths every year	*ENS*, October 6, 2006
Mongolia	Ulaanbaatar	Air pollution causes ~7000 premature deaths	*UE*, June 15, 2007
Worldwide		Anthropogenic climate change of the past 30 years already claims over 150,000 lives annually	*Nature*, November 2005

the same dispersion characteristics from the base year calculations, and exposure analysis for each of the scenarios was conducted using the equations and dose–response functions (DRFs) from health studies across the world (Lvovsky, et al., 2000; Pope, et al., 2006; CARB, 2008; Guttikunda, 2008; Mehta, 2008).

The fundamental equation utilized for estimating health impacts is as follows:

$$\delta E = \beta * \delta C * \delta P, \tag{3.15}$$

where δE is the number of estimated health effects (various end points for mortality and morbidity).

β is the DRF for a particular health end point, defined as the change in number of cases per unit change in concentration. This is established based on epidemiological studies conducted over a period of time, analyzing the trends in hospital records and air pollution monitoring. More information on DRFs is presented in the next section.

δC represents the change in concentrations; this could be the change in concentrations between two scenarios being simulated or the concentrations measured above a certain threshold value. However, WHO claims that there is no threshold over which health impacts are measured. In general, impacts are felt at the minute fluctuations in pollution.

δP is the population exposed to the incremental concentrations above; this could be on a grid-by-grid basis or for the city or region as a whole, depending on the level of information available and the goal of the analysis.

DRFs: Epidemiological studies in industrial and developing countries have shown that elevated ambient PM levels lead to an increased risk of mortality and morbidity. Health effects range from minor irritation of the eyes and the upper respiratory system to chronic respiratory disease, heart disease, lung cancer, and death. Air pollution has been shown to cause acute respiratory infections in children and chronic bronchitis in adults. It has also been shown to worsen the condition of people with preexisting heart or lung disease. Among asthmatics, air pollution has been shown to aggravate the frequency and severity of attacks. Both short-term and long-term exposures have also been linked with premature mortality and reduced life expectancy. The Health Effects Institute (HEI, USA) conducted a detailed literature survey on the impact of outdoor air pollution on human health, and the publication "Health Effects of Outdoor Air Pollution in Developing Countries of Asia: A Literature Review (2004)" includes an extensive list of references on the DRFs for various end points and methodologies to conduct epidemiological studies to develop these DRFs.

HEI's latest study under the PAPA program concludes that the "finding of a 0.6% increase in mortality for every 20 µg/m^3 of exposure to particulate air pollution is strikingly similar to comparable western results (which range from 0.4% to 0.6%) and provide increased confidence in the new Asian results. A key finding of the study is that the effect of air pollution on daily mortality remained consistent even as the degree of pollution increased to high levels, proceeding in a largely linear pattern to levels over 100 µg/m^3 (a level five times the current WHO PM$_{10}$ guideline of 20 µg/m^3)."

In other words, there is no real need to conduct epidemiological studies every time we need to assess the DRFs in a city. These studies are time consuming and constrained by budgets. Of course, if a program has enough time and resources, the

TABLE 3.5
Average DRFs for Morbidity End points

Morbidity Health End Point	DRF (β) (Effects/1 µg/m³ Change/per Capita)
Adult chronic bronchitis	0.000040
Child acute bronchitis	0.000544
Respiratory hospital admission	0.000012
Cardiac hospital admission	0.000005
Emergency room visit	0.000235
Asthma attacks	0.002900
Restricted activity days	0.038280
Respiratory symptom days	0.183000

Sources: Lvovsky et al. (2000), Bell et al. (2006), Pope et al. (2006), Ostro et al. (1998), Li et al. (2004), World Bank (1998), HEI (2004), Ostro et al. (1994), Xu et al. (1994), SAES (2000), and World Bank (2007a).

city should conduct its own epidemiological studies to investigate these functions and utilize the necessary data for impact assessment and decision making.

Based on studies conducted in the past, Table 3.5 presents an average set of DRFs for morbidity end points.

Dhaka, Bangladesh: In Dhaka city, with an estimated population of 15 million, both energy and infrastructure demands are increasing but the amenities (including environmental) cannot keep pace with the growing demand. A study in 2008 concluded that an estimated 15,000 premature deaths, as well as several million cases of pulmonary, respiratory, and neurological illness, are attributed to poor air quality in Dhaka (AQMP/World Bank, 2007). Figure 3.11 presents an overview of the measured monthly average $PM_{2.5}$ concentrations at the Sansad Bhavan (Parliament).

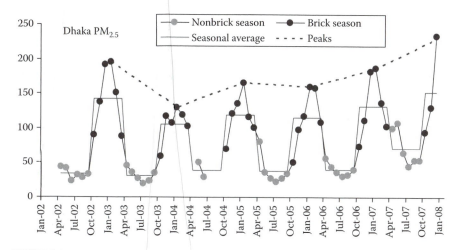

FIGURE 3.11   Monthly average $PM_{2.5}$ measurements in Dhaka city. [Data from Professor Zia Wadud, BUET (Bangladesh University of Engineering and Technology).]

The air quality in Dhaka city has deteriorated over the last decade due to a rapid change in the vehicular fleet, increased congestion, and a large increase in industrial activity (in and around the city). The annual average concentrations for $PM_{2.5}$ are ~100 μg/m³, above any of the standards for clean air and better health.

In the city, transport is the major source of air pollution. Among industries, brick kilns represent the major source, especially during the manufacturing season from October to March, depending on the monsoon rains. In Figure 3.11, $PM_{2.5}$ concentrations are split between the brick-making season and the rest, clearly presenting a distinct change in pollution trends. Seasonal averages have shifted over the years and, during the brick-manufacturing season, pollution peaks measured ~230 μg/m³ in January 2008.

Based on source apportionment studies (Begum et al., 2006a,b,c), the major sources of PM pollution during the dry season (October to March) are the following: (a) vehicular emissions, particularly motor cycles, diesel trucks, and buses (the most dominant of the sources in both fine and coarse modes); (b) soil and road dusts arising from civil construction, broken roads, and open land wind erosion; and (c) biomass burning in the brickfields and city incinerators (to the fine mode). Figure 3.12 presents a summary of the source apportionment study conducted by the Bangladesh Atomic Energy Center, Dhaka, Bangladesh for fine and coarse mode particulates at two stations: Farm gate and the Dhaka University premises (Biswas, 2009).

The analysis was conducted using "GENT" stacked filter samplers, followed by analysis of the filter samples using PIXE and receptor modeling using PMF (Guttikunda, 2009a). The study included the development of city-specific source profiles, which provide the necessary information on the biomarkers to identify sources, analyze the measured samples, and estimate the percentage contributions of various sources. It is important to note that the source apportionment results cannot be generalized to the whole city, but they do provide a basis for further analysis and an understanding of the mix of sources contributing to the air pollution, especially around the hot spots. The process of source apportionment, applied in this study, is expensive, which prohibits the inclusion of many measurement points, unlike a mobile monitoring station, which can be used to measure multiple points but cannot be used to calculate source contributions.

Motor vehicles are a known and visible source of particulate pollution in Dhaka, and require interventions ranging from technical (emission standards) to institutional (inspection and maintenance). Dust, due to resuspension on the roads, an indirect source of motor vehicle activity, is a major cause of air pollution (in the coarse mode of PM), due to lack of sufficient infrastructure (paved roads) to support the growing fleets and congestion on the roads.

The clusters of brick kilns lying north of Dhaka contribute ~40% of the measured fine PM pollution. Growing construction activity (also contributing to fugitive dust) is leading the demand for brick kilns, and the burning of biomass and low-quality coal is resulting in pollution. A majority of the brick kiln clusters are to the north of the city, as presented in Figure 3.13, and the measured peak values represent the worst-case scenario of the maximum wind blowing toward the city (the dark gray line boundary). The clusters account for ~530 brick kilns.

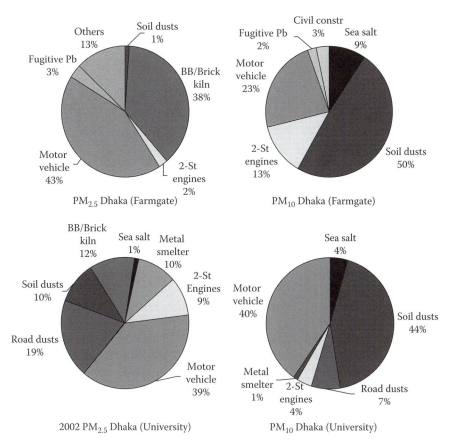

FIGURE 3.12 Source apportionment results for Dhaka.

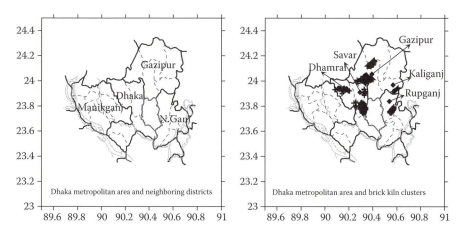

FIGURE 3.13 Dhaka metropolitan area and brick clusters.

A series of surveys conducted in Dhaka led to the establishment of an inventory of the physical location of brick kiln clusters. The survey was conducted by the Bangladesh University of Engineering and Technology, Dhaka. The details (latitudes and longitudes) are presented in AQMP/World Bank (2007).

At the brick kilns, measurements included an emission rate of 44 g/s of TSP. The emission rates were converted to PM_{10} (using a ratio of 0.3 to TSP) and to $PM_{2.5}$ (using a ratio of 0.3 to PM_{10}). This amounts to a total of 108 kilotons of PM_{10} for 180 days of operating 530 brick kilns, as presented in Figure 3.13.

Figure 3.14 (left panel) presents the seasonal (October to March) average contribution of brick kiln emissions, followed by the modeled monthly maxima over the Dhaka metropolitan area. It is important to note that the contribution of brick kilns is incremental to the daily vehicular, domestic, and industrial activities during these months. This is evident in Figure 3.11, which demonstrates the dramatic increase in $PM_{2.5}$ levels measured in the city premises.

The seasonal averages estimate a contribution of 15–60 µg/m³ of $PM_{2.5}$, which translates to 30–40% of the measured $PM_{2.5}$ concentrations in Figure 3.11, which corresponds to the estimated source contribution of brick kilns in Figure 3.12 (via source apportionment). The monthly maxima (right panel) correspond to the worst-case scenario of heavy northerly winds and a possible maximum modeled contribution of 30–100 µg/m³ of $PM_{2.5}$ over the season.

Figure 3.14 presents the monthly average contributions of brick kilns for six months of operation. The calculations presented here should not translate to undermining the influence of the other sectors on Dhaka's air quality. At the ground level, the transport sector (direct vehicle exhaust, road dust due to resuspension, and idling emissions) contributes more to the exposure levels than the long-range transport of emissions from outside the city.

The main objective of this analysis was to demonstrate the schematics of the modeling approach in understanding the contribution of brick kiln emissions in Dhaka, the characteristics of air pollution dispersion, and the physical extent of the influence of these emissions on public health.

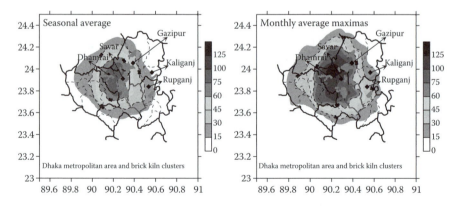

FIGURE 3.14 Seasonal average and monthly maxima of modeled $PM_{2.5}$ (µg/m³) due to brick kiln emissions.

Brick kilns contribute significantly to Dhaka air pollution problems, especially during the heightened manufacturing season, and this requires stringent interventions to reduce their incremental impact on local air quality and health. For example, the incremental pollution of 40 g/m^3 due to brick kilns translates to an increase in ~5000 premature deaths annually in Dhaka city. AQMP/World Bank (2007) outlines a series of interventions to improve combustion technologies, financial implications, and the possible benefits from brick kiln clusters on Dhaka air quality, which are currently under dialogue for implementation.

Hyderabad, India: Hyderabad, a 400-year-old city, is the state capital of Andhra Pradesh (India) and the fifth largest city in India, with a population nearing 7 million. The twin cities of Hyderabad and Secunderabad of Municipal Corporation of Hyderabad (MCH) and the neighboring 10 municipalities together form Hyderabad Urban Development Area (HUDA), a major high-tech center with increasing economic activity. The location of the city and its main industrial complexes is shown in Figure 3.15.

A multiagency study led by the Andhra Pradesh Pollution Control Board (APPCB), in collaboration with the Desert Research Institute (Reno, USA) and cofinancing from the USEPA Integrated Environmental Strategies program (Washington, DC, USA) and the World Bank (Washington, DC, USA), was designed to prepare a cobenefit action plan for air pollution control in Hyderabad, India, with the base year 2006. The program included (a) a year-long source apportionment study using a mini-vol sampler, chemical analysis, and receptor modeling using the CMB model, (b) bottom-up air pollution analysis by developing an emissions inventory for local and global air pollutants, dispersion modeling, and cobenefits analysis of the city action plan (IES, 2004, 2007, 2008). Overall, PM emissions are dominated by vehicular, industrial, and fugitive sources. A summary of the inventory is presented in Table 3.6. Garbage burning (a very uncertain source of emissions because of the lack

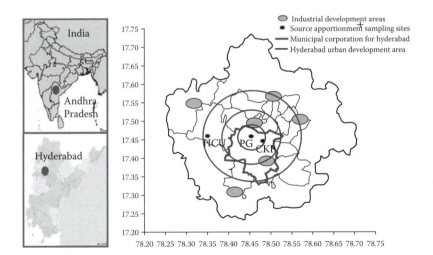

FIGURE 3.15 Geographical location of HUDA and major industrial development areas (IDAs).

TABLE 3.6

Estimated Emissions Inventory for Hyderabad in 2006 (tons/year)

Category	PM_{10}	SO_2	NO_x	CO_2
Vehicular activity	8410	6304	39,262	6,400,337
Paved road dust	3272			
Unpaved road dust	4279			
Industries	8985	4606	5070	654,717
Domestic	1845	667	545	83,485
Waste burning	810			
Total	27,599	11,577	44,877	7,138,538

of crucial information on the amount burnt and proper emission factors) is a significant unconventional source. One landfill to the southeast of the MCH border is estimated to burn on average 5% of the trash collected and, combined with domestic fuel consumption, accounts for ~10% of annual PM_{10} emissions. Emissions of PM_{10}, SO_2, NO_x, and CO_2 are estimated at 29.6 kilotons, 11.6 kilotons, 44.5 kilotons, and 7.1 million tons, respectively. For CO_2, a major GHG gas, the transport sector accounts for 90% of the emissions. Note that the inventory presented in IES (2008) includes neither the commercial sources of CO_2 emissions nor the life cycle assessments of domestic consumption.

The estimated annual average concentrations using the ATMoS modeling system, presented in Figure 3.16, include both primary PM and secondary PM due to the chemical conversion of SO_2 and NO_x emissions into sulfates and nitrates.

In the HUDA region, on average, secondary PM contributes 20–40% of the total PM_{10}. The estimated annual average concentrations were calibrated against measurements for the year 2006 with urban hot spots averaging ~200 µg/m³ on a daily basis. The highest concentrations in the first panel of Figure 3.16 represent areas with the highest industrial density; the largest density of the population is within the 10 km radius and MCH boundary lines. Within the MCH boundary, in 2006, the contributions of individual sectors ranged as 20–50% for vehicular sources, 40–70% when combined with road dust, 10–30% for industrial sources, and 3–10% for domestic and garbage burning sources.

Following air pollution modeling, the cobenefits analysis of the action plan proposed by the local authorities was conducted for 2010 and 2020. Table 3.7 summarizes the results and expected emission reductions (IES, 2008). The cobenefits of the action plan were conducted using the methodologies presented in Lvovsky et al. (2000), Pope et al. (2006), CARB (2008), Guttikunda (2008b), and Mehta (2008).

The proposed action plan includes the following: expanding the use of alternative fuels such as compressed natural gas for heavy-duty vehicles and liquefied petroleum gas (LPG) for light-duty vehicles, a program to promote the wet sweeping of paved roads to reduce fugitive dust, promoting the use of LPG for the domestic sector to replace biomass burning, introduction of stringent regulations for solid waste management garbage burning, promoting the use of public transport by introducing more

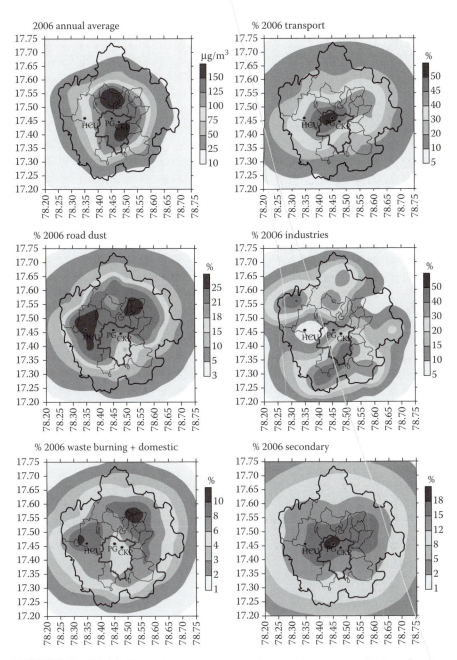

FIGURE 3.16 Modeled annual average concentrations and percent contribution of sectors to annual PM$_{10}$ in 2006.

TABLE 3.7

Summary of Total Emission Reductions under Hyderabad City Action Plan

	PM_{10}	SO_2	NO_x	CO_2
2006 Emissions BAU (tons)	29,599	11,577	44,877	7,138,538
2010 Emissions BAU (tons)	34,620	14,520	58,638	9,352,590
2010 Emissions—with controls (tons)	27,755	12,377	48,312	7,559,229
Estimated emission reductions (tons)	6864	2143	10,327	1,793,361
% Reduction from 2010 BAU	20	15	18	19
2020 Emissions BAU (tons)	43,550	18,670	63,694	10,310,520
2020 Emissions—with controls (tons)	24,110	13,365	40,059	6,968,693
Estimated emission reductions (tons)	19,440	5035	23,635	3,341,847
% Reduction from 2020 BAU	45	27	37	32

buses and introducing the bus rapid transit system along some corridors, introduction of a metro rail system to support the public transport system, and promotion of non-motorized transport by providing the necessary infrastructure.

Under these scenarios, if planned and implemented the emissions levels are expected to fall below the 2006 BAU, with reductions ranging from ~20% and ~20% in 2010 and ~45% and ~32% in 2020 for PM and CO_2 emissions, respectively, compared to their corresponding BAU scenarios.

On implementation, the estimated ambient concentration reductions are substantial in the densely populated areas, mainly due to direct reductions in the transportation sector. The ambient concentrations, when compared to the "what-if" case of the BAU scenario, are at least 40% lower in 2020. For 2020, exposure analysis was conducted and the possible number of incurred cases and estimated reductions, compared to BAU scenarios, is presented in Table 3.8.

TABLE 3.8

Estimation of Health Impacts Based on Modeling Results for Control Scenarios

Health Endpoint	Number of Incurred Cases	
	2020 BAU	2020 Control
Mortality	6347	2018
Adult chronic bronchitis	10,951	3483
Child acute bronchitis	98,650	31,373
Respiratory hospital admission	2584	822
Cardiac hospital admission	2267	721
Emergency room visit	106,720	33,939
Asthma attacks	1,314,733	418,111
Restricted activity days	17,354,479	5,519,061
Respiratory symptom days	82,964,203	26,384,226

It is important to note that these are estimated reductions and a more in-depth analysis is required, which is usually conducted as part of the feasibility studies before and after project implementation. Calculations include the control measures already in pilot implementation and assume that most will be operational by 2020, as proposed. The combined health and carbon benefits amounted to approximately US\$472 million in 2020 (IES, 2008). The health impacts are estimated using the DRFs for outdoor pollution and the methodology summarized in HEI (2004) and Guttikunda (2008b). The monetary value of the health benefits is based on the "willingness to pay" methodology (Lvovsky et al., 2000; CARB, 2008), including mortality and morbidity due to outdoor air pollution, and the carbon benefits are monetized at ~\$20 per ton of CO_2 reduced.

The combined benefits of integrated air quality and climate change policies by 2020 (or earlier, depending on feasibility and accelerated actions) are expected to make substantial improvements in the city and, given time, technical, and financial support, the implementing measures will lead to a better urban environment.

3.6.3 STATISTICAL MODEL CASE STUDIES

3.6.3.1 Linear Regression Models in Santiago, Chile

Santiago and Temuco are both cities that exceed the daily and annual PM_{10} air quality standards of 150 and 50 µg/m³, respectively. The Chilean Environmental Commission (CONAMA) has implemented a unique strategy to curb air pollution, which consists of implementing restrictions on emissions based on the results of a statistical forecast model. For example, if air quality is forecasted to be above a level of 195 µg/m³, wood-burning heaters and stoves are banned from being used. If levels are forecasted to be above 240 µg/m³, an additional 40% of cars are left out of circulation, and industries that do not meet a 32-mg/m³ PM emission standard cannot operate for 24 h. Model results are announced daily in national news outlets, as are emission restrictions for the following day. Model performance is evaluated strictly on the capability of accurately predicting these contingency levels. Model prediction failures cause widespread controversy (Global Post, 2009), as the overprediction of episodes may force people in Santiago to use public transportation, or industry to cease operations. Model underprediction of air pollution is largely overlooked by the media. The model was developed in the 1990s by Joe Cassmassi (Ulriksen and Merino, 2003). It is a statistical multivariate regression model that correlates forecasted air quality values based on current observations and on forecasted meteorological parameters, which are considered as predictors of air quality. Model output is typically maximum 24-h mean PM_{10} expressed in µg/m³ for each of the air quality measurement stations in Santiago (Perez, 2008). Meteorological parameters involved in the forecast include synoptic-scale features, results from a radiosonde located near the coast, and some forecasted parameters derived from global forecast models. These parameters are synthesized into a single parameter, the Meteorological Potential for Air Pollution (PMCA). This factor is inversely proportional to the noon ventilation factor (which is the product of mixing layer height and mean surface wind). In practice, there are five categories of PMCA, which are shown in Table 3.9. These categories are discrete parameters and are based on subjective analysis.

TABLE 3.9

Description of PCMA Categories According to Meteorological Parameters

PCMA Category	Ventilation/Dispersion Conditions	Description
1—Low	Very good	Absence of thermal inversion
		Active frontal system
		Pronounced instability
2—Regular/low	Good	Weak or elevated thermal inversion/ subsidence
		Frontal system with regular activity or light instability
		Generalized advection of humid air, low-lying clouds
		Segregated lows or cold nucleus
3—Regular	Regular	Normal wintertime anticyclonic conditions
		Absence of precipitation
		Weak advection of low-lying clouds in the west part of the basin
4—Regular/high	Bad to critical	Presence of prefrontal low-pressure system or coastal low.
		Zonal circulation index medium to low
5—High	Critical	Presence of prefrontal low or coastal low systems
		Low zonal circulation index

Source: Schmitz R. and M. Falvey. 2007. Evaluation of the PM_{10} Forecast Models in use in metropolitan region of Chile. Chilean Environmental Commision (CONAMA RM). Available at http://www.geofisica.cl/Tempo/mix/Informe_Final_PM_{10}.pdf. Accessed July 2009.

The model is a multivariate regression model, and the regression equations and coefficients were developed using 1997–1998 data. Forecasted values for each monitoring station are a function of current concentrations plus a coefficient multiplied by the PCMA. Recent model evaluation has shown that while the model is able to accurately predict goodair quality days over 85% of the time (24 h $PM_{10} < 150$ μg/m^3), it usually fails to predict bad air quality days (24 h $PM_{10} > 195$ μg/m^3). Additional information on the evaluation and use of this linear regression model is available in a report by Cassmassi (1999).

3.6.3.2 Application of ANN: A Case Study from Delhi

Application of ANN has been shown here by evaluating the performance of ANN for ambient air quality data using SO_2 data recorded at Siri Fort monitoring station, Delhi, by Minocha (2006). The ANN model used here is a back-propagation feed-forward model. The network calculates the output from each processing neuron of the ANN by starting from the input layer and propagating forward through the hidden layer to the output layer. Air quality data of the year 1998 are recorded and averaged

TABLE 3.10

Air Quality SO2 Data for Five Consecutive Weeks Recorded at 4 h Interval

Time (4-hour Interval)	$(N-2)$th Week	$(N-1)$th Week	Nth Week	$(N+1)$th Week	$(N+2)$th Week
0–4	6.33	8.62	3.42	8.28	6.01
4–8	4.50	7.30	2.80	6.42	3.70
8–12	5.72	8.67	2.99	7.51	6.99
12–16	12.44	13.07	3.11	9.33	10.67
16–20	8.82	10.37	3.09	5.61	3.90
20–24	7.33	9.66	3.46	5.53	4.44

Source: Minocha V.K. 2006. Modeling and monitoring of environmental systems, PhD thesis, University of Delhi.

for 1-h intervals, which are used for training and validation of the selected network. If the input neurons are three, data are transformed as input vector $X[x_1, x_2, x_3]$ for the longer time interval air quality data (preceding, current, and successive). If the input neurons are four, one additional neuron for the current time of the same day of the previous week's data was taken. The number of output neurons needed for disintegrating air quality data corresponds to the number of disintegrated intervals for the current interval of measurement. The duration of data that have been disintegrated is of 8, 4, 3, and 2 h intervals, causing the number of output neurons to be equal to 8, 4, 3, and 2, respectively. The number of hidden neurons is determined with a preliminary analysis of associated training errors. For the feed-forward ANN model, the selected activation function was the unipolar sigmoid function.

Table 3.10 shows the 4-h average SO_2 data for five weeks. For the 8–12 h in the Nth week, the required data consist of preceding, current, and successive data, that is, $P_1 = 2.80$, $P_2 = 2.99$, and $P_3 = 3.11$; if the input layer consists of four nodes, then the additional data will be of the same time in the previous week, that is, 8–12 h of the $(N-1)$th week, which is equal to $P_4 = 8.67$. In both cases, the data of P_2 are disintegrated to four data of 1-h intervals at output nodes as the number of output nodes depends on the time interval of the input data, which is equal to 4 h. The number of nodes in the hidden layer for which the mean square error is least is generally considered the appropriate number of neurons for that layer.

The results obtained from the ANN model are compared with the actual data available. A total of eight models are considered. $I_3H_3O_2$ indicates that the model consists of three input layers, three hidden layers, and two output layers. These models are disintegrated from 2-h duration to 1-h duration. Table 3.11 shows the mean squared errors of all the models at both the training and validation stages. Several model structures are tried for disintegration of 2-, 3-, 4-, and 8-h duration data to 1-h duration. From the several models, the following ANN structures are selected for disintegration of data into 1-h intervals.

Table 3.11 shows the value of mean square error and regression coefficient at the training and validation stages for selected ANN models. From Table 3.4, it is clear

TABLE 3.11

Comparison of Mean Square Error and Regression Coefficient for Selected ANN Models at Training and Validation Stage

Sr. No.	ANN Model Structure	Mean Square Error		Regression Coefficient		Total Number of Connection Weights of ANN	Number of Training Datasets Divided by Number of Connection Weights
		Training Stage	Validation Stage	Training Stage	Validation Stage		
1	$I_3H_4O_2$	0.12	0.14	0.99	0.90	20	8.4
2	$I_4H_4O_2$	0.11	0.13	0.99	0.95	24	7
3	$I_3H_5O_3$	0.23	0.49	0.95	0.81	30	5.6
4	$I_4H_5O_3$	0.22	0.41	0.94	0.83	35	4.8
5	$I_3H_4O_4$	0.40	0.41	0.95	0.86	28	6
6	$I_4H_4O_4$	0.32	0.30	0.96	0.85	32	5.25
7	$I_3H_3O_8$	0.91	1.06	0.82	0.44	44	3.82
8	$I_4H_4O_8$	1.21	1.47	0.77	0.38	48	3.5

Source: Minocha V.K. 2006. Modeling and monitoring of environmental systems, Ph.D thesis, University of Delhi.

that the mean square error values at the validation stage for disintegration models of the 2-, 3-, and 4-h interval data are less than 0.5 and the corresponding regression coefficients are more than 0.8. Hence ANN models can be used for the disintegration of 2-, 3-, and 4-h interval data to get 1-h data. The disintegration of 8-h interval data to 1-h data is far from satisfactory because the mean square error values are high and the regression coefficient values are low at the validation stage.

From this case study of Delhi, it is apparent that ANN is a useful tool and represents a significant advancement in the area of air quality modeling and can help reduce data collection efforts.

In the literature, there are already certain studies on ANN applications to atmospheric systems. Boznar et al. (1993) applied ANN for the short-term forecasting of atmospheric sulfur dioxide concentration in a polluted industrialized area of Slovenia. Yi and Prybutok (1996) and Comrie (1996) used the ANN model, with input as average daily meteorological data, to forecast the daily maximum ozone level. The work by Gardener and Dorling (1999) and Kolehmainen et al. (2001) indicated that the application of ANN in forecasting urban pollution gave superior results as compared to traditional regression models. Pelliccioni and Tirabassi (2006) filtered the concentration levels produced by an air pollution model with ANN to take care of disagreement between the measured and predicted values. Their study indicates that the results of the air dispersion model and the neural network can be complementary to each other and the ANN model can be combined with the air pollution model when the air pollution model gives some systematic error with respect to experimental data. Whether it is the use of dispersion models with ANN models or the combined use of other models, one is often interested in knowing the best performing modeling strategy. In this respect, the use of fuzzy theory for the ranking of modeling strategies becomes much more relevant and the description of this approach along with its use in air quality modeling becomes an important issue for consideration, as described in the next section.

3.7   SUMMARY

Air quality models, of both deterministic and statistical nature, have long been used for policy implications. Decades ago, Gaussian plume models were used to study the dispersion of large sources. Today, Eulerian and Lagrangian models are more readily available. Advances in science understanding and computational capacity, alongside decreased cost, make it possible to run a regional air quality forecast model with a single unit, double quad core processor machine; however, less than a decade ago it would have taken dozens of processors to perform the same task. Air quality models can be used in support of environmental regulators, by delimiting the areas that are in nonattainment of particular air quality standards. Models can be used in combination with observations to evaluate the contribution of specific sources to air quality measurements. Models can evaluate the contribution of a specific source to regional photochemical smog formation (such as ozone and particulates). Finally, advances in inverse modeling allow systematic improvement of model performance and estimation of emissions inventories through techniques like 4DVar data assimilation. The combination of models and policy analysis can be used in determining the benefits of implementing particular pollution abatement strategies, as shown in a case study in

India. Finally, air quality can be forecasted numerically using statistical models, such as artificial neural networks, or regression models. While these models predict maximum pollution with higher correlation coefficients than deterministic models, they do not provide information in terms of the processes involved in pollution, specifically emissions, thus limiting their use in policy making. Ideally, in future, emissions inventory estimations will improve, as will the coverage of air quality observations. At that time, deterministic models will be used to run robust and reliable air quality forecasts and diagnostics, which can be used to further develop public policy in order to reduce human exposure to pollution in a cost-effective manner.

REFERENCES

Albergel A. and Jasmin F. 1998. 3-D simulation of local-scale traffic pollution. *International Journal of Vehicle Design* 20: 79–87.

AQMP/World Bank. 2007. "Air Quality Management Program" report on the brick kiln emissions and policy implications, AQMP, Department of Energy, Dhaka, Bangladesh.

Arndt R.L., Carmichael G.R., Streets D.G., Bhatti N. 1997. Sulfur dioxide emissions and sectorial contributions to sulfur deposition in Asia, *Atmospheric Environment* 31 (10): 1553–1572.

Arndt R.L., et al. 1998. Seasonal source–receptor relationships in Asia. *Atmospheric Environment* 32: 1397–1406.

Bady M., Kato S., et al. 2009. Identification of pollution sources in urban areas using reverse simulation with reversed time marching method. *Journal of Asian Architecture and Building Engineering* 8 (1): 275–282.

Balczo M., Gromke C., and Ruck B. 2009. Numerical modeling of flow and pollutant dispersion in street canyons with tree planting. *Meteorologische Zeitschrift*. 18 (2): 197–206.

Begum et al. 2006a. Multi-element analysis and characterization of atmospheric particulate pollution in Dhaka. *Aerosol and Air Quality Research* 6 (4): 334–359.

Begum et al. 2006b. Impact of banning of two-stroke engines on airborne particulate matter concentrations in Dhaka, Bangladesh. *Journal of the Air & Waste Management Association* 56 (1): 85–89.

Begum et al. 2006c. Temporal variations and spatial distribution of ambient $PM_{2.2}$ and PM_{10} concentrations in Dhaka, Bangladesh. *Science of the Total Environment* 358 (1–3): 36–45.

Bell et al. 2006. The avoidable health effects of air pollution in three Latin American cities: Santiago, São Paulo, and Mexico City. *Environmental Research* 100: 431–440.

Benson P.E. 1992. A review of the development and application of the Caline3 and Caline4 Models. *Atmospheric Environment Part B—Urban Atmosphere* 26 (3): 379–390.

Berkowicz R. and Olesen J.R. 1986. The Danish Gaussian air pollution model (OLM): Description, test and sensitivity analysis, in view of regulatory applications. In: V.C. De Wispelaire et al., (Eds), *Air Pollution Modelling and Its Application*, Plenum Press, New York, pp. 453–481.

Binkowski F.S. and Roselle S.J. 2003. Models-3 community multiscale air quality (CMAQ) model aerosol component-1. Model description. *Journal of Geophysical Research—Atmospheres* 100 (D12): 26191–26209.

Biswas et al. 2007–2009. Atomic Energy Research Center, Dhaka, Bangladesh, personal communication.

Bond T.C., Streets D.G., et al. 2004. A technology-based global inventory of black and organic carbon emissions from combustion. *Journal of Geophysical Research—Atmospheres* 109 (D14): 1–41.

Bott A. 1989. A positive definite advection scheme obtained by non-linear re-normalisation of the advection fluxes. *Monthly Weather Review* 117: 1006–1015.

Box G.E.P. and Jenkins G.M. 1976. *Time Series Analysis—forecasting and control.* Holden-day, San Francisco.

Boylan J.W., Odman M.T., et al. 2002. Development of a comprehensive, multiscale "one-atmosphere" modeling system: Application to the Southern Appalachian Mountains. *Atmospheric Environment* 36 (23): 3721–3734.

Boznar M., Lesjak M., and Mlatkar P. 1993. A neural network-based method for the short-term predictions of ambient SO_2 concentrations in highly polluted industrial areas of complex terrain. *Atmospheric Environment B* 27 (2): 221–230.

Briggs G. 1973. Internal memo as reported by F.A. Gifford Jr. in turbulent diffusion typing schemes: A review. *Nuclear Safety* 17: 67–86.

Briggs G.A. 1975. Plume rise predictions. In: D.A. Haugen (Ed.), *Lectures on Air Pollution and Environmental Impact Analysis.* American Meteorology Society, Boston, MA, pp. 59–111.

Byun D. and Schere K.L. 2006. Review of the governing equations, computational algorithms, and other components of the models-3 Community Multiscale Air Quality (CMAQ) modeling system. *Applied Mechanics Reviews* 59 (1–6): 51–77.

Calori G. and Carmichael G.R. 1999. An urban trajectory model for sulfur in Asian megacities: Model concepts and preliminary application, *Atmospheric Environment* 33 (19): 3109–3117.

Calori G., Finardi S., et al. 2008. Long-term air quality assessment: Modeling sources contribution and scenarios in Ivrea and Torino areas. *Environmental Modeling & Assessment* 13 (3): 329–335.

Campbell J.E., Carmichael G.R., et al. 2008. Photosynthetic control of atmospheric carbonyl sulfide during the growing season. *Science* 322 (5904): 1085–1088.

CARB. 2008. *Methodology for Estimating Premature Deaths Associated with Long-term Exposure to Fine Airborne Particulate Matter in California.* California Air Resources Board, USA, http://www.arb.ca.gov/research/health/pm-mort/pm-mort.htm

Carmichael G.R., Sandu A., et al. 2008. Predicting air quality: Current status and future directions. *Air Pollution Modeling and Its Application* XIX: 481–495.

Carmichael G.R., Uno I., et al. 1998. Tropospheric ozone production and transport in the springtime in east Asia. *Journal of Geophysical Research—Atmospheres* 103 (D9): 10649–10671.

Carmichael G.R., Peters L.K., et al. 1991. The stem-Ii regional scale acid deposition and photochemical oxidant model. 1. An overview of model development and applications. *Atmospheric Environment Part A—General Topics* 25 (10): 2077–2090.

Carruthers D.J. and Holroy D.R.J. 1994. UK-ADMINS: A new approach to modeling dispersion in the earth's atmospheric boundary-layer. *Journal of Wind Engineering and Industrial Aerodynamics* 52 (1–3): 139–153.

Carter W.P.L. 2000. Implementation of the SAPRC-99 chemical mechanism into the models-3 framework, http://ftp.cert.ucr.edu/pub/carter/pubs/s99mod3.pdf

Carter W.P.L. 2003. The SAPRC-99 chemical mechanism and updated VOC reactivity scales, http://helium.ucr.edu/~ carter/reactdat.htm

Cassmassi J. 1999. Improvement of the forecast of air quality and of the knowledge of the local meteorological conditions in the metropolitan region, Informe final, CONAMA. RM, http://www.conama.cl/rm/568/articles-2581_reporte1.pdf

Chai et al., 2006. Personal communication.

Chai T., Carmichael G.R., Sandu A., Tang Y., and Daescu D.N. 2006. Chemical data assimilation of transport and chemical evolution over the Pacific (TRACE-P) aircraft measurements, *Journal of Geophysical Research* 111 (D02301).

Chai T.F., Carmichael G.R., et al. 2007. Four-dimensional data assimilation experiments with International Consortium for Atmospheric Research on Transport and Transformation ozone measurements. *Journal of Geophysical Research—Atmospheres* 112 (D12): 1–18.

Chang J.C. and Hanna S.R. 2004. Air quality model performance evaluation. *Meteorology and Atmospheric physics* 87: 167–196.

Cheremisinoff N.P. 2002. *Handbook of Air Pollution Prevention and Control* Butterworth-Heinemann Publishers, Woburn, MA.

Chock D.P., Chang T.Y., et al. 1999. The impact of an 8-h ozone air quality standard on ROG and NO_x controls in Southern California. *Atmospheric Environment* 33 (16): 2471–2485.

Cimorelli A.J., Perry S.G., et al. 2005. AERMOD: A dispersion model for industrial source applications. Part I: General model formulation and boundary layer characterization. *Journal of Applied Meteorology* 44 (5): 682–693.

Comrie A.C. 1997. Comparing neural networks and regression models for ozone forecasting. *Journal of the Air & Waste Management Association* 47 (6): 653–663.

Davidson K., Hallberg A., et al. 2007. Analysis of $PM_{2.5}$ using the Environmental Benefits Mapping and Analysis Program (BenMAP). *Journal of Toxicology and Environmental Health, Part A—Current Issues* 70 (3–4): 332–346.

de Foy et al. 2008. Basin-scale wind transport during the MILAGRO field campaign and comparison to climatology using cluster analysis. *Atmos. Chem. Phys.* 8: 1209–1224.

Diaz-Robles L., et al. 2008. A hybrid ARIMA and artificial neural networks model to forecast particulate matter in urban areas: The case of Temuco, Chile. *Atmospheric Environment* 42 (35): 8331–8340.

Dobson J.E., Bright E.A., et al. 2000. LandScan: A global population database for estimating populations at risk. *Photogrammetric Engineering and Remote Sensing* 66 (7): 849–857.

Dudhia J., Klemp J., et al. 1998. A collaborative effort towards a future community mesoscale model (WRF). 12th Conference on Numerical Weather Prediction, pp. 242–243.

Elbir T. 2003. Comparison of model predictions with the data of an urban air quality monitoring network in Izmir, Turkey. *Atmospheric Environment* 37 (15): 2149–2157.

Fast J., Aiken A.C., Allan J., et al. 2009. Evaluating simulated primary anthropogenic and biomass burning organic aerosols during MILAGRO: Implications for assessing treatments of secondary organic aerosols. *Atmos. Chem. Phys.* 9: 6191–6215.

Fast J.D., Gustafson W.I., et al. 2006. Evolution of ozone, particulates, and aerosol direct radiative forcing in the vicinity of Houston using a fully coupled meteorology-chemistry-aerosol model. *Journal of Geophysical Research—Atmospheres* 111 (D21): 1–29.

Fige. 1997. *Mobilev-Dokumentation and Benutzerhandbuch. Foschungsvorhaben 105 06 044 des Umweltbundesamts "Erarbeitun von Grundlagen fuer die Umsetzung von 40.2 des BImSchG".* Umweltbundesamt, Berlin.

Frost G.J., et al. 2006. Effects of changing power plant NO_x emissions on ozone in the eastern United States: Proof of concept. *Journal of Geophysical Research—Atmospheres* 111 (D12): 1–29.

Gardener M.W. and Dorling S.R. 1999. Neural network modeling and prediction of hourly NO_x and NO concentrations in an urban area in London. *Atmospheric Environment* 33 (5): 709–719.

Gaydos T.M., Pinder R., et al. 2007. Development and application of a three-dimensional aerosol chemical transport model, $PMCAM_x$. *Atmospheric Environment* 41 (12): 2594–2611.

Gery M.W., Whitten, G.Z., Killus J.P., and Dodge M.C. 1989. A photochemical kinetics mechanism for urban and regional scale computer modeling. *Journal of Geophysical Research* 94: 925–956.

Gidhagen L. and Johansson C. 2004. Simulation of NO_x and ultrafine particles in a street canyon in Stockholm, Sweden. *Atmospheric Environment* 38 (14): 2029–2044.

Glendinning P. and Sparrow C. 1984. Local and global behavior near homoclinic orbits. *Journal of Statistical Physics* 35: 645–696.

Global Post. 2009. Hold your breath: Santiago struggles to deal with its longtime smog woes. April 21, 2009, Available in http://www.globalpost.com/dispatch/chile/090421/smog-santiago-environment-chile. Accessed July 2009.

Grell G.A., Peckham S.E., et al. 2005. Fully coupled "online" chemistry within the WRF model. *Atmospheric Environment* 39 (37): 6957–6975.

Guttikunda S.K., et al. 2002. The contribution of megacities to regional sulfur pollution in Asia. *Atmospheric Environment* 37: 11–22.

Guttikunda S.K., Carmichael G.R., Calori G., Eck C., and Woo J.-H. 2003. The contribution of megacities to regional sulfur pollution in Asia, *Atmospheric Environment* 37 (1): 1122.

Guttikunda S.K. 2008. Estimating health impacts of urban air pollution, SIM-06-2008, SIM-Air Working Paper Series, http://www.urbanemissions.info/simair/simseries.html

Guttikunda S.K. 2009. Urban Particulate Pollution Source Apportionment: Part 1—Definition, Methodology, and Resources, SIM-16-2009, SIM-Air Working Paper Series, http://www.urbanemissions.info/simair/simseries.html

Hakami A., Seinfeld J.H., et al. 2006. Adjoint sensitivity analysis of ozone nonattainment over the continental United States. *Environmental Science & Technology* 40 (12): 3855–3864.

Hao J., Wang L., et al. 2007. Air quality impacts of power plant emissions in Beijing. *Environmental Pollution* 147 (2): 401–408.

HEI. 2004. Health Effects of Outdoor Air Pollution in Developing Countries of Asia: A Literature Review. Health Effects Institute, Boston, USA.

Holmes N.S. and Morawska L. 2006. A review of dispersion modeling and its application to the dispersion of particles: An overview of different dispersion models available. *Atmospheric Environment* 40: 5902–5928.

Holtslag A.A.M. and Nieuwstadt F.T.M. 1986. Scaling the atmospheric boundary layer. *Boundary-Layer Meteorology* 36: 201–209.

Hurley P.J., Physick W.L., et al. 2005. TAPM: A practical approach to prognostic meteorological and air pollution modelling. *Environmental Modelling & Software* 20 (6): 737–752.

IES. 2004. Emission inventory and co-benefits analysis for Hyderabad. Report prepared by EPTRI, Hyderabad, for IES India program, USEPA, Washington, DC, http://www.epa.gov/ies

IES. 2007. Particulate pollution source apportionment: A case study on Hyderabad, India. Integrated Environmental Strategies India program, USEPA, Washington, DC.

IES. 2008. Co-benefits analysis of air pollution and GHG emissions for Hyderabad, India. Integrated Environmental Strategies India program, USEPA, Washington, DC.

Jacobson M.Z. 1996. Application of a sparse-matrix, vectorized gear-type code in a new air pollution modeling system. *Zeitschrift Fur Angewandte Mathematik Und Mechanik* 76: 333–336.

Jacobson M.Z. 1997. Development and application of a new air pollution modeling system. 2. Aerosol module structure and design. *Atmospheric Environment* 31 (7): 1097.

Jacobson M.Z. 2001. GATOR-GCMM: A global-through urban-scale air pollution and weather forecast model 1. Model design and treatment of subgrid soil, vegetation, roads, rooftops, water, sea ice, and snow. *Journal of Geophysical Research—Atmospheres* 106 (D6): 5385–5401.

Jin S. and Demerjian K. 1993. A photochemical box model for urban air quality study. *Atmospheric Environment. Part B. Urban Atmosphere* 27 (4): 371–387.

Jorquera H. 2006. Estudio diagnóstico plan de gestión calidad del aire vi región. Under commission by CONAMA.

Jung Y.R., Park W.G., and Park O.-H. 2003. Pollution dispersion analysis using the puff model with numerical flow field data. *Mechanics Research Communications* 30: 277–286.

Klir J. and Foger T.A. 1988. *Fuzzy Sets, Uncertainty, and Information.* Englewood Cliffs, NJ: Prentice-Hall.

Kolehmainen M., Martikainen H., and Ruskanen J. 2001. Neural Networks and periodic components used in air quality forecasting. *Atmospheric Environment* 35: 815–825.

Korhonen H. and Lehtinen K.E.J. 2004. Multicomponent aerosol dynamics model UHMA: Model development and validation. *Atmospheric Chemistry and Physics* 4: 757–771.

Lei W., de Foy B., Zavala M., Volkamer R., and Molina L.T. 2007. Characterizing ozone production in the Mexico City metropolitan area: A case study using a chemical transport model. *Atmospheric Chemistry and Physics* 7: 1347–1366.

Lei W., Zavala M., de Foy B., Volkamer R., and Molina L.T. 2008. Characterizing ozone production and response under different meteorological conditions in Mexico City. *Atmospheric Chemistry and Physics* 8: 7571–7581.

Li G., et al. 2010. Impacts of HONO sources on the photochemistry in Mexico City during the MCMA-2006/MILAGRO Campaign. *Atmospheric Chemistry and Physics Discussion*, 10, 4143–4188.

Li J., Guttikunda S.K., Carmichael G.R., Streets D.G., Chang Y.-S., and Fung V. 2004. Quantifying the human health benefits of curbing air pollution in Shanghai, *Journal of Environmental Management* 70 (1): 49–62.

Lu R. and Turco R.P. 1997. An integrated air pollution modeling system for urban and regional scales. 1. Structure and performance. *Journal of Geophysical Research—Atmospheres* 102 (D5): 6063–6079.

Ludwig F.L., Gasiorek L.S., and Ruff R.E. 1977. Simplifications of a Gaussian Puff model for real time mini computer use. *Atmospheric Environment* 11: 431–436.

Luhar A.K. and Patil R. 1989. A general finite line source model for vehicular pollution dispersion. *Atmospheric Environment* 23: 555–562.

Lurmann F.W., Wexler A.S., Pandis S.N., Mussara S., Kumar N., and Seinfeld J.H. 1997. Modelling Urban and Regional Aerosols – II, Application to California's South Coast Air Basin, *Atmospheric Environment* 31: 2695–2715.

Lvovsky et al. 2000. Environmental costs of fossil fuels: A rapid assessment method with application to six cities. Environment Department Paper No. 78, The World Bank, Washington, DC.

Mallet V., Quelo D., et al. 2007. Technical note: The air quality modeling system Polyphemus. *Atmospheric Chemistry and Physics* 7 (20): 5479–5487.

McCubbin D., Hallberg A., et al. 2004. Assessment of urban air pollution impacts using the environmental Benefits Mapping and Analysis Program (BENMAP). *Epidemiology* 15 (4): S209.

Mediavilla-Sahagun A. and ApSimon H.M. 2006. Urban scale integrated assessment for London: Which emission reduction strategies are more effective in attaining prescribed PM_{10} air quality standards by 2005? *Environmental Modelling & Software* 21 (4): 501–513.

Mehta S. 2008. Emerging evidence on the health effects of air pollution in Asia. Presented at the 5th Better Air Quality Conference for Asian Cities, Bangkok, Thailand, http://baq2008.org/spa-mehta

Mena-Carrasco M. and Carmichael G. 2008. Final Report NASA ACMAP: Regional-Scale Modeling and Analysis Focusing on Asia. NASA Grant: NNG04GG27G.

Mena-Carrasco M., Carmichael G.R., et al. 2009a. Assessing the regional impacts of Mexico City emissions on air quality and chemistry. *Atmospheric Chemistry and Physics* 9 (11): 3731–3743.

Mena-Carrasco M., Carmichael G.R., Molina L.T., Spak S., Campos T., McNaughton C., Clarke A., and Gallardo L. 2009b. Evaluating the regional influence of Santiago de Chile on air quality and meteorology during VOCALS-REX. *Proceedings of the European Geosciences Union*, EGU2009-4743.

Mena-Carrasco M., Tang Y., et al. 2007. Improving regional ozone modeling through systematic evaluation of errors using the aircraft observations during the International Consortium for Atmospheric Research on Transport and Transformation. *Journal of Geophysical Research—Atmospheres* 112 (D12): 1–19.

Mensink C. and Colles A. 2003. Integrated air quality modeling for the assessment of air quality in streets against the council directives. *Atmospheric Environment* 37: 5177–5184.

Metternicht G. 2001. Assessing temporal and spatial changes of salinity using fuzzy logic, remote sensing and GIS foundation of an expert system. *Ecological Modelling* 144: 163–179.

Minocha V.K. 2006. Modeling and monitoring of environmental systems. PhD thesis, University of Delhi.

Moon D. and Albergel A. 1997. The use of the MERCURE CFD code to deal with an air pollution problem due to building wake effects. *Journal of Wind Engineering and Industrial Aerodynamics* 67 (8): 781–791.

Molina, L.T., et al. 2010. An overview of the MILAGRO 2006 campaign: Mexico City emissions and their transport and transformation, *Atmospheric Chemistry and Physics Discussion*, 7819–7983.

Mori A. 2000. Integration of plume and puff diffusion models/application of CFD. *Atmospheric Environment* 34: 45–49.

Morris R.E., Yarwood G., et al. 2002. Recent advances in CAM$_x$ air quality modelling. *Air Pollution Modelling and Simulation*, Proceedings: 79–88.

Nema A. and Tare V. 1989. Atmospheric dispersion under non-homogeneous and unsteady conditions. *Atmospheric Environment* 23 (4): 851–856.

Oettl D. and Sturm P.J. 2003. Dispersion from road tunnel portals: Comparison of two different modeling approaches. *Atmospheric Environment* 37: 5165–5175.

Oettl D. and Sturm P.J. 2005. Evaluation of GRAL for the pollutant dispersion from a city street tunnel portal at depressed level. *Environmental Modelling & Software* 20: 499–504.

Ojha C.S.P. and Singh V.P. 2002. A chapter on *ANN Modeling in Watershed Hydrology in Mathematical Models of Water Shed Hydrology*, V.P. Singh and D.K. Frevert (Eds). Water Resources Publishers, Colorado, USA.

Ojha C.S.P., Goyal M.K., and Kumar S. 2007. Applying fuzzy logic and the point count system to select landfill sites. *Environment Monitoring and Assessment* 135: 99–106.

Olivier J.G.J., Van Aardenne J.A., Dentener F., Ganzeveld L., and J.A.H.W. Peters. 2005. Recent trends in global greenhouse gas emissions: Regional trends and spatial distribution of key sources. In: A. van Amstel (coord.), *Non-CO$_2$ Greenhouse Gases (NCGG-4)*. Millpress, Rotterdam, ISBN 90 5966 043 9, pp. 325–330.

Ostro et al. 1994. Estimating the health effects from air pollutants: A method with an application to Jakarta. World Bank Policy Research Working Paper #1301.

Ostro et al. 1998. Estimating the health impact of air pollution: Methodology and an application to Jakarta. Working Paper Series, The World Bank, Washington, DC.

Pan L., Chai T.F., et al. 2007. Top-down estimate of mercury emissions in China using four-dimensional variational data assimilation. *Atmospheric Environment* 41 (13): 2804–2819.

Park O.H. and Seok M.G. 2007. Selection of an appropriate model to predict plume dispersion in coastal areas. *Atmospheric Environment* 41: 6095–6101.

Pasquill F. and Smith F.B. 1983. *Atmospheric Diffusion*, 3rd edition. Ellis horwood Limites, Chichester.

Pelliccioni A. and Tirabassi T. 2006. Air dispersion model and neural network: A new perspective for integrated models in the simulation of complex situations. *Environmental Modelling & Software* 21: 539–546.

Perez P. 2008. Air quality forecasting in a large city. *Air Pollution* XVI 116: 21–28.

Petersen W.B. 1980. User's Guide for HIWAY-2—a Highway Air Pollution Model, EPA-600/8-80-018, U.S. Environmental Protection Agency, Research Triangle Park, NC, 80pp.

Pilinis C. and Seinfeld J.H. 1988. Development and evaluation of an Eulerian photochemical gas-aerosol model. *Atmospheric Environment* 22 (9): 1985–2001.

Pirjola L. and Kulmala M. 2001. Development of particle size and composition distributions with a novel aerosol dynamics model. *Tellus Series B—Chemical and Physical Meteorology* 53 (4): 491–509.

Pope et al. 2006. Health effects of fine particulate air pollution: Lines that connect. *Journal of the Air & Waste Management Association* 56 (6): 709–742.

Romanowicz R., Young P., Brown P., and Diggle P. 2005. A recursive estimation approach to the spatio-temporal analysis and modeling of air quality data. *Environmental Modelling & Software* 21: 759–769.

Russell A.G., McCue K.F., and Cass G.R. 1988. Mathematical modeling of the formation of nitrogen-containing pollutants 2. Evaluation of the effects of emission controls. *Environmental Science & Technology* 22: 1336–1347.

SAES. 2000. Shanghai Energy Option and Health Impact. Report prepared by Shanghai Academy of Environmental Sciences and Shanghai Medical University.

Sandu A., Daescu D.N., Carmichael G.R., and Chai T. 2005. Adjoint sensitivity analysis of regional air quality models. *Journal of Computational Physics*, 204: 222–252.

Schell B., Ackermann I.J., et al. 2001. Modeling the formation of secondary organic aerosol within a comprehensive air quality model system. *Journal of Geophysical Research—Atmospheres* 106 (D22): 28275–28293.

Schmitz R. and Falvey M. 2007. Evaluation of the PM10 forecast models in use in metropolitan region of Chile. Chilean Environmental Commission (CONAMA RM). Available at http://www.geofisica.cl/Tempo/mix/Informe_Final_PM10.pdf. Accessed July 2009.

Schmitz R., Falvey M., and Clerc J. 2008. Optimización del modelo fotoquímico de alta resolución implementado en la fase 2007 y ampliación de su alcance a material particulado respirable y precursores de interés. Segundo Informe de Avance, Comisión Nacional de Medio Ambiente.

Scire J.S. and Yamartino R.J. 1989. *CALGRID: A Mesoscale Photochemical Grid Model*, Vol. II: User's Guide, University of Iowa.

Seigneur C., Pai P., et al. 1999. Modeling atmospheric particulate matter. *Environmental Science & Technology* 33 (3): 80A–86A.

Silibello C., Calori G., et al. 2008. Modelling of PM10 concentrations over Milano urban area using two aerosol modules. *Environmental Modelling & Software* 23 (3): 333–343.

Stern R. and Yamartino R.J. 2001. Development and first evaluation of micro-calgrid: A 3-D, urban-canopy-scale photochemical model. *Atmospheric Environment* 35: S149–S165.

Tang Y.H., Carmichael G.R., et al. 2003. Impacts of aerosols and clouds on photolysis frequencies and photochemistry during TRACE-P: 2. Three-dimensional study using a regional chemical transport model. *Journal of Geophysical Research—Atmospheres* 108 (D21): 1–25.

Tang Y.H., Carmichael G.R., et al. 2004. Three-dimensional simulations of inorganic aerosol distributions in east Asia during spring 2001. *Journal of Geophysical Research—Atmospheres* 109 (D19): 1–32.

Tie X., Madronich S., Li G., et al. 2007. Characterization of chemical oxidants in Mexico City: A regional chemical dynamical model (WRFChem) study. *Atmos. Environ.* 41: 1989–2008.

Tie X., Madronich S., Li G., et al. 2009. Simulation of Mexico City plumes during the MIRAGE-Mex field campaign using the WRF-Chem model. *Atmos. Chem. Phys.* 9: 4621–4638.

Tsimpidi A.P., Karydis V.A., Zavala M., et al. 2009. Evaluation of the volatility basis-set approach for the simulation of organic aerosol formation in the Mexico City metropolitan area. *Atmospheric Chemistry and Physics* Discussion 9: 13693–13737.

Ulriksen P. and Merino M. 2003. Air quality forecast in Santiago, Chile. GURME Expert Workshop in Santiago, Chile. http://www.cleanairnet.org/lac_en/1415/articles-51162_ulriksen.ppt (Accessed in October, 2009).

USEPA. 1995. *Screen 3 Model User's Guide*. USEPA, Research Triangle Park, NC, http://www.epa.gov/scram001/userg/screen/screen3d.pdf

USEPA. 1998. A comparison of CALPUFF modeling results to two tracer field experiments, http://www.epa.gov/scram001/7thconf/calpuff/tracer.pdf

USEPA. 2007. Guidance on the use of models and other analyses for demonstration of attainment of air quality goals for ozone, $PM_{2.5}$, and regional haze, http://www.epa.gov/scram001/guidance/guide/final-03-pm-rh-guidance.pdf EPA-454/B-07-002

USEPA. 2009. Technology Transfer Network Support Center for Regulatory Atmospheric Modeling, http://www.epa.gov/ttn/scram/

Versteeg H.K. and Malalasekera W. 2008. *An Introduction To Computational Fluid Dynamics: The Finite Volume Method*, Pearson Higher Education.

Vignati E. and Berkowicz R. 1999. Transformation of size distributions of emitted particles in the streets. *Science of the Total Environment* 235: 6479–6491.

Willis G.E. and Deardorff J.W. 1981. A laboratory study of dispersion in the middle of the convectively mixed layer. *Atmospheric Environment* 15: 109–117.

World Bank. 1997. *Clear Water & Blue Skies: China Environment 2020*. The World Bank, Washington, DC.

World Bank. 1998. *URBAIR Air Quality Management Series*. The World Bank, Washington, DC.

World Bank. 2007a. *Cost of Pollution in China*. The World Bank, Washington, DC, http://go.worldbank.org/FFCJVBTP40

World Bank. 2007b. *World Development Indicators*. The World Bank, Washington, DC, http://devdata.worldbank.org/data-query/

World Bank. 2009. *Handbook on Particulate Pollution Source Apportionment and Review of Case Studies From Around the World*. The World Bank, Washington, DC.

Xu et al. 1994. Air pollution and daily mortality in residential areas of Beijing, China. *Archives of Environmental Health* 49: 216–222.

Yamartino R.J. and Scire J.S. 1989. *CALGRID: A Mesoscale Photochemical Grid Model, Volume I: Model Formulation Document*. California Air Resources Board, Sacramento, CA.

Yamartino R.J. and Wiegand G. 1986. Development and evaluation of simple models for flow, turbulence and pollutant concentration fields within an urban street canyon. *Atmospheric Environment* 20: 2137–2156.

Yi J. and Prybutok R. 1996. A neural network model forecasting for prediction of daily maximum ozone concentration in an industrialized urban area. *Environmental Pollution* 92: 349–357.

Young P.C., Pedregal D., and Tych W. 1999. Dynamic harmonic regression. *Journal of Forecasting* 18: 369–394.

Zadeh L.A. 1965. Fuzzy sets. *Information and Controls* 8: 353–383.

Zadeh L. 1983. The role of fuzzy logic in the management of uncertainty in expert systems. *Fuzzy Sets and Systems* 11: 199–227.

Zannetti P. 1981. A new Gaussian puff algorithm for non homogeneous, non stationary dispersion in complex terrain. In: C. De Wispelaere (Ed.), *Air Pollution Modeling and Its Application-I*, Vol. 1. Plenum Press, New York.

Zannetti P. 1986. A new mixed segment-puff approach for dispersion modeling. *Atmospheric Environment* 20: 1121–1130.

Zaveri R.A., Easter R.C., et al. 2008. Model for simulating aerosol interactions and chemistry (MOSAIC). *Journal of Geophysical Research—Atmospheres* 113 (D13): 1–29.

Zawar-Reza P., Kingham S., and Pearce J. 2005. Evaluation of a year-long dispersion modeling of PM_{10} using the mesoscale model TAPM for Christchurch, New Zealand. *Science of the Total Environment* 349: 249–259.

Zeldin M.D. and Cassmassi J.C. 1979. Statistical-methods for predicting air-quality levels in the South Coast Air Basin. *Bulletin of the American Meteorological Society* 60 (5): 585.

Zhang L., Constantinescu E.M., et al. 2008. An adjoint sensitivity analysis and 4D-Var data assimilation study of Texas air quality. *Atmospheric Environment* 42 (23): 5787–5804.

Zhang Y. and Pun B. 2004. Development and application of the model of aerosol dynamics, reaction, ionization and dissolution (MADRID). *Journal of Geophysical Research—Atmospheres* 109 (D1): 1–31.

Ziomas I.C., Tzoumaka P., and Balis D. 1998. Ozone episodes in Athens, Greece. A modeling approach using data from the medcaphot-trace. *Atmospheric Environment* 32: 2313–2321.

Zia Wadud, BUET (Bangladesh University of Engineering and Technology), Personal communication.

Section II

Air Pollution and Health Effects

4 Indoor Air Pollution and Health Effects

Radha Goyal and Mukesh Khare

CONTENTS

4.1 INTRODUCTION

Through the years, man has built increasingly elaborated buildings to protect himself from the elements such as rain, snow, and warm air in summer and cool air in winter. However, such buildings do not always ensure protection of the occupants from the pollution present indoors. A person spends 90% of his time inside the buildings (Hoppe and Martinac, 1998). The indoor environment, therefore, can be viewed as a "habitat" or an "ecosystem." It is a complex "habitat" that consists of various attributes, namely occupants and their activities, the air pathways and ventilation, the building envelope and its environmental settings. Hence, it is necessary to understand the interrelationship between indoor and outdoor environments. The scientific evidence has indicated that the air within the buildings may be more polluted than the outside air causing problems associated with poor indoor air quality (IAQ). Indoor air pollution is responsible for 2.7% of the global burden of disease (WHO, 2002).

Indoor air pollution started with the use of fires in caves. The problems with indoor air were unquestionably much more apparent then than they are today. Soot found on the ceilings of prehistoric caves provides ample evidence of the high levels of pollution that are associated with inadequate ventilation of open fires (Spengler and Sexton, 1983). While chimneys first began to appear in European homes in the late twelfth century, most large medieval houses still had a central hearth in the great hall, ventilated by a louver in the roof. It was only during the sixteenth century that chimney stacks came into general use (Brimblecombe, 1987; Burr, 1997). The blackened roof timbers in many buildings that predate these innovations bear testimony to the severe pollution problems that their inhabitants faced. But prehistoric people had many more concerns than that of health hazards and little attention had been paid concerning IAQ till the early 1920s. The first systematic studies on IAQ conducted in the 1920s and 1930s focused on determining how much ventilation was needed to maintain a proper mix of metabolic gases, namely oxygen and carbon dioxide, and to control the body odors in buildings. The energy crisis period of the 1960s and early 1970s acted as the spawning ground of today's IAQ problems. Improved insulation has been accompanied by numerous other modifications to the management of indoor environments, and advances in construction technology have led to a much greater use of synthetic building materials (D'Amato et al., 1994). All these changes have undoubtedly meant that buildings are more comfortable. However, those energy conservation practices for reducing the outside air intake and enhancing airtightness in building designs have led to adverse health effects on the occupants residing in such buildings due to the production of airborne contaminants indoors and their higher concentrations indoors than outdoors (Teichman, 1995). Therefore, the research has been started to evaluate the dependence of indoor air pollution concentrations on outdoor levels of industrial and urban air pollution concentrations. In the mid-1970s, the concentration of several classes of air pollutants was found to be commonly higher indoors than outdoors. This finding has been applied to combustion by-products, volatile organic compounds (VOCs), radon and bioaerosols, which has spurred substantial interest in IAQ as an important environmental issue. Later, in the 1990s, the six-city study of the Harvard School of Public Health has discovered the "indoor air" as another dimension in determining the exposure levels of occupants, suggesting the increase in ventilation in buildings (Hensen and Burroughs, 1998). Theoretically, higher ventilation rates may reduce the indoor pollutant's concentration, but at the same time higher penetration of outdoor pollutants may also take place (Hoppe and Martinac, 1998). Therefore, greatly increased efforts in the areas of research and policy making are required to analyze and understand the complexities present in the indoor air environment and to improve the quality of air indoors. Research on the health effects of IAQ should be strengthened. A more systematic approach to the development and evaluation of interventions is desirable, with clearer recognition of the interrelationships between poverty and dependence on polluting fuels.

4.1.1 INDIAN PERSPECTIVE

In the developing world, 90% or more of our lives are dependent on the IAQ at homes, workplaces, and vehicles. In 75% of Indian households, biomass fuel (wood, dung, and crop residues) is used for cooking and heating. It accounts for 80% of India's domestic energy consumption. This fuel is typically used in open fires or simple stoves, mostly indoors, and rarely with adequate ventilation or chimneys. This situation leads to some of the highest ever recorded levels of indoor air pollution, to which young children and women are exposed daily for many hours. This exposure causes half a million premature deaths every year. According to the World Health Organization (WHO), indoor air pollution due to biomass smoke is one of the largest environmental risk factors for ill health of any kind (Bruce et al., 2000).

4.2   INDOOR AIR POLLUTION: THEIR SOURCES AND CAUSES

IAQ is defined as "air in an occupied space towards which a substantial majority of occupants express no dissatisfaction and in which there are not likely to be known contaminants at concentrations leading to exposures that pose a significant health risk" (ASHRAE, 1989). The IAQ is closely associated with the outdoor air quality. However, occupants and their activities also tend to generate contaminants in the indoor spaces. A range of conditions and the interactions of "sources," "sinks," and air movement among rooms and between the building and outside determine IAQ. Indoor air pollutants represent a complex array of constituents made up of *gases, vapors*, and *particles*. The determination of health effects related to these pollutants collectively, individually, or in certain combinations requires extensive information about the exposure of an individual to this mixture. The major indoor air pollutants that affect human health are classified broadly into three categories: *particles, vapors*, and *gases*, and their sources are broadly classified as *the activities of building occupants and other biological sources; the combustion of substances for heating or fuel*; and *emissions from building materials*. For some contaminants, infiltration from outside, either through water, air, or soil, can also be a significant source. The indoor air pollutant sources can be *external, internal, biological*, or *chemical* in nature. The *external* sources include industries and construction sources: exhausts from heating, ventilation, and air-conditioning (HVAC) equipments, vehicles, and soil gases such as radon. External biological contaminants can come from standing water that promotes mold growth. *Internal sources* of pollutants include building materials such as pressed wood board, glues, insulation, paints, stains, solvents, and other furnishings like carpet, furniture, and cabinets, HVAC systems, office equipment such as laser printers and copiers, activities like smoking and cooking, and other combustion sources such as fireplaces and furnaces, cleaning materials, both their use and storage, and pesticides. *Biological* sources in indoors are pets, plants, and humans. They produce dust mites, molds, pollen, animal dander, and bacteria. They often come from damp or water-damaged walls, floors, ceilings, and bedding and from poorly maintained air-conditioners and humidifiers. Environmental tobacco smoke (ETS) contains over 3800 compounds, including VOCs, inorganic gases, and metals, many of which are carcinogenic or can promote the carcinogenic properties of other pollutants. "Sinks" are high-surface-area or porous sites on or within which odor or other gaseous contaminants deposit. They may be located in the rooms or systems and may ultimately become secondary sources themselves. Air movement in a building consists of (a) natural air movement among rooms, sometimes fostered by the occupant's movement, (b) air movement driven by a forced air system, namely an HVAC system; air movement between the building and outside through ventilation, infiltration and exfiltration; and air movement driven by elevator piston action, the thermal stack effect, and air pressurization differentials.

The poorly designed ventilation systems and airtightness may lead to "inadequate" supply of fresh air in buildings. As a result, negative pressure develops, which may cause the outside pollutants to be drawn inside the buildings from vents, cracks, and openings. Uncontrolled temperature and humidity conditions indoors may also generate odor and bioaerosols—the *fungi, molds,* and other sickness-causing *microbes.* The indoor pollutant flow in Figure 4.1 provides an overview of the contaminant's "life" in the building. The indoor air pollutants include bioaerosols, particulates, VOCs, and inorganic and organic gases.

The *bioaerosols* are airborne microbiological particulate matters, derived from viruses, bacteria, mites, pollen, and their cellular or cell mass components. Bioaerosols are present in both indoor and outdoor environments. Floors in a hospital can be a reservoir for organisms that may subsequently be re-entrained into the air. While carpeting appears to trap microorganisms firmly, conditions within the carpet may promote their survival and dissemination. Water is a well-known source of infective agents even by aerosolization.

The *particulates* represent a broad class of chemical and physical contaminants found in the air as discrete particles. These are defined as mixtures or dispersions of solid or liquid particles. Typical examples of particulates include dust, smoke, fumes, and mists. They are broadly classified as *suspended particulate matter (SPM)* and *respirable particulate matter (RSPM)*. RSPM are generally defined as 10 μm or less in size (PM_{10}), although the Environmental Protection Agency (EPA) has expressed

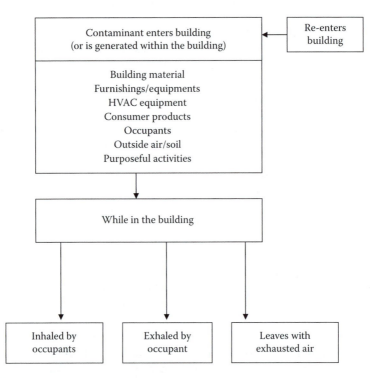

FIGURE 4.1 Indoor air pollutant flow.

concern that they are available even in less than 2.5 μm ($PM_{2.5}$) and less than 1.0 μm ($PM_{1.0}$) sizes and are the primary cause of lung cancer. Particulates found in the work environments are generated as a result of work-related activities, that is, adding batch ingredients for a manufacturing process, applying asphalt in a roofing operation, or drilling an ore deposit in preparation for blasting. The external environment is a major source of particulates because of ambient pollution. The introduction of this source is through ventilation, infiltration, and occupant traffic. In the indoor environment, particulate sources may include cleaning dirt accumulation in carpets and on other fleecing sources, construction and renovation debris, paper dust, and deteriorated insulation. ETS, kerosene heaters, humidifiers, wood stoves, and fireplaces are the common sources for RSPM indoors.

The *VOCs* exist as a gas, or can easily off-gas under normal room temperature and relative humidity (RH) conditions. A range of VOCs is always found in all nonindustrial indoor environments. After ventilation, VOCs are probably the first concern when diagnosing an IAQ problem. The list of potential sources of VOCs is lengthy and growing. Some of the major and common sources include wet emissions that have very high emission rates initially; after application, wet emissions can be present in newly constructed buildings, photocopying material, carpets, wall coverings and furnishings, refrigerants, gasoline, cosmetic products, biological matter, molded plastic containers, disinfectants, cleaning products, and ETS. While direct VOC emissions from primary sources are predominant, some materials act as sinks for emissions and then become secondary sources as they re-emit adsorbed chemicals. Floor dust, which is different from the dust in the air, has been found to be a sink and a secondary emission source for VOCs.

Inorganic gases include oxides of nitrogen, sulfur, carbon monoxide (CO), carbon dioxide (CO_2), ozone (O_3), and chlorofluorocarbons (CFCs). *Oxides of nitrogen* result mainly from cooking appliances, pilot lights, and unvented heaters. Underground or attached parking garages can also contribute to concentrations of indoor NO_x. An unvented gas stove contributes approximately 0.025 ppm of NO_2 to a home. Nitric oxide is an odorless, tasteless, colorless gas. Inhalation of NO causes the formation of methemoglobin, which adversely affects the body by interfering with oxygen transport at the cellular level. Nitrogen dioxide is a corrosive gas with a pungent odor, the odor threshold of which is reported to be between 0.11 and 0.22 ppm. NO_2 has low water solubility and therefore can be inhaled into the deep lung where it causes a delayed inflammatory response.

Sulfur dioxide (SO_2) can result from the emissions of kerosene space heaters, the combustion of fossil fuels, or burning any material containing sulfur. SO_2 is a colorless gas with a pungent odor detected at about 0.5 ppm. Because SO_2 is quite soluble in water, it can react with moisture in the upper respiratory tract to produce irritant effects on the upper respiratory mucous membrane. Contaminants' exposure to fine particulates, the depth and rate of breathing of an individual, and the presence of preexisting disease can influence the degree of SO_2 toxicity.

CO is an odorless, colorless, and tasteless gas produced by the incomplete combustion of hydrocarbons. Common indoor sources of CO include gas stove, kerosene lanterns and heaters, tobacco smoke, wood stoves and unvented or improperly vented combustion sources. CO is a chemical asphyxiate. Inhalation of CO causes a throbbing headache brought about by CO having a competitive preference for hemo-globin. Carbon monoxide inhibits oxygen transport in the blood through the formation of carboxy hemoglobin.

CO_2 is produced by human respiration. It is not normally considered to be a toxic air contaminant, but it can be a simple asphyxiate. A level of 1000 ppm has been sug-gested as being representative of delivery rates of 10 L/s per person of outside air when CO_2 is measured at equilibrium concentration and at occupant densities of 10 people per 100 m^2 floor space. CO_2 can become dangerous not as a toxic agent but as a secondary asphyxiate. At concentrations between 2500 to 5000 ppm, CO_2 can cause headache. At extremely high levels of 100,000 ppm, people lose consciousness in 10 min, and at 200,000 ppm, CO_2 causes partial or complete closure of the glottis.

O_3 arises from the electrical or coronal discharges from office equipment includ-ing laser printers and photocopiers. Ozone is a pulmonary irritant and causes changes in human pulmonary function at concentrations of approximately 0.12 ppm. Exposure to ozone at 60–80 ppb causes inflammation, bronchoconstriction, and increased air-way responsiveness.

CFCs are halogenated alkaline gases that have been used as heat transfer gases in refrigeration applications, blowing agents, and propellants in aerosol products, and as expanders in plastic foams. Inhalation exposures to CFCs can cause cardiotoxicity at chronic, low-level exposures. Chronic exposures to 1000 ppm for 8 h per day for up to 17 days caused no subjective symptoms or changes in pulmonary function.

Formaldehyde is a VOC. It is a ubiquitous chemical used in a wide variety of products and is most frequently introduced in newly constructed buildings. It is a colorless gas at room temperature and has a pungent odor at higher concentrations. The infiltration of outdoor air is one source of formaldehyde in the indoor environ-ment, but the primary sources are in the indoor environment itself: building materi-als (thermal insulation in the side walls of buildings, plywood and particle board, floor coverings, and carpet backing), combustion appliances (gas stoves and heating systems), tobacco smoke, and a large number of consumer products (paper for wax paper, facial tissues, napkins, and paper towels).

Asbestos is a widespread component of the structural environment in schools, homes, and private and public buildings. Its release in the indoor environment depends on the cohesiveness of the asbestos-containing material (ACM) and the intensity of the disturbing force. Asbestos and similar fibrous minerals have been used in construction materials, consumer products, and appliances. ACM is most frequently found in boiler insulation, pipe insulation, sprayed-on fireproofing, breaching insulation, and floor and ceiling tiles.

ETS comes from the sidestream smoke emitted from the burning end of ciga-rettes, cigars, and pipes and second-hand smoke exhaled by smokers. Breathing in ambient ETS is generally referred to as passive or involuntary smoking. The con-taminants arising from tobacco combustion can be distinguished into mainstream smoke and sidestream smoke. Both smokers and nonsmokers are exposed to side-stream smoke. Mainstream smoke is undiluted and is pulled through with the tobacco into the smoker's lungs. Sidestream smoke is directly from burning tobacco. Depending on the smoking behavior, burning temperature, and type of filter, the composition of mainstream smoke exhaled by the smoker varies substantially.

Smoking is the major source of indoor particulates (having variable composition) and numerous irritating gases. Because tobacco does not burn completely, other contaminants are given off, including sulfur dioxide, ammonia, nitrogen oxides, vinyl chloride, hydrogen cyanide, formaldehyde, radionuclides, benzene, and arsenic.

Odors are a class of contaminants in gaseous form that can bring about discomfort, irritation, stress, complaints, and even fear, panic, and mass hysteria. They arise from occupants, and their effects figure in IAQ issues predominantly on the basis of comfort rather than health. Therefore, it is very difficult to track down odor complaints, and the facility managers often relegate odor complaints to the bottom of the priority pile. Routine activities such as cooking, smoking, bathroom use, and maintenance give rise to odors that are often disagreeable and in some cases offensive. To varying degrees, almost all building materials and furnishings are sources of odor.

Radon and its progeny are the only natural airborne radionuclides. The series begins with radon-222 and ends with the alpha decay product of radon-226 and is formed from the decay of radium, which in turn results from the decay of uranium. Radon is a noble gas that can move from its site of formation, giving it a substantial opportunity to reach air that is inhaled by humans. It is an odorless and colorless gas that is always present at various concentrations in the air. The short-lived decay products of radon—polonium, lead, and bismuth—are chemically active and thus can be collected in the lungs either directly or through particles with which they attach. Approximately 90% of radon daughters attach to larger airborne particles before they can be inhaled. The most important dose arises from the alpha decay of polonium isotope. The main sources of radionuclides and radiation are building materials, soil, and ground water. It typically enters through cracks, voids, or other openings in the foundation of buildings. Conditions affecting the flow of radon are the soil factors, the building factors, and the pressure differentials.

Air temperature and humidity are two of the most important comfort variables out of six primary comfort variables, which affect the quality of the indoor environment and are important indicators of IAQ. They are also extremely important to the occupant's perception of IAQ. The hot and humid climate of India might have an adverse impact on the comfort of the occupants. Therefore, controlling the air temperature and humidity is the primary factor for achieving comfort in indoor environments. In an enclosed space, air temperature generally increases from the floor to the ceiling (vertical temperature gradient). If this temperature difference is sufficiently large, a person's head can have local warm discomfort and/or cold discomfort at the feet even though the overall average is thermally neutral. To prevent this local discomfort, the standard calls for a maximum temperature difference between the head and the feet of 3°C (5°F). Similarly, the humidity level should be below 55% or, more exactly, below a dew point of 62°F to avoid discomfort in enclosed spaces. The American Society of Heating, Refrigerating and Air-Conditioning Engineers (ASHRAE) Standard 55-2004 (ASHRAE, 2004a) notes that for thermal comfort purposes, temperature could range between approximately 67°F and 82°F. A more specific range can be determined from the standard but depends on RH, season, clothing worn, activity levels, and other factors. The standard notes that HVAC systems must be able to maintain a humidity ratio of exactly or below 0.012. This corresponds to an upper RH level as high as about 80% at low dry bulb temperatures but can be lower depending on factors such as temperature and the other factors listed above.

The standard does not specify a lower humidity limit but notes that nonthermal comfort factors may place limits on the acceptability of very-low-humidity environments. The acceptable temperature and humidity ranges for winter and summer are 68.5–76°F and 74.0–80.0°F, respectively, for a dry bulb at 30% RH, and 68.5–74.5°F and 73.0–79.0°F, respectively, for a dry bulb at 50% RH. ASHRAE Standard 62.1-2004 (ASHRAE, 2004b) recommends that RH in occupied spaces be controlled to less than 65% to reduce the likelihood of conditions that can lead to microbial growth.

The predominant sources of indoor pollutants are listed in Table 4.1.

TABLE 4.1
Sources of Indoor Air Pollutants

Principal Pollutants	Sources
Predominantly Outdoor	
SO_2, SPM/RSPM	Fuel combustion
Pollens	Trees, grass, weeds, plants
Lead, manganese	Automobiles
Calcium, chlorine, silicon, cadmium	Suspension of soils or industrial emission
VOCs, poly aromatic hydrocarbons (PAHs), etc.	Petrochemical solvents, natural sources, vaporization of unburned fuels
Both Indoor and Outdoor	
Nitrogen oxide (NO), NO_2, CO	Fuel-burning, vehicular exhaust emission
CO_2	Fuel-burning and metabolic activity
Ozone (O_3)	Photochemical reactions, photocopying machines
SPM/RSPM	ETS, resuspension, condensation of vapors and combustion products
Water vapor	Biological activity, combustion, evaporation
Organic substances (VOCs)	Volatilization, combustion, paint, metabolic action, pesticides, insecticides, fungicides
Spores	Fungi, molds
Predominantly Indoor	
Radon	Building construction, materials (concrete stone), water
Formaldehyde	Particle board, insulation, furnishings, ETS
Asbestos, minerals, and synthetic fibers	Fire-retardant, acoustic thermal, or electric insulation
Organic substances	Adhesives, solvents, cooking, cosmetics, solvents
Ammonia (NH_3)	Metabolic activity, cleaning products
PAH, arsenic, nicotine, acrolein, etc.	ETS
Mercury	Fungicides, in paints, spills in dental care facilities or laboratories, thermometer breakage
Aerosols, allergens	Consumer products, house dust, animal dander
Viable organisms	Infections

Source: National Research Council. 1981. *Indoor Pollutants.* National Academy Press, USA.

TABLE 4.2

Factors Affecting Exposure to Indoor Air Pollutants

Human activities	Working time and activities related to work	Day duties, night duties, number of hours, etc.
	Domestic work	Kitchen activities, cleaning activities
	Social life	
Geographical factors	Regional variation	Weather: RH, wind velocity, temperature
	Urban, suburban, and neighborhood variation	City size, density of buildings, population correlate with temperature, wind velocity, pressure (e.g., urban heat island)
	Variation in IAQ in buildings	Pollutant concentration in kitchen are usually greater than the rest of the rooms
Building factors	Site characteristics	Airflow around buildings, proximity to major sources of outdoor pollution
	Occupancy	Type and intensity of human activity (smoking, cleaning, etc.), spatial characteristics of a given activity, operation schedule of a building
	Design	Interior space design, envelope design (primary elements: foundations, floors, walls, and roofs; secondary elements: facing, cladding, and sheathing)

Source: National Research Council. 1981. *Indoor Pollutants*. National Academy Press, USA.

4.2.1 PARAMETERS OF IAQ

The types and quantities of pollutants found indoors vary temporally and spatially. Depending on the type of pollutant and its sources, sinks, and mixing condition, its concentration can vary a lot even in a small area. Ventilation plays an important role in deciding the IAQ of an indoor space. However, the exposure of human to indoor pollutants depends on various factors that are listed in Table 4.2.

4.2.2 VENTILATION AND IAQ

Ventilation refers to the "provision of sufficient quantities of outside air in the building for the occupants to breathe and to dilute the concentration of the pollution generated by the people, equipment and material inside the building." It plays an important role in providing good IAQ. It drives the transport of pollutants between indoor and outdoor environments and determines the residence time of pollutants indoors. Smaller the ventilation rate, longer will be the residence time of pollutants. The pollutant concentration inside the spaces is inversely proportional to the ventilation rates, that is, doubling the ventilation rate will halve the pollutant concentration.

It is necessary for both mechanical HVAC and naturally ventilated buildings (Awbi, 1991). A poorly designed HVAC system may be the culprit of IAQ problems. Biological contaminants like bacteria, virus, mold, and so on are often found in areas that provide food and moisture or water. Humidifiers, condensate pans, and poorly maintained HVAC equipment form ideal breeding grounds for these organisms. The thermal environment has been shown to affect the discomfort levels and the efficiency of the occupants. Indoor thermal conditions are controlled by the HVAC system. The energy conservation practices for reducing the outside air intake and enhancing airtightness in building designs may lead to adverse health effects on the occupants residing in such buildings. The efficient functioning of such systems depends on proper maintenance, or else it may lead to the emergence of building sickness, that is, the sick building syndrome (SBS) (Roulet and Vandaele, 1991; Allard, 2002). Therefore, a properly designed, installed, operated, and maintained HVAC system can promote IAQ. Besides, the natural ventilation is uncontrolled due to presence of operable windows, grills, vents, and unpredictable airflow in different climatic conditions. The airflow inside the naturally ventilated buildings takes place mainly because of the pressure and temperature gradients, which are further governed by two driving forces: (1) thermal buoyancy (stack pressure) and (2) the wind pressure. The airflow in naturally ventilated buildings further depends on the building shape and size, location and orientation of openings with respect to their sizes and wind direction, surrounding roughness, building configurations, topography, and so on. It also depends on the meteorological parameters, for example, the wind speed and direction, the indoor and outdoor temperatures, and the RH (Awbi, 1991; Roulet and Vandaele, 1991; ASHRAE, 2001; Allard, 2002). Theoretically, higher ventilation rates may reduce the indoor pollutant's concentration in such buildings; at the same time, higher penetration of outdoor pollutants may occur due to uncontrolled conditions, for example, uneven distribution of air or insufficient exhaust ventilation. The situation even becomes worse in cases where outdoor pollutant concentrations are higher or indoor sources of pollutants exist. Therefore, for understanding the significance of ventilation in "acceptable" IAQ, (ASHRAE revised its ventilation standard. The "acceptable" ventilation standards for IAQ are given in Table 4.3 (ASHRAE, 1989).

Although the natural ventilation is "uncontrolled" and "unpredictable" in different climatic conditions, it appears as an economically effective and attractive alternative with lower running and maintenance costs as compared to the mechanical ventilation. Moreover, the environmental concerns and the emphasis on sustainable development have also become strong driving forces for the proper use of natural ventilation. Natural ventilation as a strategy for achieving acceptable IAQ depends on the "supply of fresh air" to the space to dilute the indoor pollutant concentration.

The "volumetric airflow rates" (infiltration rates), the "air leakage" and the "airtightness," the "age of air," and the "CO_2 concentration" are the parameters that can quantify the ventilation. There are various measurement techniques available in the literature to quantify these parameters of the ventilation, namely the "tracer gas decay method," the "pulse injection method," "constant injection," the "decay method," the "pressurization test method," and so on (Persily, 1989).

4.2.3 SICK BUILDING SYNDROME

Over the years, construction techniques have significantly improved in order to make structures longer lasting and less problematic. However, not every situation

TABLE 4.3
Ventilation for Acceptable IAQ

Application		Occupancy (people/1000 ft²)	cfm/person	cfm/ft²
Food and beverage Service	Dining rooms	70	20	
	Cafeteria, fast food	100	20	
	Bars, cocktail lounges	100	30	
	Kitchen (cooking)	20	15	—
Offices	Office space	7	20	
	Reception areas	60	15	—
	Conference rooms	50	20	
Public spaces	Smoking lounge	70	60	—
	Elevator	—	—	1.00
Retail stores, sales floors, showroom floors	Basement and street	30	—	0.30
	Upper floors	20	—	0.20
	Malls and arcades	20	—	0.20
	Smoking lounge	70	60	—
Sports and amusement	Spectator areas	150	15	
	Game rooms	70	25	
	Playing floors	30	20	—
	Ballrooms and discos	100	25	
Theatres	Lobbies	150	20	
	Auditorium	150	15	—
Education	Classrooms	50	15	
	Music rooms	50	15	
	Libraries	20	15	—
	Auditoriums	150	15	
Hotels, motels, resorts dormitories	Bedrooms	—	—	
	Living rooms	—	—	30 cfm/room
	Lobbies	30	15	30 cfm/room
	Conference rooms	50	20	—
	Assembly rooms	120	15	—

Note: 1 L/s equals approximately 2 cfm.

can be predicted and sometimes events occur that lead to "problem buildings." Problem buildings can be classified either as having occupants who experience symptoms characterized as "SBS," "building-related illness (BRI)," or both. The term "SBS" is used to describe situations in which building occupants experience acute health and comfort effects that appear to be linked to duration of time spent in a building, but no specific illness or cause can be identified. The complaints may be localized to a particular room or zone or may be widespread throughout the building. In contrast, the term "BRI" is used when the symptoms of a diagnosable illness are identified and can be attributed directly to airborne building contaminants (Lane et al., 1989). Therefore, SBS may be defined as a "phenomenon that occurs in a building in which a proportion of occupants experience varying degrees of low levels of sickness or discomfort which are nonspecific in nature and are dominated by sensory reactions." The *indicators of SBS* include (1) the building occupants' complaint of symptoms associated with acute discomfort (e.g., headache; eye, nose, or throat irritation; dry cough; dry or itchy skin; dizziness and nausea; difficulty in concentrating; fatigue; and sensitivity to odors), (2) the cause of the symptoms is not known, and (3) most of the complainants report relief soon after leaving the building. The *causes of SBS* include inadequate ventilation; chemical contaminants from indoor sources; chemical contaminants from outdoor sources; biological contaminants; and psychological factors like perception of thermal, audio, and visual environments. Cases of SBS typically report vague symptoms, which cannot be objectively measured, and sufferers usually show no clinical signs of illness. SBS symptoms are linked to building occupancy because they get better on leaving the building. IAQ surveys of sick buildings often fail to find pollution problems, even though complaints are chronic and symptom prevalence among occupants is high with up to 80% of workers reporting at least one symptom (Wilson and Hedge, 1987). Table 4.4 shows the medical symptoms associated with sick buildings.

4.3   INVESTIGATING IAQ PROBLEMS

When the indoor air problems grow into full-scale concerns, the owners and operators run immediately to outside expertise. A lot can be done, however, through simple in-house keeping before such moves become necessary. The IAQ investigations can be divided into five phases for either simple in-house inspection or complex diagnosis (Hensen and Burroughs 1998):

> Phase I: *Preliminary Assessment*—a self–evaluation; data gathering; observation effort.
> Phase II: *The Qualitative Walk-Through Inspection*—conducted by trained in-house staff or as a preliminary inspection by the diagnostic team.
> Phase III: *Simple Qualitative Diagnostics*—more extensive analytical procedures conducted by the diagnostic team; limited measurements of implicated factors or surrogates.
> Phase IV: *Complex Quantitative Diagnostics*—broad in-depth testing; qualitative studies of factors in combination; medical examinations.
> Phase V: *Proactive Monitoring and Recurrence Prevention*—observation; record keeping; retesting as warranted; preventive measures.

The preliminary assessment forms or questionnaires are developed and interviews are taken for background evaluation, observational data gathering, and self-evaluation. In contrast to the preliminary investigation, the walk-through inspection involves a more thorough examination of the facility and the HVAC system by in-house personnel. In phase III, the walk-through inspection is conducted by experienced and knowledgeable personnel. It usually involves the use of temperature,

TABLE 4.4
Medical Symptoms of Sick Buildings

Syndrome	Symptoms
Sick building syndrome	Lethargy and tiredness
• Type 1	Headache
	Dry blocked nose
	Sore dry eyes
	Sore throat
	Dry skin and/or skin rashes
Sick building syndrome	Watering/itchy eyes and runny nose, that is,
• Type 2	symptoms of an allergy such as hay fever
Humidifier fever	Generalized malaise
• Flu-like symptoms	Aches and pains
	Cough
	Lethargy
	Headache
• Allergic reaction in sensitive individuals	Chest tightness
	Difficulty in breathing
	Fever
	Headache
Occupational asthma	Wheeze
	Chest tightness
	Difficulty in breathing

Sources: WHO. 1983. *Indoor Air Pollutants: Exposure and Health Effects.* WHO, Geneva;
Wilson, S. and Hedge, A. 1987. *The Office Environment Survey: A Study of Building
Sickness.* Building Use Studies Ltd., London, UK.

humidity, and airflow measurement and monitoring. CO_2 measurements are sometimes taken to assess the effectiveness of the ventilation system. In phase IV, more
precise sampling of various parameters, that is, ventilation parameters as well as
pollutants, is conducted, which is further analyzed to evaluate the intensity of the
problem and to take the preventive measures in phase V.

4.3.1 SAMPLING/MONITORING AND MODELING OF INDOOR AIR POLLUTANT

Sampling techniques can be broadly classified as *continuous*, *integrated*, and *grab* or
spot sampling. *Continuous sampling* provides real-time sampling and is used to
observe temporal fluctuations in concentrations over short periods. *Integrated sampling* provides an average sampling over a specific period. It is used when the mean
concentration is either desirable or adequate for the purpose. *Grab sampling* provides
single samples taken at specified intervals. It typically consists of admitting an air
sample into a previously evacuated vessel, drawing a sample into a deflated bag for
later analysis, or drawing (by a mechanical pump) a sample through a sample

collector to extract a contaminant from air; it is suitable when "spot" samples are adequate for the measurement of pollutants, and knowledge of temporal concentration variation over short periods is not important. Different sampling techniques are available for various pollutants. Particulates are sampled in mass according to their sizes, and gravimetric analysis techniques are in use. The major techniques developed for sampling gaseous pollutants are broadly classified as *passive* (based on membrane permeation or diffusion through a geometrically defined air space) and *nonpassive* (in which air pumping devices draw volumes through devices of known collection efficiency). Various sampling, measurement, and analysis techniques for air pollutants are listed in Table 4.5.

The concentration of a pollutant indoors depends on the relationship between the volume of air contained in the indoor space, the rate of production or release of the pollutant, the rate of removal of the pollutant from the air via reaction or settling, the rate of air exchange with the outside atmosphere, and the outdoor pollutant concentration (Maroni et al., 1995). The extent of indoor air pollution can be estimated with numerical models; mass balance equations are used to estimate the concentrations of indoor pollutants as fractions of outdoor concentrations and to estimate infiltration rates, indoor source strengths, pollutant decay rates, and mixing factors. In estimating the total exposure of humans to pollutants (exposure to pollutants encountered indoors and outdoors, in industrials sites and other workplaces, etc.), it is essential to know not only the pollutant concentration, but also the individual patterns of mobility and use of time. However, the actual human exposures are often difficult to quantify. This is largely because the behavior and activity patterns of individuals can strongly affect their levels of exposure (Harrison, 1997).

TABLE 4.5
Sampling, Measurement, and Analysis Techniques

S. No.	Pollutant	Method
1.	Carbon monoxide	Nondispersive infrared (NDIR) photometry
2.	Ozone	Chemiluminescence
3.	Nitrogen oxides	Chemiluminescence
4.	Sulfur dioxide	West Gaeke colorimetric method
5.	Formaldehyde	Spectrophotometer analysis
6.	Asbestos	Phase contrast microscopy (PCM), scanning electron microscopy (SEM), transmission electron microscopy (TEM)
7.	Radon	Thermo luminescent dosimeter (TLD)
8.	Particulate matter (PM)	Size selective samplers
9.	Biological pollutants (pollens, etc.)	Samplers with adhesive coating on slides
10.	Ventilation rates	Tracer gas technique

Sources: USEPA. 1991. *Exposure Factors Handbook.* EPA/600/8-89/043. U.S. Environmental Protection Agency, Office of Health and Environmental Assessment, Washington, DC (online publications); Persily, A. 1989. *ASHRAE Journal*, 31(6), 52–54.

The starting point in developing an IAQ model is usually a statement of the mass balance concerning the pollutant of interest (Figure 4.2). For example, consider a structure of volume V, in which makeup air enters from the outside and passes through the filter at a rate q_o. Part of the building air is recirculated through another filter at a rate q_1, and air infiltrates the structure at a rate q_2. Each filter is characterized by a factor $F \equiv (C_{inlet} - C_{outlet})/C_{inlet}$. Usually the pollutant concentration is assumed to be uniform throughout the structure. The indoor and outdoor concentrations at time t are C and C_o, respectively. The rate at which the pollutant is added to the indoor air owing to internal sources is S. The rate at which the pollutant is removed from the air owing to internal sinks is R (Wadden and Scheff, 1983). In this case, the appropriate starting equation is

$$V\frac{dC}{dt} = \underbrace{q_oC_o(1 - F_o) + q_1C\,(1 - F_1) + q_2C_o}_{\text{Input rate due to makeup, recirculated and infiltrated air}} - \underbrace{(Q_o + q_1 + q_2)C}_{\text{Output rate}} + \underbrace{S}_{\substack{\text{Source}\\\text{rate}}} - \underbrace{R}_{\substack{\text{Sink}\\\text{rate}}}. \quad (4.1)$$

Or more simply, it can be written as

$$V\frac{dC}{dt} = C_{IN}Q_{IN} - C_{OUT}Q_{OUT} + S - R, \quad (4.2)$$

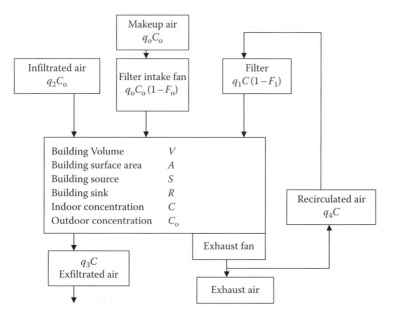

FIGURE 4.2 IAQ mass balance for an HVAC building. (Adapted from Shair, F.H. and Heitner, K.L. 1974. *Environmental Science* and *Technology*, 8, 444; Wadden, R.A. and Scheff, P.A. 1983. *Indoor Air Pollution: Characterization, Prediction, and Control.* Wiley, New York.)

where V is the volume of the room; C is the pollutant concentration in the room; C_{IN} is the concentration entering the room; Q_{IN} is the airflow into the room; C_{OUT} is the concentration leaving the room; Q_{OUT} is the airflow leaving the room; S is the source term; and R is the removal term, which includes pollutant removal by using air cleaners and sinks.

From the well-mixed assumption, C_{OUT} equals C. The equation can be rewritten as

$$V\frac{dC}{dt} = C_{IN}Q_{IN} - CQ_{OUT} + S - R. \qquad (4.3)$$

This is the case with *single compartment* mass balance modeling (NRC, 1981). It is applied to the spaces that are considered as well mixed at a given point of time and the pollutant concentration remains the same at all locations within the volume being modeled. Despite the widespread use of the well-mixed assumption, there is evidence that in many cases, indoor air concentrations are not spatially homogeneous. Well-mixed models underestimate concentration near an emission source. Recently developed models consider the "source proximate effect" (Furtaw et al., 1996), which is based on the premise that concentrations in proximity to the source are higher than those predicted by well-mixed models. *Multicompartmental modeling* is the approach to modeling nonhomogeneous indoor air concentrations. This approach divides the space into two or more well-mixed zones that are connected by interzonal airflows (Furtaw et al., 1996). A compartment is defined as a region within which spatial variations in pollutant concentrations can be neglected over the timescale of interest. Depending on the ventilation conditions, a single room, a floor, or a whole building may be adequately approximated as a single compartment. However, if either sources or sinks are not uniformly distributed throughout the region of interest and the rate of mixing throughout the region of interest is low compared with characteristic residence time, then the single-compartment model may not provide an adequate description.

4.4 INDOOR AIR POLLUTION AND HEALTH EFFECTS

The health effects of poor IAQ are dependent on several factors. A relevant consideration when determining potential health effects on a population is the effect of each *air contaminant, concentration/exposure limit, duration of exposure*, and *individual sensitivity*. The *air contaminant* may be an allergen, or it may be a carcinogenic chemical that will cause an immediate reaction with minimal long-term effects and cancer years after exposure, respectively. There are approximately 100,000 toxic substances to which building occupants are potentially exposed; however, *exposure limits* exist for less than 400 industrial chemicals. The Occupational Safety and Health Agency (OSHA), the USEPA (USEPA, 1991), and the WHO (WHO, 1983) have defined the limits of exposure to some principal indoor air pollutants. Table 4.6 shows the potential health effects of the pollutants commonly found in indoor air and their limits. The USEPA and WHO standards are for nonoccupational conditions,

TABLE 4.6
Indoor Pollutants, Their Health Effects, Standards, and Exposure Limits

Pollutant	Health Effects	EPA	OSHA	WHO
			Limits	
NO_2	Type: Immediate Causes: Irritation to the skin, eyes and throat, cough, etc.	0.09 mg/m³ (average over 1 year for 8 h exposure daily)	9 mg/m³	<0.19 mg/m³
CO	Type: Immediate Causes: Headache, shortness of breath, higher concentration. May cause sudden death.	10 mg/m³ (average over 8 h period)	55 mg/m³ (average over 8 h period)	<2.0 mg/m³
CO_2		—	9000 mg/m³	—
RSPM	Type: Cumulative Causes: Lung cancer	0.15 mg/m³ (24 h average)	5 mg/m³ (average over 8 h period)	<0.01 mg/m³
SO_2	Type: Immediate Causes: lung disorders and shortness of breath	0.13 mg/m³ (average over 1 year for 8 h exposure daily)	13 mg/m³ (average over 8 h period)	<0.5 mg/m³ (short-term exposure)
Radon	Type: Cumulative Causes: Lung cancer	4000 pCi/m³ of indoor air	≈0.1 mg/m³	—
Formaldehyde	Type: Immediate Causes: Irritation to the eyes, nose and throat, fatigue, headache, skin allergies, vomiting, etc.	0.12 mg/m³ (continuous exposure)	3 mg/m³ (average over 8 h period)	<0.06 mg/m³ (long-term and short-term exposure)
Asbestos	Type: Cumulative Causes: Lung cancer	—	200 fibers/m³ (8 h exposure period)	≈0 fiber/m³ (long-term exposure)
Pesticides	Type: Immediate Causes: Skin diseases	—	—	—
O_3	Type: Immediate Causes: Eyes itch, burn, respiratory disorders, and lowers our resistance to colds and pneumonia.	—	0.1 mg/m³ (continuous exposure)	<1800 mg/m³

Source: National Research Council. 1981. *Indoor Pollutants*. National Academy Press, USA.

whereas OSHA standards are for the occupants during their occupation. Quite recently, in 2003, the IAQ Management Group, Government of HKSAR, Hong Kong, gave the standards for "acceptable" IAQ for offices and public places, which are listed in Table 4.7.

The standards are prepared for 8-h exposure for both excellent class and good class IAQs. *Exposure duration* is of particular importance in assessing the IAQ exposures. In office buildings, exposure is generally 8–10 h a day, five days a week. In residential structures, exposure may be up to 24 h a day, seven days a week. As some substances build up in the body over time, 24-h exposure may result in an accumulation of pollutants and their subsequent impact on health effects. Thus the impact of a given concentration of air contaminant is less in office buildings than in residences. Other areas that should be considered to have potential long-duration exposures include hospital patient rooms, hotels, mental wards, and prison cells. *Individual sensitivity* contributes a huge variable to the combination of factors affecting the health of building occupants. Infants, elderly people, and sick people are the most vulnerable to the health effects of air contaminants. Immune suppressed individuals, for example, AIDS patients and organ transplant recipients, and patients of genetic diseases, for example, *Lupus erythematosus*, are particularly sensitive to common molds. Individuals who drink alcohol in excess are more susceptible to air contaminants that may affect the liver. People with dry skin are more susceptible to further drying and skin penetration by chemicals and those who smoke tobacco products

TABLE 4.7
Standards for Acceptable Indoor Air Quality

Parameters	Unit	8-h Average	
		Excellent Class	**Good Class**
CO_2	mg/m³	<1500	<1800
CO	mg/m³	<2	<10
Respiratory suspended particulate (RSP)	mg/m³	<0.02	<0.18
NO_2	mg/m³	<0.04	<0.15
Ozone (O_3)	mg/m³	<0.05	<0.12
Formaldehyde (HCHO)	mg/m³	<0.03	<0.1
Total VOC	mg/m³	<0.2	<0.6
Radon (Rn)	Bq/m³	<150	<200
Airborne bacteria	Cfu/m³	<500	<1000
Room temperature	°C	20 to <25.5	<25.5
Relative humidity	%	40 to <70	<70
Air movement	m/s	<0.2	<0.3

Note: "Excellent" class represents an excellent IAQ that a high-class and comfortable building should have.

"Good" class represents IAQ that provides protection to the public at large including the very young and the aged.

have a diminished body defense mechanism (Kousa, 2001). There is consistent evidence that exposure to indoor air pollution increases the risk of pneumonia among children under five years and chronic respiratory disease and lung cancer (in relation to coal use) among adults over 30 years old. The evidence for a link with lung cancer from exposure to biomass smoke and for a link with asthma, cataracts, and tuberculosis was considered moderate. On the basis of the limited available studies, there is tentative evidence for an association between indoor air pollution and adverse pregnancy outcomes, in particular, low birth weight, or ischemic heart disease and nasopharyngeal and laryngeal cancers (Smith, 1999).

The health effects from indoor air pollutants fall into two categories: *short-term* (immediate and/or acute) effects and *long-term* (cumulative and/or chronic) effects. *Short-term* effects may show up after a single exposure or repeated exposures. These include irritation of the eyes, nose, throat, and skin, headache, dizziness, and fatigue. These are treatable if identified. Most of the immediate effects are similar to those from cold or other viral diseases, so it is often difficult to determine if the symptoms are a result of exposure to indoor air pollution. Therefore, it is important to pay attention to the time and place where the symptoms occur. *Cumulative* effects occur only after long or repeated periods of exposure to pollutants.

Indoor air pollution has been associated with a wide range of health outcomes, and the evidence for these associations has been classified as strong, moderate, or tentative in a recent systematic review. Most of the air pollutants directly affect the respiratory and cardiovascular systems. Increased mortality, morbidity, and impaired pulmonary function have been associated with elevated levels of SO_2, and SPM or RSPM. While the precise mechanism of how exposure causes disease is still unclear, it is known that small particles and several of the other pollutants contained in indoor smoke cause inflammation of the airways and lungs and impair the immune response. *Acute* and *subacute* health effects of the inhalation of *biomass smoke* include conjunctivitis, acute respiratory irritation/inflammation, and acute respiratory infection (ARI). *Chronic* effects of the inhalation of biomass smoke are chronic obstructive pulmonary disease (COPD), chronic bronchitis, cor pulmonale, adverse reproductive outcomes and pregnancy-related problems, such as stillbirths and low birth weight, and lung cancer. The health risks associated with some of the key pollutants present in smoke and their mechanism are listed in Table 4.8.

One study in Western India found a 50% increase in stillbirths in women exposed to indoor smoke during pregnancy. Likewise, a study in Africa found that cooking with wood greatly increased the risk of stillbirth. Considerable amounts of carbon monoxide have been detected in the bloodstream of women cooking with biomass in India and Guatemala. Studies in India, Nepal, and Papua New Guinea show that nonsmoking women who have cooked on biomass stoves for many years exhibit a higher prevalence of chronic lung disease (asthma and chronic bronchitis). In Mexico, women exposed to wood smoke for many years faced 75 times more risk of acquiring chronic lung disease, about the level of risk that heavy cigarette smokers face, than women not exposed to wood smoke. One recent Colombian study found that women exposed to smoke during cooking were three times more likely to suffer from chronic lung diseases. Studies in South America and India have shown that exposure to indoor air pollution severely reduces lung function in children.

TABLE 4.8

Mechanisms by Which Some Key Pollutants in Smoke from Domestic Sources May Increase the Risk of Respiratory and Other Health Problems

S. No.	Pollutants	Mechanism	Potential Health Effects
1.	Particles (<10 μm, and particularly <2.5 μm aerodynamic diameter)	• Acute bronchial irritation, inflammation, and increased reactivity • Reduced mucociliary clearance • Reduced macrophage response and reduced immunity	• Wheezing, exacerbation of asthma • Respiratory infection • Chronic bronchitis and COPD
2.	CO	• Binding with hemoglobin to produce carboxyhemoglobin, which reduces oxygen delivery to key organs and developing fetus	• Low birth weight (carboxyhemoglobin is 2–10% higher) • Increased perinatal deaths
3.	NO$_2$	• Acute exposure increases bronchial reactivity • Long-term exposure increases susceptibility to bacterial and viral lung infections	• Wheezing, exacerbation of asthma • Respiratory infection • Reduced lung function in children
4.	SO$_2$	• Acute exposure increases bronchial reactivity • Long term: Difficult to dissociate from effects of particulates	• Wheezing, exacerbation of asthma • Exacerbation of COPD, cardiovascular disease
5.	Biomass smoke condensates including polyaromatics and metal ions	• Absorption of toxins into lungs leading to oxidative changes	• Cataract
6.	Polyaromatic HCs (benzopyrene)	• Carcinogenic	• Lung cancer • Cancer of mouth, nasopharynx, and larynx

Source: From Bruce, N., Padilla, R.P., and Albalak, R. 2000. *Bulletin of WHO*, 78(9), 1078–1092. With permission.

There is growing concern about the fact that exposure to second-hand *tobacco smoke* is a serious and substantial public health risk. Tobacco smoke contains over 3800 compounds, including VOCs, inorganic gases, and metals, many of which are carcinogenic or can promote the carcinogenic properties of other pollutants. Small children and infants raised in the presence of passive smoke are more prone to lower respiratory tract and inner ear infections.

The health effects of *airborne particulates* depend on several factors that include particle dimensions, durability, and dose. In some instances, very small exposures can cause adverse health effects (hazardous exposures), while other seemingly large exposures do not cause any adverse effects (nuisance exposures). The health risk from exposure to particulate air pollution obtained by applying the mean risk per unit ambient concentration is based on the results of some urban epidemiological studies (Smith, 1996). The range of risk was found to be 1.2–4.4% increased mortality per 10 mg/m^3 incremental increase in the concentration of respirable suspended particles (PM_{10}).

The CO emission poses a serious health problem when biomass fuels are used. Incomplete combustion of fuels produces CO. Smith (1991) estimated that about 38, 17, 5, and 2 g/meal CO is released during household cooking by using dung, crop residues, wood, and kerosene, respectively. During the use of liquid petroleum gas (LPG), a negligible amount of CO is released. A study by the National Institute of Occupational Health (NIOH), Ahmedabad, reported indoor air CO levels of 144, 156, 94, 108, and 14 mg/m^3 air during cooking by using dung, wood, coal, kerosene, and LPG, respectively. The short-term health effects of CO exposure are dizziness, headache, nausea, feeling of weakness, and so on. It also results in systemic effects by reducing the oxygen-carrying capacity of the blood. The association between a long-term exposure to CO from cigarette smoke, on the one hand, and heart disease and fetal development, on the other, has been described by Wynder (1979).

The *formaldehyde* is well recognized to be an acute irritant and long-term exposure can cause a reduction in vital capacity and chronic bronchitis. The formaldehyde is well known to form cross-links with biological macromolecules. Inhaled formaldehyde forms DNA and DNA–protein cross-links. The formaldehyde mean levels are found to be 670, 652, 109, 112, and 68 $\mu g/m^3$ of air for cattle dung, wood, coal, kerosene, and LPG, respectively (Patel and Raiyani, 1995).

The adverse health effects caused by VOCs in nonindustrial indoor environments fall under three categories: (1) irritant effects, including perception of unpleasant odors, mucous membrane irritation, and exacerbation of asthma; (2) systematic effects, such as fatigue and difficulty in concentrating; and (3) toxic, chronic effects, such as carcinogenicity.

The *bioaerosols*—fungal spores, hyphae fragment, or metabolites—can cause a variety of respiratory diseases. These range from allergic diseases including allergic rhinitis and asthma to infectious diseases such as histoplasmosis, blastomycosis, and aspergillosis. Other than these major categories of illnesses, indoor air pollution is associated with blindness and changes in the immune system. Eighteen percent of blindness in India is attributed to the use of biomass fuels. Further, a 1995 study in Eastern India found the immune system of newborns to be depressed due to the presence of indoor air pollution.

4.4.1 Disproportionate Impacts on Children and Women

Household energy practices vary widely around the world, as does the resultant death toll due to indoor air pollution. While more than two-thirds of indoor smoke-attributable deaths from acute lower respiratory infections in children occur in WHOE's African and South East Asian Regions, over 50% of the COPD deaths due to indoor air pollution occur in the Western Pacific region. In most societies, women are in charge of cooking and—depending on the demands of the local cuisine—they spend between 3 and 7 h per day near the stove preparing food. Thus 59% of all indoor air pollution-attributable deaths occur in females. Young children are often carried on their mother's back or kept close to the warm hearth. Consequently, infants spend many hours breathing indoor smoke during their first year of life when their developing airways make them particularly vulnerable to hazardous pollutants. As a result, 56% of all indoor air pollution-attributable deaths occur in children under five years of age. In addition to the health burden, fuel collection can impose a serious time burden on women and children. Alleviating this work will free women's time for productive endeavors and child care and can boost children's school attendance and time for homework.

4.4.2 IAQ and Health Effects: Indian Estimates

In India, half a million deaths each year are attributed to indoor pollution from traditional biomass fuels. India and China together account for approximately 60% of solid fuel using households in the developing world; this implies that, worldwide, about two million premature deaths each year could be attributed to household solid fuel use. Depending on the number of young children in total, indoor exposure would account for 4–6% of the global burden of disease. By comparison, urban air pollution is estimated to be responsible for 1–2% of the global disease burden. These estimates would make the health impact of indoor exposure larger than the burden from all but two of the other major preventable risk factors that have been quantified: malnutrition (15%) and lack of clean water and sanitation (7%). It surpasses the global burdens from sexually transmitted diseases, tobacco, illicit drugs, hypertension, occupational hazards, alcohol, war, vehicle accidents, or homicide. It exceeds the global burden for many diseases except total ARIs, diarrhea, and the childhood cluster of vaccine-preventable diseases (measles, diphtheria, tetanus, polio, and pertussis). If these estimates are accurate, the global burden of disease from indoor air pollution is larger than that for such well-known threats to human health as tuberculosis, AIDS/HIV, malaria, heart disease, or cancer (Murray and Lopez, 1996).

ARI is the single largest disease category worldwide (about one-twelfth of the global disease burden); it accounts for about one-eighth of the disease burden in India. Odds ratio for young children: 2–3 (10 studies in developing countries). Premature deaths: 290,000–410,000. *COPDs* such as chronic bronchitis account for about 1.5% of deaths among Indian women. Odds ratio for women cooking over biomass fires for 15 years: 2–4 (four studies in developing countries). Premature deaths: 19,000–34,000. *Lung cancer* in women is linked with cooking over open coal stoves; there is little evidence of a connection with biomass fuel. Odds ratio for

TABLE 4.9
Numbers of Deaths Attributable to Indoor Particles Air Pollution

Author	Total Deaths Attributable to Indoor Particle Air Pollution (million)	Excess Mortality by Setting (Deaths and % of Total)			
		Developed Countries		Developing Countries	
		Urban	Rural	Urban	Rural
Smith	2.8	640,000	1,800,000	250,000	30,000
		23%	67%	9%	1%
Schwela	2.7	363,000	1,849,000	511,000	Not calculated
		13%	68%	19%	

Source: From Bruce, N., Padilla, R.P., and Albalak, R. 2000. *Bulletin of WHO*, 78(9), 1078–1092. With permission.

more than 20 Chinese studies: 3–5. Premature deaths: 400–800. *Blindness* is linked with the use of biomass fuels by women in India, which has the largest burden of blindness of any region of the world. Odds ratio for this group: 1.15–1.3. Blindness does not cause premature deaths, but puts a significant burden of disability. *Tuberculosis* is responsible for 5% of the Indian burden of disease, a larger percentage than in any other region. It causes premature deaths: 50,000–130,000. *Perinatal effects* (stillbirth; low birth weight; and death or illness during the first two weeks after birth) are responsible for 8.8% of the Indian burden of disease. But there is little evidence available to make estimates about premature deaths. *Cardiovascular diseases (CVDs)* and *asthma* are. known to be related to outdoor air pollution and passive smoking in industrial countries but do not seem to have been studied in developing-country households (Smith, 1998).

The health bulletin of WHO has reported the results of studies conducted by Smith and Schwela on the number of deaths in developing and developed countries caused by indoor particle pollution for both the rural as well as urban population; the results show that in rural households the particle pollution is responsible for 67–68% of the total deaths due to indoor particle pollution. This is due to the use of biomass fuel for cooking and heating purposes. A limited number of such studies has been done till now on the rural population of developing countries. Their results are shown in Table 4.9.

4.5 INDOOR AIR POLLUTION CONTROL

In practice, there are little differences to distinguish between the control of contaminants and their treatment and mitigation. Treating an existing problem through mitigation or remediation is often the same exercise as preventive measures taken to control it in the first place. And once treated, the same action becomes a control procedure to prevent the reoccurrence. Some preventive measures can be taken that are not available in remedial treatment (e.g., building design). To accommodate the occupant needs, building owners and facility managers must recognize the multiple purposes of controlling the quality of indoor air. First of all, federal requirements for specific pollutants must be met. Furthermore, satisfying guidelines that do not have the force of law may still be legally prudent as a defense against negligence in potential lawsuits. The underlying purpose in all control procedures is to ensure that the indoor air maintains the quality needed for safety and health; satisfies comfort and productivity needs; and is as cost effective and energy effective as possible. There are six ways to control indoor air pollution:

1. Removal or substitution at the source
2. Filtration and purification of contaminants
3. Dilution or ventilation of the indoor air with the outside air or filtered recycled air
4. Encapsulation or otherwise interfering with materials' ability to give off pollutants
5. Time of use of a possible contaminant
6. Education and training of building occupants, especially operation and maintenance personnel

The EPA has consistently stated that source control is the most direct and dependable control option and is the only effective one when strong pollutant sources are present. Ventilation is a preferred way of control when the contaminant/source is unknown, when source treatment is too costly, or when the source is localized. The IAQ management may also wish to seek the advice and counsel of medical experts as a control support through medical surveillance or treatment.

REFERENCES

Allard, F. 2002. *Natural Ventilation in Buildings—A Design Handbook*. James and James (Science Publishers) Ltd., London NW1 3ER, UK.

ASHRAE. 1989. ASHRAE Standard 62-1989, *Ventilation for Acceptable Indoor Air Quality*. American Society of Heating, Refrigeration and Air Conditioning Engineers, Inc., Atlanta, GA.

ASHRAE. 2001. ASHRAE, *ASHRAE Handbook of Fundamentals*. American Society of Heating, Refrigeration and Air-Conditioning Engineers, Inc., 345 East 47th St. New York, NY.

ASHRAE. 2004a. ASHRAE Inc. Standard 55—2004, *Thermal Environmental Conditions for Human Occupancy*. ASHRAE, Atlanta, GA.

ASHRAE. 2004b. ASHRAE Inc. Standard 62.1—2004, *Ventilation for Acceptable Indoor Air Quality*. (ANSI Approved).

Awbi, H.B. 1991. *Ventilation of Buildings* (1st edition). Published by E & FN Spon—An imprint of Chapman & Hall publishers, London.

Brimblecombe, P. 1987. *The Big Smoke: A History of Air Pollution in London Since Medieval Times*. Routledge, London.

Bruce, N., Padilla, R.P., and Albalak, R. 2000. Indoor air pollution in developing countries: A major environmental and public health challenge. *Bulletin of WHO*, 78(9), 1078–1092.

Burr, M.L. 1997. Health effects of indoor combustion products. *Journal of the Royal Society of Health*, 117(6), 348–350.

D'Amato, G., Liccardi, G., and D'Amato, M. 1994. Environment and the development of respiratory allergy II: Indoors. *Monaldi Archive of Chest Disorders*, 49(5), 412–420.

Furtaw, E.J., Pandian, M.D., Nelson Dwayne R., and Behar Joseph V. 1996. Modelling indoor air concentrations near emission sources in imperfectly mixed rooms. *Journal of the Air & Waste Management Association*, 46(9), 861–868.

Harrison, P.T.C. 1997. Health impacts of indoor air pollution. *Chemistry and Industry*, 17, 677–681.

Hensen, S.J. and Burroughs, H.E. 1998. *Managing indoor air quality* (2nd edition). The Fairmont Press, Inc., Lilburn, GA.

Hoppe, P. and Martinac, I. 1998. Indoor climate and air quality. *International Journal of Biometeorology*, 42, 1–7.

HKSAR. 2003. Indoor Air Quality Assessment. *A Guide on Indoor Air Quality Certification Scheme for Offices and Public Places*. The Indoor Air Quality Management Group of the HKSAR, Hong Kong, www.aa-lab.com/air/iaq.html

Kousa, K.H. 2001. *Indoor Air Quality: Sampling Methodologies*. Lewis Publishers, Washington, DC.

Lane, C.A., Woods, J.E., and Bosman, T.A. 1989. Indoor air quality procedures for sick and healthy buildings. *ASHRAE Journal*, 31(6), 48–52.

Maroni, M., Seifert, B., and Lindvall, T. (Eds). 1995. *Indoor Air Quality—A Comprehensive Reference Book*. Elsevier, Amsterdam.

Murray, C.J. and Lopez, A.D. 1996. *The Global Burden of Disease*. Harvard University Press, Cambridge, MA.

National Research Council, 1981. *Indoor Pollutants*. National Academy Press, USA.

Patel, T.S. and Raiyani, C.V. 1995. Indoor air quality: Problems and perspectives. In *Energy Strategies and Green House Gas Mitigation*, P.R. Shukla (Ed.). Allied Publishers, New Delhi.

Persily, A. 1989. Ventilation rate in office buildings. *ASHRAE Journal*, 31(6), 52–54.

Roulet, C.-A. and Vandaele, L. 1991. Air flow patterns within buildings measurement techniques. Energy conservation in buildings and community systems programme, International Energy Agency. Technical Note, AIVC 34.

Shair, F.H. and Heitner, K.L. 1974. Theoretical model for relating indoor pollutant concentrations to those outside. *Environmental Science and Technology*, 8, 444.

Smith, K.R. 1991. Managing the risk transmission. *Toxicology and Industrial Health* 7, 319.

Smith, K.R. 1996. Indoor air pollution in developing countries: Growing evidence of its role in the global disease burden. In: *Indoor Air '96: Proceedings of 7th International Conference on Indoor Air and Climate*, K. Ikeda and T. Iwata (Eds). Institute of Public Health, Tokyo.

Smith, K.R. 1998. Indoor air pollution in India: National health impacts and the cost-effectiveness of intervention. Indira Gandhi Institute for Development Research, Goregaon, Mumbai. Discussion Note No. 4, August 1999, http://lnweb18.worldbank.org/essd/envext.nsf/51ByDocName/DiscussionNote4IndoorAirPollutionColor/$FILE/DisussionNote4August1991.pdf

Smith, K.R. 1999. Indoor air pollution. Discussion in the World Bank Report—Pollution Management in Focus.

Spengler, J.D. and Sexton, K. 1983. Indoor air pollution: A public health perspective. *Science*, 221(4605), 9–17.

Teichman, K.Y. 1995. Indoor air quality: Research needs. *Occupational Medicine*, 10(1), 217–227.

USEPA. 1991. *Exposure Factors Handbook*. EPA/600/8-89/043. U.S. Environmental Protection Agency, Office of Health and Environmental Assessment, Washington, DC.

Wadden, R.A. and Scheff, P.A. 1983. *Indoor Air Pollution: Characterization, Prediction, and Control*. Wiley, New York.

WHO. 1983. *Indoor Air Pollutants: Exposure and Health Effects*. WHO, Geneva.

WHO. 2000. Indoor air pollution in developing countries: A major environmental and public health challenge. Article in the Bulletin of the World Health Organization, http://www. who.int/indoorair/publications/bulletin/en/index.html

WHO. 2002. The health effects of indoor air pollution exposure in developing countries, http:// www.who.int/indoorair/publications/health_effects/en/index.html

Wilson, S. and Hedge, A. 1987. *The Office Environment Survey: A Study of Building Sickness.* Building Use Studies Ltd, London, UK.

Wynder, E.L. 1979. Carbon monoxide and cardiovascular disease-forum workshop. *Preventive Medicine* 8, 261.

5 Effects of Indoor Air Pollution from Biomass Fuel Use on Women's Health in India

Twisha Lahiri and Manas Ranjan Ray

CONTENTS

5.1   INTRODUCTION

Indoor air pollution (IAP) is an important cause of morbidity and mortality throughout the world. But its severity is much more in the developing countries due to burning of traditional biomass fuels such as wood, animal dung, and crop residues for
daily domestic cooking in rural households. Indeed, smoke exposure from biomass
burning is regarded as one of the most serious environmental problems facing the
developing countries throughout the world (World Bank, 1992; Smith and Mehta,
2000). It is estimated that IAP from biomass burning is responsible for 4% of the
global burden of disease (Bruce et al., 2000), and a conservative estimate indicates
that the practice of biomass fuel use endangers the health of 400–700 million people across the globe and causes 2.8 million premature deaths every year (Bruce
et al., 2000). It affects especially the women and children who spend long hours
inside the cooking areas (Larson and Rosen, 2002). As in many other developing
countries, biomass fuel is still extensively being used in rural households of India
for daily household cooking as well as for room heating in hilly areas. There have
been some studies on the effects of biomass use on respiratory health of rural
women in India, but the overall impact of biomass fuel use on public health, especially that of women and children in rural areas of the country, is yet to be eluci-
dated. In this chapter, we have collated the published reports on health impact of
IAP from biomass fuel use in India and elsewhere. In addition, we have incorporated some important findings from our ongoing investigation on respiratory and
general health impairments of women in rural areas of West Bengal, a state in
Eastern India. We have included the findings of 1260 women (median age 38 years)
who cook with biomass regularly for the past five years or more and have compared
these results with that of 650 liquid petroleum gas (LPG)-using women from similar neighborhood matched for age (median age 37 years), tobacco smoking and
chewing habits, and body mass index.

5.1.1   BIOMASS USE IN INDIAN SUBCONTINENT

About 2.4 billion people of the world, mostly in poor, developing countries of Asia,
sub-Saharan Africa, and Latin America, still rely on traditional unprocessed solid
biomass such as wood, animal dung, and crop residues as the primary source of
domestic energy (Smith et al., 2004). Globally, 52% of the households rely on biomass

as fuel for daily household cooking (WHO, 2002). Its use is much more in the Indian subcontinent. For example, biomass accounts for more than 80% of the domestic energy in India. The users are poor people and cooking is usually done in traditional chullahs (stoves), which are inefficient and emit large quantities of smoke indoors (Figure 5.1). Coupled with these, the cooking areas are often poorly ventilated, which results in greater exposure to combustion fumes (IIPS, 1995). A survey conducted jointly by the National Council of Applied Economic Research and World Bank in 1996 has reported that about 578 million tons of biomass is used annually in India as a source of domestic energy of which wood constitutes 52%, animal dung 21%, and agricultural refuse 20%. In general, rural domestic sector in India uses 1.2–2.1 kg of biomass/capita/day (Smith, 1987).

FIGURE 5.1   Photographs showing (a) collection of biomass fuel and (b) cooking with it in open chullah (oven) in rural India.

In West Bengal, where the study has been carried out, 45.6 million tons of biomass is used annually for cooking and room heating (Reddy and Venkataraman, 2002). Firewood remains the mainstay of biomass fuel in the state with an annual usage of 23.3 million tons/year, followed by agricultural wastes (11.9 million tons/year) and dung cakes (10.3 million tons/year) (Reddy and Venkataraman, 2002). Biomass fuel use as a source of domestic energy is even more in Bangladesh and Nepal where 88% and 80% of the population, respectively are dependent on biomass fuels for household cooking. The percentage of users is marginally lower in Pakistan (72%) and Sri Lanka (67%; WHO, 2006).

Poverty is a strong predictor of IAP from biomass fuel use in developing countries. People living in rural areas of these countries are generally poor who can ill afford cleaner fuels such as LPG with their modest income. Instead, they rely on biomass that is cheaper (or free of cost) and readily available. No wonder, there are only a handful of LPG users in rural India, and in most of the cases the LPG owners also use biomass (mixed user) to reduce fuel cost. Besides developing nations, poor people of some industrialized countries who earlier used cleaner fuels are now reverting to biomass because of political change and economic downturn. Notable examples include people residing in Tajikistan and Kyrgyz Republic after the breakdown of the former Soviet Union.

5.1.2 IAP FROM BIOMASS BURNING

Biomass fuels are at the high end of the fuel ladder in terms of pollution emissions and at the low end in terms of combustion efficiency (Smith et al., 1994). Smoke emitted during biomass burning contains a wide range of pollutants including particulates, carbon monoxide (CO), oxides of nitrogen and sulfur, polycyclic aromatic hydrocarbons (PAHs), volatile organic compounds (VOCs), and trace metals including Fe, Cu, Ni, Cr, and Pb (Zhang and Smith, 1996; Bruce et al., 2000). Airborne particles with a diameter of less than 10 μm (PM_{10}) are hazardous because they can be inhaled deep inside the lungs and thus serve as a vehicle for toxic matters that usually adsorb onto their surface. Burning biomass emits more PM_{10} than LPG or electricity. For example, particle concentrations in the kitchens vary from 200 to 5000 μg/m^3 of air during cooking with biomass fuels (Smith, 1993; Ezzati and Kammen, 2002) in contrast to 200–380 μg/m^3 during cooking with LPG or electricity (Ellegard, 1996). Moreover, fine (aerodynamic diameter < 2.5 μm, $PM_{2.5}$) and ultrafine particles (diameter < 0.1 μm) constitute a bulk of the biomass smoke (Tesfaigzi et al., 2002) and most of the mass is due to the presence of fine particles, that is, $PM_{2.5}$, which is more harmful for human health than is PM_{10} (WHO, 1999). Biomass is considered even more harmful than diesel, because biomass smoke contains 10 times more mass concentration of respirable particles having a diameter of 0.5–0.8 μm compared with diesel (Venkataraman and Rao, 2001).

In a typical biomass-using Indian kitchen, the concentration of PM_{10} ranged from 500 to 2000 μg/m^3 during cooking (Balakrishnan et al., 2002). Similarly, in Nepalese households, PM_{10} level in kitchens using biomass fuels was about three times higher than that in those using cleaner fuels such as kerosene, LPG, and biogas (Srestha and Srestha, 2005). The type of fuel, oven type, and ventilation have a significant impact on indoor air quality. Compared with wood, combustion of dung cakes and briquette emitted significantly higher CO and particulate matter (Venkataraman and Rao, 2001). In rural India, the ovens are not vented outside, kitchens are often poorly ventilated, and half of the households do not have separate kitchens (Mishra et al., 1999). These factors contribute significantly to high particulate level in cooking areas.

5.1.3 HEALTH IMPACT OF BIOMASS SMOKE EXPOSURE: EXCESS MORBIDITY AND MORTALITY

Chronic exposure to biomass smoke is one of the most important environmental and public health problems in developing countries, especially in women who cook with these fuels and their children who attend to the fires or stay close to their mothers during cooking (Smith and Mehta, 2000; Smith, 2002). Cumulative exposures to high levels of particulate pollutants could be hazardous for the biomass users, because 1% increase in total daily mortality occurs for every 20 $\mu g/m^3$ increase in PM_{10} level in breathing air (Samet et al., 2000). It has been estimated that IAP from biomass use in developing countries is responsible for 4–5% of global burden of disease for both deaths and disability adjusted lost life years (DALYs) from acute respiratory infections (ARI), chronic obstructive pulmonary disease (COPD), tuberculosis, asthma, lung cancer, ischemic heart disease, and blindness (Smith and Mehta, 2003).

Globally, over 1.2 million (Smith and Mehta, 2003) to about 2.8 million premature deaths per year (Bruce et al., 2000) have been attributed to biomass fuel use. World Health Organization (WHO) has put the mortality figure to 2.2–2.5 million annually (WHO, 1997). The problem is escalating with the passage of time and the number of annual premature deaths due to IAP from biomass fuel use has been projected at 9.8 million by the year 2030 (Bailis et al., 2005). In India, IAP from biomass combustion contributes to 4–6% of the national burden of disease and 400,000–550,000 premature deaths every year (Smith, 2000).

5.1.4 SCOPE OF THE WORK AND OBJECTIVE OF THE STUDY

Many PAHs and VOCs present in biomass smoke, such as benzo(a)pyrene [B(a)P] and benzene, are potential human carcinogens (Zhang and Smith, 1996). It has been estimated that women in India who cook regularly with biomass are on average exposed to a B(a)P that is equivalent to smoking 2–20 packs of cigarettes per day (Smith and Liu, 1994). It is reasonable to assume therefore that the womenfolk of the developing countries who are chronically exposed to air contaminated with chemicals emitted from combustion of biomass are at a considerable health risk. However, the level of IAP from biomass fuel use in rural households in India, its effects on respiratory and systemic health of the women who cook with these fuels, and the possible mechanism of such effects at the cellular and subcellular levels are largely unknown as very little attention has been paid to this important public health issue by the clinicians, scientists, and administrators. In view of this, the present work was undertaken to evaluate the health impact of IAP from biomass fuel use among rural women in India who used to cook with these fuels. The objective was to prepare a database on respiratory and systemic health impairments in association with biomass use in the country and to search for biomarkers of pollution exposure and effect suitable for biomonitoring of health impact of air pollution in the country.

5.2 STUDY PROTOCOL

5.2.1 SUBJECTS

The study was conducted among 1260 women aged 24–56 years (median age 38 years) of rural West Bengal, a state in Eastern India, who cooked exclusively with unprocessed solid biomass such as wood, dung cake, and agricultural wastes like hay, jute stick, rice bran, dried leaves, and so on. For comparison, 650 age-matched nonsmoking women from the same neighborhood who used to cook with relatively clean fuel LPG were enrolled. Apparently, healthy nonsmoking women who cook regularly preparing family meal for 3–5 h per day, 6–7 days a week for the past five years or more were enrolled. Pregnant and lactating women and those currently under medication were excluded. The study protocol was approved by the Ethics Committee of Chittaranjan National Cancer Institute, Kolkata, India.

5.2.2 EVALUATION OF RESPIRATORY HEALTH

Data on prevalence of respiratory symptoms were collected through personal interview using validated, structured questionnaire. Pulmonary function test (PFT) was done using a portable spirometer following the procedure recommended by American Thoracic Society (ATS, 1995). COPD was diagnosed following the criteria of the Global Initiative for Chronic Obstructive Lung Disease (GOLD; Pauwels et al., 2001). Cellular lung response to air pollution was evaluated by examining the cytology of exfoliated airway cells in Papanicolaou-stained sputum.

5.2.3 MEASUREMENT OF IAP

Real-time measurement of airborne pollutant (PM_{10} and $PM_{2.5}$) concentration in cooking areas was measured by portable, laser photometer (DustTrak TM Aerosol monitor, model 8520, TSI Inc., MN, USA).

5.2.4 HEMATOLOGICAL ASSAYS

Hematological parameters like total and differential counts of white blood cells (WBC) and total count of platelets were done on freshly collected venous blood without anticoagulant using a hemocytometer under a light microscope. β2 Mac-1 integrin (CD11b/CD18) expression on circulating neutrophils and monocytes, P-selectin (CD 62P) expression on platelet surface, a measure of platelet activation, and circulating leukocyte-platelet aggregates were determined by flow cytometry.

5.2.5 REPRODUCTIVE TOXICITY OF BIOMASS SMOKE EXPOSURES

The impact of chronic exposure to smoke from biomass burning during household cooking on five menstrual cycle characteristics, namely (i) short cycles, (ii) long cycles (oligomenorrhea), (iii) irregular cycles, (iv) missed periods, and (v) intermenstrual bleeding were evaluated by self-reporting questionnaire through personal interview by trained female members of the research team following the protocol of Farr et al. (2004).

5.2.6   GENOTOXICITY AND ANTIOXIDANT DEFENSE

Chromosomal breaks were examined by micronucleus (MN) assay (Tolbert et al., 1991) and DNA damage by single cell gel electrophoresis (Comet assay, Singh et al., 1988), and antioxidant enzyme superoxide dismutase (SOD) was measured in plasma by spectrophotometry.

5.2.7   QUESTIONNAIRE SURVEY FOR NEUROBEHAVIORAL PROBLEMS

A neurobehavioral symptom questionnaire, adopted from the subjective symptom questionnaire accompanying the WHO Neurobehavioral Core Test Battery, Wechsler's memory scale (Wechsler, 1945), and 21-item Beck's depression inventory (BDI; Beck et al., 1961) was administered to them. In addition, the questionnaire focused on other symptoms like burning sensation in extremities (feeling of burn in distal and terminal portions of the body such as hand and foot), tingling (repetitive moving pin prick-like sensation), numbness (temporary loss of sensation), vertigo (an illusionary sensation that the body or surrounding environment is revolving), and dizziness (sensation of unsteadiness with a feeling of movement within the head, giddiness). The concentrations of dopamine (DA), epinephrine (E), and norepinephrine (NE) in blood plasma were measured by high-performance liquid chromatography while plasma cholinesterase was determined spectrophotometrically following the procedure developed by Ellman et al. (1961).

5.3   SALIENT FINDINGS

5.3.1   PARTICULATE POLLUTION IN COOKING AREAS

The concentration of PM_{10} in kitchens of biomass-using households during cooking varied between 479 and 2156 $\mu g/m^3$ with a mean of 625 $\mu g/m^3$. In contrast, a mean PM_{10} level of 169 $\mu g/m^3$ (ranged 97–254 $\mu g/m^3$) was recorded during cooking with LPG. The corresponding $PM_{2.5}$ levels were 312 $\mu g/m^3$ (ranged 262–915 $\mu g/m^3$) and 77 $\mu g/m^3$ (ranged 46–117 $\mu g/m^3$), respectively during cooking with biomass and LPG. This suggests that women who cooked regularly with biomass fuels were several times more exposed to particulate pollutants than LPG users.

5.3.2   EFFECTS ON RESPIRATORY HEALTH

5.3.2.1   Rise in Respiratory Symptoms

Biomass users suffer more from respiratory symptoms that indicate the presence of underlying respiratory illness. In this study, 45% of biomass users had experienced

TABLE 5.1

Prevalence (%) of Respiratory and Other Symptoms among Rural Women

Symptom	LPG-Users ($n = 650$)	Biomass Users ($n = 1260$)
Upper respiratory symptom, overall	26.2	45.1*
Sinusitis	6.3	11.7*
Runny or stuffy nose	12.6	21.2*
Sore throat	9.2	20.3*
Common cold and fever	12.3	16.3*
Lower respiratory symptom, overall	24.7	48.3*
Dry cough	15.6	16.7*
Cough with phlegm	18.3	32.0*
Wheezing breath	21.2	30.3*
Chest discomfort or tightness	19.7	53.4*
Breathlessness	19.7	64.0*
Other symptoms		
Headache	40.7	82.3*
Physician-diagnosed asthma	2.7	4.6*

*$p < 0.05$ compared with control in χ^2 test; more than one symptoms were present in many individuals.

one or more upper respiratory symptoms like sinusitis, recurrent runny or stuffy nose, frequent sore throat, and common cold with fever in the past three months compared with 26% of LPG users (Table 5.1). Similarly, lower respiratory symptoms such as recurrent dry cough or sputum-producing cough, chest discomfort or chest tightness, wheeze, and breathlessness were present in 48% of biomass users compared with 25% of LPG users.

Higher prevalence of respiratory symptoms among biomass users may suggest underlying respiratory infections. IAP from combustion of biomass fuels has been linked to ARI (Pandey et al., 1989; Ezzati and Kammen, 2001; Mishra, 2003) and tuberculosis (Gupta et al., 1997; Mishra et al., 1999; Smith and Mehta, 2003). Overall, 51% of the prevalence of active tuberculosis among Indians aged 20 years and above is attributable to cooking smoke from biomass fuels (Mishra et al., 1999). The Indian National Survey found an adjusted odds ratio of 2.7 for tuberculosis in women in association with biomass use (Mishra et al., 1999).

5.3.2.2 Asthma

Asthma is a chronic respiratory disease characterized by sudden attacks of labored breathing, chest tightness, and coughing. It is a complex multifactorial disease with both genetic and environmental components. Physician-diagnosed asthma was more prevalent in biomass users compared with LPG-using women (4.6% vs. 2.7% in LPG users). Since asthma has a genetic predisposition, biomass smoke alone is unlikely to be responsible for the observed rise in the prevalence of the disease in biomass users in some studies. However, when asthmatics are exposed to smoke, the symptoms are aggravated (Uzun et al., 2003).

TABLE 5.2

Spirometric Measurements of Lung Function in Biomass Users Following ATS (1995) Criteria

Parameter	LPG Users	Biomass Users
FVC (L)	2.3 ± 0.8	1.6 ± 0.6*
FEV_1 (L)	1.9 ± 0.8	1.3 ± 0.5*
FEV_1/FVC	80.9 ± 2.6	82.3 ± 4.2
PEFR (L/s)	2.7 ± 1.4	1.9 ± 0.9*
$FEF_{25-75\%}$ (L/s)	2.1 ± 1.1	1.7 ± 1.3*

Results are mean $\pm$SD; *$p < 0.05$ compared with LPG users.

5.3.2.3 Reduction in Lung Function

Biomass users had significantly lower forced vital capacity (FVC), forced expiratory volume in one second (FEV_1), peak expiratory flow rate (PEFR), and forced expiratory flow during the middle half of FVC (FEF25—75%) than LPG users ($p < 0.05$, Table 5.2), suggesting a significant reduction in lung function. Overall, pulmonary function was reduced in 70.3% of biomass users in contrast to 34.7% of LPG-using controls ($p < 0.05$; Figures 5.2 and 5.3). Restrictive type of impairment (FVC <80% predicted) was predominant. Reduced pulmonary function was more frequent (75%) in women without kitchens than in those who possessed separate kitchens (62%, $p < 0.05$).

Like the present finding, reduction in lung function has been observed in women who are chronically exposed to biomass smoke (Pandey, 1984; Behera et al., 1994; Dutt et al., 1996; Perez-Padilla et al., 1996; Amoli, 1998; Albalak et al., 1999; Laffon et al., 1999; Pauwels et al., 2001; Golshan et al., 2002; Arslan et al., 2004). As in Eastern India, the predominant type of lung function impairment in biomass users in North India was a restrictive type of ventilatory defect characterized by FVC <80% of the predicted FVC (Behera et al., 1994). The fall in FVC may be due to reduced inspiratory force producing a reduction in total lung capacity, subclinical edema,

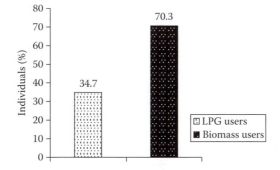

FIGURE 5.2 Effects of chronic biomass smoke exposures on lung function. Results are expressed as percentage of individuals with reduced lung function.

FIGURE 5.3 Prevalence of different types of lung function deficits in biomass and LPG-using women from Eastern India.

and an increase in pulmonary blood volume or changes in airway closure (Mason et al., 2000). Fibrosis of the lung following infections such as pulmonary tuberculosis can lead to restrictive pattern of spirometry. Besides, CO, PM_{10}, and $PM_{2.5}$ present in biomass smoke could accomplish injury to the lung tissue by their ability to form free radicals in bronchial mucosa and alveoli (Haponik et al., 1988). High-resolution computed tomography of thorax of Turkish women who used biomass as cooking fuel has revealed a variety of structural changes including thickening of interlobular septa (Ozbay et al., 2001). The probable mechanism of reduction in FVC could be that neutrophils, once activated, are capable of releasing collagenase, elastase, neutral protease, and various oxidants, all of which markedly derange the alveolar structure. Gradually, there is a loss of type I alveolar cells and capillary endothelial cells with the proliferation of type II cells and interstitial fibroblasts and accumulation of type I collagen, which may ultimately lead to fibrosis (Raghu et al., 1999). As a consequence, the alveolar wall loses elasticity, and vital capacity of the lung is markedly reduced. Therefore, lung inflammation with associated restrictive lung function impairment in biomass fuel users may indicate increased risk of interstitial lung fibrosis, sarcoidosis, or adult respiratory distress syndrome. Children living in homes that used biomass fuel also had reduced lung function characterized by lowered FVC and FEV_1 (Rinne et al., 2006).

5.3.2.4 Adverse Cellular Lung Reaction

Lungs are the main entry point of airborne pollutants. An adult human being inhales about 10,000–15,000 L of air everyday that contains life-sustaining oxygen as well as various harmful gases, chemicals, and bioaerosols. Therefore, lungs and the airways are endowed with a strong cellular defense mechanism to protect the gas exchange area from the inhaled harmful agents. The most important defense cells in the lungs are the alveolar macrophages (AMs). They readily engulf airborne particulates and microorganisms to ensure sterility and cleanliness of the lung epithelium for effective gas exchange. It has been estimated that the total surface area of inhaled carbon within an AM is 13 times higher in women who cook with biomass fuels than a nonuser and 7.5 times higher in children from biomass-using households (Kulkarni et al., 2005).

We found increased number of AMs and inflammatory cells such as neutrophils, eosinophils, and lymphocytes in sputum (Table 5.3 and Figure 5.4). The changes are generally interpreted as adverse lung response to challenges from airborne bacterial, allergic, and viral agents. Airway neutrophilia plays a key role in the pathogenesis of pulmonary fibrosis and its magnitude correlates with lung function impairment

TABLE 5.3
Differential Cell Count of Sputum of Nonsmoking, Biomass Fuel Using Women of Eastern India

Parameter	LPG Users	Biomass Users
Alveolar macrophage/hpf	6.7 ± 3.1	$12.9 \pm 0.9^*$
Multinucleated AM (% of cells)	2.5 ± 1.2	$8.2 \pm 3.5^*$
Neutrophil/hpf	36.6 ± 0.8	$57.9 \pm 0.6^*$
Eosinophil/hpf	0.7 ± 0.3	$2.1 \pm 0.5^*$
Lymphocyte/hpf	2.4 ± 0.2	$5.2 \pm 0.4^*$
Ciliocytophthoria (% individual)	0.5	3.0
Goblet cell hyperplasia (% individual)	1.3	5.5
Curschmann's spiral (% individual)	0	4.4
Charcot–Leyden crystal (% individual)	0	0.9
Koilocyte (% individual)	2.9	6.0
Siderophage/hpf	0.4 ± 0.2	$4.2 \pm 1.8^*$

Results are mean ±SD; hpf, high power field (40× objective, 10× eye piece) of microscope; $^*p < 0.05$ compared with LPG users.

(Beeh et al., 2003; Fujimoto et al., 2003). On the other hand, sputum eosinophilia (eosinophil >3% of total cells) is associated with eosinophilic bronchitis, chronic nonproductive (dry) cough, and wheeze (Gibson et al., 2001; Ayik et al., 2003; Fujimoto et al., 2003). The presence of both neutrophilia and eosinophilia in sputum may indicate chronic obstructive lung disease in nonsmokers (Birring et al., 2002). Considering these reports, several respiratory symptoms found in excess in biomass users of this study may be attributed to neutrophilic and eosinophilic inflammation of the airways.

In addition, the number of multinucleated AMs was significantly increased in biomass users, suggesting altered AM activity. Biomass users had greater prevalence of ciliocytophthoria, that is, injury to the ciliated epithelial cells of the airways, which is a common feature of respiratory virus infections; goblet cell hyperplasia and excess mucous production, presumably to facilitate disposal of excess inhaled pollutants; Curschmann's spiral, suggesting underlying airway obstruction; Charcot–Leyden crystal, implying eosinophilia and accompanying hypersensitivity; koilocytes, cellular changes associated with human papilloma virus infection; and abundance of iron-laden macrophages (siderophages) indicating covert pulmonary hemorrhage (Table 5.3).

Airway inflammation in biomass users could be attributed to fine and ultrafine particles present in biomass smoke that can induce production and release of proinflammatory mediators (cytokines, chemokines, etc.) from AMs, triggering recruitment of inflammatory cells from circulation to the lungs. Incidentally, indoor-generated particulates are more bioactive on AMs than are outdoor particulate matter (Long et al., 2001). Besides, organs other than the lungs can generate these inflammatory mediators, because lung inflammation was found to be mediated by production of tumor necrosis factor-alpha in the liver (Folch-Puy et al., 2003). Smoke inhalation has been

(a)   (b)

(c)   (d)

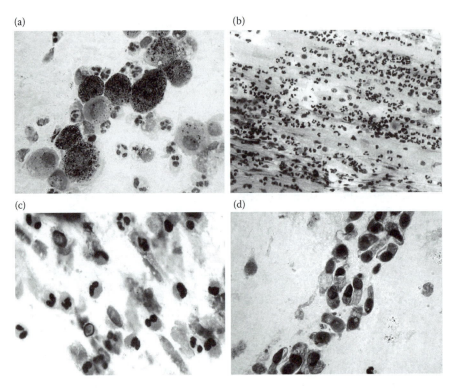

FIGURE 5.4 Photomicrographs of sputum samples of biomass users showing abundance of particle-laden alveolar macrophages (a), neutrophilia (b), increased eosinophil number (stained orange, c) and goblet cell hyperplasia (d), indicating high particulate pollution exposure, inflammation of the airways, bronchial allergy, and mucus hypersecretion, respectively. Pap-stained ×200 (b) ×1000 (a,c,d).

shown to facilitate traffic of inflammatory cells and fluid from pulmonary capillaries to the airway spaces by increasing alveolo-capillary permeability (Laffon et al., 1999) causing pulmonary edema. These changes predispose the lower respiratory tract to bacterial infection by interfering with mucociliary clearance and reducing bacterial killing by AMs (Hogg, 2000). Indeed, chronic exposures to smoke have been linked to pulmonary edema (Soejima et al., 2001), respiratory infections, and COPD (Smith, 2000; Bruce et al., 2000; Ezzati and Kammen, 2001).

5.3.2.5 Rise in COPD

COPD is characterized by symptoms of chronic bronchitis (presence of cough and expectorations on most of the days for at least three months in a year for two consecutive years or more) and reduced lung function (FEV_1 and FEV_1/FVC). COPD is a major public health concern around the globe. It is ranked 12th as a burden of disease worldwide and is projected to rank 5th by the year 2020 in terms of number of deaths and loss of quality of life (Varkey, 2004). COPD was found in 6.3% of biomass users compared with 2.1% in controls (Table 5.4).

TABLE 5.4

Prevalence (%) and Severity of COPD among Biomass Using Women of West Bengal

	Prevalence (%)	
COPD Severity	LPG Users ($n = 650$)	Biomass Users ($n = 1260$)
Mild	1.1	1.6*
Moderate	0.5	2.1*
Severe	0.5	1.6*
Very severe	0.1	1.0*
Overall	2.2	6.3*

*$p < 0.05$.

5.3.3 Hematological, Immunological, and Cardiovascular Changes

Biomass-using women had increased WBC, neutrophil and eosinophil counts, and lowered hemoglobin and erythrocyte levels than LPG-using controls. However, no significant difference was recorded in platelet, monocyte, and lymphocyte counts between biomass users and controls (Table 5.5). The blood smear of 18% of biomass users displayed anisocytosis (change in red cell size) and 23% had poikilocytosis (change in red cell shape) against 9% and 12% of age-matched LPG-using controls. Abundance of "target" cells was found in 17% of biomass users compared with 5% of controls. Toxic granulation of neutrophils, an indication of bacterial infection and inflammation, was found in 81% of biomass users in contrast to 29% of LPG users.

Chronic exposure of women to biomass smoke may alter the body's immune defense. We found alterations in lymphocyte subsets in biomass-using women of Eastern India. They had 17% reduction in absolute number of CD4+ T-helper cells and 14% fall in CD19+ B-lymphocytes but 31% increase in CD16 + CD56+ natural

TABLE 5.5

Hematological Changes in Biomass Users

	LPG Users	Biomass Users
Hemoglobin (g/dL)	11.5 ± 1.2	10.9 ± 1.1*
RBC ($10^6/\mu L$)	4.1 ± 0.3	3.8 ± 0.4*
Platelet ($10^5/\mu L$)	2.3 ± 0.4	2.2 ± 0.4
WBC/μL	7524 ± 482	8350 ± 518*
Neutrophil/μL	4244 ± 311	4701 ± 367*
Lymphocyte/μL	2528 ± 205	2488 ± 203
Monocyte/μL	211 ± 27	219 ± 29
Eosinophil/μL	497 ± 54	919 ± 105*

Results are mean $\pm$SD; *$p < 0.05$ compared with LPG users.

TABLE 5.6

Lymphocyte Subsets in Peripheral Blood of Biomass and LPG-Using Women

Lymphocyte Subtype	LPG Users	Biomass Users
CD4 + T cell (%)	45.4 ± 3.8	38.3 ± 3.5*
CD4 + T cell/μL	1148 ± 98	953 ± 87*
CD8 + T cell (%)	26.6 ± 2.7	30.6 ± 3.2
CD8 + T cell/μL	672 ± 68	761 ± 73*
CD19 + B cell (%)	15.7 ± 1.7	13.7 ± 1.8
CD19 + B cells/μL	397 ± 46	341 ± 50*
CD16 + CD56 + NK cell (%)	11.8 ± 1.7	17.3 ± 1.8*
CD 16 + 56 + NK-cells/μL	298 ± 44	430 ± 52*

Results are mean ±SD; *$p < 0.05$ compared with LPG users.

killer cells and 12% rise in CD8+ T-lymphocytes in peripheral blood compared with LPG users, suggesting changes in immune defense (Table 5.6). Air pollutants commonly found in biomass smoke have been shown to cause hematological alterations (Ray et al., 2003) and suppression of the lung's immunity in animal and human studies (Chang et al., 1990; Fujii et al., 2001; Mukae et al., 2001). Lymphocytes are the key players in maintaining the body's surveillance over infiltrating pathogens. Among all the air pollutants, B(a)P, found in large quantities in biomass smoke, is particularly important because it causes immune suppression by downgrading B cell-mediated humoral immunity (Hardin et al., 1992; Szezeklik et al., 1994).

The expression of CD11b/CD18, the activation markers of neutrophils and monocytes, was significantly increased in women who were chronically exposed to biomass smoke. The mean fluorescence intensity (MFI) for CD 11b on neutrophils was elevated from 683.5 ± 85.9 (SD) in controls to 1150.2 ± 143.2 ($p < 0.001$) in biomass users (Figure 5.5). MFI of CD18 on neutrophils was also increased from 385.1 ± 56.4 to 568.7 ± 64.4 ($p < 0.001$). Similarly, the MFI of CD11b and CD18 on peripheral

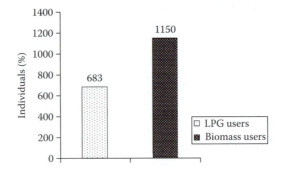

FIGURE 5.5   Marked increase in mean fluorescence intensity (MFI) of CD11b expression on neutrophil surface of biomass users suggesting activation of these cells.

blood monocytes of biomass users was increased by 50% and 62%, respectively when compared with that of controls. The present findings along with that of earlier ones (Ray et al., 2006) suggest functional activation and migration of inflammatory cells from airways to the tissues of women chronically exposed to biomass smoke.

Cardiovascular disease (CVD) is well known for its association with active and passive smoking and outdoor air pollution in the Western world. In comparison, little is known about its association with IAP from biomass burning in the developing countries. We found remarkable increase in platelet activity, a risk factor for CVD, in women who cook regularly with biomass fuels. Compared with 2.2% of P-selectin-positive circulating platelets of LPG users, biomass users had 6.3% P-selectin-expressing activated platelets in peripheral blood (Figure 5.6). Similarly, expression of soluble P-selectin (GMP-140) in plasma was significantly increased in biomass users (Figure 5.6). Platelet activation was accompanied by a sharp rise in the percentage of neutrophil–platelet (9.1% vs. 3.8%) and monocyte–platelet aggregates (6.6% vs. 2.7%). Platelet–leukocyte complexes enable leukocytes to be involved in hemostasis and thrombosis. CD11b expression is intimately related to the levels of plasma markers of coagulation activation and increased circulating leukocyte–platelet aggregates represent an additional prothrombotic mechanism, and a risk factor for thrombosis, because they promote the formation of fibrin via induction of tissue factor expression on monocytes, release of superoxide anions, and production of proinflammatory cytokines from polymorphonuclear leukocytes (PMN), which, in turn, augment platelet activation and modulate the coagulant properties of the endothelium. Taken together, the findings suggest a greater risk factor of CVD in biomass users.

5.3.4 ADVERSE EFFECTS ON FEMALE REPRODUCTIVE SYSTEM

5.3.4.1 Alterations in Menstrual Cycle Length

Long-term exposure to biomass smoke was associated with adverse effects on the reproductive system of premenopausal women of this study. Menstrual cycle length

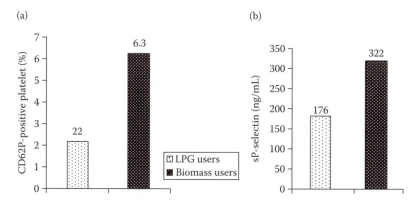

FIGURE 5.6 Increase in the percentage of P-selectin (CD62P)—expressing activated platelets in circulation (a) and the level of soluble P-selecin (sP-selectin) in plasma (b) of women chronically exposed to smoke from biomass burning during daily household cooking. The changes are usually associated with greater risk of cardiovascular diseases.

TABLE 5.7

Prevalence (%) of Abnormal Menstrual Cycle Characteristics in Premenopausal Nonsmoking Women of West Bengal

Menstrual Cycle Characteristics	LPG Users ($n = 523$)	Biomass Users ($n = 1010$)
Irregular cycle length	16.2	29.7*
Short cycle (≤24 days)	6.1	7.5
Long cycle (≥36 days)	10.1	22.2*

*$p < 0.05$ compared with LPG users in χ^2 test.

was altered in 29.7% of women who cook with biomass compared with 16.2% of LPG users. About 22% of biomass users had oligomenorrhea, that is, prolonged cycles (≥36 days) compared with 10.1% of LPG users. However, we did not find any change in the prevalence of short cycle between biomass and LPG users (Table 5.7). Long cycles could be detrimental for the reproductive health of the women, because women with oligomenorrhea had greater risk of abortions, stillbirth, and preterm delivery (Krassas, 2000). Indeed, we had found more instances of spontaneous abortion and stillbirth among biomass users.

We found long cycles in a large number of biomass users. It is not clear whether the shift toward prolonged cycles among biomass users is due to lengthening of a specific phase of the menstrual cycle or of experiencing missed periods and reporting long cycles. The menstrual cycle is controlled by the endocrine system and several factors may influence its length and regularity (Harlow and Ephross, 1995). The effect of biomass could be on the ovary, the estrous cycle, or reproductive hormones like estrogen, follicle stimulating hormone (FSH), and leuteinizing hormones (LH).

Delay in ovulation, reduction in the number of corpora lutea and disruption of the estrous cycle may lead to extended cycles. Long menstrual cycle length in many biomass users may reflect anovulation and relatively low levels of estrogen exposure. In conformity with this, we found a fall in plasma estradiol and progesterone levels in biomass users (Table 5.8). In addition, the concentration of LH was doubled although the level of FSH remained relatively unaffected. Change in plasma LH level in follicular and luteal phases has been reported in female traffic police chronically exposed to vehicular exhausts (Monti et al., 2006). Similarly, smoking and ETS affect female reproductive hormones partly via their antiestrogen effect (Chen et al., 2005). Thus, cumulative exposures to biomass smoke have been shown to alter the levels of female reproductive hormones. Biomass users are usually poorer than LPG users and women from these families have the additional stressful task of fuel collection (wood and crop residues) and preparation (dung cake). Therefore, physical exhaustion and stress associated with poverty are much more in these women. Both these factors can also be an important predictor of prolonged menstrual cycle length (Thurston et al., 2000). Indeed, we found a sharp rise in cortisol level among biomass users (Table 5.8) that indicate greater level of stress in these women.

TABLE 5.8
Hormone Levels in Blood Plasma of Premenopausal Women Participants of the Study

Hormone	LPG Users	Biomass Users
Estradiol (pg/mL)	219.6 ± 34.7	182.5 ± 32.8*
Progesterone (ng/mL)	4.4 ± 1.3	2.1 ± 0.7*
Leuteinizing hormone (mIU/mL)	18 ± 2.4	36.4 ± 5.6*
Follicle stimulating hormone (mIU/mL)	9.4 ± 2.6	10.2 ± 2.7
Prolactin (ng/mL)	14.3 ± 3.6	24.1 ± 4.7*
Cortisol (ng/mL)	95.5 ± 21.3	143.8 ± 33.7*

*$p < 0.05$ compared with LPG users in Student's "t" test.

5.3.4.2 Spontaneous Abortion

The instances of spontaneous abortions were very high among biomass users compared with their neighbors who used LPG as the cooking fuel (Figure 5.7). About 14% of biomass-using women experienced spontaneous abortion once in their life compared with 7.3% of LPG users, and 7.5% of biomass users of this study had experienced this painful episode more than once compared with 1.7% of LPG users (Table 5.9).

5.3.4.3 Low Birth Weight and Stillbirth

We found low birth weight of the baby (body weight less than 2.5 kg) in 42.5% cases of child delivery among biomass users, which was 3 times more than that of LPG users (Figure 5.8). Like the present finding, higher episodes of stillbirth and low birth weight have been reported in women who cook with wood, dung, or straw than those who cook with cleaner fuels (Mishra et al., 2004, 2005). On an average, the reduction in birth weight varies from 50 to 300 g with a mean of 175 g when compared with that of users of cleaner fuel like LPG, natural gas, and electricity (Mishra et al., 2004). Pregnant women tending fires with biomass for domestic

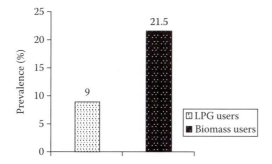

FIGURE 5.7   Comparison of the prevalence of spontaneous abortion between biomass and LPG-using women showing 2.3 times more risk among the former.

TABLE 5.9

Prevalence (%) of Reproductive and Associated Problems in Indian Women Using Biomass Fuel for Domestic Cooking

Finding	LPG Users	Biomass Users
Spontaneous Abortion		
Once	7.3	13.9*
Twice	1.7	6.2*
Thrice	0	1.3*
Premature delivery of the baby	1.4	1.5
Stillbirth	1.4	3.5*

*$p < 0.001$ compared with LPG users.

cooking have high CO level in their blood, and they have a greater chance of delivering baby with low birth weight (<2.5 kg), which is an important risk factor in infant mortality and morbidity (Boy et al., 2002). Exposure of pregnant women to air pollution may lead to other adverse birth outcomes such as preterm delivery, intrauterine growth retardation, small head circumference, and increased rate of malformations (Dejmek et al., 1999; Bobak, 2000).

5.3.5 DEPLETION OF ANTIOXIDANT DEFENSE

Air pollutants generally cause cellular injury via oxidative stress and consequent production of free radicals, which are toxic for cellular macromolecules including DNA. To combat oxidative stress, the body has a strong antioxidant defense system. An important member of this defense system is the enzyme SOD. We found 51% reduction (118 vs. 244 U/mL in LPG users) in erythrocyte SOD level in women chronically exposed to biomass smoke, suggesting down regulation of antioxidant defense (Figure 5.9). Total antioxidant level was also reduced from 1.5 ± 0.3 mmol/L in LPG users to 0.67 ± 0.27 mmol/L in women who used biomass fuels. Thus, it

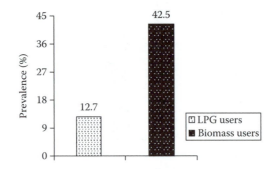

FIGURE 5.8 Comparison of the prevalence of low birth weight of the new born between LPG and biomass fuel users showed 3.3-fold higher risk in the latter group.

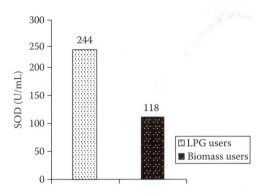

FIGURE 5.9   Depletion of superoxide dismutase enzyme in plasma of women who cook exclusively with wood, dung cake and agricultural wastes.

appears that chronic biomass smoke exposures reduce the body's antioxidant defense that may lead to enhanced cellular injury. Like the present study, exposure to biomass smoke have been shown to be associated with increased lipid peroxidation and a significant decrease in antioxidant enzyme activities in Turkish women (Gani et al., 2000). Short-course exposure to biomass smoke has also been shown to decrease plasma antioxidant levels in rabbits (Sezer et al., 2006).

5.3.6   GENOTOXICITY AND CARCINOGENIC CHANGES

5.3.6.1   Chromosomal Breakage

Biomass smoke contains a host of mutagens and potential carcinogens including B(a) P, 1,2 butadiene, and benzene (Wafula et al., 1990; Zhang and Smith, 1996). Therefore, chronic exposure to biomass smoke may result in genetic toxicity. We have examined this possibility by undertaking the MN test and the Comet assay, which are biomarkers of damage at the chromosomal and DNA levels, respectively. We found three times more MNs in exfoliated buccal and airway epithelial cells of biomass users compared with those of LPG users (Figure 5.10) implying an increased rate of chromosomal breakage in cells directly at the pathway of smoke exposure (Table 5.10).

5.3.6.2   DNA Damage

Comet morphology (Figure 5.10) was found in 18.6% peripheral blood lymphocytes of biomass users in contrast to 11.7% in LPG-using controls. Moreover, the extent of damage in terms of tail length and olive tail moment was much more in biomass users (Table 5.11). Elevated frequency of MN formation in peripheral blood lymphocytes has also been reported among women who cook with biomass fuels in North India (Musthapa et al., 2004).

5.3.6.3   Cancer Risk

Chromosomal and DNA breakage are associated with increased risk for many diseases including cancer (Stich et al., 1982; Tolbert et al., 1991). MN is regarded as an important biomarker to predict the relative risk of cancer in the upper aerodigestive

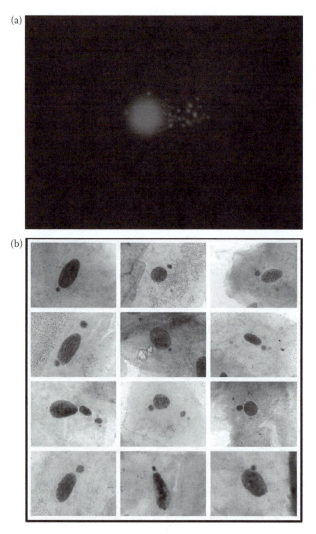

FIGURE 5.10 Photomicrographs showing "comet" formation in peripheral blood lymphocytes (a) and micronucleus formation in buccal epithelial cells (b) of biomass using women, suggesting DNA damage and chromosomal breakage, respectively.

tract (Bloching et al., 2000). Therefore, elevated MN numbers in biomass users implies a higher risk of cancer among them. Carcinogenesis is a multistep process and it takes years to manifest itself. Cancer development usually starts with genetic changes leading to activation of oncogenes, inactivation of tumor suppressor genes, methylation of the promoter, and so on. The genetic changes are paralleled by progressive changes in a cell's morphology from normality to progressively more severe abnormalities (metaplasia–dysplasia–carcinoma *in situ* to malignancy). Metaplasia is thus the initial morphological change in this journey toward neoplasia. This usually develops as an adaptive response to toxic insults and the cells behave differently from that of normal. Metaplasia often leads to dysplasia when cell turnover

TABLE 5.10

Micronucleus (MN) Frequency in Exfoliated Epithelial Cells

Cell Type	MN per 1000 Cells	
	LPG Users	Biomass Users
Airway epithelial cell	1.2 ± 0.4	$3.7 \pm 0.1^*$
Buccal mucosal cell	1.1 ± 0.2	$2.9 \pm 0.5^*$

Results are mean $\pm$SD; $^*p < 0.05$ compared with LPG users.

becomes more rapid (Snead et al., 2003). Accordingly, we examined the prevalence of metaplasia and dysplasia in airway epithelial cells in Pap-stained sputum samples. We found a remarkable 7.5-fold rise in the prevalence of metaplasia in airway epithelial cells. More importantly, 3.1% of biomass users displayed dysplasia of airway cells compared with 0.3% of LPG users (Figure 5.11).

Although it is rarely seen in the sputum samples of nonsmokers, we found metaplasia with atypia, a hallmark of cytological change in the lower respiratory tract after carcinogen exposure (Kamei et al., 1993), and a risk factor for lung cancer (Vine et al., 1990), in women exposed to biomass smoke. Therefore, the presence of these cytological changes in the sputum of biomass-using women indicates extremely harmful effects of biomass smoke on the airways. Indeed, biomass smoke has been implicated as a major risk factor for lung cancer among women in India (Smith, 2000; Smith and Mehta, 2003; Behera and Balamugesh, 2005). In West Bengal, lung cancer is the 5th leading cancer type among nonsmoking women with an annual incidence of five per 100,000 females. Besides lung cancer, IAP from biomass fuel use has been recognized as a risk factor for nasopharyngeal (Clifford, 1972) and head and neck cancer (Dietz et al., 1995). Moreover, biomass smoke activates human papilloma virus type-16 and -18 and significantly increases the risk of cancer of the uterine cervix (Velema et al., 2002). Incidentally, cervical cancer is the most prevalent cancer type in women in rural areas of the developing world. Thus, it will be worthwhile to investigate

TABLE 5.11

DNA Damage in Peripheral Blood Lymphocyte in Biomass and LPG-Using Women

Comet Parameters	LPG Users	Biomass Users
Lymphocyte with "comet" morphology (%)	11.7 ± 2.5	$18.6 \pm 3.6^*$
Tail length of comet (μm)	31.4 ± 3.4	$45.5 \pm 5.5^*$
Olive tail moment (arbitrary unit)	1.4 ± 0.5	$4.1 \pm 0.9^*$

Results are mean $\pm$SD; $^*p < 0.05$ compared with LPG users.

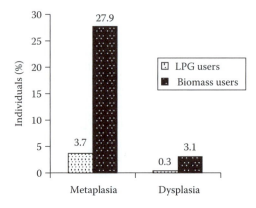

FIGURE 5.11 Prevalence of metaplasia and dysplasia of airway epithelial cells in sputum of women in Eastern India who cook regularly either with biomass fuels or LPG.

whether biomass use is causally related to the development of the most prevalent cancer type in poor, rural women in India and other developing nations.

5.3.7 DEPRESSION AND OTHER NEUROBEHAVIORAL CHANGES

Exposures to high level of air pollution for long may cause brain damage (Calderon-Garciduenas et al., 2002). We have explored that aspect in biomass users by evaluating the prevalence of neurobehavioral symptoms. Compared with LPG users, biomass-using women frequently complained of fatigue and feet numbness. Moreover, they had greater prevalence of anxiety, depression, abnormal smell, taste and vision, and transient loss of memory (Table 5.12). The symptoms were mainly psychoneurological or due to abnormal reactions of the cranial nerves (abnormal smell, taste, and vision), peripheral nerves (feet numbness), and mucous membrane irritations (cough, shortness of breath, and eye and nose irritations). Controlling for education, family income, environmental tobacco smoke as possible confounders, multivariate logistic regression analysis showed a significant positive association between PM_{10} and $PM_{2.5}$ levels in indoor air, particularly the latter, and depression (OR = 1.26 and 1.33; 95% CI and; 1.10–1.55 and 1.16–1.72 for PM_{10} and $PM_{2.5}$ respectively). Similarly, PM_{10} and $PM_{2.5}$ levels were positively associated with transient loss of memory, burning sensation in extremities, reduced sense of smell, and blurred vision. Lifetime exposure to biomass smoke, calculated as exposure hour-years, correlated positively in Spearman rank correlation test with depression ($\rho = 0.535$, $p < 0.001$) and other symptoms (ρ values varied between 0.322 and 0.621, $p < 0.01$).

5.3.7.1 Changes in Neurotransmitter Level

Behavior is the outcome of multiple mechanisms within the central nervous system (CNS), and changes are sensitive indicators of nervous system dysfunction. Ultrafine particles present in combustion products cross the alveolar-capillary barrier, reach the blood stream (Nemmar et al., 2002), and may influence the activities of all vital organs including the CNS. Animal study has shown impairment of brain function

TABLE 5.12
Prevalence of Neurobehavioral Symptoms

	LPG Users	Biomass Users
Depression[a]	24.6	70.6*
Burning sensation in extremities	7.7	15.8*
Vertigo/dizziness	10.6	9.4
Palpitation	11.7	20.9*
Anxiety	10.6	29.9*
Feeling of drunkenness	2.6	7.0*
Inability to concentrate	8.6	32.7*
Transient loss of memory	14.3	29.0*
Reduced sense of taste	2.4	5.9*
Reduced sense of smell	1.8	9.1*
Numbness in feet	12.1	33.1*
Blurred vision	2.0	16.8*
Eye irritation	7.3	44.0*
Tears while cooking	7.0	46.6*

*$p < 0.05$ compared with controls in χ^2 test.

[a] Assessed by Beck's 21-point depression score; *$p < 0.05$ compared with LPG users.

following exposure to high levels of air pollution (Calderon-Garciduenas et al., 2002). Therefore, it is an interesting proposition to explore whether cumulative exposures to combustion products mediate changes in CNS function that, in turn, results in neurobehavioral symptoms. Accordingly, we measured neurotransmitter levels in plasma and found significantly elevated ($p < 0.05$) levels of E, NE, and serotonin in women who cook with biomass fuels. The increase was 64.5% above the mean control level in the case of E, while 58.5% and 94.3% increase was recorded for NE and serotonin, respectively. In contrast, appreciable fall in dopamine and acetylcholinesterase levels was found in these women (Table 5.13).

TABLE 5.13
Plasma Catecholamine and Acetylcholinesterase Levels

	LPG Users	Biomass Users
Epinephrine (ng/mL)	0.16 ± 0.12	1.1 ± 0.26*
Norepinephrine (ng/mL)	10.2 ± 0.82	13.3 ± 0.89*
Dopamine (ng/mL)	6.5 ± 0.45	4.4 ± 0.36*
Serotonin (ng/mL)	10.6 ± 2.8	20.6 ± 4.6*
Acetylcholinesterase (U/mL)	9.7 ± 1.2	6.8 ± 0.7*

Results are mean $\pm$SD; *$p < 0.05$ compared with rural women.

5.4 CONCLUSION

Biomass fuels are extensively used in rural India for domestic cooking and room heating because they are cheap and readily available. When biomass burns in traditional ovens in poorly ventilated kitchens, a common feature in rural India, it emits particulates and gaseous pollutants causing indoor pollution at least three times more than that in LPG-using households. Women who cook with these fuels everyday suffer from a multitude of health problems, both physical and mental. These include lung function reduction, asthma, COPD, inflammation of the airways, covert pulmonary hemorrhage, platelet hyperactivity and associated higher cardiovascular risk, immune alteration, reproductive failures, underweight baby and hormonal imbalance, chromosomal and DNA damage, airway cell metaplasia and dysplasia and consequent higher risk of cancer in the airways and lungs, and depression with changes at the level of neurotransmitters. Even after controlling potential confounders like socioeconomic conditions and environmental tobacco smoke, the changes remained positively associated with PM_{10} and $PM_{2.5}$ levels in indoor air. Since biomass remains the main source of domestic energy in developing countries throughout the globe, health impairments among biomass users as observed in this study may represent the health conditions of women of other developing nations as well. Thus the magnitude of the problem seems enormous and it warrants immediate attention of all concerned to this public health problem. Exploitation of alternative energy source including solar power and/or supply of cleaner fuels such as LPG to the rural people at an affordable price could be the long-term solution to the problem. As an immediate measure, we recommend an initiative to introduce user-friendly smokeless ovens and better kitchen ventilation in all biomass-using households of the country.

ACKNOWLEDGMENTS

The authors are grateful to the Central Pollution Control Board, Delhi and the WHO (India) for their financial support to carry out this study and to the people of West Bengal for their active interest and cooperation.

REFERENCES

Albalak, R., Frisancho, A.R., and Keeler, G.J. 1999. Domestic biomass fuel combustion and chronic bronchitis in two rural Bolivian villages. *Thorax* 54: 1004–1008.
Amoli, K. 1998. Bronchopulmonary disease in Iranian housewives chronically exposed to indoor smoke. *Eur Respir J* 11: 659–663.
Arslan, M., Akkurt, I., Egilmez, H., Atalar, M., and Salk, I. 2004. Biomass exposure and the high resolution computed tomographic and spirometric findings. *Eur J Radiol* 52: 192–199.
ATS (American Thoracic Society). 1995. Standardization of spirometry. *Am J Respir Crit Care Med* 152: 1107–1136.
Ayik, S.O., Basoglu, O.K., Erdinc, M., Bor, S., Veral, A., and Bilgen, C. 2003. Eosinophilic bronchitis as a cause of chronic cough. *Respir Med* 97: 695–701.
Bailis, R., Ezzati, M., and Kammen, D. 2005. Mortality and greenhouse gas impacts and petroleum energy futures in Africa. *Science* 308: 98–103.

Balakrishnan, K., Sankar, S., Parikh, J., et al. 2002. Daily average exposures to respirable particulate matter from combustion of biomass fuels in rural households of southern India. *Environ Health Perspect* 110: 1069–1075.

Beck, A.T., Ward, C.H., Mendelson, M., Mock, J., and Erbaugh, J. 1961. An inventory for measuring depression. *Arch Gen Psychiatry* 4: 561–571.

Beeh, K.M., Beier, J., Kornmann, O., and Buhl, R. 2003. Neutrophilic inflammation in induced sputum of patients with idiopathic pulmonary fibrosis. *Sarcoidosis Vasc Diffuse Lung Dis* 20: 138–143.

Behera, D. and Balamugesh, T. 2005. Indoor air pollution as risk factor for lung cancer in women. *J Assoc Physicians India* 53: 190–192.

Behera, D., Jindal, S.K., and Malhotra, H.S. 1994. Ventilatory function in non-smoking rural Indian women using different cooking fuels. *Respiration* 61: 89–92.

Birring, S.S., Brightling, C.E., Bradding, P., et al. 2002.Clinical, radiologic, and induced sputum features of chronic obstructive pulmonary disease in non-smokers: A descriptive study. *Am J Respir Crit Care Med* 166: 1078–1083.

Bloching, M., Hofmann, A., Lautenschlager, C., Berghaus, A., and Grummt, T. 2000. Exfoliative cytology of normal buccal mucosa to predict the relative risk of cancer in the upper aerodigestive tract using the MN-assay. *Oral Oncol* 36: 550–555.

Bobak, M. 2000. Outdoor air pollution, low birth weight, and prematurity. *Environ Health Perspect* 108: 173–176.

Boy, E., Bruce, N.G., and Delgado, H. 2002. Birth weight and exposure to kitchen wood smoke during pregnancy in rural Guatemala. *Environ Health Perspect* 110: 109–114.

Bruce, N., Perez-Padilla, R., and Albalak, R. 2000. Indoor air pollution in developing countries: A major environmental and public health challenge for the new millennium. *Bull World Health Organ* 78: 1078–1092.

Calderon-Garciduenas, L., Azzarelli, B., Acuna, H., et al. 2002. Air pollution and brain damage. *Toxicol Pathol* 30: 373–389.

Chang, J.C., Distler, S.G., and Kaplan, A.M. 1990. Tobacco smoke suppresses T-cells but not antigen-presenting cells in the lung-associated lymph nodes. *Toxicol Appl Pharmacol* 102: 514–523.

Chen, C., Wang, X., Wang, L., et al. 2005. Effects of environmental tobacco smoke on levels of urinary hormone markers. *Environ Health Perspect* 113: 412–417.

Clifford, P. 1972. Carcinogenesis in the nose and throat: Nasopharyngeal carcinoma in Kenya. *Proc R Soc Med* 65: 682–686.

Dejmek, J., Selevan, S., Benes, I., Solansky, I., and Sram, R. 1999. Fetal growth and maternal exposure to particulate matter during pregnancy. *Environ Health Perspect* 107: 475–480.

Dietz, A., Senneweld, E., and Maier, H. 1995. Indoor air pollution by emissions of fossil fuel single stoves: Possibly a hitherto underrated risk factor in the development of carcinomas in the head and neck. *Otolaryngol Head Neck Surg* 112: 308–315.

Dutt, D., Srinivasa, D.K., Rotti, S.B., Sahai, A., and Konar, D. 1996. Effect of indoor air pollution on the respiratory system of women using different fuels for cooking in an urban slum of Pondicherry. *Natl Med J India* 9: 113–117.

Ellegard, A. 1996. Cooking smoke and respiratory symptoms among women in low-income areas of Maputo. *Environ Health Perspect* 104: 980–985.

Ellman, G.L., Courtney, K.D., Andres, V. Jr., and Featherstone, R.M. 1961. A new and rapid colorimetric determination of acetylcholinesterase activity. *Biochem Pharmacol* 7: 88–95.

Ezzati, M. and Kammen, D.M. 2001. Indoor air pollution from biomass combustion and acute respiratory infections in Kenya: An exposure response study. *Lancet* 358: 619–624.

Ezzati, M. and Kammen, D.M. 2002. The health impacts of exposure to indoor air pollution from solid fuels in developing countries: Knowledge, gaps, and data needs. *Environ Health Perspect* 110: 1057–1068.

Farr, S.L., Cooper, G.S., Cai, J., Savitz, D.A., and Sandler, D.P. 2004. Pesticide use and menstrual cycle characteristics among premenopausal women in the agricultural health study. *Am J Epidemiol* 160: 1194–1204.

Folch-Puy, E., Garcia-Movtero, A., Lovanna, J.L., et al. 2003. The pancreatitis-associated protein induces lung inflammation in the rat through activation of TNF-α expression in hepatocytes. *J Pathol* 199: 398–408.

Fujii, T., Hayashi, S., Hogg, J.C., Vincent, R., and Van Eeden, S.F. 2001. Particulate matter induces cytokine expression in human bronchial epithelial cells. *Am J Respir Cell Mol Biol* 25: 265–271.

Fujimoto, K., Yamaguchi, S., Urushibata, K., Koizumi, T., and Kubo, K. 2003. Sputum eosinophilia and bronchial responsiveness in patients with chronic non-productive cough responsive to anti-asthma therapy. *Respirology* 8: 168–174.

Gani, H., Seyfikli, Z., Celik, V.K., Akkurt, I., and Abadoglu, O. 2000. The effect of biomass on lipid peroxidation and antioxidant activities on Turlish female groups. *Eur Respir J* 16: 108–112.

Gibson, P.G., Simpson, J.L., Chalmers, A.C., et al. 2001. Airway eosinophilia is associated with wheeze but is uncommon in children with persistent cough and frequent chest colds. *Am J Respir Crit Care Med* 164: 977–981.

Golshan, M., Faghihi, M., Roushan-Zamir, T., et al. 2002. Early effects of burning rice farm residues on respiratory symptoms of villagers in suburbs of Isfahan, Iran. *Int J Environ Res* 12: 125–131.

Gupta, B., Mathur, N., Mahendra, P., Srivastava, A., Swaroop, V., and Agnihotri, M. 1997. A study of household environmental risk factors pertaining to respiratory diseases. *Energy Environ Monitor* 13: 61–67.

Haponik, E.F., Crapo, R.O., Herndon, D.N., Traber, D.L., Hudson, L., and Moylan, J. 1988. Smoke inhalation. *Am Rev Respir Dis* 138: 1060–1063.

Hardin, J.A., Hinoshita, F., and Sherr, D.H. 1992. Mechanisms by which benzo[a]pyrene, an environmental carcinogen, suppresses B cell lymphopoiesis. *Toxicol Appl Pharmacol* 117: 155–164.

Harlow, S.D. and Ephross, S.A. 1995. Epidemiology of menstruation and its relevance to women's health. *Epidemiol Rev* 17: 265–286.

Hogg, J.C. 2000. Chronic bronchitis: The role of viruses. *Semin Respir Infect* 15: 32–40.

IIPS. 1995. National family health survey (MCH and Family Planning): India 1992–1993. International Institute for Population Sciences (IIPS), Bombay.

Kamei, T., Kohno, T., Ohwada, H., Takeuchi, Y., Hayashi, Y., and Fukuma, S. 1993. Experimental study of the therapeutic effects of folate, vitamin A, and vitamin B12 on squamous metaplasia of the bronchial epithelium. *Cancer* 71: 2477–2483.

Krassas, G.E. 2000. Thyroid disease and female reproduction. *Fertil Steril* 74: 1063–1070.

Kulkarni, N.S., Rudon, B., Panditi, S.L., Abebe, Y., and Grigg, J. 2005. Carbon loading of alveolar macrophages in adults and children exposed to biomass smoke particles. *Sci Total Environ* 345: 23–30.

Laffon, M., Pittet, J.F., Modelska, K., Matthay, M.A., and Young, D.M. 1999. Interleukin-8 mediates injury from smoke inhalation to both the lung endothelial and the alveolar epithelial barriers in rabbits. *Respir Crit Care Med* 160: 1443–1448.

Larson, B.A. and Rosen, S. 2002. Understanding household demand for indoor air pollution control in developing countries. *Social Sci Med* 55: 571–584.

Long, C.M., Suh, H.H., Kobzik, L., Catalano, P.J., Ning, Y.Y., and Koutrakis, P. 2001. A pilot investigation of the relative toxicity of indoor and outdoor fine particles: *In vitro* effects of endotoxin and other particulate properties. *Environ Health Perspect* 109: 1019–1026.

Mason, N.P., Barry, P.W., Pollard, A.J., et al. 2000. Serial changes in spirometry during an ascent to 5333 m in the Nepalese Himalayas. *High Alt Med Biol* 1: 185–195.

Mishra, V. 2003. Indoor air pollution from biomass combustion and acute respiratory illness in preschool age children in Zimbabwe. *Int J Epidemiol* 32: 847–853.

Mishra, V., Dai, X., Smith, K.R., and Mika, L. 2004. Maternal exposure to biomass smoke and reduced birth weight in Zimbabwe. *Ann Epidemiol* 14: 740–747.

Mishra, V., Retherford, R.D., and Smith, K.R. 2005. Cooking smoke and tobacco smoke as risk factors for stillbirth. *Int J Environ Health Res* 15: 397–310.

Mishra, V.K., Retherford, R.D., and Smith, K.R. 1999. Biomass cooking fuels and prevalence of tuberculosis in India. *Int J Infect Dis* 3: 119–129.

Monti, C., Ciarrocca, M., Cardella, C., et al. 2006. Exposure to urban stressor and effects on luteinizing hormone (LH) in female outdoor workers. *J Environ Sci Health A Tox Hazard Subst Environ Eng* 41: 1437–1448.

Mukae, H., Vincent, R., Quinlan, K., et al. 2001. The effect of repeated exposure to particulate air pollution (PM_{10}) on the bone marrow. *Am J Respir Crit Care Med* 163: 201–209.

Musthapa, M.S., Lohani, M., Tiwari, S., Mathur, N., Prasad, R., and Rahman, Q. 2004. Cytogenetic biomonitoring of Indian women cooking with biofuels: Micronucleus and chromosomal aberration tests in peripheral blood lymphocytes. *Environ Mol Mutagen* 43: 243–249.

Nemmar, A., Hoet, P.H., Vanquickenborne, B., et al. 2002. Passage of inhaled particles into the blood circulation in humans. *Circulation* 105: 411–414.

Ozbay, B., Uzun, K., Arslan, H., and Zehir, I. 2001. Functional and radiological impairment in women highly exposed to indoor biomass fuels. *Respirology* 6: 255–258.

Pandey, M.R., Boleij, J.S.M., Smith, K.R., and Wafula, E.M. 1989. Indoor air pollution in developing countries and acute respiratory infections in children. *Lancet* 1: 424–429.

Pandey, M.R. 1984. Domestic smoke pollution and chronic bronchitis in a rural community of the hill region of Nepal. *Thorax* 39: 337–339.

Pauwels, R.A., Busse, W.W., O'Byrne, P.M., et al. 2001. The inhaled steroid treatment as regular therapy in early asthma (START) study: Rationale and design. *Control Clin Trials* 22: 405–419.

Perez-Padilla, R., Regalado, J., and Vedal, S. 1996. Exposure to biomass smoke and chronic airway disease in Mexican women: A case-control study. *Am J Respir Crit Care Med* 154: 701–706.

Raghu, G., Mageto, Y.N., Lockhart, D., Schmidt, R.A., Wood, D.E., and Godwin, J.D. 1999. The accuracy of the clinical diagnosis of new-onset idiopathic pulmonary fibrosis and other interstitial lung disease. A prospective study. *Chest* 116: 1168–1174.

Ray, M.R., Basu, C., and Lahiri, T. 2003. Haematological changes and up-regulation of P-selectin expression in circulating platelets of Indian women chronically exposed to emissions from biomass fuels. *Comp Clin Path* 12: 151–154.

Ray, M.R., Mukherjee, S., Roychoudhury, S., et al. 2006. Platelet activation, upregulation of beta$_2$ Mac-1 integrin expression on leukocytes and increase in circulating leukocyte–platelet aggregates in Indian women chronically exposed to biomass smoke. *Hum Exp Toxicol* 25: 627–635.

Reddy, M.S. and Venkataraman, C. 2002. Inventory of aerosol and sulphur dioxide emissions from India: II. Biomass combustion. *Atmos Environ* 36: 699–612.

Rinne, S.T., Rodas, T.J., Bender, B.S., et al. 2006. Relationship of pulmonary function among women and children to indoor air pollution from biomass use in rural Ecquador. *Respir Med* 100: 1208–1215.

Samet, J.M., Dominici, F., Curriero, F.C., Coursac, I., and Zeger, S.L. 2000. Fine particulate air pollution and mortality in 20 US Cities, 1987–1994. *N Engl J Med* 34: 1742–1749.

Sezer, M., Fidan, F., Koken, T., Serteser, M., and Unlu, M. 2006. Effects of cigarette and biomass smoke exposure on oxidant–antioxidant status in rabbits. *Toxicol Environ Chem* 88: 169–174.

Singh, N.P., McCoy, M.T., Tic, R.R., and Schneider, E.L. 1988. A simple technique for quantization of low levels of DNA damage in individual cells. *Exp Cell Res* 175: 184–191.

Smith, K. and Mehta, S. 2000. The burden of disease from indoor air pollution in developing countries: Comparison of estimates. USAID/WHO Global Consultation on the Health Impact of Indoor Air Pollution and Household Energy in Developing Countries: Setting the Agenda for Action, May 3–4, Washington, DC.

Smith, K.R. 1987. *Biomass Fuels, Air Pollution and Health: A Global Review*. New York: Plenum Press.

Smith, K.R. 1993. Fuel combustion, air pollution exposure, and health: Situation in developing countries. *Ann Rev Energy Environ* 18: 529–566.

Smith, K.R. 2000. Inaugural article: National burden of disease in India from indoor air pollution. *Proc Soc Natl Acad Sci USA* 97: 13286–13293.

Smith, K.R. 2002. Indoor air pollution in developing countries: Recommendations for research. *Indoor Air* 12: 198–207.

Smith, K.R., Apta, M.G., Yoqing, M., Wongsekiarttirat, W., and Kulkarni, A. 1994. Air pollution and the energy-ladder in Asian cities. *Energy* 19: 587–600.

Smith, K.R. and Liu, Y. 1994. Indoor air pollution in developing countries. In: J.M. Samet (ed.) *Epidemiology of Lung Cancer*. New York: Marcel Dekker, pp. 151–184.

Smith, K.R. and Mehta, S. 2003. The burden of disease from indoor air pollution in developing countries: Comparison of estimates. *Int J Hyg Environ Health* 206: 279–289.

Smith, K.R., Mehta, S., and Maeusezahl-Feuz, M. 2004. Indoor smoke from household solid fuels. In: M. Ezzati, A.D. Rodgers, A.D. Lopez, and C.J.L. Murray (eds) *Comparative Quantification of Health Risks: Global and Regional Burden of Disease due to Selected Major Risk Factors*, Vol. 2, Geneva: World Health Organisation, pp. 1437–1495.

Snead, D.R.J., Perunovic, B., Cullen, N., Needham, M., Dhillon, D.P., Satoh, H., and Kamma, H. 2003. Hn RNP B1 expression in benign and malignant lung disease. *J Pathol* 200: 88–94.

Soejima, K., Traber, L.D., Schmalstieg, F.C., Traber, L.D., Szabo, C., Salzman, A., and Taber, D.L. 2001. Role of nitric oxide in vascular permeability after combined burns and smoke inhalation injury. *Respir Crit Care Med* 163: 745–752.

Srestha, I.L. and Srestha, S.L. 2005. Indoor air pollution from biomass fuels and respiratory health of the exposed population in Nepalese households. *Int J Occup Environ Health* 11: 150–160.

Stich, H.F., Stich, W., and Parida, B.B. 1982. Elevated frequency of micronucleated cells in the buccal mucosa of individuals at high risk for oral cancer: Betel quid chewers. *Cancer Lett* 17: 125–134.

Szezeklik, A., Szezeklik, J., Galuszka, Z., Musial, J., Kolaryzk, E., and Targosz, D. 1994. Humoral immunosuppression in men exposed to polycyclic aromatic hydrocarbons and related carcinogens in polluted environments. *Environ Health Perspect* 102: 302–304.

Tesfaigzi, Y., Singh, S.P., and Foster, J.E., et al. 2002. Health effects of subchronic exposure to low levels of wood smoke in rats. *Toxicol Sci* 65: 115–125.

Thurston, S.W., Ryan, L., Christiani, D.C., et al. 2000. Petrochemical exposure and menstrual disturbances. *Am J Ind Med* 38: 555–564.

Tolbert, P., Shy, C.M., and Allen, J.W. 1991. Micronucleus and other nuclear anomalies in buccal smears: Methods development. *Mutat Res* 271: 69–77.

Uzun, K., Ozbay, B., Ceylan, E., Gencer, M., and Zehir, I. 2003. Prevalence of chronic bronchitis-asthma symptoms in biomass fuel exposed females. *Environ Health Prev Med* 8: 13–17.

Varkey, A.B. 2004. Chronic obstructive pulmonary disease in women: Exploring gender differences. *Curr Opin Pulm Med* 10: 98–103.

Velema, J.P., Ferrera, A., Figueroa, M., et al. 2002. Burning wood in the kitchen increases the risk of cervical neoplasia in HPV-infected women in Honduras. *Int J Cancer* 97: 536–541.

Venkataraman, C. and Rao, G.U. 2001. Emission factors of carbon monoxide and size-dissolved aerosols from biofuel combustion. *Environ Sci Technol* 15: 2100–2107.

Vine, M.F., Schoenobach, V.J., Hulka, B.S., Koch, G.G., and Samsa, G. 1990. Atypical meta-plasia and incidence of bronchogenic carcinoma. *Am J Epidemiol* 131: 781–793.

Wafula, E.M., Onyango, F.E., Mirza, W.M., et al. 1990. Epidemiology of acute respiratory tract infections among young children in Kenya. *Rev Infect Dis* 12: S1035–S1038.

Wechsler, D. 1945. A standardized memory scale for clinical use. *J Psychol* 19: 87–95.

WHO. 1997. *Health and Environment for Sustainable Development*. Geneva: World Health Organization.

WHO. 1999. *Global Air Quality Guidelines*. Geneva: World Health Organization.

WHO. 2002. *The World Health Report 2002: Reducing Risks, Promoting Healthy Life*. Geneva: World Health Organization .

WHO. 2006. *Fuels for life: Household Energy and Health*, E. Rehfuess (ed.). Geneva: WHO Library Cataloguing-in-Publication Data.

World Bank. 1992. *World Development Report*. New York: Oxford University Press.

Zhang, J. and Smith, K.R. 1996. Hydrocarbon emissions and health risks from cookstoves in developing countries. *J Expos Anal Environ Epidemiol* 6: 147–161.

6 Health Effects of Urban Air Pollution in India

Manas Ranjan Ray and Twisha Lahiri

CONTENTS

6.1   INTRODUCTION

6.1.1   Air Pollution Scenario in Indian Cities

In recent times, India has been experiencing a phase of rapid growth and economic development reflected by industrialization, urbanization, rise in income, and motor vehicle use. A parallel rise in air pollution is evident from the survey conducted by the Central Pollution Control Board (CPCB, 1997) showing that only 2 out of the 70 cities monitored had clean air (Table 6.1).

Delhi, the political capital of India, is one of the 10 most polluted cities of the world. An estimate of respirable suspended particulate matter (RSPM) levels in ambient air among 10 large Indian cities in 2002 has put Delhi in the top position, followed closely by Ahmedabad and Kanpur (Table 6.2). On the other hand, Chennai had the cleanest air among all other cities (Figure 6.1).

TABLE 6.1
Air Quality Status Based on Measurements of Total Suspended Particulate Matter in Indian Cities in 1997

Pollution Status	Number of Towns	Names
Critical pollution (above 1.5 times of the standard)	29	Agra, Ahmedbad, Alwar, Ankleshwar, Anpara, Bhopal, Chandigarh, Dehradun, Delhi, Faridabad, Ghaziabad, Gajraula, Gobindgarh, Guwahati, Haldia, Howrah, Indore, Jaipur, Jharia, Jodhpur, Kanpur, Lucknow, Ludhiana, Mumbai, Patna, Pune, Surat, Vadodara, Vapi
High pollution (between 1 and 1.5 times the standard)	18	Bangalore, Bhilai, Chittor, Dhanbad, Hyderabad, Jalandhar, Kota, Korba, Nagpur, Nasik, Paonta Sahib, Pondicherry, Rourkela, Raipur, Satna, Sholapur, Sindri, Talcher, Thiruvanathapuram, Udaipur, Visakhapatnam, Yamunanagar
Moderate pollution (between 0.5 and 1 times the standard)	17	Angul, Chanderpur, Chennai, Coimbatore, Damtal, Dombivali, Jabalpur, Jamshedpur, Kochi, Kozhikode, Mysore, Ponda, Parwanoo, Nagda, Rayagada, Shimla, Vasco
Clean air (below 0.5 times the standard)	2	Shillong, Tuticorin

Source: CPCB (Central Pollution Control Board). 1997. Air quality data. Central Pollution Control Board, Delhi, India.

6.1.2 SOURCES OF AIR POLLUTION IN URBAN INDIA

There are three important sources of ambient air pollution in Indian cities: (i) emissions from motor vehicles due to combustion and evaporation of automotive fuels, (ii) emissions from industrial units and construction of buildings, infrastructure in and around the cities, and (iii) emissions from domestic sources. Of these, the contribution vehicular source is increasing over the years (Figure 6.2). In contrast,

TABLE 6.2
Five Most Polluted Cities in India in 2002

City	Annual Average RSPM Level ($\mu g/m^3$)	% Change Over National Standard for Residential Areas (60 $\mu g/m^3$)
Delhi	168 ± 122	+180
Kanpur	164 ± 102	+173
Ahmedabad	164 ± 68	+173
Jaipur	118 ± 93	+97
Kolkata	97 ± 51	+62

Note: The values are expressed as mean ± standard deviation (SD).

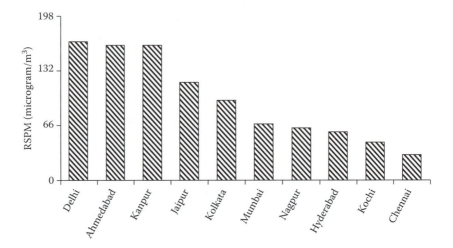

FIGURE 6.1   Annual average concentration of RSPM (RSPM, PM_{10}) in important Indian cities.

emissions from industrial and domestic sources are declining because coal-based industries are increasingly using cleaner fuel, highly polluting industrial units are being shifted outside the city areas, and liquefied petroleum gas (LPG), a relatively cleaner fuel, is replacing coal and biomass as the cooking fuel in urban households.

6.1.3   Air Pollution from Vehicular Sources

6.1.3.1   Exponential Growth in the Number of Motor Vehicles

One of the major reasons for high levels of vehicular pollution in Indian cities is the exponential rise in the number of motorized vehicles in the country in the recent past. The number of motor vehicles in India has increased 29 times in the past three decades, from 1.9 million in 1971 to 55.0 million in 2001 (Badami, 2005, Figure 6.3).

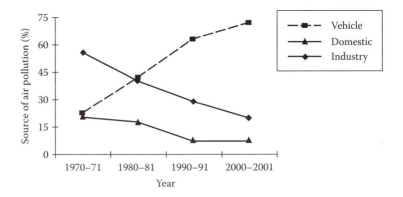

FIGURE 6.2   Relative contributions of the three chief sources of urban air pollution in Delhi during 1970–2001.

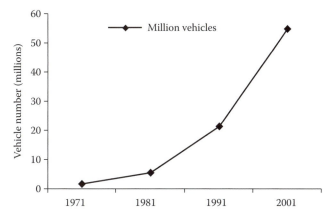

FIGURE 6.3   Increase in the number of motor vehicles in India during 1971–2001.

The increase was sevenfold for buses, ninefold for trucks, 10-fold for cars, jeeps, and taxis, but a remarkable 67-fold for two-wheelers (Badami, 2005).

Delhi has the highest number of vehicles in the country—more than three other metros, Mumbai, Kolkata, and Chennai, put together. The population of Delhi has increased from 3.53 million in 1970 to 13.80 million in 2001, registering a rise of 3.9-fold over a period of 30 years. In contrast, the number of registered motor vehicles in the city in 2001 was 3.42 million against 0.2 million during 1970–1971, illustrating a rise of 17-fold that far surpassed population growth. Similarly, the total number of registered vehicles in Kolkata in January 2000 was 721,775, which is about three times that of the 1982–1983 number. This extensive rise in vehicle numbers is reflected in the air pollution scenario of the cities.

6.1.3.2   Traffic Congestion and Low Traffic Speed

Traffic congestion and low traffic speed increase fuel consumption and enhance emissions of carbon monoxide (CO) and hydrocarbon per vehicle-km by 200% or more. The problems are common in Indian cities due to inadequate road space, poor road conditions, indiscriminate use of speed breakers, waterlogging during monsoons, political or religious processions on busy roads, and encroachment of roads by pavement dwellers and street hawkers.

6.1.3.3   Poor Fuel Quality and Adulteration

Motor vehicles have internal combustion engines, which burn a mixture of air and fuel to produce energy that propels the vehicle. The type and quantity of the pollutants released during this combustion is influenced by several factors. The type of fuel—petrol, diesel, or compressed natural gas (CNG)—is one of them. Although petrol and diesel are more polluting fuels than CNG or LPG, these fuels are largely used in India. But the benefit of using cleaner automotive fuel has been proved beyond doubt in Delhi where the level of ambient pollution has substantially declined following the introduction of CNG as fuel for public transport vehicles.

Fuel and lubricating oil quality have also contributed significantly to transport air pollution in India. Indian transport fuels are still inferior to those presently available in Europe and the United States (CSE, 2002). Indian gasolines have a high volatility, and the vast majority of gasoline vehicles are carburated, not fuel injected. These facts coupled with India's high ambient temperatures increase the potential for evaporating hydrocarbons to generate ground-level ozone. Until recently, benzene was not controlled in Indian gasoline. As a consequence, ambient benzene level in Delhi in the late 1990s was several folds higher than the maximum allowable limit by the European Union (CSE, 2002). Yet another important factor is the adulteration of fuel and lubricating oil. This is particularly rampant when the operators do not own the vehicles (Badami, 2005).

6.1.3.4 Poor Vehicle Maintenance

Poor vehicular maintenance is another important contributing factor in vehicular pollution in India. Several studies have shown that maintenance is a significant factor in vehicular emissions. For instance, particulates can increase 10-fold in poorly maintained two-stroke motorcycles and autorickshaws using poor-quality lubricating oil and by 20 times in diesels with damaged fuel-injection systems.

6.1.4 Industrial Emissions

The share of industries as a source of air pollution is rapidly declining in Indian cities for the past 30 years. Industrial pollution was the biggest contributor (56%) to Delhi's air pollution during 1970–1971. In contrast, only 20% of Delhi's air pollution is now generated by the industries, especially by the three coal-based thermal power plants.

6.1.5 Emissions from Domestic Sources

Pollution from household sources is mainly due to the use of coal, kerosene, biomass (wood, cow dung, and agricultural refuse), and LPG for daily household cooking and room heating. Like the industries, the contribution of household sector to urban air pollution is steadily declining in India. For example, only 8% of Delhi's current air pollution is contributed by household sources, compared with 21% in 1970–1971 and 18% in 1980–1981. Similarly, only 2% of Kolkata's air pollution is contributed by domestic sector. The type of fuel used for domestic cooking has significantly changed in Indian cities from the early 1970s to the late 1990s. Biomass fuel has been largely replaced by cleaner fuels like LPG. However, there are a large number of pavement and slum dwellers in many cities. These people generally use biomass as cooking fuel that contributes to a substantial amount of smoke and particulate in urban air.

6.2 CONSTITUENTS OF POLLUTED AIR

6.2.1 Particulate Matter: The Major Toxic Component of Urban Air

Particulate matter (PM) is a complex mixture of suspended solid and liquid particles (Brook et al., 2003) classified into primary (particles emitted directly by emission sources) and secondary (particles formed through the atmospheric reaction of gases) particles. The PMs vary greatly in size, composition, concentration, depending on origin and age. The size distributions are important for health impacts. Particles larger than 10 μm in diameter are deposited almost exclusively in the nose and throat, whereas those smaller than 1 μm reach the lower regions of the lung. The intermediate size range gets deposited between these two extremes of the respiratory tract. Outdoor (ambient) PM size ranges from approximately 0.001 to 100 μm in aerodynamic diameter. There are three main size categories for PM measured in urban air.

6.2.2 Coarse Particles (PM$_{10}$)

The particles in this category are larger than 1 μm and extend up to 100 μm. However, for toxicity studies, the most important particles are those that have a diameter of less than10 μm (PM$_{10}$) because they are respirable whereas the larger particles are not. PM$_{10}$ deposit relatively quickly with a lifetime of less than 2 days, and exposure may lead to adverse responses in the lungs triggering an array of problems in the lungs and heart (Brunekreef and Forsberg, 2005; Harrabi et al., 2006). PM$_{10}$ has also been associated with emergency hospital admission for asthma, bronchitis, and pneumonia in older people (Ye et al., 2001). For every 10 μg/m^3 increase of PM$_{10}$, mortality from all causes increases by 0.51% (Samet et al., 2000).

6.2.3 Accumulation Mode or Fine Particles (PM$_{2.5}$)

They consist of PM with a diameter between 0.1 and 2.5 μm. Airborne particles smaller than 2.5 μm (PM$_{2.5}$) are usually called fine particles. They account for the majority of the mass of suspended particles and deposit slowly leading to a long atmospheric lifetime of 5–10 days. These particles may penetrate deep inside the airways and are more strongly linked with adverse health effects (USEPA, 1996). Fine particles are composed mainly of carbonaceous materials (organic and elemental), inorganic compounds (sulfate, nitrate, and ammonium), and trace metal compounds (iron, aluminum, nickel, copper, zinc, and lead). There are potentially thousands of different compounds adsorbed on fine particles that may exert harmful biological effects. A modest rise in PM$_{2.5}$ level has been shown to be associated with changes in cardiac function (Mar et al., 2005). The toxicity of fine particles is mediated by generation of oxidative stress and production of free radicals (Furuyama et al., 2006).

6.2.4 Nuclei Mode or Ultrafine Particles

The particles in this category are smaller than 0.1 μm. They are also known as ultrafine particles (UFPs). They do not last long in the air since they deposit or rapidly form fine particles by coagulation. UFPs are present in great numbers in polluted urban air (Jaques and Kim, 2000). They have a carbonaceous core with attached inorganic and organic materials that can cause adverse health effects (Oberdorster, 2000). The UFPs have less mass than coarse particle fractions but they are much greater in number and have a relatively large surface area-to-mass ratio, making them potential carriers of harmful gaseous compounds. UFPs escape alveolar macrophage (AM) surveillance in the lungs, which is otherwise very efficient for larger particles (Hahn et al., 1977). Exposure to high doses of UFPs can cause severe pulmonary inflammation and hemorrhage, high degree of alveolar and interstitial edema, disruption of epithelial and endothelial cell layers, and death (Oberdorster et al., 1992; Peters et al., 1997; Oberdorster, 2000). Even modest exposure to UFPs from vehicular exhausts can lead to cardiovascular problems, pulmonary diseases, and cancer (Vinzents et al., 2005).

6.2.5 SO_2 AND NO_x

Coal burning is a major source of sulfur dioxide (SO_2) in air. It is an acidic gas, which combines with water vapor in the atmosphere to produce acid rain. SO_2 in ambient air can affect human health (Routledge et al., 2006), particularly in those suffering from asthma and chronic lung diseases and exacerbates respiratory symptoms and impaired breathing in sensitive individuals (Lipfert, 1994). Nitrogen oxides (NO_x) are formed during combustion processes at high temperatures from the oxidation of nitrogen in air or fuel. The major types of oxides of nitrogen are nitric oxide (NO) and nitrogen dioxide (NO_2). The main source of NO is road traffic, emitted from both petrol and diesel engine motor vehicles. NO_x is a precursor of ozone formed in the troposphere. Oxides of nitrogen are immunotoxic and increase the susceptibility to respiratory tract infection. Continued or frequent exposures to high concentrations of NO_x in breathing air may cause irritation of the lungs and consequent acute respiratory illness (Hasselblad et al., 1992).

6.2.6 CARBON MONOXIDE

CO is a toxic gas emitted into the atmosphere as a result of combustion processes and by oxidation of hydrocarbons and other organic compounds. CO is produced almost entirely from road traffic and remains in the atmosphere for approximately one month before being oxidized to CO_2. The largest contributors of CO are petrol-fueled vehicles. CO binds strongly to hemoglobin in red blood corpuscles resulting in the production of carboxyhemoglobin. This impairs the transport of oxygen within the blood and can result in adverse effects on tissues with high oxygen needs such as the cardiovascular and nervous systems. A recent study has shown that chronic exposures to CO may cause adverse birth outcomes such as reduced birth weight and intrauterine growth retardation (Salam et al., 2005).

6.2.7 POLYCYCLIC AROMATIC HYDROCARBONS

About 200 different kinds of hydrocarbons are emitted from the combustion of petrol and diesel. Of these, the polycyclic aromatic hydrocarbons (PAHs) are of particular interest due to their carcinogenic (cancer-causing) potential. PAHs are usually adsorbed on the PM and enter the body through inhalation. These compounds are semivolatile in nature. Several PAHs like benzo(a)pyrene [B(a)P] are highly carcinogenic (Hrudkova et al., 2004). Incidence of lung cancer has been reported in persons directly exposed to B(a)P from automobile exhausts and biomass fuel burning during household cooking).

6.2.8 Volatile Organic Compounds

Volatile organic compounds (VOCs) consist of various classes of carbon-containing chemicals that are gaseous at room temperature. They are released into the environment from petrol and diesel, especially the former, by evaporation or as combustion products. Some VOCs (e.g., benzene) are hematotoxic, neurotoxic, leukemogenic, and carcinogenic (Wallace, 1984, 1989; Farris et al., 1993).

6.2.9 Air Toxics of Biological Origin

Biological agents present in polluted air may cause several diseases. There are many sources of these pollutants. Pollens originate from plants; people and animals transmit viruses; bacteria are carried by man, animal, soil, and plant debris; and household pets are sources of saliva and animal dander. The protein in urine from rats and mice is a potent allergen. When it dries, it can become airborne. Along with particulate pollution, numerous airborne bacteria enter the body during respiration. Several of these are pathogenic to humans. For example, *Mycobacterium tuberculosis* causes tuberculosis while *Streptococcus pneumoniae* causes bacterial pneumonia. Globally, pneumonia causes 2 million deaths in children (20% of the total number of child deaths) every year and 70% of them occur in Africa and Southeast Asia.

Like bacteria, viral infections have been linked to air pollution. *Mumps* virus (mumps), *Myxovirus influenza* (influenza), *Poliovirus* (poliomyelitis), *Rhinovirus* causing common cold, *Rubella* virus (measles), *Varicella* virus (chicken pox), *Variola pox* virus causing small pox, *Haemophilus influenzae*, respiratory syncytial virus (RSV), influenza, parainfluenza, and adenoviruses are some of the viruses that spread through polluted air.

Increased presence of fungi in indoor environment due to poor housing conditions, older house age, relative lack of sun exposure, and absence of insulation enhances the risk of respiratory illnesses including oral toxicosis and airway allergy (Howden-Chapman et al., 2005). The problems may turn serious requiring hospitalization (Khalili et al., 2005). Airborne pollens and other allergens are major causative agents of bronchial hypersensitivity and asthma. Asthma exacerbation is the most common cause of hospital admission in children. Airborne bacteria and virus infections in allergic asthmatics further increase the risk of hospitalization (Murray et al., 2006). Pollen exposure is usually associated with respiratory tract allergy and eosinophil accumulation in the nasopharynx and the airways (Onbasi et al., 2005).

6.3 HEALTH EFFECTS OF AIR POLLUTION

Harmful effects of air pollution on human health are recognized for centuries. Globally, 1.1 billion people breathe polluted, unhealthy air (UNEP, 2002). The consequence can be disastrous because epidemiological studies have established a direct relationship between air pollution and health hazards ranging from morbidity (illness) to mortality (death from illness). It may be recalled that about 8000 people die every day from diseases related to air pollution exposure across the globe. In addition, air pollution is responsible for 4.6 million lost life-years every year (WHO,

2005). This burden of ill health is not equally distributed, as approximately two-thirds occur in developing countries owing to high levels of outdoor air pollution in many Asian cities. For instance, each year 500,000 deaths occur in China due to air pollution against 60,000 deaths in the United States.

6.3.1 Excess Mortality

The London fog incident in 1952 conclusively established an association between air pollution and increased mortality (Logan, 1952). Since then, several epidemiological studies in the USA and Europe have established a clear relationship between air pollution exposure and excess mortality (Samet et al., 1981; Dockery et al., 1982; Wichmann et al., 1989; Archer, 1990; Ostro et al., 1991; Ponka, 1991; Pope and Dockery, 1992; Bobak and Leon, 1992; Lipfert, 1994; Thurston, 1996). Air pollution is associated with increased risk of acute respiratory infections, the principal cause of infant and child mortality in the developing countries (Bendahmane, 1997). Each 10 $\mu g/m^3$ increase in annual average $PM_{2.5}$ level may lead to 4%, 6%, and 8% rise in the risk of all-cause, cardiopulmonary, and lung cancer mortality, respectively (Pope et al., 2002). An increase in PM_{10} by 10 $\mu g/m^3$ has been reported to cause 0.76% excess deaths from cardiovascular causes and 0.58% excess mortality from respiratory diseases (Analitis et al., 2006). Dockery and his coworkers (1993) showed an association between mortality rates and PM_{10} levels not only from lung cancer but also from cardiopulmonary diseases. They estimated 3.4% excess deaths from respiratory diseases and 1.4% from cardiovascular diseases (CVDs) for every 10 $\mu g/m^3$ increases in PM_{10}. The overall increase in mortality was calculated as 1% for every 10 $\mu g/m^3$ rise in PM_{10} (Dockery et al., 1999; Viegi et al., 1999, 2000). Samet et al. (2000) reviewed the subject and concluded that for every 10 $\mu g/m^3$ rise in PM_{10}, there was an increase in mortality from all causes by 0.51% and from cardiopulmonary diseases by 0.68%. It has been calculated that if the pollution level is brought down to National Ambient Air Quality Standards, 10,647 premature deaths could be avoided in Kolkata alone and another 9859 in Delhi every year.

Mortality and morbidity associated with air pollution are primarily due to the toxic effects of the PM (Morgan et al., 1998; Hong et al., 1999; Peters et al., 2000; Arena et al., 2006). Associations have also been reported with gaseous air pollutants, namely, ozone (Anderson et al., 1996), NO_2 (Anderson et al., 1996), SO_2 (Gouvea and Fletcher, 2000), and CO (Gouvea and Fletcher, 2000). Compared with particulates, however, the relationship between gaseous pollutants and mortality is less consistent.

6.3.2 Increased Morbidity

Besides mortality, air pollution could initiate and/or aggravate several diseases. Excess morbidity is often reflected in absenteeism from school and work, restricted activity at home, more attendance to outpatient medical services, and emergency visits to clinics and hospitalization (Shy et al., 1978). Air pollution-related pulmonary diseases for which hospital admissions are usually required are acute bronchitis, pneumonia, emphysema, bronchiectasis, chronic airway obstruction, and attacks of asthma. Besides lung diseases, air pollution is significantly associated with cardiac and vascular problems (von Klot et al., 2005; Maheswaran et al., 2005; Mills et al., 2005).

6.3.3 ADDITIVE AND SYNERGISTIC EFFECTS OF AIRBORNE POLLUTANTS

Following inhalation, air pollutants act on the target tissues in unison rather than individually. The pollutants may also react with each other and some of the compounds generated in the process may be more toxic than the primary pollutants.

The additive or cumulative response to a mixture is the sum of the effects induced by the individual components of the mixture. Conceptually, the additive effect occurs only when the action of each pollutant is independent. When a pollutant does not elicit a response when acting alone but increases the effect of another co-occurring pollutant, the effect is called potentiation. Synergism refers to any combination of action in which the result is more than which would be attained if the actions were entirely independent of each other. In other words, in a synergistic process, the whole is greater than the sum of its parts. As, for example, smoking and exposure to vehicular emission or air pollution result in a greatly increased probability of lung cancer compared to the risk of either smoking or asbestos exposure alone.

Human exposure to complex mixtures of air pollutants is a challenge to the toxicologists and epidemiologists because of the enormous range of variations and confounding factors making exposure assessment, study design, and data interpretation difficult. Therefore, it is debatable whether the observed changes in human subjects could be attributed to benzene alone. To explore these points, parallel experiments need to be conducted in experimental animals under controlled laboratory conditions where the animals are exposed to measured doses of benzene in drinking water and also through inhalation. Comparing the health response following controlled benzene exposure with those obtained from vehicular emission-exposed population can give an insight into the possible health effects of benzene from vehicular emission.

6.4 MATERIALS AND METHODS

During 2000–2006, we conducted epidemiological studies on the health impact of urban air pollution among 6862 nonsmoking residents of Kolkata (former Calcutta, capital of the state of West Bengal in Eastern India) and Delhi having a median age of 43 years and 3715 age- and sex-matched nonsmokers as controls from relatively less polluted rural areas of West Bengal where particulate pollution (PM_{10}) level was considerably lower. In addition, 12,688 school-going children (age 8–17 years) of these two cities and 5649 from rural West Bengal were examined. The study protocol was approved by the Ethics Committee of Chittaranjan National Cancer Institute, Kolkata, India.

Data on the concentration of PM having an aerodynamic diameter of less than 10 μm (PM_{10}) in ambient air were obtained from the Central Pollution Control Board, Delhi and the West Bengal State Pollution Control Board, Kolkata. The prevalence of respiratory symptoms was estimated from a questionnaire survey using a validated, structured questionnaire. Pulmonary function test (PFT) was carried out following the procedure of the American Thoracic Society (ATS, 1995). Chronic obstructive pulmonary disease (COPD) was diagnosed following the criteria of the Global Initiative for Chronic Obstructive Lung Disease (GOLD; Pauwels et al., 2001). Cellular lung response to air pollution was evaluated by examining cytology of exfoliated airway cells in Papanicolaou-stained sputum. Hematological parameters like total and differential counts of white blood cells (WBC) and total count of platelets were done on freshly collected venous blood without anticoagulants by using a hemocytometer under a light microscope. β_2 Mac-1 integrin (CD11b/CD18) expression on circulating neutrophils and monocytes, P-selectin (CD 62P) expression on platelet surface, a measure of platelet activation, and circulating leukocyte–platelet aggregates were determined by flow cytometry. Chromosomal breaks were examined by micronucleus (MN) assay and DNA damage by single cell gel electrophoresis (Comet assay) and antioxidant enzyme superoxide dismutase (SOD) was measured in plasma by spectrophotometry.

6.5   SALIENT FINDINGS

6.5.1   Effects on Respiratory Health

Since airborne pollutants generally enter the body through inhalation, the lungs and airways are the primary target organs. The airways of the lungs are represented by the trachea (windpipe), and beyond it are bronchi (with cartilage cover) and bronchioles (without cartilage). The bronchioles lead to air spaces called alveoli, which have an average diameter of 200 μm each. A recent study has demonstrated that there are approximately 480 million alveoli in both lobes of an adult human lung, and men have more alveoli and larger lung volume than women (Ochs et al., 2004). The mean size of a single alveolus is 4.2×10^6 μm^3. Alveoli make up approximately 64% of the lung space (Ochs et al., 2004). Human lungs have a total surface area of 1400 m^2, and everyday we inhale approximately 15 m^3 of air (that is, 15,000 liters). The weight of this inhaled air is greater than the food we consume and the water we drink in a day. The lung volume and breathing frequencies of healthy adults at rest are 400–500 mL and 15–17 breaths per minute, respectively (Tobin et al., 1983). Recent study has documented that a constant number of respiratory units is maintained from childhood to adulthood while both the smallest bronchioles and alveoli expand in size to produce the increased lung volume with increased age and height (Zeman and Bennett, 2006).

6.5.1.1   Rise in the Prevalence of Respiratory Symptoms

Symptoms are a form of signals that act as indicators of any underlying illness or disease. Epidemiological studies on respiratory health are generally based on collection of data on the prevalence of respiratory symptoms. Respiratory symptoms are usually classified into two broad groups: (a) upper respiratory symptoms (URS), which include runny and stuffy nose, sinusitis, sore throat, wet cough, dry cough, cold head, fever, burning or red eyes and (b) lower respiratory symptoms (LRS), which include chronic cough, wheeze, sputum production, shortness of breath, and chest discomfort. Most of the respiratory diseases underlying these symptoms are

TABLE 6.3

Prevalence (%) of Respiratory Symptoms in Children

Symptom	Rural	Delhi
Upper respiratory symptoms	14.6	23.1*
Lower respiratory symptoms	8.0	17.0*
Overall	18.2	32.1*

Note: Many children had more than one symptoms; $*p < 0.05$ compared with rural subjects.

caused by bacterial, fungal, or viral infections, or structural or functional damage to the respiratory system mediated by air pollutants.

During 2000–2006, we conducted epidemiological studies on health impacts of urban air pollution among 6862 nonsmoking residents of Kolkata and Delhi (median age 43 years) and 3715 age- and sex-matched nonsmokers as controls from relatively less polluted rural areas of West Bengal, where the particulate pollution (PM_{10}) level was considerably lower. In addition, 12,688 school-going children (8–17 years) of these two cities and 5649 from rural West Bengal were examined.

We found one or more respiratory symptoms present in 32.1% children of Delhi compared with 18.2% of age- and sex-matched children of rural areas (Lahiri et al., 2006b). In Delhi, URS and LRS were present in 23% and 17% children, respectively, compared with 14.6% and 8% in their rural counterparts ($p < 0.05$) (Table 6.3). The respiratory symptoms were more prevalent in girls than in boys (36.3% versus 30%). After controlling potential confounders like parental smoking and socioeconomic status (SES), multivariate logistic regression analysis revealed a positive association between the PM_{10} level in ambient air and the prevalence of respiratory symptoms. A parallel study among adults revealed one or more respiratory symptoms in the past 3 months in 33.2% of the residents of Delhi ($n = 6005$) compared with 19.6% of age- and sex-matched subjects ($n = 1046$) of rural West Bengal, indicating 1.7 times more prevalence in urban subjects (Lahiri and Ray, 2006a). URS were present in 21.5% of the residents of Delhi compared with 14.7% of the rural subjects (Figure 6.4) Common symptoms were sore throat (8.1% versus 4.5% in rural areas), runny or stuffy nose (7.8% versus 5%), and common cold with fever (7.4% versus 4.1%). We found LRS in 22.3% of the citizens of Delhi compared with 12.7% of rural subjects ($p < 0.001$, Figure 6.4).

The frequent symptoms were breathlessness (9.9% versus 4.8%), sputum-producing cough (8.8% versus. 4.6%), and dry cough (7.3% versus. 4.2%). As in Delhi, the high prevalence of cough, sinusitis, bronchitis, and asthma has been found in association with traffic-related air pollution in Kolkata (Basu et al., 2001). In Kolkata, the prevalence of URS in the past 12 months was 41.3% compared with 13.5% in rural subjects, while LRS were found in 47.8% of urban people in contrast to 35% of rural controls (Lahiri et al., 2000a). In general, the respiratory symptoms were most prevalent during winter when the air pollution level was the highest, and lowest in monsoon when the air was least polluted. However, sinusitis was most prevalent during

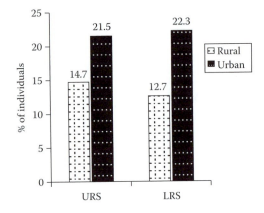

FIGURE 6.4   Prevalence (%) of upper and lower respiratory symptoms in rural and urban (Delhi) subjects.

monsoon when the level of fungal aeroallergen was higher. Except for wheeze, all other symptoms were most prevalent in low SES and in women.

6.5.1.2 Rise in Bronchial Asthma

The problem of asthma results from intermittent narrowing of the airways and consequent shortness of breath. Its early symptoms are wheeze, tightening of chest, shortness of breath, and persistent dry cough. We found current asthma (dyspnea and wheeze at any time in the last 12 months) in 4.6% schoolchildren of Delhi compared with 2.5% of age- and sex-matched rural children. Physician-diagnosed asthma was recorded in 1.7% of the children of Delhi against 0.9% in rural children. Asthma was more prevalent in large-sized families (>6 members), and multivariate logistic regression analysis revealed a positive association between particulate air pollution (PM_{10} level) and asthma attacks (OR = 1.28, 95% CI, 1.07–1.42), but not with asthma prevalence. In adults, current asthma and physician-diagnosed asthma were present in 7.6% and 3.6% of the residents of Delhi, respectively against 3.9% and 2.1% in rural subjects. In Kolkata, we found physician-diagnosed asthma in 5.8% of the citizens against 3.5% in rural subjects. Although asthma has genetic predisposition, exposure to air pollution exacerbate asthma attacks (Cakmak et al., 2004). In agreement with our observations, strong associations between severe asthma symptoms and cumulative exposures to exhaust from diesel-fueled engines (Riedl and Diaz-Sanchez, 2005) along with occupational exposures to benzene have been reported (Cakmak et al., 2004).

6.5.1.3 Reduction of Lung Function

The measurement of mechanics of breathing is collectively known as PFT. An instrument called spirometer measures it. The two major patterns of abnormal PFT are restrictive and obstructive type of lung function deficits. In restrictive lung function deficit, lung volume (FVC, Forced vital capacity) decreases below 80% of the predicted value. The subjects inhale reduced volume of air due to reduction in total lung capacity. The condition is often caused by infection and inflammation that leave

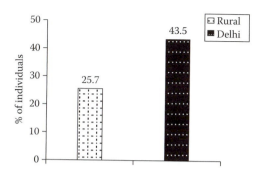

FIGURE 6.5 Prevalence (%) of lung function deficits among schoolchildren in rural and urban India.

scars on the lung tissue leading to the loss of elasticity of the lung, such as in the case of pneumonia or tuberculosis. Obesity and neuromuscular problems could also lead to restrictive lung function deficit. On the other hand, obstructive type of lung function deficit is indicated by a decrease in the forced expiratory volume in one second (FEV_1)/FVC ratio below 70%. Decline in FEV_1/FVC usually results from obstruction in large airways while fall in Forced expiratory flow between 25% and 75% of FVC, mid expiratory force ($FEF_{25-75\%}$) signifies small airways obstruction (Dassen et al., 1986; Vedal et al., 1987). Obstructive lung function is common in case of asthma and chronic bronchitis. A large number of studies have shown a decline in FVC and FEV_1 with increasing concentration of particulate air pollution (Johnson et al., 1982; Lebowitz et al., 1985; Brunekreef et al., 1991; Kilburn, 2000; Ibald-Mulli et al., 2002; Golshan et al., 2002; Frye et al., 2003; Asero et al., 2005).

We found reduction in lung function in 43.5% of the schoolchildren of Delhi compared with 25.7% of matched controls from rural areas of West Bengal and Uttaranchal (Figure 6.5; Lahiri et al., 2006b). Delhi's children had increased prevalence of the restrictive (20.3% versus 14.3% in rural), obstructive (13.6% versus 8%), as well as combined (both restrictive and obstructive) type of lung functions deficits (9.6% versus 3.5%) (Figure 6.6).

Besides higher prevalence, the magnitude of lung function deficits was much more in Delhi. For example, 7.3% of the schoolchildren of the city had severe lung function deficits compared with 2.2% of the children from rural areas where the ambient air pollution level was much lower. As in the case of children, 40.3% of adult individuals of Delhi had reduced lung function against 20.1% of rural subjects. Their mean FVC, FEV_1, $FEF_{25-75\%}$ and peak expiratory flow rate (PEFR) values were decreased by 9.4%, 13.3%, 10.4%, and 9.3%, respectively (Table 6.4). Restrictive type of lung function deficits was more prevalent (22.5%) than the obstructive type (in 10.7%), and 7.1% of the citizens of Delhi had the combined type of lung function deficits (Figure 6.6). The magnitude of lung function reduction was also much more in Delhi. For instance, 6.7% of the residents of Delhi had severe lung restriction (FVC <40%) against 1.3% of rural subjects ($p < 0.001$), and 2.7% of the citizens of Delhi had severe lung obstruction (FEV_1/FVC <30%) against 0.8% of their rural counterparts ($p < 0.001$). In Kolkata too, we found that 46.9% of the citizens had reduced

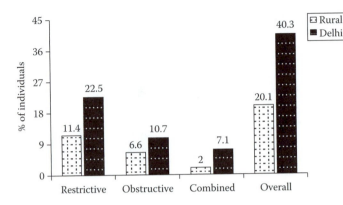

FIGURE 6.6 Prevalence of different types of lung function deficits in the residents of Delhi.

TABLE 6.4

Comparison of Spirometric Lung Function Measurements between Rural Subjects and Residents of Delhi

Spirometric Measurements	Rural Control	Delhi	p Value
FVC (L)	3.38 ± 0.52	3.06 ± 0.43	<0.05
FEV$_1$(L)	2.94 ± 0.54	2.55 ± 0.53	<0.05
FEF$_{25-75\%}$ (L/s)	3.27 ± 0.65	2.93 ± 0.47	<0.05
PEFR (L/s)	3.65 ± 0.83	3.31 ± 0.66	<0.05

Note: Results are mean ± SD; *$p < 0.05$ compared with rural control.

lung function against 22% of rural subjects, and the restrictive type of lung function deficit was the most prevalent.

The risk factors of lung function deficits in urban children and adults were female gender, winter season, exposure to environmental tobacco smoke (ETS), and low SES. After controlling potential confounders, the restrictive (OR = 1.35, 95% CI 1.07–1.58), obstructive (OR = 1.45, 95% CI, 1.16–1.82), and combined type of lung function deficits (OR = 1.74, 95% CI, 1.37–2.71) were found to be positively associated with the PM$_{10}$ level in ambient air. A strong negative correlation was found in Spearman's correlation test between the PM$_{10}$ level and different lung function measurements. The correlation was strongest for FVC ($\rho = -0.74$, $p < 0.0005$) and FEV$_1$ ($\rho = -0.62$, $p < 0.0005$).

6.5.1.4 Rise in COPD Prevalence

COPD represented by chronic bronchitis and emphysema was diagnosed on the basis of bronchitis symptoms (presence of cough and expectorations on most of the days for at least 3 months in a year for two consecutive years or more) and spirometric characteristics as defined in GOLD criteria (Pauwels et al., 2001). In Delhi, the prevalence of COPD was 3.9% against 0.8% in rural controls ($p < 0.001$). Even 3.1% of never-smokers of the city had this disease compared with only 0.2% of nonsmoking controls (Lahiri and Ray, 2006a).

6.5.1.5   Adverse Cellular Lung Response

Both chronic and acute exposures to air pollution have been shown to directly affect the structural integrity of the respiratory system. Continued chemical exposures can cause necrosis and subsequent sloughing off of ciliated epithelial cells. Microscopic examination of these exfoliated cells in spontaneously expectorated sputum provides important information regarding the pathophysiological changes in the lung tissue and development of lung disease including malignancy (Roby et al., 1990).

6.5.1.6   Airway Inflammation

Compared with rural controls, sputum samples of the residents of Delhi were 1.6 times more cellular and contained significantly increased number ($p < 0.05$) of inflammatory cells such as neutrophils, eosinophils, lymphocytes, and AMs, indicating airway inflammation (Table 6.5, Figure 6.7).

Likewise, residents of Kolkata had 3.6 times more cells in their sputum than their rural counterparts. Highest rise in neutrophils and eosinophils was observed in persons occupationally exposed to vehicular emissions such as traffic policemen and street hawkers of the city (Lahiri et al., 2000a). The findings suggest inflammation and allergic lung disease. As in the case of adults, an elevated number of sputum eosinophils and neutrophils along with a marked rise in the prevalence of respiratory symptoms has been reported in school-going children of the city (Lahiri et al., 2000b). The changes could be attributed to cumulative exposures to the city's vehicular pollution, because inhalation of diesel exhausts significantly increases the number of inflammatory cells in the airways and alveoli (Salvi et al., 1999; Takizawa et al., 2000). Like the present findings, Saccomanno and his coworkers (1970) demonstrated cellular changes in the respiratory epithelium following chronic air pollution exposures.

Mucus plugs, goblet cell hyperplasia, and nuclear anomaly of columnar epithelial cells were frequent, and metaplasia and dysplasia of epithelial cells were found in 16% and 2.6% of urban subjects, respectively, compared with 3% and 0.6% of controls ($p < 0.01$, Figure 6.8). Air pollution exposure was associated with the presence

TABLE 6.5
Sputum Cytology of Nonsmoking Adults of Delhi

Cell Type	Rural Control	Delhi
Cells/hpf	46.2 ± 11.9	76.2 ± 24.9*
Neutrophil/hpf	29.8 ± 5.6	48.1 ± 7.6*
Eosinophil/hpf	0.6 ± 0.2	3.3 ± 1.6*
Lymphocyte/hpf	2.9 ± 1.3	4.7 ± 2.5*
AM/hpf	6.9 ± 1.6	12.9 ± 2.6*

Note: The results are expressed as mean ± SD; hpf, high power field of microscope (×400); *$p < 0.05$ compared with controls.

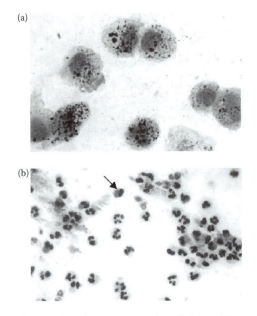

FIGURE 6.7 Photomicrographs of sputum samples of the residents of Kolkata showing abundance of particle-laden AMs (a) and eosinophils (b, arrow). Papanicolaou-staining ×400 (b), 1000 (a).

of abnormal columnar epithelial cells and squamous metaplasia in sputum of young adults and children (Plamenac et al., 1973, 1978).

6.5.1.7 Changes in Alveolar Macrophages, the First Line of Cellular Lung Defense

Macrophages are the principal defense cells in the airways and alveoli. Pulmonary macrophages include AMs, airway macrophages, and interstitial macrophages. AMs are the dominant phagocytic cells that act as the first line of cellular defense in the lungs. They play a pivotal role in lung defense through their activities like particle clearance from the inner airways by endocytosis and phagocytosis followed by killing of invading pathogens via generation of oxygen radicals and release of degradative enzymes (Becker, 1995). Besides, AMs actively participate in inflammation, wound healing, and tissue repair through their vast array of secretory products (Laskin and Pendino, 1995). In adult human lungs, there are approximately 480 million alveoli and each alveolus is defended by about 73 macrophages. Therefore, human lungs contain around 35.0 billion macrophages for its defense against inhaled pollutants.

We found a remarkable increase in the AM number in the residents of Delhi and Kolkata, and the cells were larger (mean diameter 27.8 μm against 16.2 μm in rural), often multinucleated, and heavily loaded with phagocytosed particles. Several-fold rise in the AM number was found in sputum of adults and children of highly polluted Kolkata compared with the inhabitants of Sunderban islands in Bay of Bengal where

(a)

(b)

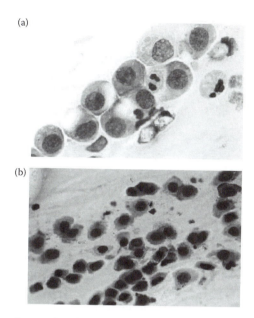

FIGURE 6.8 Photomicrographs of sputum samples of the residents of Delhi showing goblet cell hyperplasia (a) and metaplasia of airway epithelial cells (b). Papanicolaou-staining ×1000.

the air pollution level was significantly lower (Lahiri et al., 2000a,b). Similar changes were observed in schoolchildren and adult citizens of Delhi (Lahiri et al., 2006b). The AM number showed close parallelism with the PM_{10} level: highest in winter, lowest in monsoon, and intermediate in summer. AMs also showed a direct relationship with the degree of exposure to urban air pollution. Persons occupationally exposed to high levels of vehicular emission like traffic policeman, drivers, garage workers, and street hawkers of Kolkata and Delhi had significantly higher AM count compared with relatively less-exposed office employees (Basu et al., 2001; Lahiri et al., 2006b). Thus, the AM number in sputum appears to be a sensitive biomarker of air pollution exposure. Despite the rise in cell number, air pollution exposures can affect AM function because particle overloading impairs the phagocytic activity of AMs, particularly after infection that induce an increased production of interferon γ (Lundborg et al., 2001). Moreover, UFPs in urban dust and diesel exhaust cause cytoskeletal toxicity leading to impaired function of the macrophages, compromising lung defense (Moller et al., 2002).

6.5.1.8 Covert Pulmonary Hemorrhage

In case of lung hemorrhage, AMs rapidly engulf and disintegrate erythrocytes to form degradation products including heme iron (Perez-Arellano et al., 1992). These iron-containing macrophages are known as siderophages. Their presence in sputum in high numbers is indicative of either past intrathoracic bleeding or extravasations of red blood cells into the alveoli due to a sluggish blood flow (Grubb, 1994).

TABEL 6.6

Iron Deposition in Alveolar Macrophages in Sputum of Residents of Delhi

Siderophage in Sputum	Rural Control	Delhi	Kolkata
Number/hpf	0.6 ± 0.2	3.7 ± 1.1*	5.4 ± 2.3*
Golde score	12 ± 8	44 ± 32*	52 ± 33*
% of subjects with Golde score ≥100	1.0	3.9*	4.5*

Note: The results are mean ± SD; *$p < 0.001$ compared with controls.

Siderophages were abundant in sputum of the residents of Delhi and Kolkata and the city people had a higher Golde score, a measure of heme iron deposition in lungs, implying microscopic hemorrhage in the lungs (Table 6.6 and Figure 6.9). Siderophage numbers were higher in winter when air pollution levels were high, in persons belonging to low SES, and in persons occupationally exposed to vehicular pollution such as traffic policemen, and autorickshaw and taxi drivers (Roy et al., 2001). Abundance of siderophages in sputum possibly due to microscopic hemorrhage in the inner airways is recognized as an indicator of adverse lung reaction to air

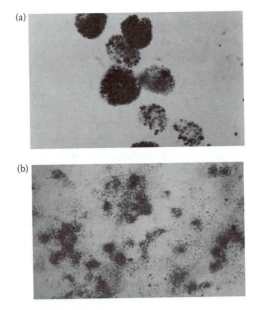

FIGURE 6.9 Photomicrographs of sputum samples of the residents of Kolkata showing cytochemical localization of iron (a) and elastase (b). Note the abundance of siderophages in "a" and release of excess elastase (blue-stained) in "b." Cytochemical-staining ×1000.

pollution (Roy et al., 2001). Like the present findings, siderophages were in excess in sputa of traffic policemen in Italy (Giovagnoli et al., 1999), automobile service station workers and hawkers of Kolkata (Roy et al., 2001), and in steel plant workers who had pulmonary hemorrhage. Deposition of iron causes oxidative stress, inflammation, and neutrophilic lung injury (Ghio et al., 2000).

6.5.1.9 Upregulation of Elastase Activity

Elastin is a fibrous protein present in the elastic tissues of lungs. Elastase, a proteolytic enzyme found in the lysosomes of neutrophils and AMs, is capable of destroying elastin. Release of excess elastase from these cells promotes development of emphysema as it leads to degradation of alveolar wall (Laskin and Pendino, 1995; Churg et al., 2002; Shapiro et al., 2003). In cytochemical analysis, we found overexpression of this enzyme in AMs and neutrophils of the residents of Kolkata and Delhi (Figure 6.9). Elastase-positive AMs were 9.4 ± 1.9/hpf in the residents of Delhi against 2.9 ± 0.8/hpf in rural controls. Kolkata's residents had 58% elastase-positive AMs out of which nearly 52% displayed high enzyme activity. In contrast, only 34.9% of AMs were positive for this enzyme, out of which 16% exhibited high elastase activity in rural individuals (Table 6.7). Enzyme activity was confined to the cell in controls whereas a considerable amount of the enzyme was released into the extracellular matrix in urban individuals. Since the enzyme is active only when released from the cell of origin, higher tissue degradation is likely in the urban group. Among all the individuals examined in Kolkata, the highest number of elastase-containing AMs was found in automobile service station workers followed

---

TABLE 6.7

Elastase Activity of Alveolar Macrophages Following Occupational and Environmental Exposures to Air Pollution in Kolkata

Group	Elastase-Positive AM (%)	Elastase-Positive AM/hpf	% of AM Showing High Enzyme Activity
Occupationally exposed	67.2 ± 0.4*	18.9 ± 0.2*	56.9 ± 0.9*
Driver	56.8 ± 1.5*	15.1 ± 0.9*	54.4 ± 1.7*
Traffic policeman	67.3 ± 1.3*	18.9 ± 0.7*	59.4 ± 1.6*
Roadside hawker	68.3 ± 0.7*	18.6 ± 0.6*	51.9 ± 1.3*
Environmentally exposed	44.3 ± 0.3*	7.0 ± 0.3*	45.0 ± 0.5*
Office employee	45.3 ± 0.8*	7.3 ± 0.5*	50.1 ± 1.3*
Housewife	49.4 ± 1.1*	10.2 ± 0.6*	41.6 ± 0.9*
Student	41.5 ± 0.7*	5.5 ± 0.5*	42.5 ± 1.1*
Rural control	34.9 ± 0.7	1.2 ± 0.4	16.0 ± 0.7

Note: Results are mean $\pm$ S.E.; *$p < 0.05$ compared with the corresponding rural control.

by traffic policemen and roadside hawkers (Basu et al., 2001). High elastase activity indicates greater risk of damage to the bronchial and alveolar walls among the residents of Delhi and Kolkata who were chronically exposed to high levels of particulate pollution.

6.5.1.10   TNF-α, iNOS, and p53 Overexpression, AM Apoptosis

Immunocytochemical study showed overexpression of tumor necrosis factor alpha (TNF-α) and inducible nitric oxide synthase (iNOS) in AMs and p53 protein overexpression in epithelial cells. TUNEL assay showed upregulation of apoptosis in AMs but downregulation in sputum neutrophils.

6.6   SYSTEMIC EFFECTS OF AIR POLLUTION IN URBAN INDIA

6.6.1   CARDIOVASCULAR CHANGES

Air pollution exposure and CVDs are intimately related. CVDs associated with air pollution such as angina, cardiac insufficiency, hypertension, and myocardial infarction (MI), that is, heart attack, are growing concern worldwide (Zareba et al., 2001; Dockery, 2001; Goldberg et al., 2001; Donaldson et al., 2001). Only 8.5% of air pollution-related deaths in the United States in 1997 were from respiratory diseases (COPD, pneumonia, influenza, etc.) while cardiovascular deaths accounted for 39.5% of all deaths (Greenle et al., 2000). Thus, CVDs, rather than respiratory ailments, are the most important causes of death from air pollution exposures. Compared with Europe and the United States, the risk of CVDs and deaths from these diseases occur at a younger age among urban Indians (Cropper et al., 1997).

6.6.2   HYPERTENSION

Our study in Delhi during 2002–2005 has recorded hypertension in 36.1% of the residents of the city who were lifetime nonsmokers compared with 9.5% of rural controls (Table 6.8). The prevalence of hypertension increased progressively with age: it was 24.8% in the 21–30-year age group, 28.1% in the 31–40-year age group, 45.3% in the 41–50-year age group, 52.5% in the 51–60-year age group, and 63.9% in the 60 + -year age group.

TABLE6.8
Prevalence (%) of Hypertension in Nonsmoking Citizens of Delhi

Hypertension	Rural Control	Delhi
Systolic	4.4	2.7
Diastolic	2.5	16.7*
Systolic + diastolic	2.6	16.7*
Overall	9.5	36.1*

Note: *$p < 0.05$ compared with controls.

Besides having greater prevalence, the severity of hypertension was much more in urban subjects. In Delhi, 15.4% of the citizens had relatively less severe Stage 1 hypertension with systolic blood pressure (SBP) between 140 and 159 mm Hg and 4.0% had more severe Stage 2 hypertension (SBP $\geq$160 mm Hg) in contrast to 6.1% and 0.9% of rural controls who had Stage 1 and Stage 2 systolic hypertension, respectively. Similarly, Stage 1 (DBP 90–99 mm Hg) and Stage 2 diastolic hypertension (DBP $\geq$100 mm Hg) were present in 23.4% and 10.0% of the citizens of Delhi compared with 4.4% and 0.8% of the control subjects, respectively.

6.6.2.1   Air Pollution and Other Risk Factors for Hypertension

A significant positive association was found between the PM_{10} level in Delhi's air and SBP and DBP of the citizens in Spearman's rank correlation test. The correlation was stronger for DBP ($\rho = 0.350$, $p < 0.005$). Conditional logistic regression analysis revealed that the risk factors for hypertension were high SES, elevated RSPM level, and overweight/obesity. Spearman's correlation illustrated a significant positive correlation between body mass index (BMI) with systolic ($\rho = 0.297$, $p < 0.01$) and diastolic hypertension ($\rho = 0.327$, $p < 0.005$). Therefore, particulate air pollutants along with lifestyle and SES significantly contributed to the greater prevalence of hypertension in Delhi. Like the present report, studies conducted from the late 1990s have consistently shown a positive association between the PM_{10} level and hospital admissions for CVDs (Schwartz, 1999; Burnett et al., 2001; Moolgavkar, 2000; Lin and Kou, 2000) and sudden deaths in patients with stable angina and MI (Lind et al., 2001; Peters et al., 2001, 2004). In combined analyses across six eastern U.S. cities, each 10 $\mu g/m^3$ increase in $PM_{2.5}$ was found to be associated with 2.1% increase in total mortality from ischemic heart diseases (Schwartz et al., 1996). People at the highest risk tend to be those with preexisting CVDs and the elderly. In addition, persons with diabetes are 2 times more susceptible to CVDs mediated by airborne pollutants (Singh et al., 2000; Zanobetti and Schwartz, 2001).

6.6.3   Relationship between Blood Pressure and Lung Function

SBP was negatively correlated with FVC ($\rho = -0.574$, $p < 0.001$), FEV_1 ($\rho = -0.586$, $p < 0.001$), $FEF_{25-75\%}$ ($\rho = -0.561$, $p < 0.001$), and PEFR ($\rho = -0.411$, $p < 0.002$) values. Similarly, DBP was negatively correlated with FVC ($\rho = -0.583$, $p < 0.001$), FEV_1 ($\rho = -0.536$, $p < 0.001$), $FEF_{25-75\%}$ ($\rho = -0.631$, $p < 0.001$), and PEFR ($\rho = -0.433$, $p < 0.001$) values. In essence, lung function and blood pressure are inversely correlated.

6.6.4   Mechanism of Cardiovascular Effects of Air Pollution: Upregulation of Platelet Activity

The mechanism by which air pollution affects the cardiovascular system is largely unknown. Seaton et al. (1999) first proposed that pollutant exposure induces a transient increase in blood coagulability, and this was subsequently corroborated by the study of Schwartz (2002), who demonstrated an association between PM_{10} and all three markers of cardiovascular risk: higher level of fibrinogen in blood plasma, greater number of platelets in circulation, and elevated WBC count. Several other workers have linked air pollution, especially the PM_{10} level, with the markers of adverse cardiovascular events such as increase in peripheral white cell counts (Salvi et al., 1999), elevation of interleukin-6 level (Ernst and Resch, 1993; Ridker et al., 2000), upregulation of C-reactive protein (CRP; Seaton et al., 1999; Peters et al., 2001), rise in plasma viscosity due to increase in fibrinogen level (Ghio et al., 2000; Peters et al., 2001; Schwartz, 2001), and alteration in cardiovascular autonomic control (Ibald-Mulli et al., 2001). Our flow cytometric study showed $3.4 \pm 0.6\%$ P-selectin-expressing activated platelets in circulation of the residents of Delhi compared with $1.6 \pm 0.4\%$ in age- and sex-matched rural subjects ($p < 0.001$). Moreover, the absolute number of P-selectin expressing platelets in circulation was $10,268 \pm 1232$ (mean $\pm$ SE) per μL of blood in urban subjects versus $3682 \pm 960/\mu$L in rural controls, indicating 2.8 times more activated platelets in circulation of the former group. Since platelet hyperactivity is a major risk factor for CVDs, the changes could be causally associated with greater risk of the diseases in urban subjects. The proposed mechanism of air pollution-related cardiovascular events (Frampton, 2001; Donaldson et al., 2001) is as follows (see Box 6.1).

**BOX 6.1   THE MECHANISM OF AIR POLLUTION-
RELATED CARDIOVASCULAR EVENTS**

Air pollution exposure and deposition of PM in the airways
↓
Generation of reactive oxygen species (ROS)
↓
Activation of transcription factors and gene expression
↓
Upregulation of fibrinogen, platelets, CRP, cytokines, and adhesive molecules
↓
Hypercoagubality, oxidation of LDL cholesterol, and lung inflammation
↓
Generation of foam cells, plaque formation in arteries, and atherosclerosis
↓
Rupture of plaques leading to angina and myocardial infarction (CVD)

6.6.5   HEMATOLOGICAL AND IMMUNOLOGICAL CHANGES

Chronic exposures to urban air pollution in Delhi led to increase in hemoglobin, erythrocyte, total leukocyte, and platelet levels (Lahiri et al., 2006b). In addition, the citizens had several morphological abnormalities of blood cells such as abundance of "target" cells (signifying liver problem related to cholesterol metabolism), increased number of metamyelocytes and band cells (infection and inflammation), toxic granulation in neutrophil (bacterial infection) and giant platelets in circulation (platelet activation) (Table 6.9).

Delhi's air pollution was also associated with a significant reduction ($p < 0.05$) in the percentage of CD4+ T-helper cells and concomitant increase in the percentage of

TABLE 6.9

Prevalence (%) of Abnormal Cell Types in Peripheral Blood

Cell Type	Rural Control	Delhi
"Target" RBC	3.3	11.7*
Toxic granulation in neutrophil	15.8	34.5*
Metamyelocyte/band cell >20%	4.5	8.1*
Giant platelets	2.6	4.8*

Note: Results are expressed as % of individuals with abnormal cell morphology; $*p < 0.05$ compared with rural subjects in χ^2 test.

TABLE 6.10

Absolute Number of Different Lymphocyte Subtypes in Circulation of the Residents of Delhi

Cell Type	Rural Control	Delhi
CD4+/µL	939 ± 65	$795 \pm 63*$
CD8+cell/µL	562 ± 51	$719 \pm 68*$
CD19+cell/µL	456 ± 45	$329 \pm 60*$
CD16+56+ cell/µL	243 ± 44	$475 \pm 65*$

Note: Results are mean $\pm$ SD; $*p < 0.05$ compared with rural subjects.

CD8+ T-cytotoxic cells ($p < 0.05$) (Table 6.10). As a result, the CD4:CD8 cell ratio was reduced. The citizens of Delhi also demonstrated a significant fall in CD19+ B cells and rise in CD56+ natural killer (NK) cells in peripheral blood. Statistical analysis showed a negative correlation between the PM_{10} level and the number of CD4+ and CD19+ cells (ρ values −0.429 and −0.325, respectively, $p < 0.001$), while the correlation was positive for CD8+ and NK cells (ρ values 0.531 and 0.785, respectively, $p < 0.001$), suggesting that the particulate air pollutants have played a significant role in mediating alterations in lymphocyte subtypes in the residents of Delhi.

6.6.6 Metabolic Changes

6.6.6.1 Liver Function

Liver function was assessed through biochemical measurements of four serum parameters—ALT, AST, albumin and total protein in Delhi, and control groups who were never-smokers and nonusers of alcoholic drinks. Moreover, they did not have a history of jaundice in the past one year. Results showed that the concentrations of all four parameters were higher than the upper limit of the respective normal range in 4.6% of the subjects in Delhi against 1.2% in controls (Lahiri et al., 2006b). Thus, it seems that liver function was altered in 4.6% of the residents of Delhi compared with 1.2% of controls.

6.6.6.2 Kidney Function

Kidney function was simultaneously assessed through biochemical measurements of serum urea and creatinine subjects whose liver enzymes were measured. It is evident that the concentrations of serum urea and creatinine were higher in 2.6% subjects of Delhi against 1.2% of controls, suggesting a greater prevalence of kidney function impairment in Delhi (Lahiri et al., 2006b).

6.6.7 NEUROTOXICITY OF AIR POLLUTION

Besides physical health, air pollution exposure may lead to impairment of mental health, because toxic effects of PMs on the peripheral nervous system and the central nervous system (CNS) have been reported (Kilburn, 2000). We found depression in 69% of the citizens of Delhi compared with 35.4% of rural controls (Lahiri et al., 2006b). Besides, the residents of Delhi had higher prevalence of several other neurobehavioral symptoms such as burning sensation in extremities (7% versus 3% in control), anxiety (17.8% versus 6.6%), feeling of drunkenness (5.6% versus 3%), inability to concentrate (23.2% versus 9.2%), transient loss of memory (21.6% versus 9.2%), reduced sense of smell (3.4% versus 0.7%), blurred vision (4.9% versus 2.2%), and sleep disturbance (7% versus 3%).

Greater prevalence of these symptoms was accompanied by marked elevation of plasma epinephrine (E) and norepinephrine (NE) levels with concomitant fall in plasma dopamine (DA), suggesting alterations of plasma catecholamine levels. Moreover, plasma acetyl cholinesterase (AChE) concentration was reduced by 43% in the residents of Delhi ($p < 0.001$), suggesting changes in cholinergic neurotransmission. Controlling for age and passive smoking as possible confounders, logistic regression analysis showed positive association between the PM_{10} level and the concentration of E (OR = 1.33; 95% CI, 1.08–1.67) and NE in plasma (OR = 1.47; 95% CI, 1.10–2.12). Similarly, a positive association was found between *trans,trans*-muconic acid (*t,t*-MA) excretion (a measure of benzene exposure) and plasma E (OR = 1.23; 95% CI, 1.05–1.52), and NE (OR = 1.34; 95% CI, 1.12–1.71). Thus, the changes in catecholamines could be attributed, at least in part, to chronic inhalation of polluted air with respect to particulates and benzene (Lahiri and Ray, 2006a). Like the present findings, changes in the CNS have been linked to vehicular emission (Kilburn, 2000). Benzene present in polluted air produced discrete changes in NE and DA turnover in certain areas of the hypothalamus (Andersson et al., 1983). A recent study from this laboratory has demonstrated short-term memory loss in mice exposed to benzene (Banik and Lahiri, 2005).

6.6.8 CHANGE IN ANTIOXIDANT STATUS

Oxidative stress is the injury caused to cells resulting from increased formation of ROS and/or decreased antioxidant reserve. Reactive oxygen metabolites (O_2^-, $OH^\cdot$, H_2O_2) are products of aerobic metabolism and are continuously produced *in vivo* which, if not neutralized with the endogenous antioxidants, may endanger the cellular integrity. These free radicals may cause oxidative DNA damage, lipid peroxidation, and enzymatic oxidation leading to cellular injury. PM containing a variety of toxic substances on their surface induces production of ROS by AMs and neutrophils and induces inflammation. Living organisms have a well-orchestrated machinery of antioxidant defense mechanism for their survival against oxidative stress. SOD is a member of the family of metalloenzymes and is the most important antioxidant present in the body to act against superoxide (O_2^-) radicals. It accelerates the dismutation of superoxide anion into hydrogen peroxide (H_2O_2) and molecular oxygen.

$$2O_2^- + 2H^+ \xrightarrow{\text{SOD}} H_2O_2 + O_2$$

Other antioxidant enzymes are catalase and glutathione peroxidase. In addition, there are several antioxidants present in food such as vitamins E, C, and A, and flavonoids.

6.6.8.1 Depletion of SOD and Total Antioxidant Status

A significant depletion ($p < 0.05$) of erythrocyte SOD was found among the residents of Delhi (Table 6.11) as compared to rural controls (Lahiri and Ray, 2006a). The reduction of SOD was 23% for male and 52% for females. Overall, a 30% decline in the concentration of SOD in blood was found among the nonsmokers of Delhi, implying a significant deficit in antioxidant activity, especially in the city's women. The normal reference value of the total antioxidant level in blood is 1.3–1.77 mmol/L of plasma. The control male subjects of this study had a higher value than the normal limit, but the females had a lower value. Overall, the control group had a normal total antioxidant level. In contrast, the residents of Delhi had a drastically reduced level of total antioxidant. The urban mean was one-fourth of the control mean, and 65% lower than the lower limit of normal range.

Particulate air pollution could be an important contributing factor to depleted SOD and total antioxidant level, because a negative correlation (ρ values -0.257, and -0.470, respectively, $p < 0.05$) was found between the PM_{10} level and SOD and total antioxidant status. In conformity with our findings, $PM_{2.5}$ has been shown to significantly increase lipid peroxidation levels and decrease SOD, catalase, and glutathione peroxidase activities in laboratory animals. A fall in SOD level in the liver has been documented following benzene poisoning (Pan et al., 2003). In a recent study, Holovska et al. (2005) have shown a decrease in glutathione peroxidase (GSHPx) and glutathione-S-transferase (GST) activities, but not in SOD and thiobarbituric acid reactive substances (TBARS), in the liver of rats after benzene exposure.

TABLE 6.11
Concentration of Superoxide Dismutase in Erythrocytes

Parameter	Control	Delhi
Erythrocyte SOD (U/mL)	255.3 ± 38.4	$179.4 \pm 27.8*$
Total antioxidant in plasma (mmol/L)	1.93 ± 0.62	$0.46 \pm 0.23*$

Note: Results are mean ± SD; *$p < 0.05$ compared with the respective control group.

6.6.9 GENOTOXIC EFFECTS OF AIR POLLUTION

Chronic exposures to vehicular emission may cause genetic changes. Urban air contains a wide spectrum of mutagens and carcinogens such as benzene, diesel soot, heavy metals, and PAHs (Klumpp et al., 2006). The pollutants, their metabolites, or ROS generated by the pollutants interact with DNA causing strand breaks (Moller and Wallin, 1998), impairment of DNA repair system, dysregulation of the cell cycle, and induction of programmed cell death, that is, apoptosis (Eastman and Barry, 1992; Tovalin et al., 2006).

6.6.9.1 Micronucleus Formation and Other Nuclear Anomalies

MN is defined as microscopically visible, round, or oval chromatin masses in the cytoplasm next to the nucleus (Schmid, 1975). An MN consists of a part of the chromosome or chromatid or a whole chromosome that has not been incorporated in the spindle apparatus due to aberrant mitosis (Schmid, 1975). Assessment of the number of MNs is widely used to identify the genotoxic damages and its formation is considered to be a simple biomarker of mutagenic effects of environmental pollutants (Stich et al., 1982). Since the air pollutants enter the body via the nasal-oropharyngeal route, the epithelial cells of the buccal mucosa are in constant contact with foreign particles. A comparison of MN count between rural and urban people without tobacco smoking and chewing habits showed 2.3 times higher value in the latter group (3.5 versus 1.5 micronucleated cells/1000 cells, $p < 0.001$, Figure 6.10). Since MN formation is associated with chromosomal breakage, a higher frequency of chromosomal damage in buccal and airway epithelial cells that are at the direct route of exposure of inhaled pollutants can be envisaged among the citizens of Kolkata and Delhi. Besides MNs, Comet assay showed DNA damage in 33.8% lymphocytes of the citizens of Delhi and Kolkata compared with 18.5% in controls ($p < 0.05$), and the extent of damage was much more in urban subjects ($p < 0.05$). Therefore, MNs and Comet assays showed significantly increased genotoxicity among urban subjects. PM_{10} levels of these two cities were positively correlated with MN formation in buccal ($\rho = 0.40$ in Delhi and 0.43 in Kolkata, $p < 0.01$) and airway epithelial cells ($\rho = 0.44$ in Delhi and 0.47 in Kolkata, $p < 0.01$).

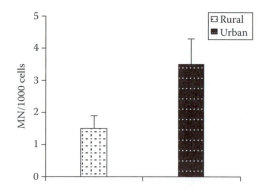

FIGURE 6.10 Micronucleated cell per 1000 exfoliated buccal epithelial cells in nonsmokers and nonchewers. Bars represent the standard deviation of mean values.

6.6.9.2   Benzene Exposure and Genotoxic Changes

We found elevated levels of urinary t,t-MA, a measure of benzene exposure, in the residents of Kolkata and Delhi. For example, mean t,t-MA levels of 218 and 326 µg/g creatinine were found in office workers and autorickshaw drivers of Delhi, respectively, compared with a t,t-MA level of 102 µg/g creatinine in rural controls. A strong correlation ($\rho = 0.77$, $p < 0.001$) was found between the benzene level in ambient air and the MN frequency in buccal and airway epithelial cells (Lahiri et al., 2006b). Likewise, benzo(a)pyrene concentration in the breathing air of Delhi showed a significant, positive correlation with MN count, but the strength of this correlation was weaker than what was elicited by benzene ($\rho = 0.33$ and 0.41 for buccal and airway cells, $p < 0.05$) (Lahiri and Ray, 2006a).

6.6.10   AIR POLLUTION AND CANCER

The incidence of cancer is on the rise among the population living in the Indian cities. Of particular importance is the fact that age-adjusted rate of lung cancer per 100,000 population is highest in Kolkata compared to all other cities studied (ICMR, 2004). Air pollution is believed to be an important contributor to the disease, since large numbers of carcinogens are present in automobile exhausts and in industrial and household emissions. The notable environmental carcinogens are benzene and benzo(a)pyrene. Benzene is a Class 1 carcinogen (confirmed as a human carcinogen) while benzo(a)pyrene and diesel exhaust particles belong to Class 2A (probably carcinogenic to humans) according to the International Agency for Research on Cancer (IARC, 1982). Diesel exhaust exposure has been linked to increased incidence of lung cancer (Garshick et al., 2004; Laden et al., 2006) while reactive benzene metabolites, especially p-benzoquinone, induce acute myeloid leukemia (Descatha et al., 2005), chronic lymphocytic leukemia (Glass et al., 2003), aplastic anemia, and myelodysplastic syndromes (Kang et al., 2005). Besides leukemia, benzene exposure is believed to be responsible for cancers of the brain, lung, paranasal cavity, and esophagus (Forni, 1996; Beach and Burstyn, 2006). In general, sustained exposures to air pollution can cause cancers of the lung (Kuper et al., 2002; Vineis and Husgafvel-Pursiainen, 2005; Chen et al., 2005), larynx (Zheng et al., 2001; Kuper et al., 2002), nasopharynx (Kuper et al., 2002; Sasco et al., 2004), esophagus (Kuper et al., 2002; Guo et al., 2004), oral cavity (Kuper et al., 2002), urinary bladder (Mastrangelo et al., 1996; Guo et al., 2004), uterine cervix, breast (Johnson, 2005; Bonner et al., 2005; Grant and Garland, 2006), and leukemia and lymphoma (Kasim et al., 2005; Fritschi et al., 2005).

6.7   CONCLUSION

The levels of particulate pollutants in ambient air of all major Indian cities have been far above the permissible standards for long. Our study has revealed that chronic exposures to these pollutants through breathing air are adversely affecting the physical as well as mental health of the citizens. The worst sufferers are the children and the elderly. While a majority of air pollution-related diseases such as respiratory infections are transient and curable, some are not. For instance, COPD is not fully irreversible, and cure for lung cancer is still elusive. Therefore, air pollution can be viewed as a serious public health concern, and concerted efforts should be made by all concerned to mitigate the problem by every conceivable way.

ACKNOWLEDGMENTS

The authors gratefully acknowledge the financial assistance received from Department of Environment, Government of West Bengal, Central Pollution Control Board, Delhi, and World Health Organization, India, for carrying out this study.

REFERENCES

Analitis, A., Katsouyanni, K., Dimakopoulou, K., et al. 2006. Short-term effects of ambient particles on cardiovascular and respiratory mortality. *Epidemiology* 17: 230–233.

Anderson, H.R., de Leon, A.P., Bland, J.M., Bower, J.S., and Strachan, D.P. 1996. Air pollution and daily mortality in London: 1987–92. *BMJ* 312: 665–669.

Andersson, K., Fuxe, K., Eneroth, P., Isaksson, O., Nyberg, F., and Roos, P. 1983. Rat growth hormone and hypothalamic catecholamine nerve terminal systems evidence for rapid and discrete reductions in dopamine and noradrenaline levels and turnover in the median eminence of the hypophysectomized male rat. *Eur J Pharmacol* 95: 271–275.

Archer, V.E. 1990. Air pollution and fatal lung disease in three Utah counties. *Arch Environ Health* 45: 325–334.

Arena, V.C., Mazumdar, S., Zborowski, J.V., et al. 2006. A retrospective investigation of PM10 in ambient air and cardiopulmonary hospital admissions in Allegheny County, Pennsylvania: 1995–2000. *J Occup Environ Med* 48: 38–47.

Asero, V., Mistretta, A., Arcidiacono, G., et al. 2005. The puzzling relationship between cigarette smoking, reduced respiratory function, and systemic inflammation. *Chest* 128: 3772–3773.

ATS (American Thoracic Society). 1995. Standardization of spirometry. *Am J Respir Crit Care Med* 152: 1107–1136.

Badami, M.G. 2005. Transport and urban air pollution in India. *Environ Manage* 36: 195–104.

Banik, S. and Lahiri, T. 2005. Decrease in brain serotonin level and short-term memory loss in mice: A preliminary study. *Environ Toxicol Pharmacol* 19: 367–370.

Basu, C., Ray, M.R., and Lahiri, T. 2001. Traffic-related air pollution in Calcutta associated with increased respiratory symptoms and impaired alveolar macrophage activity. *J Environ Poll* 8: 187–195.

Beach, J. and Burstyn, I. 2006. Cancer risk in benzene exposed workers. *Occup Environ Med* 63: 71–72.

Becker, S. 1995. Fc-mediated macrophage phagocytosis. *Methods Immunol Toxicol* 2: 27–38.

Bendahmane, D.B. 1997. Air pollution and child health: Priorities for action. Environmental Health Project, Activity. Report No. 38. US AID, Washington, DC.

Bobak, M. and Leon, D. 1992. Air pollution and infant mortality in the Czech Republic, 1986–1988. *Lancet* 340: 1010–1014.

Bonner, M.R., Han, D., Nie, J., et al. 2005. Breast cancer risk and exposure in early life to polycyclic aromatic hydrocarbons using total suspended particulates as a proxy measure. *Cancer Epidemiol Biomarkers Prev* 14: 53–60.

Brook, R.D., Brook, J.R., and Rajagopala, S. 2003. Air pollution: The 'heart' of the problem. *Curr Hypertens Rep* 5: 32–39.

Brunekreef, B. and Forsberg, B. 2005. Epidemiological evidence of effects of coarse airborne particles on health. *Eur Respir J* 26: 309–318.

Brunekreef, B., Kinney, P.L., Ware, J.H., et al. 1991. Sensitive subgroups and normal variation in pulmonary function response to air pollution episodes. *Environ Health Perspect* 90: 189–193.

Burnett, R.T., Smith-Doiron, M., Stieb, D., et al. 2001. Association between ozone and hospitalization for acute respiratory diseases in children less than 2 years of age. *Am J Epidemiol* 153: 444–452.

Cakmak, A., Ekici, A., Ekici, M., et al. 2004. Respiratory findings in gun factory workers exposed to solvents. *Respir Med* 98: 52–56.

CPCB (Central Pollution Control Board). 1997. Air quality data. Central Pollution Control Board, Delhi, India.

Chen, Y., Yang, Q., Krewski, D., Burnett, R.T., Shi, Y., and McGrail, K.M. 2005. The effect of coarse ambient particulate matter on first, second, and overall hospital admissions for respiratory disease among the elderly. *Inhal Toxicol* 17: 649–655.

Churg, A., Zay, K., Shay, S., et al. 2002. Acute cigarette smoke-induced connective tissue breakdown requires both neutrophils and macrophage metalloelastase in mice. *Am J Respir Cell Mol Biol* 27, 368–374.

Cropper, M.L., Simon, N.B., Alberini, A., and Sharma, P.K. 1997. The health effects of air pollution in Delhi, India. *Policy Research Working Paper 1860*. The World Bank Development Research Group.

CSE (Center for Science and Environment). 2002. A report on the independent inspection of fuel quality at fuel dispensing stations, oil tanks and tank lorries. Available at http:// www.cseindia.org/html/cmp/air/Fnladul.pdf.

Dassen, W., Brunekreef, B., and Hoek, G. 1986. Decline in children's pulmonary function during an air pollution episode. *J Air Pollut Control Assoc* 36: 1223–1227.

Descatha, A., Jenabian, A., Conso, F., and Ameille, J. 2005. Occupational exposures and haematological malignancies: Overview on human recent data. *Cancer Causes Control* 16: 939–953.

Dockery, D.W. 2001. Epidemiologic evidence of cardiovascular effects of particulate air pollution. *Environ Health Perspect* 109: 483–486.

Dockery, D.W., Pope, C.A., III, Xu, X., et al. 1993. An association between air pollution and mortality in six U.S. cities. *N Eng J Med* 329: 1753–1759.

Dockery, D.W., Pope, C.A., III, Kanner, R.E., Martin Villegas, G., and Schwartz, J. 1999. Daily changes in oxygen saturation and pulse rate associated with particulate air pollution and barometric pressure. *Health Effects Inst Rep* 83: 1–19.

Dockery, D.W., Ware, J.H., and Ferris, B.G., Jr. 1982. Change in pulmonary function in children associated with air pollution episodes. *J Air Pollut Control Assoc* 32: 937–942.

Donaldson, K., Stone, V., Seaton, A., and MacNee, W. 2001. Ambient particle inhalation and the cardiovascular system: Potential mechanism. *Environ Health Perspect* 109(Suppl. 4): 523–527.

Eastman, A. and Barry, M.A. 1992. The origin of DNA breaks: A consequence of DNA damage, DNA repair, or apoptosis? *Cancer Invest* 10: 229–240.

Ernst, E. and Resch, K.L. 1993. Fibrinogen as a cardiovascular risk factor: A meta-analysis and review of the literature. *Ann Intern Med* 118: 956–963.

Farris, G.M., Everitt, J.I., Irons, R.D., and Popp, J.A. 1993. Carcinogenicity of inhaled benzene in CPA mice. *Fundam Appl Toxicol* 20: 503–507.

Forni, A. 1996. Benzene-induced chromosome aberrations: A follow-up study. *Environ Health Perspect* 104: 1309–1312.

Frampton, M.W. 2001. Systemic and cardiovascular effects of airway injury and inflammation: Ultrafine particle exposure in humans. *Environ Health Perspect* 109: 529–532.

Fritschi, L., Benke, G., and Hughes, A.M., et al. 2005. Occupational exposure to pesticides and risk of non-Hodgkin's lymphoma. *Am J Epidemiol* 162: 849–857.

Frye, C., Hoelscher, B., Cyrys, J., Wjst, M., Wichmann, H.E., and Heinrich, J. 2003. Association of lung function with declining ambient air pollution. *Environ Health Perspect* 111: 383–388.

Furuyama, A., Hirano, S., Koike, E., and Kobayashi, T. 2006. Induction of oxidative stress and inhibition of plasminogen activator inhibitor-1 production in endothelial cells following exposure to organic extracts of diesel exhaust particles and urban fine particles. *Arch Toxicol* 80: 154–162.

Garshick, E., Laden, F., Hart, J.E., et al. 2004. Lung cancer in railroad workers exposed to diesel exhaust. *Environ Health Perspect* 112: 1539–1543.

Giovagnoli, M.R., Alderisio, M., Cenci, M., Nofroni, I., and Vecchione, A. 1999. Carbon and hemosiderin-laden macrophages in sputum of traffic policeman exposed to air pollution. *Arch Environ Health* 54: 284–290.

Ghio, A.J., Kim, C., and Devlin, R.B. 2000. Concentrated ambient air particles induce mild pulmonary inflammation in healthy human volunteers. *Am J Respir Crit Care Med* 162: 981–988.

Glass, D.C., Gray, C.N., Jolley, D.J., et al. 2003. Leukemia risk associated with low-level benzene exposure. *Epidemiology* 14: 569–577.

Goldberg, M.S., Burnett, R.T., Bailar, J.C., III, et al. 2001. Identification of persons with cardiorespiratory conditions who are at risk of dying from the acute effects of ambient air particles. *Environ Health Perspect* 109: 487–494.

Golshan, M., Faghihi, M., Roushan-Zamir, T., et al. 2002. Early effects of burning rice farm residues on respiratory symptoms of villagers in suburbs of Isfahan, Iran. *Int J Environ Health Res* 12: 125–131.

Gouvea, N. and Fletcher, T. 2000. Time series analysis of air pollution and mortality: Effects by cause, age and socioeconomic status. *J Epidemiol Community Health* 54: 750–755.

Grant, W.B. and Garland, C.F. 2006. The association of solar ultraviolet B (UVB) with reducing risk of cancer: Multifactorial ecologic analysis of geographic variation in age-adjusted cancer mortality rates. *Anticancer Res* 26: 2687–2699.

Greenle, R.T., Murray, T., Bolden, S., and Wingo, P.A. 2000. Cancer statistics, 2000. *CA: Cancer J Clin* 50: 7–33.

Grubb, C. 1994. *Diagnostic Cytopathology, a Text and Colour Atlas*. Churchill Livingstone, London. pp. 65–112.

Guo, H., Lee, S.C., Chan, L.Y., and Li, W.M. 2004. Risk assessment of exposure to volatile organic compounds in different indoor environments. *Environ Res* 94: 57–66.

Hahn, F.F., Newton, G.J., and Bryant, P.L. 1977. *In vitro* phagocytosis of respirable-sized monodisperse particles by alveolar macrophages. In: C.L. Sanders, R.P. Schnider, G.E. Dagle, and H.A. Ragen (eds) *Pulmonary Macrophages and Epithelial Cells*. VA: Technical Information Center, Energy research and development administration symposium series, pp. 424–435.

Harrabi, I., Rondeau, V., Dartigues, J.F., Tessier, J.F., and Filleul, L. 2006. Effects of particulate air pollution on systolic blood pressure: A population-based approach. *Environ Res* 101: 89–93.

Hasselblad, V., Kotchmar, D.J., and Eddy, D.M. 1992. Synthesis of environmental evidence: Nitrogen dioxide epidemiology studies, EPA/600/8-91/049A, U.S. Environmental Protection Agency, Research Triangle Park, NC.

Holovska, K., Jr., Sobekova, A., Holovska, K., et al. 2005. Antioxidant and detoxifying enzymes in the liver of rats after subchronic inhalation of the mixture of cyclic hydrocarbons. *Exp Toxicol Pathol* 56: 377–383.

Hong, Y.C., Leem, J.H., Ha, E.H., and Christiani, D.C. 1999. PM(10) exposure, gaseous pollutants, and daily mortality in Inchon, South Korea. *Environ Health Perspect* 107: 873–878.

Howden-Chapman, P., Saville-Smith, K., Crane, J., and Wilson, N. 2005. Risk factors for mold in housing: A national survey. *Indoor Air* 15: 469–476.

Hrudkova, M., Fiala, Z., Borska, L., Novosad, J., and Smolej, L. 2004. The effect of polycyclic aromatic hydrocarbons to bone marrow. *Acta Med. (Hradec Kralove)* 47: 75–81.

IARC. 1982. "Some industrial chemicals and dyestuffs" in Monographs on the *Evaluation of Carcinogenic Risk of Chemicals to Humans*, Vol. 29. International Agency for Research on Cancer, Lyon, pp. 345–389.

ICMR. 2004. National Cancer Registry Programme. In: A. Nandakumar, P.C. Gupta, P. Gangadharan, and R.N. Visweswara (eds) *Mapping Patterns of Cancer*, Vol. I. Indian Council of Medical Research, Bangalore, pp. 13–18.

Ibald-Mulli, A., Stieber, J., Wichmann, H.E., Koenig, W., and Peters, A. 2001. Effects of air pollution on blood pressure: A population-based approach. *Am J Public Health* 91: 571–577.

Ibald-Mulli, A., Wichmann, H.E., Kreyling, W., and Peters, A. 2002. Epidemiological evidence on health effects of ultrafine particles. *J Aerosol Med* 15: 189–201.

Jaques, P.A. and Kim, C.S. 2000. Measurement of total lung deposition of inhaled ultrafine particles in healthy men and women. *Inhal Toxicol* 12: 715–731.

Johnson, K.C. 2005. Accumulating evidence on passive and active smoking and breast cancer risk. *Int J Cancer* 117: 619–628.

Johnson, K.G., Loftsgarden, D.O., and Gideon, R.A. 1982. The effects of Mount St. Helens Volcanic Ash on the pulmonary function of 120 elementary schoolchildren. *Am Rev Respir Dis* 126: 1066–1069.

Kang, S.K., Lee, M.Y., Kim, T.K., Lee, J.O., and Ahn, Y.S. 2005. Occupational exposure to benzene in South Korea. *Chem Biol Interact* 153: 65–74.

Kasim, K., Levallois, P., Abdous, B., Auqe, R.P., and Johnson, K.C. 2005. Lifestyle factors and the risk of adult leukemia in cancer. *Cancer Causes Control* 16: 489–500.

Khalili, B., Montanaro, M.T., and Bardana, E.J., Jr. 2005. Inhalational mold toxicity: Fact or fiction? A clinical review of 50 cases. *Ann Allergy Asthma Immunol* 95: 239–246.

Kilburn, K.H. 2000. Effects of diesel exhaust on neurobehavioral and pulmonary function. *Arch Environ Health* 55: 11–17.

Klumpp, A., Ansel, W., Klumpp, G., et al. 2006. Tradescantia micronucleus test indicates genotoxic potential of traffic emissions in European cities. *Environ Pollut* 139: 515–522.

Kuper, H., Boffetta, P., and Adami, H.O. 2002. Tobacco use and cancer causation: Association by tumour type. *J Intern Med* 252: 206–224.

Laden, F., Hart, J.E., Eschenroeder, A., Smith, T.J., and Garshick, E. 2006. Historical estimation of diesel exhaust exposure in a cohort study of U.S. railroad workers and lung cancer. *Cancer Causes Control* 17: 911–919.

Lahiri, T. and Ray, M.R. 2006a. Assessment of air pollution related respiratory problem in children of Delhi. Report submitted to Central Pollution Control Board, Delhi.

Lahiri, T., Ray, M.R., and Lahiri, P. 2006b. Health effects of air pollution in Delhi. Report submitted to Central Pollution Control Board, Delhi.

Lahiri, T., Ray, M.R., Mukherjee, S., Basu, C., and Lahiri, P. 2000a. Marked increase in sputum alveolar macrophages in residents of Calcutta: Possible exposure effect of severe air pollution. *Curr Sci* 78: 399–304.

Lahiri, T., Roy, S., Basu, C., Ganguly, S., Ray, M.R., and Lahiri, P. 2000b. Air pollution in Calcutta elicits adverse pulmonary reaction in children. *Ind J Med Res* 112: 21–26.

Laskin, D.L. and Pendino, K.J. 1995. Macrophages and inflammatory mediators in tissue injury. *Annu Rev Pharmacol Toxicol* 35: 655–677.

Lebowitz, M.D., Holberg, C.J., Boyer, B., and Hayes, C. 1985. Respiratory symptoms and peak flow associated with indoor and outdoor air pollutants in the southwest. *J Air Pollut Control Assoc* 35: 1154–1158.

Lin, Y.S. and Kou, Y.R. 2000. Acute neurogenic airway plasma exudation and edema induced by inhaled woodsmoke in guinea pigs: Role of tachykinins and hydroxyl radical. *Eur J Pharmacol* 394: 139–148.

Lind, P., Hedblad, B., Stavenow, L., Janzon, L., Eriksson, K.F., and Lindgarde, F. 2001. Influence of plasma fibrinogen levels on the incidence of myocardial infarction and death is modified by other inflammation-sensitive proteins: A long-term cohort study. *Arterioscler Thromb Vasc Biol* 21: 452–458.

Lipfert, F.W. 1994. *Air Pollution and Community Health: A Critical Review and Data Sourcebook.* Van Nostrand Reinhold, New York, p. 437.

Logan, W.P.D. 1952. Mortality in the London fog incident. *Lancet* 264: 336–338.

Lundborg, M., Johard, U., Lastbom, L., Gerde, P., and Camner, P. 2001. Human alveolar macrophage phagocytic function is impaired by aggregates of ultrafine carbon particles. *Environ Res* 86: 244–253.

Maheswaran, R., Haining, R.P., Brindley, P., et al. 2005. Outdoor air pollution, mortality, and hospital admissions from coronary heart disease in Sheffield, UK: A small-area level ecological study. *Eur Heart J* 26, 2543–2549.

Mar, T.F., Koenig, J.Q., Jansen, K., et al. 2005. Fine particulate air pollution and cardiorespiratory effects in the elderly. *Epidemiology* 16: 681–687.

Mastrangelo, G., Fadda, E., and Marzia, V. 1996. Polycyclic aromatic hydrocarbons and cancer in man. *Environ Health Perspect* 104: 1166–1170.

Mills, N.L., Amin, N., Robinson, S.D. et al. 2005. Do inhaled carbon nanoparticles translocate directly into the circulation in humans? *Am J Respir Crit Care Med* 173: 426–431.

Moller, P. and Wallin, H. 1998. Adduct formation, mutagenesis and nucleotide excision repair of DNA damage produced by reactive oxygen species and lipid peroxidation products. *Mutat Res* 410: 271–290.

Moller, W., Hofer, T., Ziesenis, A., Karg, E., and Heyder, J. 2002. Ultrafine particles cause cytoskeletal dysfunctions in macrophages. *Toxicol Appl Pharmacol* 182: 197–207.

Morgan, G., Corbett, S., Wlodarczyk, J., and Lewis, P. 1998. Air pollution and daily mortality in Sydney, Australia, 1989 through 1993. *Am J Public Health* 88: 759–764.

Moolgavkar, S. 2000. Air pollution and daily mortality in three US counties. *Environ Health Perspect* 108: 777–784.

Murray, C.S., Poletti, G., Kebadze, T., et al. 2006. A study of modifiable risk factors for asthma exacerbations: Virus infection and allergen exposure increase the risk of asthma hospitalization in children. *Thorax* 61: 376–382.

Oberdorster, G. 2000. Toxicology of ultrafine particles; *in vivo* studies. *Phil Trans R Soc Lond A* 358: 2719–2740.

Oberdorster, G., Ferin, J., and Morrow, P.E. 1992. Volumetric loading of alveolar macrophages (AM): A possible basis for diminished AM-mediated particle clearance. *Exp Lung Res* 18: 87–104.

Ochs, M., Nyengaard, J.R., Jung, A., et al. 2004. The number of alveoli in the human lung. *Am J Respir Crit Care Med* 169: 120–124.

Onbasi, K., Sin, A.Z., Doganavsargil, B., Onder, G.F., Bor, S., and Sebik, F. 2005. Eosinophil infiltration of the oesophageal mucosa in patients with pollen allergy during the season. *Clin Exp Allergy* 35: 1423–1431.

Ostro, B.D., Lipsett, M.J., Wiener, M.B., and Selner, J.C. 1991. Asthmatic responses to airborne acid aerosols. *Am J Public Health* 81: 694–602.

Pan, H.Z., Na, L.X., and Tao, L. 2003. DNA damage and changes of antioxidative enzymes in chronic benzene poisoning mice. *Zhonghua Lao Dong Wei Sheng Zhi Ye Bing Za Zhi* 21: 423–425.

Pauwels, R.A., Buist, A.S., Calverely, P.M.A., Jenkins, C.R., and Hurd, S.S. 2001. Global strategy for the diagnosis, management, and prevention of chronic obstructive pulmonary disease. *Am J Respir Crit Care Med* 163: 1256–1276.

Perez-Arellano, J.L., Losa, Garcia, J.E., Garcia Macias, M.C., et al. 1992. Hemosiderin-laden macrophages in bronchoalveolar lavage fluid. *Acta Cytol* 36: 26–30.

Peters, A., Doring, A., Whichmann, H.E., and Koenig, W. 1997. Increased plasma viscosity during an air pollution episode: A link to mortality? *Lancet* 349: 1582–1587.

Peters, A., Dockery, D.W., Muller, J.E., and Mittleman, M.A. 2001. Increased particulate air pollution and the triggering of myocardial infarction. *Circulation* 103: 2810–2815.

Peters, A., Liu, E., Verrier, R.L., et al. 2000. Air pollution and incidence of cardiac arrhythmia. *Epidemiology* 11: 11–17.

Peters, A., von Klot, S., Heier, M., et al. 2004. Exposure to traffic and the onset of myocardial infarction. *N Engl J Med* 351: 1721–1730.

Plamenac, P., Nikulin, A., and Pikula, B. 1973. Cytologic changes of the respiratory tract in young adults as a consequence of high levels of air pollution exposure. *Acta Cytol* 17: 241–244.

Plamenac, P., Nikulin, A., Pikula, B., and Markovic, Z. 1978. Cytologic changes in the respiratory tract in children smokers. *Acta Cytol* 23: 389–391.

Ponka, A. 1991. Asthma and low level air pollution in Helsinki. *Arch Environ Health* 46: 262–270.

Pope, C.A., III and Dockery, D.W. 1992. Acute health effects of PM_{10} pollution on symptomatic and asymptomatic children. *Am Rev Respir Dis* 145: 1123–1128.

Pope, C.A., III, Barnett, R.T., Thun, M.J., et al. 2002. Lung cancer, cardiopulmonary mortality, and long term exposure to fine particulate air pollution. *JAMA* 287: 1132–1141.

Ridker, P.M., Hennekens, C.H., Buring, J.E., and Rifai, N. 2000. C-reactive protein and other markers of inflammation in the prediction of cardiovascular disease in women. *N Engl J Med* 342: 836–843.

Riedl, M. and Diaz-Sanchez, D. 2005. Biology of diesel exhaust effects on respiratory function. *J Allergy Clin Immunol* 115: 221–228.

Roby, T.J., Swan, G.E., Sorensen, K.W., Hubbard, G.A., and Schumann, G.B. 1990. Discriminant analysis of lower respiratory tract components associated with cigarette smoking, based on quantitative sputum cytology. *Acta Cytol* 34: 147–154.

Routledge, H.C., Manney, S., Harrison, R.M., Ayres, J.G., and Townend, J.N. 2006. Effect of inhaled sulphur dioxide and carbon particles on heart rate variability and markers of inflammation and coagulation in human subjects. *Heart* 92: 220–227.

Roy, S., Ray, M.R., Basu, C., Lahiri, P., and Lahiri, T. 2001. Abundance of siderophages in sputum. Indicator of an adverse lung reaction to air pollution. *Acta Cytol* 45: 958–964.

Saccomanno, G., Saunders, R.P., Klein, M.J., Archer, V.E., and Brennan, L. 1970. Cytology of the lung in reference to irritant, individual sensitivity and healing. *Acta Cytol* 14: 377–381.

Salam, M.T., Millstein, J., Li, Y.F., Lurmann, F.W., Margolis, H.G., and Gilliland, F.D. 2005. Birth outcomes and prenatal exposure to ozone, carbon monoxide, and particulate matter: Results from the Children's Health Study. *Environ Health Perspect* 113: 1638–1644.

Salvi, S., Blomberg, A., Rudell, B., et al. 1999. Acute inflammatory responses in the airways and peripheral blood after short-term exposure to diesel exposure in healthy human volunteers. *Am J Respir Crit Care Med* 159: 702–709.

Samet, J.M., Domonici, F., Frank, C., Curriero, F.C., Coursac, I., and Zeger, S.L. 2000. Fine particulate air pollution and mortality in 20 U.S. cities, 1987–1994. *N Engl J of Med* 343: 1742–1749.

Samet, J.M., Speizer, F.E., and Bishop, Y. 1981. The relationship between air pollution and emergency room visits in an industrial community. *J Air Pollut Control Assoc* 31: 236–240.

Sasco, A.J., Secretan, M.B., and Straif, K. 2004. Tobacco smoking and cancer: A brief review of recent epidemiological evidence. *Lung Cancer* 45: S3–S9.

Schmid, W. 1975. The micronucleus test. *Mutat Res* 31: 9–15.

Schwartz, J., Ballester, F., Saez, M., et al. 2001. The concentration–response relation between air pollution and daily deaths. *Environ Health Perspect* 109: 1001–1006.

Schwartz, J., Laden, F., and Zanobetti, A. 2002. The concentration–response relation between PM(2.5) and daily deaths. *Environ Health Perspect* 110: 1025–1029.

Schwartz, J., Spix, C., and Touloumi, G. 1996. Methodological issues in studies of air pollution and daily counts of deaths or hospital admissions. *J Epidemiol Comm Health* 50: 3–11.

Schwartz, J.S. 1999. Economics and cost-effectiveness in evaluating the value of cardiovascular therapies. Comparative economic data regarding lipid-lowering drugs. *Am Heart J* 137: S97–S104.

Seaton, A., Soutar, A., Crawford, V., et al. 1999. Particulate air pollution and the blood. *Thorax* 54: 1027–1032.

Shapiro, S.D., Goldstein, N.M., Houghton, A.M., Kobayashi, D.K., Kelley, D., and Belaaouag, A. 2003. Neutrophil elastase contributes to cigarette smoke-induced emphysema in mice. *Am J Pathol* 163: 2329–2335.

Shy, C.M., Kleinbaum, D.G., and Morgenstern, H. 1978. The effect of misclassification of exposure status in epidemiological studies of air pollution health effects. *Bull N Y Acad Med* 54: 1155–1165.

Singh, J.P., Larson, M.G., O'Donnell, C.J., et al. 2000. Association of hyperglycemia with reduced heart rate variability (The Framingham Heart Study). *Am J Cardiol* 86: 309–312.

Stich, H.F., Stich, W., and Parida, B.B. 1982. Elevated frequency of micronucleated cells in the buccal mucosa of individuals at high risk for oral cancer: Betel quid chewers. *Cancer Lett* 17: 125–134.

Takizawa, H., Abe, S., Ohtoshi, T., et al. 2000. Diesel exhaust particles up-regulate expression of intercellular adhesion molecules-1 (ICAM-1) in human bronchial epithelial cells. *Clin Exp Immunol* 120: 356–362.

Thurston, G.D. 1996. A critical review of PM10-mortality time-series studies. *J Expos Anal Environ Epidemiol* 6: 3–21.

Tobin, M.J., Chada, T.S., Jenouri, J., Birch, S.J., Gazeroglu, H.B., and Sackner, M.A. 1983. Breathing patterns. 1. Normal subjects. *Chest* 84: 202–205.

Tovalin, H., Valverde, M., Morandi, M.T., Blanco, S., Whitehead, L., and Rojas, E. 2006. DNA damage in outdoor workers occupationally exposed to environmental air pollutants. *Occup Environ Med* 63: 230–236.

UNEP (United Nations Environment Programme). 2002 Environmental threats to children. In: *Children in the New Millennium*. United Nations Environment Programme, United Nations Children's Fund and World Health Organization, Geneva, pp. 43–86.

USEPA. 1996. *Air Quality Criteria for Particulate Matter*, Vol. III. EPA/600/P-95/001cF, Washington, DC.

Vedal, S., Schenker, M.B., Munoz, A., Samet, J.M., Batterman, S., and Speizer, F.E. 1987. Daily air pollution effects on children respiratory symptoms and peak expiratory flow. *Am J Public Health* 77: 94–98.

Viegi, G., Pedreschi, M., Baldacci, S., et al. 1999. Prevalence rates of respiratory symptoms and diseases in general population samples of North and Central Italy. *Int J Tuberc Lung Dis* 3: 1034–1042.

Vieg, G., Pedreschi, M., Pistelli, F., et al. 2000. Prevalence of airways obstruction in a general population: European Respiratory Society vs American Thoracic Society definition. *Chest* 117: 339S–345S.

Vineis, P. and Husgafvel-Pursiainen, K. 2005. Air pollution and cancer: biomarker studies in human populations. *Carcinogenesis* 26: 1846–1855.

Vinzents, P.S., Moller, P., Sorensen, M., et al. 2005. Personal exposure to ultrafine particles and oxidative DNA damage. *Environ Health Perspect* 113: 1485–1490.

von Klot, S., Peters, A., Aalto, P., et al. 2005. Health Effects of Particles on Susceptible Subpopulations (HEAPSS) Study Group. Ambient air pollution is associated with increased risk of hospital cardiac readmissions of myocardial infarction survivors in five European cities. *Circulation* 112: 3073–3079.

Wallace, L.A. 1989. Major sources of benzene exposure. *Environ. Health Perspect* 82: 165–169.

Wallace, L.A. 1984. Organic chemicals in indoor air: A review of human exposure studies and indoor air quality studies. In: R.B. Gammage, S.V. Kaye, and V.A. Jacobs (eds) *Indoor Air and Human Health, Part V*. Lewis Publishers, Chelsea, MI, pp. 361–378.

WHO (World Health Organization). 2005. Air quality guidelines global update 2005. Report on a Working Meeting, Bonn, Germany, October 18–20. Wichmann, H.E., Hubner, H.R., Malin, E., et al. 1989. The significance of health risks caused by outdoor pollution, demonstrated by cross-sectional studies of pseudocroup in Baden-Wurttemberg. *Offentl Gesundheitswes* 51: 414–420.

Ye, F., Piver, W.T., Ando, M., and Portier, J. 2001. Effects of temperature and air pollutants on cardiovascular and respiratory diseases for males and females older than 65 years of age in Tokyo, July and August 1980–1995. *Environ Health Perspect* 109, 355–359.

Zanobetti, A. and Schwartz, J. 2001. Are diabetics more susceptible to the health effects of airborne particles? *Am J Respir Crit Care Med* 164: 831–833.

Zareba, W., Nomura, A., and Couderc, J.P. 2001. Cardiovascular effects of air pollution: What to measure in ECG? *Environ Health Perspect* 109: 533–538.

Zeman, K.L. and Bennett, W.D. 2006. Growth of the small airways and alveoli from childhood to the adult lung measured by aerosol-derived airway morphometry. *J Appl Physiol* 100: 965–971.

Zheng, Z., Park, J.Y., Guillemette, C., Schantz, S.P., and Lazarus, P. 2001. Tobacco carcinogen-detoxifying enzyme UGT1A7 and its association with orolaryngeal cancer risk. *J Natl Cancer Inst* 93: 1411–1418.

7 Air Pollutants Exposure and Health Effects during the MILAGRO– MCMA2006 Campaign

Horacio Tovalin, Olf Herbarth, Martha P. Sierra-Vargas, Bo Strandberg, Salvador Blanco, Libia Vega, Constantinos Sioutas, Juan J. Hicks, Rubén Marroquín, Gustavo Acosta, Marco Guarneros, Vicente Hernández, Elizabeth Estrada-Muñiz, Ivonne M. Olivares, Dora A. Pérez, Yessica Torres-Ramos, Frank Ulrich, Robyn Hudson, Ernesto Reyes, Tracy Rodríguez, Guillermo Elizondo, and Eliseo Cantellano

CONTENTS

7.1 INTRODUCTION

Mexico City Metropolitan Area (MCMA) is one of the most densely populated cities in the world, with 18 million people according to the 2000 census (INEGI, 2001). MCMA is at an altitude of 2240 m above sea level, surrounded by mountains on the south, west, and east. Air pollution driven by local emissions can affect large areas within closed valleys, where restricted air movement concentrates the pollutants. Due to the altitude and latitude, MCMA receives intense solar radiation, a condition that added to a less efficient combustion promotes the photochemical formation of secondary pollutants such as ozone and particulate matter (PM) (Molina and Molina, 2002).

In Mexico City, the concentrations of some of the criteria air pollutants have declined during the last decade. However, ozone and PM concentration remain above the Mexican standard for many days in certain city zones. Forty percent of the ozone (O_3) measurements in 2006 were above the 0.11 ppm 1-hour standard and 50% were above the 0.08 ppm 8-hour standard. Twenty percent of the time, PM smaller than 10 μm (PM_{10}) concentrations was above the 120 μg/m³ 24-hour standard and more than 4 million of children lived in areas where PM smaller than 2.5 μm ($PM_{2.5}$) concentrations was above the 15 μg/m³ annual standard (SMA, 2007). Furthermore, some substances emitted by mobile sources are recognized as carcinogenic or genotoxic, such as benzene, 1,3-butadiene, formaldehyde, cadmium, and others (IARC, 2009); they are not monitored regularly, but have high concentrations outdoors and indoors (Serrano-Trespalacios et al., 2004). Observations from the intensive MCMA-2003 Campaign showed that MCMA motor vehicles produce high levels of primary PM, particle-bound polycyclic aromatic hydrocarbons (PAHs), and a wide range of air toxics, including formaldehyde, acetaldehyde, benzene, toluene, and xylenes (Molina et al., 2007).

7.2 AIR POLLUTION-RELATED HEALTH EFFECTS IN MCMA

Cohen et al. (2005) reported that air pollution in developed and developing countries causes about 3% of mortality from cardiopulmonary disease, about 5% of mortality from cancer of the trachea, bronchus, and lung, and about 1% of mortality worldwide from acute respiratory infections in children younger than 5 years. These effects represent about 0.8 million (1.2%) premature deaths and 6.4 million (0.5%) years of life lost. Other epidemiological studies have demonstrated health impacts for total suspended matter (TSP), PM_{10} and $PM_{2.5}$, the latter being the most suitable measure for health effects (Téllez-Rojo et al., 2000). Furthermore, newer epidemiological studies indicate the importance of smaller than 1-μm fine particles, including so-called ultrafine (UF) particles (<100 nm). Fine and UF particle masses are insignificant compared to the total particle mass, but their number density is considerable, as are their reactivity and toxic effects. Therefore, the particle number density (number/cm³) is recommended in order to complete the characterization of UF particles (Téllez-Rojo et al., 2000).

A variety of air pollution-related health outcomes have been reported in Mexico City. In the respiratory tract, O_3 exposure can cause inflammation and a higher susceptibility to infections (Castillejos et al., 1995), worsening of asthma (English, 1994; Romieu et al., 1996), decrements in peak expiratory flow (PEF) in association with PM exposures (Gold et al., 1999), a deficiency in FVC and FEV1 growth among schoolchildren living in Mexico City (Rojas-Martinez et al., 2007), and increased mortality (Borja-Aburto et al., 1997). Studies of the relationship between PM exposure and mortality have also shown a significant increase in morbidity and mortality related to chronic exposure to photochemical pollutants and PM (Bobak and Leon, 1999; Loomis et al., 1999). Among the different damage mechanisms involved, studies in Mexico City have described E-selectin expression by PM_{10} insoluble components and endotoxins that can induce an increase in monocytic cell adhesion (Alfaro-Moreno et al., 2007). Other studies have explored the impact of air pollution in the olfactory function that appears to be substantial in young and middle-aged Mexico City residents (Hudson et al., 2006).

Moreover, exposure to Mexico City air pollution has also been associated with cell culture genetic damage. For example, PM extracts induce DNA damage (Gutiérrez-Castillo et al., 2006); similar results have been reported in human samples (Calderon-Garcidueñas et al., 1996; Valverde et al., 1997; Rojas et al., 2000).

Many of these air pollution-related health effects have, as a common damage mechanism, the production or worsening of oxidative stress. Oxidative stress represents a serious imbalance between production of reactive species (oxidants) and antioxidant defense in favor of the former, leading to potential cell damage (Halliwell and Whiteman, 2004). Oxidative stress has been implicated in the pathogenesis of several chronic diseases such as asthma, chronic obstructive pulmonary disease, arteriosclerosis, myocardial acute infarction, and diabetes mellitus (Hicks et al., 1996; Medina-Navarro et al., 1997; Sierra–Vargas et al., 2004). To counteract the severe damage caused by oxidative stress, the organism has a battery of enzymatic antioxidants such as superoxide dismutase (SOD), catalase (CAT), glutathione peroxidase (GPx), and paraoxonase (PON1), and nonenzymatic antioxidants such as β-carotene, vitamin E, vitamin C, and sulfhydryl groups. SOD is a specific superoxide anion scavenger and one of the principal antioxidant enzymes in mammalian cells while GPx and CAT remove the hydrogen peroxide (Chang et al., 2001).

Other oxidative stress-related markers involve the presence of injury in polyunsaturated free fatty acids, proteins, and DNA and the activity level of antioxidant enzymes. Myeloperoxidase (MPO) activity is a marker of cardiovascular risk (Aldus et al., 2003), nitroblue tetrazolium dye reduction (NTE) evaluates protein damage, and thiobarbituric acid reactive substances (TBARS) and PON1 activity are used to evaluate polyunsaturated free fatty acid damage (Medina-Navarro et al., 1997).

Other related oxidative stress markers are C-reactive protein (CRP), ceruloplasmin (CP), and nitric oxide (NO). CRP is an inflammation marker that increases in response to acute injury, infection, and other inflammatory stimuli. CRP is also a predictor of cardiovascular disease production as well as of diabetes mellitus and other metabolic disorders. CP, a serum α2-glycoprotein that carries more than 95% of the copper present in plasma, is an acute-phase protein released in response to infection and inflammation. It plays an important anti-inflammatory role as a superoxide anion scavenger. NO release is associated with a number of pathological processes such as septic shock, inflammation, diabetes, and so on (Cousins, 1985; Koening et al., 1999).

Air pollutants can also induce an immunological response in exposed individuals, expressed by cytokine differential response. Interleukin IL-6 initiates inflammatory reactions, protecting against viral and bacterial infections (Shi et al., 2006). IL-2 is the major autocrine growth factor for T lymphocytes, a determinant of the magnitude of T cell-dependent immune responses. It stimulates the synthesis of IFN-γ and lymph toxin, the growth of natural killer (NK) cells, and the production of lymphokine-activated killer (LAK) cells (Kidd, 2003).

IL-4 is a regulator of allergic reactions caused by activated mast cells and basophils, and CD8+ T cells, and is required for the production of IgE. IL-4 inhibits macrophage activation and blocks the effects of IFN-γ, including increased production of cytokines such as IL-1, NO, and prostaglandins (Lucey et al., 1996). IFN-γ induces an antiviral state, is antiproliferative, and is a potent activator of mononuclear phagocytes to kill phagocytosed microbes. The two major activities of IL-10 are the inhibition of cytokine (i.e., TNF, IL-1, chemokine, and IL-12) production by macrophages and the inhibition of the accessory functions of macrophages in T cell activation and stimulation on B cells (O'Garra and Vieira, 2007).

IL-12 presence is related to chronic inflammation; it is produced by T and B lymphocytes, NK, and monocytes and is responsible for the differentiation pathway to Th1-type T cells and a precursor of the inflammatory response in cellular immunity (Trinchieri, 2003). The macrophage colony stimulating factor (GM-CSF) can enhance both cellular and humoral immune responses; it acts directly on T and B lymphocytes to promote their differentiation (Gattoni et al., 2006).

The effects of pollutants on immune system responses are not well understood. For instance, a recent report did not indicate that differences in IL-6, TNF-α, IFN-γ, and IL-1β levels are related to different pollutant exposure levels (Moreno-Ramírez et al., 2006); another study reported that the percent of NK cells and secretion of IL-2 were increased in women exposed to urban pollutants (Ciarrocca et al., 2006); and a third study found that proinflammatory cytokines and suppressive cytokines were produced in significant amounts for each 100 ppb increment of sulfur dioxide concentration (Vega et al., 2007).

7.3   HIGH-RISK POPULATIONS

Some MCMA population subgroups are more susceptible to chronic exposure to air pollutants and have a higher risk than others; these include children, elderly people, and those who work long hours in outdoor locations (Forastiere et al., 2007).

In the MCMA, 2.7 million people are younger than 18 years, 50% of whom live below the poverty level (INEGI, 2005). They are simultaneously exposed to poverty, deficient nutrition, and high concentrations of environmental pollutants (CCA, 2003). Some children's activities make them more susceptible to pollutant insult; children breathe more air, drink more liquids, and eat more food per unit of body weight because they are more active than adults. MCMA children physically engage in many hours of outdoor activities throughout the year, due to mild weather (Landrigan and Anjali Garg, 2002). Therefore, children inhale higher amounts of air pollutants, which could impair new alveoli (Etzel and Balk, 1999). Toxic pollutant effects can also permanently damage children's development [e.g., PAHs, carbon monoxide (CO), and lead] (Bearer, 1995). These conditions can retard the child's development and increase the risk for the later development of different chronic diseases (Chaudhuri, 1998).

Exposure to traffic (Herbarth et al., 1992) and indoor exposures have to be considered when children are evaluated for pollution-related health problems. In all cases, dust and volatile organic compounds (VOCs) (Herbarth et al., 2001; Wolkoff and Nielsen, 2001) may play an important role in the observed health outcomes. Important indoor VOC sources are smoking and emissions from building materials and furnishings. These exposures have been definitely found to affect inflammatory processes as well as allergic disorders in children (Wieslander et al., 1997; Diez et al., 2000). Romieu et al. (2002) reported that a 20 $\mu g/m^3$ increase in PM_{10} was related to an 8% increase in lower respiratory illnesses among children on the same day, and a 10 $\mu g/m^3$ weekly mean of $PM_{2.5}$ was related to a 21 $\mu g/m^3$ increase in lower respiratory tract infection. Recently, Calderon-Garciduenas et al. (2007) found that MCMA children have elevated plasma endothelin-1, which is associated with an increase in pulmonary arterial pressure, a condition that may influence their health in the future.

Elderly populations also have susceptibility to air pollutants due to a decline of their organic defenses, even though they have fewer outdoor activities. In highly polluted areas of the MCMA, air pollutants, mainly PM, can reach them indoors by infiltration, producing a significant exposure risk (Pope and Dockery, 1992; Tovalin-Ahumada et al., 2007a). Oxidative stress due to poor nutrition and genetic factors, among others, may increase their susceptibility to pollutant exposures due to differential antioxidant capacity, possibly promoting the progressive development of neurodegenerative diseases (Sanchez-Rodriguez et al., 2005). In addition, several reports suggest that ambient $PM_{2.5}$ and O_3 can reduce the high-frequency component of heart rate variability in both elderly subjects (Holguín et al., 2003) and young healthy adults (Vallejo et al., 2006). This may explain, in part, results such as those reported by Borja-Aburto et al. (1997), who found an excess MCMA mortality rate of 6% per 100 $\mu g/m^3$ increment of total suspended particles and an excess mortality rate of 3% per 100 ppb increment in 1 hour maximum O_3.

In less developed megacities, such as Mexico City, a significant proportion of the labor force works in informal markets, where many of them spend long hours outdoors. Many workers in the service and transport sectors also experience similar conditions. In Mexico City, about 200,000 people work as taxi and bus drivers and more than 100,000 work as street vendors (SETRAVI, 2007); they have direct exposures to mobile source emissions on high-traffic-density streets (Ortiz et al., 2002), producing significant exposure to air pollutants. Outdoor workers in Mexico City have higher exposures to $PM_{2.5}$, above the 65 $\mu g/m^3$ Mexican standard, and two or more times higher exposures to ozone and benzene, toluene, methyl *tert*-butyl ether, and *n*-pentane than indoor workers (Tovalin-Ahumada and Whitehead, 2007b). In Mexico City, a survey among outdoor workers found a relationship between their exposure to select VOCs, O_3, and $PM_{2.5}$ and the presence of severe DNA damage (Tovalin et al., 2006).

Occupational activities and in-home environments cause many individuals to be exposed to PM of vegetative, animal, or microbial origin. Inhalation of endotoxins, a major component of the outer membrane of Gram-negative bacteria, carried by airborne PM, leads to known adverse effects on human health. Endotoxins may initiate a cascade of biochemical and cellular events giving rise to multiple dysfunctions associated with sick building syndrome or to acute chronic lung diseases (Jacobs,

1997; Rylander, 1999). On the other hand, fungal spores, mycelium, and metabolic products are all allergenic substances that increase the incidence of asthma and rhinitis in the exposed population (Beaumont, 1988).

7.4   THE MCMA-2006 CAMPAIGN

Megacities are major sources of air pollutants that influence the air quality at a regional and global scale, and may also affect the health conditions of the large populations residing in them. Some studies indicate that policies directed at reducing air pollution would prevent numerous adverse health outcomes, including tens of thousands of deaths, children's medical visits, cases of chronic bronchitis, and millions of asthma attacks: events with an economic value of billions of dollars (Evans et al., 2002; Molina and Molina, 2004; Bell et al., 2006).

In 2006, a multinational team of investigators performed the Megacity Initiative: Local and Global Research Observations (MILAGRO) Campaign in and around Mexico City. This campaign had four components, involving a range of geographic scales. The MCMA-2006 Campaign examined pollutant emissions and boundary layer concentrations within the Mexico City Basin. The Megacity Aerosol Experiment in Mexico City (MAX-Mex) campaign investigated the evolution of aerosols and gas–aerosol interactions in the MCMA plume. The Megacity Impacts on Regional and Global Environments Mexico City (MIRAGE-Mex) campaign studied the evolution of the Mexico City plume on larger regional scales. The INTEX-B campaign observed the evolution and transport of pollution on continental scales (Molina et al., 2008).

MCMA-2006 featured a combination of a central fixed site, mobile laboratories, fixed mobile units, and personal exposure monitoring deployed throughout the metropolitan area to measure aerosol PM, VOCs, O_3, and other pollutant gas concentrations as well as meteorological and solar radiation parameters. In addition, this project studied children's and their parents' exposure to air pollutants and health conditions in three different downward sites intercepting the Mexico City air pollutant plume toward the northern neighboring states of Mexico and Hidalgo. The first receptor site was urban, Iztapalapa, in SE downtown Mexico City (T0)*; the second was a suburban site, Tecamac, in the State of Mexico, north of Mexico City (T1); and the third was a semirural site, San Pedro, in Hidalgo State, NE to Mexico City (T2) (Figure 7.1). The project also evaluated correlations among these pollutants and the oxidative stress status and health conditions in sampled site populations.

The volunteers investigated were children 9–12 years old and their parents; the children were recruited from elementary schools in the three areas. Before the campaign started, printed and verbal information about the project was distributed to the students and their parents asking for their participation and parents' authorization. Teachers were provided with pertinent information about the project and asked to collaborate. After the project was completed, several information meetings were organized to present the results to the participants.

---

* Due to logistic problems, the T0 site of this subproject was not the same as the T0 site referred to in the MCMA-2006 study, which was located at Instituto Mexicano del Petróleo in NW of MCMA.

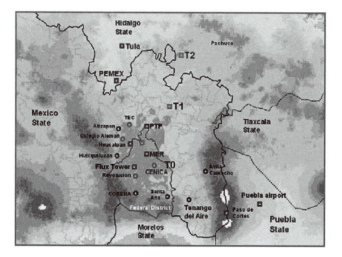

FIGURE 7.1 Monitoring sites.

The microenvironmental and personal exposure patterns to O_3, VOCs, $PM_{2.5}$, UF particles, CO, and endotoxins were characterized at these sites. The air pollution-related respiratory and olfactory health outcomes, oxidative stress status, and cytokine expression levels of participating children and their parents were studied. In this section, some general results of the project are presented.

7.4.1 OZONE

During the campaign, O_3 outdoor levels at homes and schools were three to six times higher than personal exposure and indoor concentrations. O_3 concentrations were higher at the semirural and urban sites than at the suburban site. The highest mean home outdoor concentrations were detected at T2 and T0 (189 and 166 ppb) than at T1 (80 ppb); many measurements were above the 110 ppb 1-hour Mexican standard. Indoors, T2 and T0 concentrations were much lower than the corresponding outdoor levels (18 and 28 ppb). Due to outdoor activities, personal exposures for both children and adults tended to be higher than indoor ones but smaller than outdoor ones. Adult ozone personal exposures at T0 and T2 (33 ppb) were similar. These exposures were influenced by local production, plume oxidants transported to the site, transportation time exposure, and work areas. For children, personal exposure was higher at T0 (31 ppb), probably impacted by local O_3 levels at the site and time outdoors (Figure 7.2). This pollutant level would still represent a significant impact on the health of the local population.

7.4.2 PM$_{2.5}$

A different pattern of home $PM_{2.5}$ concentrations both indoors and outdoors was observed for the three sites. The mean $PM_{2.5}$ indoor versus outdoor concentrations at T2 (34 and 25 $\mu g/m^3$) and T1 (38 and 30 $\mu g/m^3$) showed higher values indoors, while

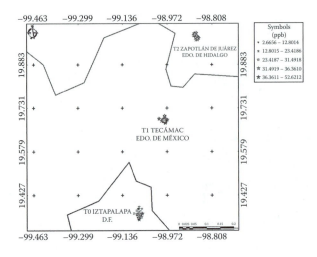

FIGURE 7.2 Ozone levels at homes outdoors, and children and parent personal exposure by site (ppb).

at T0 (43 and 42 μg/m³) these concentrations were similar. The PM$_{2.5}$ concentration pattern at T0 may be due to this area's higher PM concentration, which increases the home infiltration rates. There are no other significant differences between the sites' building characteristics or indoor combustion sources.

Parents' PM$_{2.5}$ personal exposure was higher at T1 (66 μg/m³) and similar at T0 and T2 (52 and 48 μg/m³) due to larger transportation exposures (longer commutes), high heavy vehicle density, and a higher secondary PM formation rate in the area (Figure 7.3). The PM personal exposures to air pollutants tended to be higher than home indoor pollutant concentrations. Many measurements were around the 65 μg/m³ 24-hour Mexican standard and may significantly impact the health of the susceptible local population. Measurements of PM$_{2.5}$ mass and number were performed at homes with direct reading monitors [Dust-Trak and Condensation Particle Counting (CPC)]. Mass concentrations measured directly were those from filter samples, varying between 28.6 and 60.6 μg/m³. The indoor concentrations were higher at T2 and T1 and similar to outdoor concentrations at T0 (Figure 7.4).

Fine particle number densities in homes were higher when the human activity level was high, when there was an open kitchen in the apartment, and/or when outdoor concentrations were low (ranging from 20,000 to 100,000 particle number concentration). In schools, with children present, indoor concentrations were higher than those outdoors, and higher than those indoors after classes. The number concentrations were much lower in rooms with closed windows and doors and reduced activity.

The mean outdoor/indoor concentration ratios during the campaign were 0.85–0.88 (day) and 0.93 (night); these ratios were considerably higher than the concentration ratios measured indoors with closed windows in Germany (~0.30) (Wolkoff and Nielsen, 2001). This higher number and ratio concentration of indoor particles in homes in comparison with more developed countries can be explained by warmer weather conditions, a higher human activity factor, large rooms with low

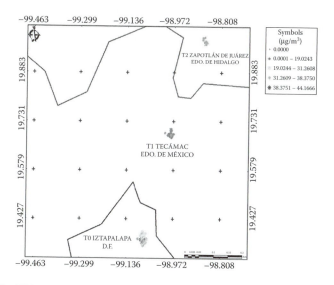

FIGURE 7.3 PM$_{2.5}$ levels at homes indoors, outdoors, and parent personal exposure by site (µg/m^3).

area-to-volume ratios (minimizing precipitation on the walls), less tight buildings with higher air exchange rates, and windows open for long hours during the spring season (producing higher outdoor PM infiltration).

7.4.3 UF Particles

UF particle concentrations, mostly the smallest ones, were elevated in the three studied sites and showed different patterns related to specific local sources in

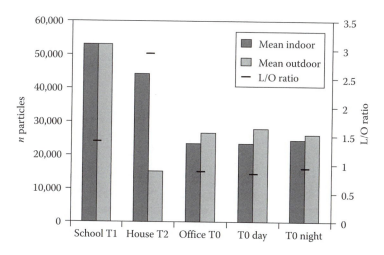

FIGURE 7.4 Mean indoor and outdoor particle concentration ratios (µg/m^3).

each site. For all sites, the PM_1 concentrations were higher outdoors than indoors (9.48 versus 4.3 µg/m³ at T2). The $PM_{0.5}$ concentrations were almost the same outdoors and indoors (4.31 versus 4.06 µg/m³ at T1). The $PM_{0.25}$ outdoor concentrations were three times the indoor values (9.50 versus 3.20 µg/m³ at T0 and T1). The outdoor and indoor $PM_{0.25}$ concentrations were the most abundant fraction in all the sites (50.67 versus 23.74 µg/m³ at T1) (Figure 7.5). Production of a small fraction of PM is mainly related to secondary PM formation and diesel vehicle emissions. The smallest UF fractions in the MCMA may be associated with some health problems like the oxidative stress effects observed in the volunteers.

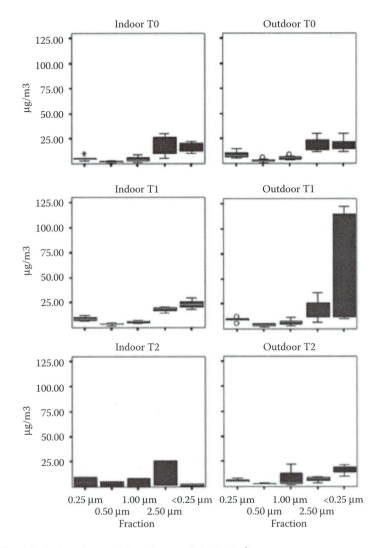

FIGURE 7.5 UF particles by size, place, and site (µg/m³).

7.4.4 ENDOTOXINS

The mean endotoxin outdoor concentrations in T0, T1, and T2 were 6, 8, and 19 U/mg, respectively (Figure 7.6). Endotoxin indoor concentrations were higher at T2 than at T0 and T1, that is, 67.5, 14.11, and 2.94 U/mg, respectively. The endotoxin concentration in personal filters was higher at T0 compared with the other sites. The high incidence of outdoor endotoxins indicates significant airborne levels of human and animal fecal coliforms and endotoxins due to entrainment of solid organic material residues. The endotoxin indoor values suggest poor sanitation and ventilation, mainly in rural area homes. The higher endotoxin level in T0 personal samples may be due to the longer outdoor time for these volunteers. These findings are consistent with previous studies that reported the range of airborne endotoxin air concentrations (Bötcher et al., 2003; Bouillard et al., 2005).

7.4.5 FUNGI

The occurrence of outdoor fungi colonies was highest at T0 with the largest quantity of isolated colonies [645 colony-forming units (CFU) isolated]. For indoor samples T1 and T0 showed a similar number of isolations, and for the personal samples T2 showed the highest isolation of fungi colonies. The most isolated genus at outdoor, indoor, and personal levels was *Penicillium* sp. (41.2%, 20.68%, and 28.8%, respectively). The low frequency of fungi colonies may be related with the low humidity present during the campaign. In rural areas, there are many potential emission sources, including plants, soil, and PM produced by harvest activities, with the wind as a medium for spreading spores (Chen et al., 2004). The geneses *Alternaria*, *Aspergillus*, *Cladosporium*, *Fusarium*, *Penicillium*, and *Rhizopus* have been reported as potent allergy inductors (Nadel, 1980); many of these were isolated in this study. Thus, future studies should correlate the concentrations of these geneses with the incidence of asthma and allergies.

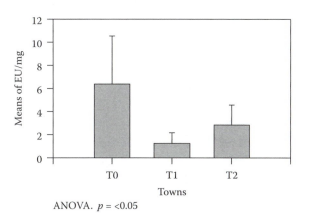

ANOVA. $p = <0.05$

FIGURE 7.6 Endotoxin concentrations in $PM_{2.5}$ personal samples by site ($\mu g/m^3$).

7.4.6   VOCs

The level of exposure to VOCs decreased with distance from the city. For each measurement site, the exposure ranking was typically adults > children > apartments/homes > schools. Indoor exposure to VOCs was usually higher than outdoor exposure (Figure 7.7). Outdoor VOC patterns at the different sampling sites were similar. The main components were toluene, limonene, and xylene. In most of the cases, T0 was the highest polluted site for every component, followed by T1 and T2, except for limonene, which was higher at T1.

Benzene and 1,3-butadiene exposures followed the same pattern of indoor > outdoor > personal levels. Depending on sampling sites, the concentrations ranged typically from 0.3–4 µg/m^3 and 2–10 µg/m^3 for 1,3-butadiene and benzene (Figure 7.8), respectively. A comparison between indoor and outdoor levels showed higher indoor concentrations, possibly due to cigarette smoke in those homes. In general, the personal sample levels reflect those obtained by corresponding stationary indoor or outdoor sampling results. VOC levels in the MCMA were higher than those at other urban sites (Rehwagen et al., 2003).

It is important to note that MCMA VOC emissions apparently influence VOC levels at suburban (T1) and semirural (T2) sites and their surrounding areas. This MCMA influence is supported by the observation that T1 local VOC emissions represent less than 0.04% of the State of Mexico's urban counties since it has only seven industrial/commercial units reported in a recent emissions inventory (SMAEM, 2004). There is no VOC emission inventory available for T2, but it can be expected that local VOC emissions at that site are low.

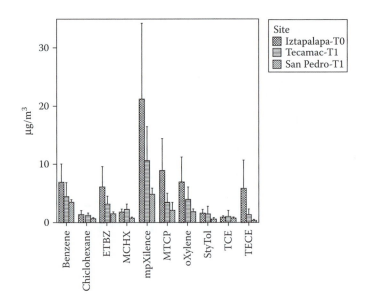

FIGURE 7.7   VOC indoor levels by site (µg/m^3).

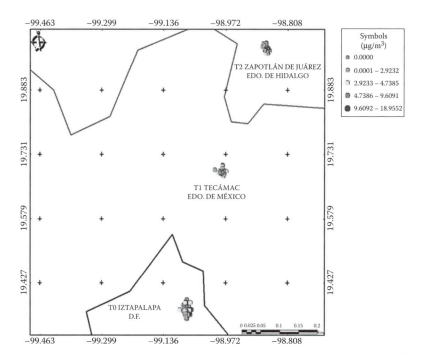

FIGURE 7.8 Benzene levels at homes indoors and outdoors, and personal exposure by site ($\mu g/m^3$).

7.4.7 CARBON MONOXIDE

CO levels for each microenvironment, as well as personal exposure levels, were higher at the urban sites. For T1 and T0 homes, outdoor CO concentrations were higher (2.39 and 2.36 ppm) than at T2; the same was true for indoor CO concentrations (2.87 and 2.86 ppm). Personal exposure to CO was higher at T1 (3.42 ppm), followed by T0 and T2 (2.55 and 2.41 ppm, respectively). In all the sites, CO concentrations were higher inside homes and schools than outside (from 1.21 to 1.64 times). Higher concentrations were recorded inside public and private motor vehicles and kitchens. However, most measurements were lower than the 11 ppm 8-hour Mexican standard.

7.4.8 RESPIRATORY AND OLFACTORY CONDITION

The frequency of respiratory diseases was high; during the campaign parents reported that 55% of their children had respiratory diseases at T2, with 20% at T1 and T0. Spirometry results showed that median flow (FEF25-75) was smaller at T1 and T0. Children's observed health conditions were probably impacted by the local levels of air pollutants, as noticeable by the different reported levels at suburban and urban sites.

Olfactory threshold results indicate that children from T0 have the poorest olfactory sensitivity. Children from the semirural site, T2, presented the lowest thresholds. These thresholds represented a 2.8-fold difference for T0 and a 1.6-fold difference for

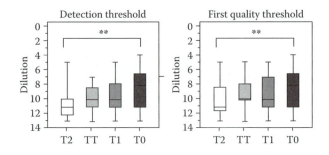

FIGURE 7.9   Children olfactory threshold by site.

T1 (Figure 7.9). The differences in olfactory thresholds might also indicate that the adverse effects of air pollution on the olfactory system appear early in life.

7.4.9   Oxidative Stress in Children

MPO enzyme plasma activity was significantly elevated in children living in T1 in comparison with those in T2 (57.8 versus 47.7 U/mg protein), and the related lipids damage expressed as lipoperoxidation products (TBARS) concentration was higher at T2 than at T0 and T1 (13.00, 9.01, and 9.44 µM, respectively). PON1 enzyme activity was higher at T2 than at T1 and T0 (0.122, 0.103, 0.099 nmol p-nitrofenol/mg protein, respectively), and this correlates with the greater presence of carbonylation in T2 children than at the other sites. Finally, the T1 site children showed low antioxidant status expressed by the lowest NTE reduction (Figure 7.10).

A multivariate analysis showed significant associations among carbonylation and ozone and nitrogen oxide exposures as an expression of pollutant-related protein damage. NBT reduction was related to benzene exposure and having a mother who smokes, expressing a reduction of the antioxidant state associated with these exposures. TBARS production was associated with PM_{10} exposure and a mother who smokes. PON1 activity was associated with ozone and parents who smoke, and MPO activity was associated with methylcyclopentane and trichloroethylene exposures (Table 7.1).

This study result raises the possibility that oxidative damage to lipids, proteins, and other biological effects in children may be related to local air pollution levels. These levels could be influenced by the MCMA pollutant PM plume, which is transported to long distances and thus affects a wide area of neighboring states (including T2 site children).

7.4.10   Oxidative Stress in Adults

The results for adults indicate that CRP did not differ significantly among adults living in the three sites (2.05, 1.33, and 1.38 for T0, T1, and T2, respectively). CRP

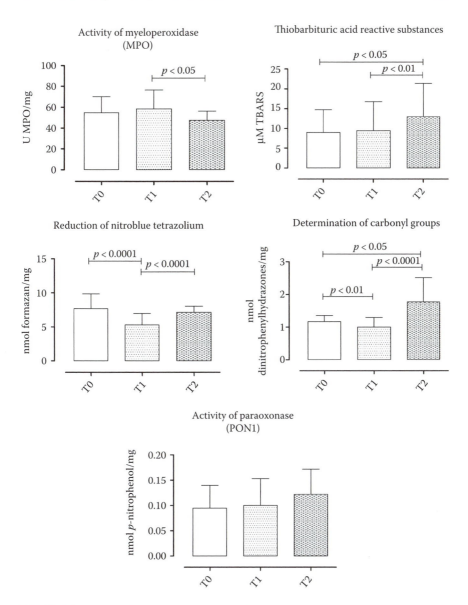

FIGURE 7.10 Oxidative stress markers in children by site.

is a sensitive, although nonspecific, marker for inflammation; high CRP levels constitute a risk factor for cardiovascular events in apparently healthy individuals (Tracy et al., 1997). In this case, CRP was not sensitive enough to express a differential inflammatory condition related to air pollution exposure.

It is well known that serum CP and Cu vary, responding to inflammation and infectious events. The CP mean levels were higher at T0 than at T1 and T2 (91.4, 62.94,

TABLE 7.1

Children's Oxidative Stress Biomarkers and Air Pollutants Regression Models[a]

Marker	Related Covariate	R	R^{2b}	p
MPO	Methylcyclopentane	0.26	0.07	0.02
	Trichloroethylene			
TBARS	Mother smoker	0.36	0.13	0.02
	PM_{10}			
Carbonylation	Ozone	0.57	0.32	≤0.00
	Nitrous oxide			
PON1	Ozone	0.30	0.09	0.01
	Mother smoker			
	Father smoker			
NBT	Benzene	0.47	0.22	≤0.00
	Mother smoker			

[a] Linear regression.
[b] R^2 corrected.

and 57.24, respectively). This indicator may suggest either a greater inflammation response from the T0 population and/or higher Cu levels in comparison with the other studied sites (Mendez et al., 2004).

The activities of the antioxidant enzymes GPx (77.10, 47.11, and 50.97 U/g of hemoglobin at T0, T1, and T2, respectively) and SOD (1625, 1124, and 1510 U/g of hemoglobin at T0, T1, and T2, respectively) were highest at T0 (Figure 7.11). This may be an acute response to the higher production of air pollution-related hydrogen peroxide and superoxide anions. Thus, it may not be surprising that these individuals developed protective antioxidant defenses.

Nitrite levels were highest at T2 (71.59). Increased nitrite levels at T2 might be attributed to the activation of inducible isoforms of NO synthesis in leukocytes (neutrophils, eosinophils, basophils, and mononuclear phagocytes) (Abou-Seif and Youssef, 2004), because NO synthetase is expressed at very high levels in macrophages activated by exposures to bacterial lipopolysaccharide (LPS) (Sheffler et al., 1995), irritant air pollutants, or the consumption of nitrite-rich food common in the Mexican diet.

For adults, the multivariate analysis showed significant associations among GPx and benzene, ozone, NO_2, SO_2, and living near a factory. CRP was associated with NO_2, SO_2, and the presence of new furniture in homes. Nitrites were correlated with 1,3-butadiene, insecticides, and PM_{10} exposures, and CP was correlated with 1,3-butadiene, dodecane, styrene, and new furniture (Table 7.2).

7.4.11 CYTOKINES

In the case of cytokines, mixed reactions to air pollutants were observed, since not only were proinflammatory cytokines increased but also suppressive cytokines

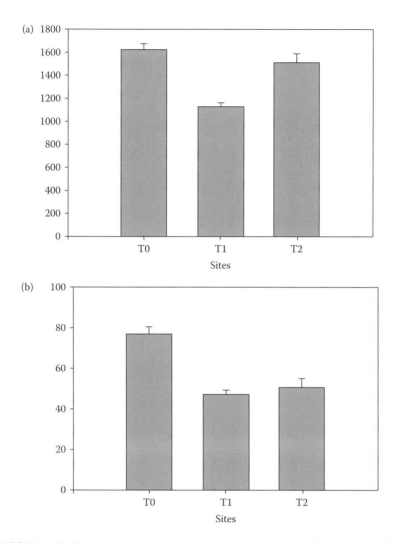

FIGURE 7.11 Oxidative stress markers in adults by site. Superoxide dismutase u/g of hemoglobin. T0 = Iztapalapa Distrito Federal, T1 = Tecamac Mexico, T2 = Tizayuca Hidalgo ANOVA. T0 vs T1 $p = 0.001$, and T0 vs T2 $p = 0.539$. (b) Gluthatione Peroxidase u/g of hemoglobin T0 = Iztapalapa Distrito Federal, T1 = Tecamac Mexico, T2 = Tizayuca Hidalgo ANOVA. T0 vs T1, and T2. $p = 0.001$.

were produced in significant amounts (Vega et al., 2007). Children from T2 showed higher levels of IL-6, IL-4, IFN-γ, GM-CSF, and IL-2 in their serum samples, while those from T0 showed the lowest levels of all cytokines (Figure 7.12). The parents' cytokines were highest at T1, except for IL-10, IFN-γ, and GM-CSF, which were highest at T0.

Children's cytokine regression models showed an association between IL-4 and $PM_{2.5}$, and NO_2 exposures; and IL-6 was correlated with $PM_{2.5}$, SO_2, and PM_{10}

TABEL 7.2
Adults' Oxidative Stress Biomarkers and Air Pollutants Regression Models[a]

Marker	Related Covariate	R	R^{2b}	p
GPx	Benzene	0.68	0.43	≤0.00
	O_3			
	NO_2			
	SO_2			
	Factory close			
CRP	NO_2	0.33	0.08	0.01
	SO_2			
	New furniture			
Nitrites	1,3-Butadiene	0.42	0.14	0.01
	Insecticides			
	PM_{10}			
CP	1,3-Butadiene	0.38	0.11	≤0.00
	Dodecane			
	Styrene			
	New furniture			

[a] Linear regression.
[b] R^2 corrected.

exposures and living near a factory (Table 7.3). For adults, an important association between IL-4 and trichloroethylene and smoking was observed. GM-CSF and IL-10 were associated with ozone and smoking, and IL-12 was associated with home traffic density and limonene (Table 7.4).

It is possible that some of the observed changes on seric cytokines may be related to contaminants other than those evaluated in this study, such as PAHs. PAHs are known to modulate cytochrome expression and immune responses (Elizondo and Vega, 2007), which may produce immunosuppression (Duramad et al., 2007), or asthma and allergy (Chung, 2001).

7.5   CONCLUSIONS

A wide range of potentially harmful airborne pollutants emitted or produced by atmospheric processes in megacities makes it necessary to assess the exposure profiles of high-risk population subgroups. This approach is also necessary to study different interactions between pollutants and their early effects on biomarkers and other health conditions. The results obtained during the MILAGRO Campaign reported in this chapter illustrate the high frequency of many subclinical expressions related to air pollution, even though a significant reduction of MCMA air pollution levels has been achieved in recent times. It is also important to keep in mind that children are the most sensitive individuals in a population (Chung, 2001). More

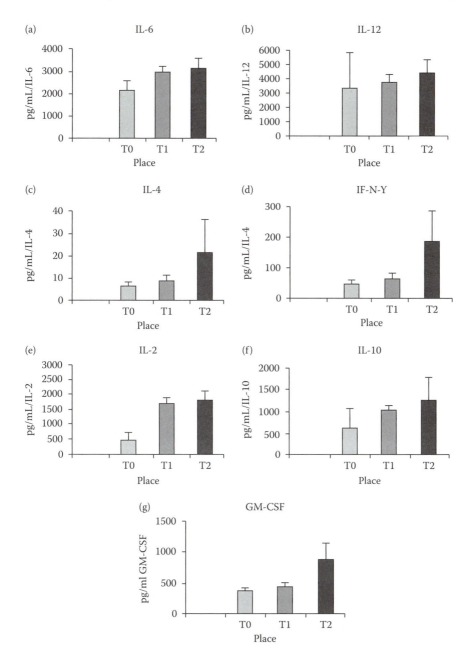

FIGURE 7.12   Children serum cytokine levels by site.

research should be focused on how children are affected by air pollution and other xenobiotics. The air pollution-related indicators presented here may lead to future development or worsening of chronic diseases if more stringent controls and preventive programs are not established in the MCMA.

TABLE 7.3
Children's Cytokines and Air Pollutants Regression Models[a]

		R	R[2][b]	p
IL-4	$PM_{2.5}$	0.24	0.04	0.06
	NO_2			
IL-6	$PM_{2.5}$	0.86	0.78	≤0.00
	Factory near			
	SO_2			
	PM_{10}			

[a] Linear regression.
[b] R^2 corrected.

TABLE 7.4
Adults' Cytokines and Air Pollutants Regression Models[a]

	Covariates	R	R[2][b]	p
IL-4	Trichloroethylene	0.74	0.52	≤0.00
	Smoker			
GM-CSF	Ozone	0.42	0.13	0.02
	Smoker			
IL-10	Ozone	0.55	0.26	≤0.00
	Smoker			
IL-12	Traffic density	0.58	0.29	≤0.00
	Limonene			

[a] Linear regression.
[b] R^2 corrected.

ACKNOWLEDGMENTS

This project was partially funded by the Comisión Ambiental Metropolitana-Mexico, the Department of Occupational and Environmental Medicine, Göteborg University, the Human Exposure Research and Epidemiology, UFZ-Leipzig, NIH/Fogarty International Center 5 D43TW00644, the Centro Nacional de Investigación y Capacitación Ambiental-INE, and the Fondazione Salvatore Maugeri.

The authors acknowledge Luisa T. Molina, Lars Barregard, René Lugo, Henry Wöhrnschimmel, Alejandro Treviño, José Samudio, Beatriz Cárdenas, Rosa Maria Berbabé, Felipe Angeles, Francisco Mandujano, Meike Schilde, and Paolo Sacco for their support and commentaries, and Jephte Cruz, Martha A. Hernández, Amilcar Torres, Martha Hernández, Gonzalo González, and Yazmín Affif for their collaboration during the campaign.

REFERENCES

Abou-Seif MA and Youssef AA. 2004. Evaluation of some biochemical changes in diabetic patients. *Clin Chim Acta* 346: 161–170.

Aldus S, Heeschen C, Meinertz T, et al. 2003. Myeloperoxidase serum levels predict risk in patients with acute coronary syndromes. *Circulation* 108: 1440–1445.

Alfaro-Moreno E, López-Marure R, Montiel-Dávalos A, Symonds P, Osornio-Vargas A, Rosas I, and Murray C. 2007. E-Selectin expression in human endothelial cells exposed to PM_{10}: The role of endotoxin and insoluble fraction. *Environ Res* 103(2): 221–228.

Bearer C. 1995. How are children different from adults? *Environ Health Perspect.* 103(6): 7–12.

Beaumont F. 1988. Clinical manifestations of pulmonary Aspergillus infections. *Mycoses* 31: 15–20.

Bell ML, Davis DL, Gouveia N, Borja-Aburto VH, and Cifuentes LA. 2006. The avoidable health effects of air pollution in three Latin American cities: Santiago, Sao Paulo, and Mexico City. *Environ Res* 100(3): 431–440.

Bobak M and Leon DA. 1999. The effect of air pollution on infant mortality appears specific for respiratory causes in the postneonatal period. *Epidemiology* 10(6): 666–669.

Borja-Aburto VH, Loomis DP, Bangdiwala SI, Shy CM, and Rascón-Pacheco RA. 1997. Ozone, suspended particulates, and daily mortality in Mexico City. *Am J Epidemiol* 145(3): 258–268.

Bötcher M, Björksten B, Gustafson S, Voor T, and Jenmalm M. 2003. Endotoxin levels in Estonian and Swedish houses dust and atopy in infancy. *Clin Exp Allergy* 33: 295–300.

Bouillard L, Michel O, Dramaix M, and Devleeschouwer M. 2005. Bacterial contamination of indoor air, surfaces, and settled dust, and related dust endotoxin concentration in healthy office building. *Ann Agric Environ Med* 12: 187–192.

Calderon-Garcidueñas L, Osnaya-Brizuela N, and Ramirez-Martinez L. 1996. DNA strand breaks in human nasal respiratory epithelium are induced upon exposure to urban pollution. *Environ Health Perspect* 104: 160–168.

Calderon-Garciduenas L, Vincent R, Mora-Tiscareno A, et al. 2007. Elevated plasma endothelin-1 and pulmonary arterial pressure in children exposed to air pollution. *Environ Health Perspect* 115(8): 1248–1253.

Castillejos M, Gold DR, Damokosh AI, Serrano P, Allen G, and McDonnell WF. 1995. Acute effects of ozone on the pulmonary function of exercising schoolchildren from Mexico City. *Am J Respir Crit Care Med* 152: 1501–1507.

CCA. 2003. Comisión para la Cooperación Ambiental de América del Norte. Informe del Taller de América del Norte sobre Indicadores de Riesgo y Salud Ambiental de la Infancia. CCA, http://www.cec.org/pubs_docs/documents/index.cfm?varlan=espanol&ID=840 (accessed March 10, 2007).

Chang SC, Kao MC, and Lin CT. 2001. Modulation of NO and cytokines in microglial cells by CUZN-superoxide dismutase. *Free radic Biol Med* 31: 1084–1089.

Chaudhuri N. 1998. Child health, poverty and the environment: The Canadian context. *Can J Public Health* 89: S26–S30.

Chen J, Shi J, Wang S, Yang S, Lou J, and Liu Z. 2004. Environmental mycological study and respiratory disease investigation in tussah silk processing workers. *J Occup Health* 46: 418–422.

Chung F. 2001. Anti-inflammatory cytokines in asthma and allergy: Interleukin-10, interleukin-12, interferon-gamma. *Mediat Inflamm* 10(2): 51–59.

Ciarrocca M, Tomei F, Bernardini A, et al. 2006. Immune parameters in female workers exposed to urban pollutants. *Sci Total Environ* 370(1): 17–22.

Cohen AJ, Ross Anderson H, Ostro B, et al. 2005. The global burden of disease due to outdoor air pollution. *J Toxicol Environ Health A* 68(13–14): 1301–1307.

Cousins RJ. 1985. Absorption, transport, and hepatic metabolism of copper and zinc: Special reference to metallothionein and ceruloplasmin. *Physiol Rev* 65: 238–309.

Diez U, Kroessner T, Rehwagen M, et al. 2000. Effects of indoor painting and smoking on airway symptoms in atopy risk children in the first year of life: Results of the LARS-study. Leipzig Allergy High-Risk Children Study. *Int J Hyg Environ Health* 203: 23–28.

Duramad P, Tager IB, and Holland NT. 2007. Cytokines and other immunological biomarkers in children's environmental health study. *Toxicol Lett* 172(1–2): 48–59.

Elizondo G and Vega L. 2007. Interactions between cytochrome P450s and cytokines modify inflammatory responses and parasitosis outcome. In: *Advances in the Immunobiology of Parasitic Diseases*, LI Terrazas (Ed.). Research Signpost, India, pp. 53–72.

English P. 1994. Examining associations between childhood asthma and traffic flow using a geographic information system. *Arch Environ Health* 49: 223–227.

Etzel R, and Balk YS (Eds). 1999. *Handbook of Pediatric Environmental Health*. American Academy of Pediatrics, Nueva York, p. 420.

Evans J, Levy J, Hammitt J, et al. 2002. Health benefits of air pollution control. In: *Air Quality in the Mexico Megacity: An Integrated Assessment*, LT Molina and MJ Molina (Eds). Kluwer Academic Publishers, Dordrecht, the Netherlands, pp. 103–136.

Forastiere F, Stafoggia M, Tasco C, et al. 2007. Socioeconomic status, particulate air pollution, and daily mortality: Differential exposure or differential susceptibility. *Am J Ind Med* 50(3): 208–216.

Gattoni A, Parlato A, Vangieri B, Bresciani M, and Derna R. 2006. Interferon-gamma: Biologic functions and HCV therapy (type I/II) (1 of 2 parts). *La Clínica Terapéutica* 157(4): 377–386.

Gold DR, Damokosh AI, Pope CA III, Dockery DW, McDonnell WF, and Serrano P. 1999. Particulate and ozone pollutant effects on the respiratory function of children in southwest Mexico City. *Epidemiology* 10: 8–16.

Gutiérrez-Castillo ME, Roubicek DA, Cebrián-García ME, De Vizcaya-Ruíz A, Sordo-Cedeño M, and Ostrosky-Wegman P. 2006. Effect of chemical composition on the induction of DNA damage by urban airborne particulate matter. *Environ Mol Mutagen* 47(3): 199–211.

Halliwell B and Whiteman M. 2004. Measuring reactive species and oxidative damage *in vivo* and in cell culture: How should you do it and what the results mean? *Brit J Pharmacol* 142: 231–255.

Herbarth O, Fritz GJ, Behler JC, et al. 1992. Epidemiologic risk analysis of environmental attributed exposure on airway diseases and allergies in children. *Centr Eur F Publ Health* 7(2): 72–76.

Herbarth O, Fritz GJ, Krumbiegel P, Diez U, Franck U, and Richter M. 2001. Effect of sulfur dioxide and particulate pollutants on bronchitis in children—a risk analysis. *Environ Toxicol* 16(3): 269–276.

Hicks JJ, Medina-Navarro R, Guzman-Grenfell A, Wacher N, and Lifshitz A. 1996. Possible effect of air pollutants (Mexico City) on superoxide dismutase activity and serum lipoperoxides in the human adult. *Arch Med Res* 27(2): 145–149.

Holguín F, Téllez-Rojo MM, Hernández M, et al. 2003. Air pollution and heart rate variability among the elderly in Mexico City. *Epidemiology* 14(5): 521–527.

Hudson R, Arriola A, Martinez-Gomez M, and Distel H. 2006. Effect of air pollution on olfactory function in residents of Mexico City. *Chem Senses* 31(1): 79–85.

IARC. 2009. Overall Evaluations of Carcinogenicity to Humans Group 1: Carcinogenic to humans, http://monographs.iarc.fr/ENG/Classification/crthgr01.php (accessed January 10, 2009).

INEGI. 2001. *Censo de Población 2000*. INEGI. México.

INEGI. 2005. *II Conteo de Población y Vivienda*. INEGI, México.

Jacobs R. 1997. Endotoxin in the environments. *Int J Occup Environ Health* 3: 5.

Kidd P. 2003. Th1/Th2 balance: The hypothesis, its limitations, and implications for health and disease. *Altern Med Rev* 8(3): 223–246.

Koening W, Sund M, Frohlich M, Fischer HG, Lowel H, and Doring A. 1999. C-reactive protein, a sensitive marker of inflammation, predicts future risk of coronary heart disease in initially healthy middle-aged men: Results from the MONICA (Monitoring Trends and Determinants in Cardiovascular Disease) Augsburg cohort study, 1984 to 1992. *Circulation* 99: 237–242.

Landrigan PJ and Anjali Garg A. 2002. Chronic effects of toxic environmental exposures on children's health 1998. *Clin Toxicol* 40(4): 449–456.

Loomis D, Castillejos M, Gold DR, McDonnell W, and Borja-Aburto VH. 1999. Air pollution and infant mortality in Mexico City. *Epidemiology* 10(2): 118–123.

Lucey DR, Clerici M, and Shearer GM. 1996. Type 1 and type 2 cytokine dysregulation in human infectious, neoplastic, and inflammatory diseases. *Clin Microbiol Rev* 9(4): 532–562.

Medina-Navarro R, Lifshitz A, Wacher N, and Hicks JJ. 1997. Changes in human serum antioxidant capacity and peroxidation after four months of exposure to air pollutants. *Arch Med Res* 28(2): 205–258.

Mendez MA, Araya M, Olivares M, Pizarro F, and Gonzalez M. 2004. Sex and ceruloplasmin modulate the response to cooper exposure in healthy individuals. *Environ Health Perspect* 112: 1654–1657.

Molina LT, Kolb CE, de Foy B, et al. 2007. Air quality in North America's most populous city—overview of MCMA-2003 Campaign. *Atmos Chem Phys* 7: 2447–2473.

Molina LT, Madronich S, Gaffney JS, and Singh HB 2008. Overview of MILAGRO/INTEX-B Campaign, *IGAC Newslett* (38): 2–15.

Molina LT and Molina MJ. 2002. Cleaning the air: A comparative overview. In: *Air Quality in the Mexico Megacity. An integral Assessment.* Molina LT and Molina MJ (Eds). Kluwer Academic, The Netherlands.

Molina LT and Molina MJ. 2004. Improving air quality in megacities: Mexico City case study. *Ann N Y Acad Sci* 1023: 142–158.

Moreno-Ramírez E, Hernández-Urzúa MA, González-Villegas AC, Casas-Solís J, and Zaitseva G. 2006. Estudio de inmunoglobulinas, citocinas proinglamatorias, linfoproliferación y fagocitosis en sangre periférica de jóvenes sanos expuestos a diferentes niveles de contaminación atmosférica. *Revista Alergia México* 53(1): 3–8.

Nadel JA. 1980. Mechanisms of airway responses to inhaled substances. *Arch Environ Health* 16: 171.

O'Garra A and Vieira P. 2007. Th1 cells control themselves by producing interleukin-10. *Nat Rev Immunol* 7(6): 425–428.

Ortiz E, Alemon E, and Romero D. 2002. Personal exposure to benzene, toluene and xylene in different microenvironments at the México City metropolitan zone. *Sci Total Environ* 287: 241–248.

Pope CA III and Dockery DW. 1992. Acute health effects of PM_{10} pollution on symptomatic and asymptomatic children. *Am Rev Respir Dis* 145: 1123–1128.

Rehwagen M, Schlink U, and Herbarth O. 2003. Seasonal cycle of VOCs in apartments. *Indoor Air* 13(3): 283–291.

Rojas-Martinez R, Perez-Padilla R, Olaiz-Fernandez G, Mendoza-Alvarado L, Moreno-Macias H, Fortoul T, McDonnell W, Loomis D, and Romieu I. 2007. Lung function growth in children with long-term exposure to air pollutants in Mexico City. *Am J Respir Crit Care Med* 176(4): 377–384.

Rojas E, Valverde M, and Lopez MC. 2000. Evaluation of DNA damage in exfoliated tear duct epithelial cells from individuals exposed to air pollution assessed by single cell gel electrophoresis assay. *Mutat Res Genet Toxicol Environ Mutagen* 468: 11–17.

Romieu I, Meneses F, Ruiz S, et al. 1996. Effects of air pollution on the respiratory health on asthmatic children living in Mexico City. *Am J Respir Crit Care Med* 154: 300–307.

Romieu I, Samet JM, Smith KR, and Bruce N. 2002. Outdoor air pollution and acute respiratory infections among children in developing countries. *J Occup Environ Med* 44(7): 640–649.

Rylander R. 1999. Health effects among workers in sewage treatment plants. *Occup Environ Med* 56: 354–357.

Sanchez-Rodriguez MA, Retana-Ugalde R, Ruiz-Ramos M, Munoz-Sanchez JL, Vargas-Guadarrama LA, and Mendoza-Nunez VM. 2005. Efficient antioxidant capacity against lipid peroxide levels in healthy elderly of Mexico City. *Environ Res* 97(3): 322–329.

SMA. 2007. Secretaria del Medio Ambiente del DF (Environment Department). La calidad del aire en la Zona Metropolitana del Valle de México 1986–2006. Informe del estado y tendencias de la contaminación atmosférica. Gobierno del Distrito Federal, México D.F. 2007.

SMAEM. 2004. Secretaria del Medio Ambiente del Estado de México. Inventario de Emisiones de la Zona Metropolitana del Valle Cuautitlán-Texcoco. Gobierno del Estado de México, Toluca.

Serrano-Trespalacios PI, Ryan L, and Spengler JD. 2004. Ambient, indoor and personal exposure relationships of volatile organic compounds in Mexico City Metropolitan Area. *J Expo Anal Environ Epidemiol* 14(Suppl. 1): S118–S132.

SETRAVI. 2007. Informe SETRAVI. Enero-Agosto de 2007. Gobierno del Distrito Federal. México D.F.

Sierra–Vargas MP, Guzmán-Grenfell AM, Olivares-Corichi IM, Torres-Ramos JD, and Hicks-Gómez JJ. 2004. Participación de las especies reactivas del oxígeno en las enfermedades pulmonares. *Rev Inst Nal Enf Resp Mex* 17(2): 35–148.

Shi Y, Liu CH, Roberts AI, et al. 2006. Granulocyte–macrophage colony-stimulating factor (GM-CSF) and T-cell responses: What we do and don't know. *Cell Res* 16(2): 126–133.

Sheffler LA, Wink DA, Melillo G, and Cox GW. 1995. Exogenous nitric oxide regulates IFN-γ plus lipopolysaccharide-induced nitric oxide synthase expression in mouse macrophages. *J Immunol* 155: 886–894.

Téllez-Rojo MM, Romieu I, Ruiz-Velasco S, Lezana MA, and Hernández-Avila MM. 2000. Daily respiratory mortality and PM_{10} pollution in Mexico City: Importance of considering place of death. *Eur Respir J* 16: 391–396.

Tovalin H, Valverde M, Morandi MT, Blanco S, Whitehead L, and Rojas E. 2006. DNA damage in outdoor workers occupationally exposed to environmental air pollutants. *Occup Environ Med* 63: 230–236.

Tovalin-Ahumada H, Whitehead L, and Blanco S. 2007a. Personal exposure to $PM_{2.5}$ and element composition—a comparison between outdoor and indoor workers from two Mexican cities. *Atmos Environ* 41: 7401–7413.

Tovalin-Ahumada H and Whitehead L. 2007b. Personal exposures to volatile organic compounds among outdoor and indoor workers in two Mexican cities. *Sci Total Environ* 376(1–3): 60–71.

Tracy RP, Lemaitre RN, Psaty BM, Ives DG, Evans RW, and Cushman M. 1997. Relationship of C-reactive protein to risk of cardiovascular disease in the elderly: Results from the cardiovascular health study and the rural health promotion project. *Arterioscler Thromb Vasc Biol* 17: 1121–1127.

Trinchieri G. 2003. Interleukin-12 and the regulation of innate resistance and adaptive immunity. *Nat Rev Immunol* 3(2): 133–146.

Valverde M, del Carmen Lopez M, Lopez I, et al. 1997. DNA damage in leukocytes and buccal and nasal epithelial cells of individuals exposed to air pollution in Mexico City. *Environ Mol Mutagen* 30(2): 147–152.

Vallejo M, Ruiz S, Hermosillo AG, Borja-Aburto VH, and Cardenas M. 2006. Ambient fine particles modify heart rate variability in young healthy adults. *J Expo Sci Environ Epidemiol* 16(2): 125–130.

Vega L, Rodríguez-Sosa M, García-Montalvo EA, Del Razo LM, and Elizondo G. 2007. Non-optimal levels of dietary selenomethionine alter splenocyte response and modify oxidative stress markers in female mice. *Food and Chem Toxicol* 45(7): 1147–1153.

Wieslander G, Norback D, Bjornsson E, Janson C, and Boman G. 1997. Asthma and the indoor environment: The significance of emission of formaldehyde and volatile organic compounds from newly painted indoor surfaces. *Int Arch Occup Environ Health* 69: 115–124.

Wolkoff P. and Nielsen GD. 2001. Organic compounds in indoor air—their relevance for perceived indoor air quality. *Atmos Environ* 35(26): 4407–4417.

8 Polycyclic Aromatic Hydrocarbons
Sources, Distribution, and Health Implications

Nirat Rajput and Anita Lakhani

CONTENTS

8.1 INTRODUCTION

The runway development of modern technology has provided us with a comfortable, safe, and an affluent way of life, but in a much less conspicuous way, it has also presented us with potential health hazards. Smoking, urban and industrial environmental pollution, and the indiscriminate introduction of inadequately listed chemicals into various products contribute to the exposure of the general population to chemical carcinogens. Hence, it is becoming apparent that the problem of chemical carcinogens has escaped its traditional setting as an occupational hazard, advancing to the stage of becoming a potential threat to the population at large. In recent years, there has been considerable concern regarding the possible adverse effects of toxic compounds in the atmospheric environment on human health. Attention has been focused particularly on chemical carcinogens in ambient air. Polycyclic aromatic hydrocarbons (PAHs) were one of the first atmospheric pollutants to be identified as carcinogenic and mutagenic (IARC, 1984). PAHs are ubiquitous environmental contaminants present in air, water, soil, and vegetation. They are also found in remote and pristine areas such as the Arctic. Belonging to a group of compounds commonly known as persistent organic pollutants (POPs), they have attracted much attention in recent years because of their inherent toxicity and ability to disperse in the environment by direct emissions to the air and consequently by long-range transport. Many PAHs are potentially carcinogenic and mutagenic. There is growing evidence that priority PAHs and their derivatives have a dioxin-like potency. They are multi-aromatic ring systems composed of carbon and hydrogen atoms arranged in the form of fused aromatic rings (linear, cluster, or angular arrangement). The PAH family includes 660 substances indexed by the National Institute of Standards and Technology (Sander and Wise, 1997) and approximately 30–50 of them commonly occur in the environment. They are generally produced in incomplete combustion processes, and their occurrence and emissions have therefore been substantial during the past centuries because of the abundant use of fuels for industrial applications, heating, transport, and many other purposes. Thus, PAHs are ubiquitous contaminants in both the general environment and in certain working environments. They are semivolatile compounds present in the atmosphere in both the vapor phase and the particulate phase as well as dissolved or suspended in precipitation. Most of the more potent carcinogens in this group have more than three rings, hence the name PAHs (or polynuclear aromatic hydrocarbons).

8.1.1 STRUCTURE AND PROPERTIES

PAHs consist of two or more fused benzene rings in linear, angular, or cluster arrangements such as naphthalene, acenaphthene, anthracene, pyrene, chrysene, and benzo(a)pyrene (BaP) that are called alternant PAHs. They may also be composed of unsaturated four-, five-, and six-membered rings, referred to as nonalternant PAHs (Bostrom et al., 2002). The basic structural units are phenanthrene, anthracene, and pyrene. Fusion of a benzene ring to face "a" of anthracene and pyrene gives rise to the most widely studied benz(a)anthracene and benz(a)pyrene. These compounds exist in many structural forms and also have a large number of isomers. The structures of PAHs and their structural interrelationships are depicted in Figure 8.1.

The properties of PAHs are correlated to the number of rings, while minor differences within each ring homologue can be attributed to the arrangement of rings. Table 8.1 lists the physical properties of PAH compounds commonly found in urban environmental samples. The general characteristics common to the class are high melting and boiling points and low vapor pressures. PAHs are sparingly soluble or practically insoluble in water. The solubility of BaP, for example, in water is of the

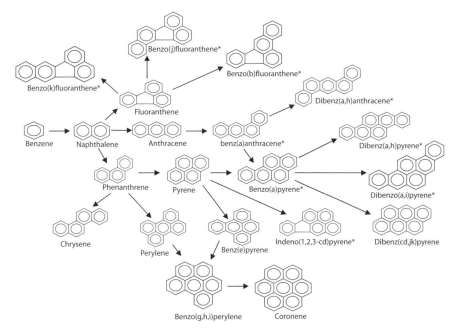

FIGURE 8.1 Structural interrelationships of PAHs in the environment. * Signifies structural forms and isomers.

order of 10^{-8} to 10^{-7} mol/L; surfactants such as detergents can substantially increase its solubility by formation of micelles. PAHs are soluble in many organic solvents: benzene, acetone, hexane, and tetrahydrofuran readily dissolve PAHs. They are also lipophilic; the lipophilicity increases with increasing complexity. The PAHs have a wide range of molecular weights from 128 to 276 with boiling points ranging from 218°C to 525°C, while some of the isomers have small differences in their boiling points. Vapor pressure generally tends to decrease with the increase in molecular weight. Low-molecular-weight (LMW) PAHs containing two or three fused rings are more volatile than high-molecular-weight (HMW) PAHs containing more than three fused rings, which are primarily associated with particles.

Carcinogenic PAHs of environmental interest readily vaporize at higher temperatures and have a tendency to adsorb to particulate matter such as dust particles. Most of the PAHs are fluorescent in ultraviolet light and are photosensitive, forming endoperoxides that then undergo ring cleavage and dealkylation. There is some evidence on the fact that PAHs adsorbed to particles have a greater susceptibility to photooxidation. There is also evidence on the fact that some PAHs may be biodegradable.

8.2 FORMATION AND SOURCES

PAHs and their heteroatom analogues are formed by incomplete combustion or high-temperature pyrolytic processes involving fossil fuels, for example coal, oil, wood, gasoline, and diesel fuel or more generally materials containing C and H (Tavares

TABLE 8.1
Physical Properties of Selected PAHs

PAHs	Molecular Formula	Molecular Weight	Vapor Pressure (Pa at 25°C)	Boiling Point (°C)	Melting Point (°C)	Solubility in Water (μg L^{-1})
Naphthalene	$C_{10}H_8$	128	1.4	217.9	81	3.10×10^4
Phenanthrene	$C_{14}H_{10}$	178	1.6×10^2	340	100.5	1.29×10^3
Anthracene	$C_{14}H_{10}$	178	8.0×10^4	342	216.4	73
Fluoranthene	$C_{16}H_{10}$	202	1.2×10^{-3}	375	108.8	260
Pyrene	$C_{16}H_{10}$	202	6.0×10^{-4}	393	150.4	135
Chrysene	$C_{18}H_{12}$	228	8.4×10^{-5}	448	253.8	2
Benzo(a)anthracene	$C_{18}H_{12}$	228	2.8×10^{-5}	400	160.7	14
Benzo(b)fluoranthene	$C_{20}H_{12}$	252	6.7×10^{-5}	481	168.3	1.2
Benzo(k)fluoranthene	$C_{20}H_{12}$	252	1.3×10^{-8}	480	215.7	2.5
Acenaphthylene	$C_{10}H_8$	152	8.8×10^{-1}	265–275	92–93	—
BaP	$C_{20}H_{12}$	252	7.3×10^{-7}	496	178.1	3.8
Acenaphthene	$C_{12}H_{10}$	154	2.9×10^{-1}	279	85	3.93×10^3
Benzo(ghi)perylene	$C_{22}H_{12}$	276	1.4×10^{-8}	545	278.3	0.26
Dibenzo(a,h)anthracene	$C_{22}H_{14}$	278	1.3×10^{-8}	524	266.6	0.5
Indeno(cd)pyrene	$C_{22}H_{12}$	276	1.3×10^{-8}	536	163.6	62
Fluorene	$C_{13}H_{10}$	166	8.0×10^{-2}	293–295	115–116	1.98×10^3

Source: Adapted from Pitts-Finlayson, B.J. and Pitts, J.N., Jr. 1986. *Atmospheric Chemistry: Fundamentals & Experimental Techniques*, pp. 877–892. Wiley, New York.

et al., 2004). They are formed by mechanisms involving pyrolysis and pyrosynthesis. The higher alkanes present in fuels and plant materials form PAHs by pyrolysis. During pyrolysis, reactive free radicals are produced at a temperature of 500–800°C in the chemically reducing zone of a flame burning with an insufficient supply of oxygen. The mechanism of function is illustrated in Figure 8.2. At the temperature of pyrolysis, aromatic ring systems are the most stable among the structural types present while aliphatic carbon–carbon bonds and carbon–hydrogen bonds readily break down to yield molecular fragments of free radical character, which then undergo recombination in the reducing atmosphere to form partially condensed aromatic molecules. The relatively greater stability of the aromatic rings thus allows the gradual accumulation of condensed ring structures as the duration of pyrolysis progresses. Under these conditions, acetylene and butadiene undergo chain-lengthening leading first to vinylcyclohexene to yield the *n*-butylbenzene radical. A molecule of *n*-butylbenzene radical can then join a molecule of tetralin to form 1-phenyl-4-(1,2,3,4-tetrahydro-5-naphthyl)butane, which in turn yields BaP and benzo(j)fluoranthene following ring closure and dehydrogenation. Acetylene is actually formed during the thermal decomposition of many organic materials. However, it is not necessary to assume that the pyrolytic formation of BaP invariably involves initial cracking down to acetylene. Long-chain alkanes may be split to C10 units, which

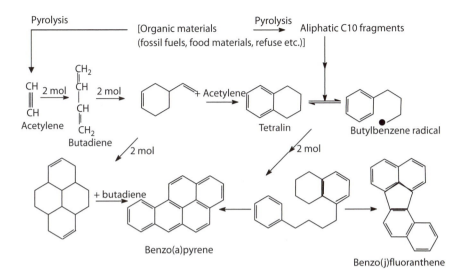

FIGURE 8.2 Mechanism of pyrolytic formation of BaP and benzo(j)fluoranthene from organic materials. (Adapted from Badger, G.M., 1962. *The Chemical Basis of Carcinogenic Activity.* Springfield III., Charles C. Thomas; Badger, G.M., Kimber, R.W.L., and Novotny, J., 1964. *Aust. J. Chem.* 17, 778–786.)

then undergo cyclization to form the C6–C4 units. Dimerization of two vinylcyclohexene to dodecahydropyrene also occurs, which, followed by the addition of a molecule of butadiene and aromatization, yields BaP. BaP is probably formed by both mechanisms as shown in Figure 8.2.

The low hydrocarbons form PAHs by pyrosynthesis; the tendency of hydrocarbons to form PAHs by pyrosynthesis increases in the following order: aromatics > cycloolefins > olefins > parafins. It has also been suggested that PAHs are formed during combustion by mechanisms involving slow Diels–Alder condensations, rapid radical reactions, and ionic reactions (Haynes, 1991). Within the internal combustion engines, PAH formation is favored by the radical formation mechanism and the production is affected by factors like the type of fuel, amount of oxygen, and temperature (Lima et al., 2005).

They are initially generated at the source in the gaseous state; at higher temperatures of combustion sources, larger proportions are present in the vapor phase. On cooling the reaction mixture, these PAHs condense from the vapor phase onto coexisting particulate substrates. PAHs on reaction with other atmospheric pollutants, namely, NO_x, SO_2, and so on, may form hetero-PAHs. The carcinogenicity and mutagenicity of these hetero-PAH compounds is greater than their parent compounds.

8.2.1 Sources

8.2.1.1 Natural Sources

Natural sources of PAHs include combustion (forest fires and volcanoes) and biosynthesis (sediment diagenesis, tar pits, and biological conversion of biogenic precursors) (Oros and Simoneit, 2001a,b). They are found in coal tar, crude oil, creosote, and roofing tar and also have a cosmic origin (Beegle et al., 2001). Burning of biomass in wildfires has been a natural process on this planet and is an important primary source of soot and organic particulate matter, which influences atmospheric chemical, optical, and radiative properties through direct and indirect mechanisms (Oros and Simoneit, 2001a; Simoneit, 2002). Fires can produce both "solid" and "liquid" aerosol particles. The particle formation process in flames begins with the creation of condensation nuclei such as PAHs from ejected gases as well as from a variety of "soot-like" species (Turns, 1996). The formation of cyclic molecules and PAHs as nuclei in the flame zone is linked to the availability of double- and triple-bonded hydrocarbons in the biomass, and is rate limited by the formation of the first aromatic ring. As the PAH molecules grow between 3000 and 10,000 amu through chemical and coagulative processes, these microparticles become the condensation nuclei for other pyrolized species and may experience considerable growth. Subsequently, many of these particles may be reduced in size through further oxidation in the interior of the flame zone if temperatures exceed 1100 K. If insufficient oxygen is transported into the flame, or if the temperature is not high enough to complete oxidation ($T < 1100$ K), many of these particles may undergo a secondary condensation growth phase and be emitted in the form of smoke (Reid et al., 2005).

8.2.1.2 Anthropogenic Sources

Anthropogenic sources are the major contributors of the more hazardous PAH species and include mobile as well as stationary categories. Mobile categories are mainly vehicular (petrol and diesel engines) (Baek and Jenkins, 2004). Stationary categories include domestic heating, refuse burning, and agricultural and industrial activities such as metallurgical enterprises, foundries, timber treatment plants, as well as industries focusing on the carbonization, distillation, and gasification of coal, coke, and aluminum (Mastral et al., 2000).

8.2.1.2.1 Mobile Sources

Mobile sources include a variety of vehicles such as automobiles, aircraft, shipping, railways and off-road vehicles, engines, and equipment that generate air pollution— that move or can be moved from place to place. Vehicular emissions contribute to the majority of PAH emissions in urban areas (Lim et al., 2005, 2007; Yang et al., 2005; Ravindra et al., 2006a,b; Marr et al., 2006). Exhaust emissions of PAHs from vehicular sources originate by three distinct mechanisms:

  i. Synthesis from simpler molecules in the fuel, particularly from aromatic compounds (Mitchell et al., 2000)
  ii. Storage in engine deposits and subsequent emission of PAHs already present in the fuel (Lin et al., 2006)
  iii. Pyrolysis of the lubricant (Westerholm and Lin, 1994)

PAH emission from vehicles depends on several factors such as fuel type, fuel parameters like fuel additives and lubricating oil (Sjogren et al., 1996), engine type, engine efficiency and combustion conditions (Vaaraslati et al., 2004; Lim et al., 2005), driving conditions including cold start conditions (Rijkeboer and Zwalve, 1990), ambient temperature (Laurikko and Nylund, 1993), exhaust after treatment devices, engine adjustment (Rijkeboer and Zwalve, 1990), and emission control.

The effect of engine type is largely determined by the design of the combustion system, fuel–air mixture, temperature within the combustion chamber, and manufacturing quality. It has been established that different vehicles operating under the same conditions with the same fuel can emit significantly different amounts of PAH (Velasco et al., 2004). The main engine-operating parameters that affect exhaust PAH content are engine load, air-to-fuel ratio, and engine coolant temperature (Lim et al., 2005, 2007; Yang et al., 2005). The starting conditions of vehicles' hot or cold engines have a minor influence on the partitioning of PAHs in gaseous or particulate phases, with particularly higher emissions of PAHs under cold start conditions for gasoline vehicles than for diesel vehicles (Devos et al., 2006). Diesel engines have higher particulate emissions than gasoline-fueled vehicles and are the major sources of LMW PAHs, whereas light-duty gasoline engines are the principal source of HMW PAHs such as BaP and benzo(a,h)anthracene.

Marr et al. (1999) reported a significant correlation between the chemical composition and emission rate of PAHs for gasoline-powered vehicles and indicated that unburnt fuel is a significant source of PAH. The production of PAHs from gasoline vehicles also depends on the air/fuel ratio; the amount of PAHs decreases with a leaner mixture. Jones et al. (2004) found that the proportion of HMW PAHs decreased with an increase in the air/fuel ratio. Similarly, the composition of a fuel and the PAH content in the fuel also influences PAH emissions for diesel-powered vehicles (Ravindra et al., 2006b). In addition to vehicle exhaust, resuspended road dust from wear of tires, asphalt, and brake linings may also contribute to the PAH levels in ambient air (Boulter, 2005). PAHs emitted from these sources are mainly associated with coarse particles.

PAHs are also emitted by trains, aircraft, and ships. The use of diesel and diesel/electric locomotives is the main source of PAHs in rail transportation. In the developing countries, the use of coal-fired locomotives may also contribute to PAH emission. PAHs are also emitted from aircraft exhaust and depend on the volatility of the fuel and its composition as well as on the power setting of the engine (Chen et al., 2006). PAHs are also reported to be emitted from shipping operations (Cooper, 2003).

8.2.1.2.2 Stationary Sources

Coal combustion: Coal, as a combustible rock rich in carbon, is a crucial component of the energy mix that fuels our society. In addition, coal is the fuel of choice for most heat-intensive industrial processes, such as production of steel, aluminum, and concrete. Coal is mainly composed of a wide variety of organic structures such as aromatic clusters, aliphatic bridges and rings, side chains, and oxygen functional groups. Upon heating, coal structures undergo major physical and chemical changes and release volatile organic compounds. PAH formation during coal combustion may occur through complex pathways involving both pyrolysis and pyrosynthesis (Ni et al., 2003). There are two important factors influencing PAH emission in coal combustion: (1) the unburned organic materials because of the poor combustion conditions (Mastral et al., 2000), and (2) the initial stage of pyrolysis process in the combustion (Mastral et al., 2000). Different types of free radicals, aromatic structures, and alkyl structures undergo association through cyclization that lead to PAH formation.

PAH emission from various coals is found to have a relationship with their volatile content, and the complete combustion of coals with a high volatile content is more difficult to achieve (Oanh et al., 1999). The burning conditions also affect PAH emission. Total PAH emission depends mainly on the pyrolytic process and to a lesser degree on the combustion efficiency. The combustion temperature, amount of air, and coal rank also affect PAH emissions; emission initially increases and then decreases as the temperature increases (Mastral et al., 2000). Further, there is a specific PAH emission trend that follows the combustion temperature. PAHs in the gas phase are higher than those in the solid phase. The higher the percentage of excess air, the lower the total PAHs emitted. A lower percentage of excess air could favor the deposition of PAHs on the more stable particulate matter, while a higher percentage of excess air could favor the shift of PAHs to the gas phase (Mastral et al., 2000).

Biofuel combustion: The smoke from biomass fuels is a complex mixture of aerosols containing significant amounts of carbon monoxide (CO), suspended particulate matter, hydrocarbons, and NO_x (Naeher et al., 2005). In general, each individual biomass emits "chemical fingerprint" of natural (unaltered) and thermally altered (pyrolysis) organic constituents, which is source specific and unique in composition upon burning; thus, they can be utilized as specific indicators for identifying fuel source inputs, transport mechanisms, and receptor fate in samples of atmospheric fine particulate matter (Oros and Simoneit, 2001a,b). Four-ring PAHs (e.g., fluoranthene and pyrene) and five- to six-ring PAHs (e.g., BaP and benzo(ghi) perylene) have been found to dominate over two- to three-ring (e.g., phenanthrene) compounds during the combustion of kulim (*Scorodocarpus* spp.) and seraya (*Shorea* spp.) wood. Quantitative estimation of Indonesian forest fires shows that wood burning contributes to 25–35% of the total PAH emission in the Malaysian atmosphere (Okuda et al., 2002). Small-scale combustion such as that of domestic cook stoves, which have a low burning rate (a few hundreds of grams to a few kilograms per hour), leads to a greater formation of PAHs, resulting in their higher emission factors. This is related to the high volatile content of biofuels, which leads to a higher possibility of incomplete burning.

Smoking: Cigarette smoke is a complex aerosol with multiple classes of chemical compounds, 400–500 gaseous components and approximately 1010 particles ml-1 including numerous PAHs, and is a major source of particle-bound PAHs (PPAHs) in the indoor environment (Velasco et al., 2004). PAHs can be formed by degradation of organic cigarette components to simpler fragments followed by recombination of the simpler fragments during the pyrolysis processes in the burning cigarette. Guerin et al. (1987) observed that sidestream smoke is the major contributor while Baek and Jenkins (2004) observed the predominance of chrysene followed by benzo(a)anthracene and BaP in cigarette smoke. A commission by the California Air Resources Board found that the levels of PAHs associated with cigarette smoking were 1.5–4 times higher than other indoor combustion sources (Sheldon et al., 1993).

8.3 CHEMICAL TRANSFORMATIONS

PAHs are chemically inert compounds. When PAHs react, they undergo two types of reactions that is, electrophilic substitution and addition reaction. The former is preferred since it does not destroy the aromatic character of the PAHs, while addition is often followed by elimination resulting in a net substitution.

Several studies have demonstrated that many PAHs are susceptible to photochemical and chemical oxidation under simulated atmospheric conditions. PAH compounds undergo chemical transformation by gas particle interactions in emission plumes, exhaust systems, and even during atmospheric transport. PAHs on reaction with other atmospheric pollutants such as O_3, NO_2, SO_2, HNO_3, and peroxyacetyl nitrate (PAN) are concurrently exposed to sunlight and molecular oxygen and may form hetero-PAHs like oxy, hydroxy, nitro, and hydroxylnitro PAHs. Nitro and oxy PAH reaction products may be present in the gas phase as well as in the particulate phase. Important chemical reactions undergone by PAH compounds are nitration, ozonolysis, and photooxidation.

PAHs react with O_3 to form direct acting mutagens such as diones, quinines, and epoxides. They undergo photooxidation by ultraviolet rays from the sun to yield mutagenic and carcinogenic reaction products, namely aldehydes, ketones, and acids. Photochemical transformations have generally been considered to be the most important mode of atmospheric decomposition of PAHs of both phases (Kamens et al., 1990). However, the reactivity of PAHs in the gas phase is significantly larger than when associated with carbonaceous particulate substrates (Esteve et al., 2006). Thus, PAHs appear to be more stable when adsorbed on naturally occurring particles such as soot or fly ash than when present in pure form or in solution, adsorbed on silica gel or alumina, or coated on the glass surface. Transformations of PAHs on particles can affect the particles' toxicity through the formation of species that are more toxic, for example nitro-PAHs, or less toxic. Heterogeneous reactions of PAHs on particles can change the particles' hydrophilicity and thus alter their potential to act as cloud condensation nuclei (Jones et al., 2004). Further, secondary aerosol components formed by gas-phase reactions can coat over freshly emitted particles by condensation. Through this type of transformation, PAHs that are initially present on the surface of particles may become less accessible for heterogeneous reactions and less bioavailable.

8.3.1 Nonphotochemical Degradation

Although irradiation appears to play a major role in the degradation of airborne PAHs, light is not necessarily required for the atmospheric decomposition of PAHs. PAHs may also degrade by nonphotochemical pathways such as evaporative or oxidative reactions with gaseous pollutants. Atmospheric degradation of PAHs may also occur due to reaction with SO_x or their acid derivatives, particularly in aerosols or when adsorbed on particles.

8.4   ENVIRONMENTAL SIGNIFICANCE OF PAHs AND THEIR EFFECTS ON HUMAN HEALTH

PAHs are hydrophobic compounds and their persistence in the environment is mainly due to their low water solubility and electrochemical stability. Evidence suggests that the lipophilicity, environmental persistence, and genotoxicity of PAHs increase up to four or five fused benzene rings (CPCB, 2002). There is now ample evidence to suggest that the metabolically activated intermediates of carcinogenic hydrocarbons initiate carcinogenesis by binding covalently to DNA in the target tissues and form protein adducts and activate aryl hydrocarbon receptor (AhR)-mediated activity, and may interfere with estrogen receptor (ER)-mediated signaling (Hilscherova et al., 2001). Subsequent DNA replication results in mutations leading to carcinogenesis. Covalent binding of carcinogenic hydrocarbons to DNA has been demonstrated under both *in vitro* and *in vivo* conditions. With a few exceptions, there is good correlation between the carcinogenic potency of a hydrocarbon and binding to DNA.

Concerning carcinogenicity, the International Agency for Research on Cancer (IARC) has classified three PAHs into "group 2A (probably carcinogenic to humans)," and nine PAHs into "group 2B (possibly carcinogenic in humans; inadequate evidence in humans; sufficient evidence in animals)" (IARC, 1987). The carcinogenicity of PAHs is about 1/103–1/104 of that of 2,3,7,8-tetrachlorodibenzodioxin (TCDD) (Machala et al., 2001); on the other hand, the PAH concentration in the atmosphere is about 104–106 times higher than that of TCDD (Mandalakis et al., 2002).

Human exposure to PAHs occurs principally by direct inhalation, ingestion, or dermal contact as a result of the widespread presence and persistence of PAHs in the urban environment. BaP and pyrene are the most important carcinogenic PAHs and are components of combustion processes, coke oven and foundry emissions, cigarette smoke, and charcoal-grilled meats. Most of the data collected on the human health effects of PAH exposure arise from epidemiological studies conducted in the occupational setting. There is a large body of evidence supporting an excess risk of lung cancer in workers exposed to mixtures of PAHs at coke ovens, coal gasification plants, petroleum refineries, aluminum smelters, iron and steel foundries, and with bitumen, diesel, and asphalt. The highest PAH levels are probably found in coke-ovens, although the levels to which workers are exposed are not described in most epidemiological studies.

Given the long-term evidence suggesting that PAHs elevate the risk of various cancers, immunotoxic and respiratory problems, a number of recommendations have been made for the protection of human health. These recommendations include eliminating or minimizing emissions in occupational settings, improved monitoring of urban air pollution, and public education. The complexity of investigating the health effects associated with PAH exposure in the ambient air is reflected in the absence of research addressing this issue to date. Improved understanding of their sources and atmospheric risks posed by PAH exposure is therefore necessary.

8.5   TOXICOKINETICS

PAHs are highly lipid soluble and are absorbed from the lung, gut, and skin of mammals. Inhaled PAHs are predominantly adsorbed on soot particles. After deposition in the airways, the particles can be eliminated by bronchial clearance. PAHs might be partly removed from the particles during transport on the ciliated mucosa and may penetrate into the bronchial epithelium cells where metabolism takes place. BaP and other PAHs are readily adsorbed from the gastrointestinal tract when present as solutes in various dietary lipids. Their adsorption is facilitated by the presence of bile salts in the intestinal lumen.

BaP and other PAHs stimulate their own metabolism by inducing microsomal cytochrome P-450 monooxygenases and epoxide hydrolases. BaP is initially oxidized to several arene oxides and phenols (Figure 8.3). The arene oxides may rearrange spontaneously to phenols (3-OH-, 6-OH-, 7-OH-, and 9-OH-BaP), undergo hydration (catalyzed by microsomal epoxide hydrolases) to the corresponding *trans*-dihydrodiols (4,5-, 7,8-, or 9,10-dihydrodiol), or react covalently with glutathione, either spontaneously or when catalyzed by cytosolic glutathione-*S*-transferases. The phenols can be further oxidized to quinones (1,6-, 3,6-, or 6,12-quinone). In addition, secondary epoxides derived from the phenols and dihydrodiols (resulting in diol epoxides) are formed following further oxidation by the cytochrome P-450 monooxygenase system (Graslund and Jernstrom, 1989). PAHs exert their mutagenic and carcinogenic activities through biotransformation to chemically reactive intermediates, which bind covalently to cellular macromolecules (*inter alia*, DNA). Extensive and systematic studies on the tumorigenicity of individual PAH metabolites in animals have led to the conclusion that vicinal or the so-called bay-region diol epoxides are the ultimate mutagenic and carcinogenic species of alternant PAHs, although not necessarily the only ones (Graslund and Jernstrom, 1989). These diol epoxides are easily converted by epoxide ring opening into electrophilic carbonium ions, which are alkylating agents that covalently bind to nucleophilic sites in the DNA bases and in proteins.

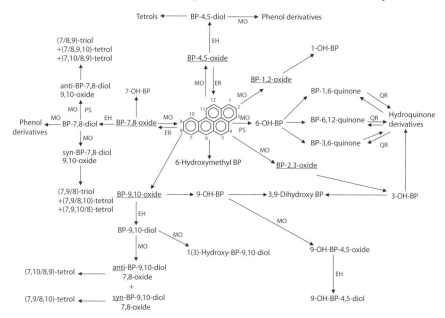

FIGURE 8.3 Toxicokinetics of PAHs.

8.6 STRUCTURE–ACTIVITY RELATIONSHIP

The study of structure–activity relationships was initiated in the 1930s in an attempt to delineate the structural features of chemical compounds responsible for their carcinogenic activity. Although thousands of compounds have been tested for carcinogenicity to date, it is evident that the structure–activity relationships remain largely unresolved and are meaningful only within a class of chemical agents. For example, some general structural features common to carcinogenic PAHs are known. Carcinogenic PAHs are essentially planar molecules. Unsubstituted PAHs with only three rings are inactive, and a few with more than six rings still show some activity; there is a correlation between the molecular size and the carcinogenicity of these compounds. In general, the presence of an angular ring pattern (phenanthrene) in the molecule is necessary for carcinogenicity.

The angular ring structure is of particular interest because the central ring contains a bond (e.g., 9,10 of phenanthrene; 5,6 of benz(a)anthracene; and 4,5 of pyrene) with the highest double bond character, that is, it is more reactive than an ordinary aromatic double bond. This bond was termed as the K-region. In addition to the K-region, some PAHs have another region (L-region) of high reactivity. These carbon atoms are the most reactive sites for substitution reactions. The theory of chemical carcinogenesis (the K- and L-region theory) was postulated by Pullman and Pullman (1955), according to which for a PAH to be carcinogenic, it must have a reactive K-region and an inactive L-region (inactivated by substitution or ring fusion); the K-region was concluded to be the site of critical reaction between the carcinogen and the cell constituents involved in the carcinogenesis.

Boyland (1950) conceived that the reactive intermediate of PAHs might be epoxides (Figure 8.4). The results of various studies consistently indicated that the K-region epoxides are not more carcinogenic than their parent compounds, suggesting that the K-region is not involved in the reactions leading to carcinogenesis. Instead, a new region (the "bay" region, Figure 8.5) is now considered to be the

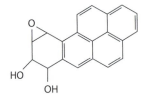

FIGURE 8.4   Reactive intermediate epoxide.

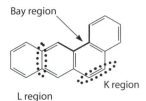

FIGURE 8.5   Bay region in PAHs.

critical molecular site. A bay region in a PAH exists when two nonadjacent benzene rings, one of which is a benzo-ring, are in proximity. Using BaP as a model compound, several groups of investigators have independently arrived at the conclusion that 7,8-dihydrodiol-9,10-epoxide of BaP, a bay-region epoxide, is the ultimate carcinogen of BaP (Marquardt et al., 1977). Based on perturbation molecular orbital calculation, it can be predicted that diol epoxides are more reactive than K-region epoxides and that bay-region epoxides on saturated, angular benzo-rings of unsubstituted PAHs are the ultimate carcinogens. Substitution of the ring hydrogen of PAHs by substituent groups such as alkyl, hydroxyl, halogen, and so on can yield compounds of enhanced or reduced carcinogenicity, depending on the nature of the substituent group and the position of substitution.

8.7 HEALTH IMPACTS

8.7.1 RISK CHARACTERIZATION

The occurrence of PAHs in urban air has caused particular concern because of the continuous nature of the exposure and the size of the population at risk.

A number of approaches have been developed for evaluating the potencies of the various PAHs with regard to the possible inhalation cancer risk to humans (Nisbet and Lagoy, 1992; MOE, 1997; WHO, 1998). One approach commonly referred to as the toxic equivalence factor (TEF) calculates the inhalation risk for excess lung cancer over the risk posed by BaP for each of its copollutant carcinogenic PAH in the polluted ambient air (Papageorgopoulou et al., 1999). The TEF approach has been extensively used for hazard assessment of different classes of toxic chemical mixtures. As shown in Equation 8.1, the overall toxicity or toxic equivalents (TEQs) of a mixture are defined by the concentration of individual compounds (C_i) in a mixture multiplied by their relative potencies or TEFs.

$$TEQ = \Sigma[C_i] \times TEF_i. \tag{8.1}$$

The assumptions implicit in utilization of the TEF approach include the following: all the individual compounds act through the same biologic or toxic pathway; the effects of individual chemicals in a mixture are essentially additive at submaximal levels of exposure; the dose–response curves for different congeners should be parallel; and the organotropic manifestations of all congeners must be identical over the relevant range of doses. TEF_i values are either derived for a species-specific response or are a composite value obtained from TEFs for several responses, and individual TEFs are usually determined relative to the activity of a standard or reference compound. The TEF approach has been applied to different structural classes of compounds, including PAHs, halogenated aromatic hydrocarbons (HAHs), and endocrine disruptors. Chu and Chen and Clement Associates were the first to develop TEFs for PAHs (Nisbet and Lagoy, 1992). Their work was eventually synthesized and adapted by Nisbet and Lagoy into a set of commonly referenced order-of-magnitude estimates. The utility of this approach was demonstrated in studies using relatively simple reconstituted PAH mixtures in rodent carcinogenicity models. It has been demonstrated that there are a number of important factors that can significantly modulate the genotoxicity of PAH mixtures, including the presence or absence of BaP, the dose, and the solvents used in carcinogen administration.

8.7.2 MUTAGENICITY ASSAYS

Over 100 *in situ*, short-term bioassays, using a variety of cell types from bacteria and phage to human cells, now exist for detecting potential chemical mutagens and carcinogens. Some of these have been employed to evaluate the mutagenicity of ambient fine particles and of primary polycyclic organic matter (POM) from motor vehicle exhaust and coal and wood combustion. The most widely employed test to date is the Ames/Salmonella Test, which is a short-term bacterial reverse mutation assay, specifically designed to detect a wide range of chemical substances that can produce genetic damage that leads to gene mutations. It is a rapid and relatively inexpensive assay and employs special strains of *Salmonella typhimurium* to detect frameshift or base pair substitution reverse mutations.

This assay procedure utilizes variants of *S. typhimurium* that have already undergone a mutation to render it incapable of synthesizing its own histidine as normal bacteria can. For growth, these bacteria therefore need to be provided with exogenous histidine. On exposure to mutagenic agents, these bacteria will mutate back (revert) to the wild type and regain the ability to produce histidine. Therefore, on exposure of these bacteria to a chemical, and culturing the exposed bacteria on plates containing minimal histidine, only those organisms that have undergone the back mutation or reversion are able to survive, and the unmutated ones starve due to lack of histidine. The larger the number of bacterial colonies formed, the more mutagenic the compound. Some chemicals can induce mutations all by themselves or after undergoing metabolic activation by the bacterial enzymes. These are the "direct-acting" mutagens. Other compounds, typically PAHs, require metabolism by the cytochrome P-450 system to be active. These are termed "indirect-acting" mutagens. We therefore test compounds both with and without an enzyme preparation containing the cytochrome P-450.

Direct mutagenicity of extracts of ambient POM has been confirmed in major cities throughout the world. Most of the activity is in submicron particles, as with BaP, and is almost exclusively due to frameshift-type chemical mutagens. The activities without (−S9) and with (+S9) metabolic activation are generally about the same, but there are some significant differences. For example, ambient POM samples taken near major heavily traveled streets or freeways during rush-hour traffic show enhanced activity when S9 is added.

This mutagenic effect has been reported for ambient POM collected in Stockholm, Sweden (Alfheim and Lindskog, 1984), Santiago, Chile (Adonis and Gil, 2000), Denmark (Fielberg et al., 2002), Alaska (Watts and Wallace, 1988), and California (Archado and Archer, 1993). Intense research on the chemical composition and mutagenic and carcinogenic activities of exhaust emissions from diesel engines, motor vehicles, wood stoves, hot water boilers, and power plants have also been reported.

8.8 STANDARD LIMITS FOR PAH IN THE AIR

The U.S. Environmental Protection Agency has classified PAHs with BaP as an indicator species as a B-2 pollutant, which means they are probable human carcinogens with sufficient evidence from animal studies but inadequate evidence from human studies (CPCB, 2002). The World Health Organization (WHO) has added PAHs into the list of priority pollutants both in air and in water. WHO has prescribed a guide value of 1.0 ng m^{-3} (as BaP) for a measuring period of one year while the Netherlands have prescribed a guide value of 0.5 ng m^{-3} (as BaP) and a limit value of 5 ng m^{-3} (CPCB, 2002). The USSR Ministry of Health has set up the maximum allowable concentration (MAC) for BaP as 15 µg/100 m^3 in the air of industrial working zones and 0.1 µg/100 m^3 in ambient air. The European Union has proposed to reach ambient levels of 6 ng m^{-3} for PAH indicator parameter (BaP) by January 1, 2010 (CPCB, 2002).

The recommended ambient air quality standards for BaP in India was 5 ng m^{-3} till January 2005 and the concentrations are proposed to be reduced to 1 ng m^{-3} by January 1, 2010, at a reduction rate of 1 ng m^{-3} every year (CPCB, 2002).

8.9 CONTROL MEASURES

Complete removal of PAHs is impossible; they can only be controlled. Simple measures such as removing the superpollutant vehicles and arranging the chaotic movements of buses in transfer stations would be highly effective in improving air quality. Special attention must be taken in underground parking lots, as those in commercial centers, where ventilation is low, and cars idle in long queues. In the same sense, it is necessary that diesel and heavy-duty trucks also undergo periodical exhaust emissions and engine tests. These tests must be effective and focused in the use of particle traps on the diesel vehicles and in the exhaust catalysts on gasoline cars.

Presence of smokers, poorly adjusted pilot stoves, insufficient ventilation, and badly functioning air-conditioning systems have proved to be major PPAH sources in places such as in residences, restaurants, and shopping and entertainment centers, among others. Therefore, smoking must be banned in workplaces and hospitality venues such as clubs and pubs in an attempt to protect the public from exposure to second-hand smoke (Repace, 2003).

Emission control requirements for PAHs are more challenging and require an integrated approach of emission controls components/systems, engine technology, and fuel formulation. Vehicular pollution contributes significantly to the global PAH emissions and a number of strategies have been adopted that focus on controlling the effectiveness of combustion processes, capture, and control of emissions of particles. The catalytic converters for gasoline engines have a marked effect on the reduction of PAHs that usually varies between 80% and 90%, except BaP for which a reduction of 94% has been observed (Schauer et al., 2002). Catalytic converters for diesel engines also reduce total PAH emissions; however, the reductions are not as high as that for gasoline engines (CONCAWE, 1998). In heavy-duty diesel engines, the use of turbocharging and intercooling use of trap oxidizers and filters are currently being developed to reduce particulate emissions.

The emissions of PAHs from agriculture sources are difficult to quantify and control because of the uncertainty in the emission factors and the wide occurrence of these activities. However, the open burning of the agricultural residue/waste can be regulated. Industrial sources, on the other hand, are comparatively well understood and are increasingly being regulated in developed countries. Indeed, improved energy management can lead to improved combustion, which in turn leads to lower emissions. Domestic sources of PAHs are geographically widespread, and the PAH emissions are largely unregulated. Some countries have regulations for controlling the general emission from some domestic heating systems. Modern gas and oil burners, used for circulatory heating systems and hot water systems, have relatively low PAH emissions. Similarly, solid fuel systems (e.g., wood, coal, and peat), which are automatically controlled and fed, are thermally more efficient (and have lower PAH emissions) than those that are hand-fed.

8.10 SUMMARY

PAHs belong to the group of POPs. They are composed of carbon and hydrogen atoms arranged in the form of fused benzene rings having a linear cluster or angular arrangement. There are thousands of PAH compounds in the environment and are mainly of anthropogenic origin. However, individual PAHs differ substantially on the basis of their physical and chemical properties. Generally, the HMW compounds are less water soluble, less volatile, and more lipophilic than LMW compounds. These are resistant to degradation, can remain in the environment for long periods, and have the potential to cause adverse environmental and health effects. PAHs are known for their strong mutagenic, carcinogenic, and toxic properties and are therefore listed by the U.S. Environment Protection Agency and the European Commission as priority pollutants. The best-known moral compound from this group is the highly carcinogenic BaP.

ACKNOWLEDGMENTS

The authors are grateful to the Director, Dayalbagh Educational Institute, Agra, for providing necessary help and the Department of Science and Technology (Project No.: SR/S4/AS:207/02), New Delhi, for financial assistance.

REFERENCES

Adonis, M. and Gil, L. 2000. Polycyclic aromatic hydrocarbon levels and mutagenicity of inhalable particulate matter in Santiago, Chile. *Inhalation Toxicology* 12, 1173–1183.
Alfheim, I. and Lindskog, A. 1984. A comparison between different high volume sampling systems for collecting ambient airborne particles for mutagenicity testing and for analysis of organic compounds. *Science of the Total Environment* 34, 203–222.
Archado, A. and Archer, V.E. 1993. Air pollution and fatal lung disease in three Utah counties. *Archives of Environmental Health* 45, 325–334.
Badger, G.M., 1962. *The Chemical Basis of Carcinogenic Activity*. Springfield III., Charles C. Thomas.

Badger, G.M., Kimber, R.W.L., and Novotny, J., 1964. The formation of aromatic hydrocarbons at high temperatures from 300 to 900°C at 50°C intervals. *Aust. J. Chem.* 17, 778–786.

Baek, S.O. and Jenkins, R.A. 2004. Characterization of trace organic compounds associated with aged and diluted side stream tobacco smoke in a controlled atmosphere–volatile compounds and polycyclic aromatic hydrocarbons. *Atmospheric Environment* 38, 6583–6599.

Beegle, L.W., Wdowiak, T.J., and Harrison, J.G. 2001. Hydrogenation of polycyclic aromatic hydrocarbons as a factor affecting the cosmic 6.2 micron emission band. *Spectrochimica Acta Part A* 57, 737–744.

Bostrom, C.E., Gerde, P., Henberg, A., et al. 2002. Cancer risk assessment, indicators and guidelines for polycyclic aromatic hydrocarbons in the ambient air. *Environmental Health Perspectives* 110, 451–488.

Boulter, P. 2005. A review of emission factors and models for road vehicle non-exhaust particulate matter. TRL Project Report for DEFRA, PPR065, pp. 25–26.

Boyland, E. 1950. The biological significance of metabolism of polycyclic aromatic compounds. *Biochemical Society Symposium* 5, 40–54.

Chen, Y.-C., Lee, W.-J., Uang, S.-N., Lee, S.-H., and Tsai, P.-J. 2006. Characteristics of polycyclic aromatic hydrocarbon (PAH) emissions from a UH-1H helicopter engine and its impact on the ambient environment. *Atmospheric Environment* 40, 7589–7597.

CONCAWE. 1998. Polycyclic aromatic hydrocarbons in automotive exhaust emissions and fuels. Report No. 98/55, Brussels.

Cooper, D.A. 2003. Exhaust emissions from ships at berth. *Atmospheric Environment* 37, 3817–3830.

CPCB. 2002. *Parivesh: Benzene in Air and Its Effects on Human Health.* Central Pollution Control Board, Ministry of Environment and Forest, Delhi, available at www.cpcb. nic.in.

Devos, O., Combet, E., Tassel, P., and Paturel, L. 2006. Exhaust emissions of PAHs of passenger cars. *Polycyclic Aromatic Compounds* 26, 69–78.

Esteve, W., Budzinski, H., and Villenave, E. 2006. Heterogeneous reactivity of OH radicals with phenanthrene. *Polycyclic Aromatic Compounds* 23, 441–456.

Fielberg, A.S., Longwell, J.P., and Sarofim, A.F. 2002. Metal enhanced soot and PAH formation. *Combustion and Flame* 92, 241–253.

Graslund, A. and Jernstrom, B. 1989. DNA–carcinogen interaction: Covalent DNA-adducts of benzo(a)pyrene 7,8-dihydrodiol 9,10-epoxides studied by biochemical and biophysical techniques. *Quarterly Review of Biophysics* 22, 1–37.

Guerin, M.R., Higgins, C.E., and Griest, W.H. 1987. The analysis of the particulate and vapour phases of tobacco smoke. *IARC Science Publication* 81, 115–139.

Haynes, B.S. 1991. Soot and hydrocarbons in combustion. In: W. Bartock and A.F. Sarofim (Eds), *Fossil Fuel Combustion: A Source Book*, pp. 261–326. Wiley, New York.

Hilscherova, K., Kannan, K., Kang, Y.S., et al. 2001. Characterization of dioxine-like activity of sediments from a Czech river basin. *Environmental Toxicology and Chemistry* 20, 2768–2777.

IARC. 1984. *Polynuclear Aromatic Compounds—Part I. Chemicals, Environmental and Experimental Data.* Monograph 32. International Agency for Research on Cancer (IARC), Lyon.

IARC. 1987. *IARC Monographs on the Evaluation of the Carcinogenic Risk of Chemicals to Humans.* IARC Suppl. 7, Lyon, France.

Jones, C.C., Chughtai, A.R., Murugaverl, B., and Smith, D.M. 2004. Effects of air/fuel combustion 30 ratio on the polycyclic aromatic hydrocarbon content of carbonaceous soots from selected fuels. *Carbon* 42, 2471–2484.

Kamens, R.M., Guo, J., Guo, Z., and McDow, S.R. 1990. Polycyclic aromatic hydrocarbon degradation by heterogeneous reactions with N2O5 on atmospheric particles. *Atmospheric Environment* 24(5), 1161–1173.

Laurikko, J. and Nylund, N.O. 1993. Regulated and unregulated emissions from catalyst vehicles at low ambient temperatures. SAE Technical Papers Series No. 930946.

Lim, M.C.H., Ayoko, G.A., and Morawska, L. 2005. Effect of fuel composition and engine operating conditions on polycyclic aromatic hydrocarbon emissions from a fleet of heavy-duty diesel buses. *Atmospheric Environment* 39, 7836–7848.

Lim, M.C.H., Ayoko, G.A., Morawska, L., Ristovski, Z.D., and Jayaratne, E.R. 2007. Influence of fuel composition on polycyclic aromatic hydrocarbon emissions from a fleet of in-service passenger cars. *Atmospheric Environment* 41, 150–160.

Lima, A.L.C., Farrington, J.W., and Reddy, C.M. 2005. Combustion-derived polycyclic aromatic hydrocarbons in the environment—a review. *Environmental Forensics* 6, 109–131.

Lin, Y.C., Lee, W.J., and Hou, H.C. 2006. PAH emissions and energy efficiency of palm biodiesel blends fueled on diesel generator. *Atmospheric Environment* 40, 3930–3940.

Machala, M., Ciganek, M., Blaha, L., Minksova, K., and Vondrack, J. 2001. Aryl hydrocarbon receptor-mediated and estrogenic activities of oxygenated polycyclic aromatic hydrocarbons and azaarenes originally identified in extracts of river sediments. *Environmental Toxicology and Chemistry* 20, 2736–2743.

Mandalakis, M., Tspakis, M., Tsoga, A., and Stephanou, E.G. 2002. Gas-particle concentrations and distribution of aliphatic hydrocarbons, PAHs, PCBs and PCDD/Fs in the atmosphere of Athens (Greece). *Atmospheric Environment* 36, 4023–4035.

Marquardt, H., Baker, S., Grover, P.L., and Sims, P. 1977. Malignant transformation and mutagenesis in mammalian cells induced by vicinal diol epoxides derived from benzo(a) pyrene. *Cancer Letter* 3, 31–36.

Marr, L.C., Dzepina, K., Jimenez, J.L., et al. 2006. Sources and transformations of particle bound PAHs in Mexico City. *Atmospheric Chemistry and Physics Discussions* 5, 12741–12773.

Marr, L.C., Kirchstetter, T.W., Harley, R.A., Miguel, A.H., Hering, S.V., and Hammond, S.K. 1999. Characterization of polycyclic aromatic hydrocarbons in motor vehicles fuels and exhaust emissions. *Environmental Science and Technology* 33, 3091–3099.

Mastral, T.A., Maria, A., and Callen, S. 2000. A review of polycyclic aromatic hydrocarbon emissions from energy generation. *Environmental Science and Technology* 34, 3051–3057.

Mitchell, K., Steere, D.E., Taylor, J.A., et al. 2000. Impact of diesel fuel aromatics on particulate, PAH and nitro-PAH emissions. SAE Emission Technology Collection, No. 942053.

MOE. 1997. *Scientific Criteria Document for Multimedia Standards Development. Polycyclic Aromatic Hydrocarbons (PAH). Part I: Hazard Identification and Dose–Response Assessment.* Ministry of the Environment (MOE), Toronto, Ontario.

Naeher, L.P., Smith, K.R., Brauer, M., et al. 2005. *Critical Review of the Health Effects of Woodsmoke.* Berkley: University of California. Available at http://ehs.sph.berkeley.edu/krsmith/publications/HC%20woodsmoke%20report%20Mar%2031%2005(rev).pdf

Ni, M.J., You, X.F., Li, X.D., et al. 2003. Study of PAHs formation from different kinds of coal combustion process. *Power Engineering* 11, 78–89.

Nisbet, C. and LaGoy, P. 1992. Toxic equivalency factors (TEFs) for polycyclic aromatic hydrocarbons (PAHs). *Regulatory Toxicology and Pharmacology* 16, 290–300.

Oanh, N.T.K., Reutergardh, L.B., and Dung, N.T. 1999. Emission of polycyclic aromatic hydrocarbons and particulate matter from domestic combustion of selected fuels. *Environmental Science and Technology* 33, 2703–2709.

Okuda, T., Kumata, H., Zakaia, M., Naraoka, H., Ishiwatari, R., and Takada, H. 2002. Source identification of Malaysian atmospheric polycyclic aromatic hydrocarbons nearby forest firs using molecular and isotopic compositions. *Atmospheric Environment* 36, 611–618.

Oros, D.R. and Simoneit, B.R.T. 2001a. Identification and emission factors of molecular tracers in organic aerosols from biomass burning. Part 1. Temperate climate conifers. *Applied Geochemistry* 16, 1513–1544.

Oros, D.R. and Simoneit, B.R.T. 2001b. Identification and emission factors of molecular tracers in organic aerosols from biomass burning. Part 2. Deciduous trees. *Applied Geochemistry* 16, 1545–1565.

Papageorgopoulou, A., Manoli, E., Touloumi, E., and Samara, C. 1999. Polycyclic aromatic hydrocarbons in the ambient air of Greek towns in relation to other atmospheric pollutants. *Chemosphere* 39, 2183–2199.

Pitts-Finlayson, B.J. and Pitts, J.N., Jr. 1986. *Atmospheric Chemistry: Fundamentals & Experimental Techniques*, pp. 877–892. Wiley, New York.

Pullman, A. and Pullman, B. 1955. Electronic structure and carcinogenic activity of aromatic molecules: New development. *Advances in Cancer Research* 3, 117–167.

Ravindra, K., Bencs, L., Wauters, E., et al. 2006a. Seasonal and site specific variation in vapor and aerosol phase PAHs over Flanders (Belgium) and their relation with anthropogenic activities. *Atmospheric Environment* 40, 771–785.

Ravindra, K., Wauters, E., Tyagi, S.K., Mor, S., and Van, G.R. 2006b. Assessment of air quality after the implementation of CNG as fuel in public transport in Delhi, India. *Environmental Monitoring and Assessment* 115, 405–417.

Reid, J.S., Koppmann, R., Eck, T.F., and Eleuterio, D.P. 2005. A review of biomass burning emissions part II: Intensive physical properties of biomass burning particles. *Atmospheric Chemistry and Physics* 5, 799–825.

Repace. 2003. An air quality survey of respirable particles and particulate carcinogens in Delawere hospitality venues before and after a smoking ban. Report prepared before the IMPACT Delawere Tobacco Prevention Coalition. Delawere, USA.

Rijkeboer, R. and Zwalve, W. 1990. Car emissions in the field. *Science of the Total Environment* 93, 159–166.

Sander, L. and Wise, S. 1997. *Polycyclic Aromatic Hydrocarbon Structure Index*. National Institute of Standards and Technology, Special publication 922, US Government Printing Office Washington, http://ois.nist.gov/pah/sp922_Result.cfm

Schauer, J.J., Kleeman, M.J., Cass, G.R., and Simoneit, B.R.T. 2002. Measurement of emissions from air pollution sources. 5. C1–C32 organic compounds from gasoline-powered motor vehicles. *Environmental Science and Technology* 36, 1169–1180.

Sheldon, L., Clayton, A., Keever, J., Perritt, R., and Whitaker, D. 1993. Indoor concentrations of polycyclic aromatic hydrocarbons in California residences. Contract A033-132, Final Report, Air Resources Board: Sacramento, CA.

Simoneit, B.R.T. 2002. Biomass burning—a review of organic tracers for smoke from incomplete combustion. *Applied Geochemistry* 17, 129–162.

Sjogren, M., Li, H., Banner, C., Rafter, J., Westerholm, R., and Rannug, U. 1996. Influence of physical and chemical characteristics of diesel fuels and exhaust emissions on biological effects of particle extracts: A multivariate statistical analysis often diesel fuels. *Chemical Research and Toxicology* 9(1), 197–207.

Tavares Jr. Moacir, Pinto Juandir, P., Souza Alexandre, L., Scarminio Ieda, S., and Solci Maria Cristina. 2004. Emission of polycyclic aromatic hydrocarbons from diesel engine in a bus station, Londrina, Brazil. *Atmospheric Environment* 38, 5039–5044.

Turns, S.R. 1996. *An Introduction to Combustion, Concepts and Applications*, pp. 291–297. McGraw-Hill, New York.

Vaaraslati, K., Virtanen, A., Ristimaki, J., and Keskinen, J. 2004. Nucleation mode formation in heavy-duty diesel exhaust with and without a particulate filter. *Environmental Science and Technology* 38, 4884–4890.

Velasco, E., Siegmann, P., and Siegmann, H.C. 2004. Exploratory study of particle bound polycyclic aromatic hydrocarbons in different environments of Mexico City. *Atmospheric Environment* 38, 4957–4968.

Watts, L.A. and Wallace, L. 1988. Major sources of benzene exposure. *Environmental Health Perspectives* 82, 165.

Westerholm, R. and Lin, H. 1994. A multivariate statistical analysis of fuel-related polycyclic aromatic hydrocarbon emission from heavy-duty diesel vehicles. *Environmental Science and Technology* 28, 962–972.

WHO/IPCS, 1998. *Environmental Health Criteria 202, Selected non-Heterocyclic PAHs*. WHO, Geneva.

Yang, H., Hsieh, L.T., Liu, H.C., and Mi, H.H. 2005. Polycyclic aromatic hydrocarbon emissions from motorcycles. *Atmospheric Environment* 39, 17–25.

9 Cellular Mechanisms behind Particulate Matter Air Pollution– Related Health Effects

*Ernesto Alfaro-Moreno, Claudia García-Cuellar,
Andrea De-Vizcaya-Ruiz, Leonora Rojas-Bracho,
and Alvaro R. Osornio-Vargas*

CONTENTS

9.1   INTRODUCTION

Urban air pollution represents a major problem affecting millions of people around the world. It has been considered by the World Health Organization (WHO) as responsible for 865,000 yearly deaths and 1.1 years of life lost due to premature mortality (DALYs)/1000 capita per year in urban settings worldwide (WHO, 2007). Epidemiological studies have provided most of the evidence linking air pollution to human disease (Samet and Krewski, 2007). Adverse health conditions and diseases affecting the respiratory and cardiovascular systems of vulnerable populations, such as children, the elderly, and those with preexisting cardiopulmonary illnesses, represent the main health concerns related to air pollution. Included among these health endpoints are increased cardiopulmonary mortality, asthma exacerbation, chronic obstructive pulmonary disease (COPD), decreased heart rate variability, hypercoagulability state, impaired lung function growth in children, and lung cancer, among others (Kunzli and Tager, 2005; Pope and Dockery, 2006; Samet and Krewski, 2007). Recently, attention has been paid to health outcomes affecting other systems, manifested as low birth weight (Bell et al., 2007) and the induction of chromosomal abnormalities in children (Pedersen et al., 2006). All these effects have been associated predominately with at least one of the so-called criteria pollutants that are routinely measured to assess air quality in most cities around the world. Criteria pollutants include airborne particulate matter (PM), ozone (O_3), sulfur dioxide (SO_2), nitrogen dioxide (NO_2), carbon monoxide (CO) and lead (Pb) (US-EPA, 2006). Given that this chapter focuses on PM, a description of this pollutant will follow.

PM comprises a complex mixture of solid and liquid components, which can vary significantly in size, composition, and shape. PM ranges in size from 0.005 to 100 μm in aerodynamic diameter (total suspended particles, TSP). PM is monitored in most urban settings as PM_{10} (particles with mean aerodynamic diameter $\leq$10 μm) and/or $PM_{2.5}$ (particles with mean aerodynamic diameter $\leq$2.5 μm). PM_{10} are inhalable particles, composed of four distinct fractions: coarse, fine, ultrafine, and nanoparticles. The coarse or thoracic fraction has an aerodynamic diameter between 10 and 2.5 μm ($PM_{10-2.5}$); fine particles are referred as respirable fraction ($PM_{2.5}$); ultrafine particles (UFPs) are smaller than 0.1 μm; and nanoparticles are 50 nm or less in diameter. Size of particles translates into their potential for deposition and clearance in the respiratory tract, which is highly relevant for triggering deleterious health effects.

Particle composition varies with size and emission source. Particles may contain metals, organic compounds, ions, condensed reactive gas products, microorganisms and microorganism products (e.g., endotoxins), nitrates, sulfates, and elemental carbon, among others (Pope and Dockery, 2006). A basic carbon structure, metals, hydrocarbons, and secondary particles compose fine particles and UFPs. These latter species include sulfates, nitrates, and organic compounds that are formed by condensation of high-temperature vapors or by atmospheric chemical reactions of nitrogen oxides, SO_2, volatile organic compounds (VOCs), and other reactive molecules (Hinds et al., 1985; HEI 2002). In contrast, the coarse fraction is formed by mechanical processes and is mainly composed of insoluble mineral oxides, hydroxides, sulfates, carbonates, halides, and so on; it may also contain biological matter, for instance, pollen and spores. Bacteria and virus may be found in both, the fine and coarse fractions (Spengler and Wilson, 1996).

There are multiple emission sources of atmospheric particles, driven by both natural and anthropogenic activities. Natural sources include dust storms, volcanic activity, forest fires, erosion, and so on. Anthropogenic sources include high-temperature metallurgic and combustion processes, involving wood, paper, oil, coal, or other fossil fuels. In urban settings, the main emission sources are automobiles and buses, as well as power plants and industrial activities (Molina et al., 2004; Molina and Molina, 2004).

Current understanding of PM health effects is rudimentary. It is likely that PM toxicity depends on their physicochemical properties, including size, shape, composition, and reactivity. Current evidence is not conclusive on the specific PM characteristics responsible for adverse health responses and toxicity. Multiple epidemiological studies conducted in urban settings have shown strong associations between mortality and atmospheric concentrations of fine particle mass, and between mortality and particle emissions associated with motor vehicles and other fossil and biofuel combustion processes (Hinds et al., 1985; Spengler and Wilson, 1996; US-EPA, 2000; HEI, 2002; Dominici et al., 2007). However, other studies have concluded that coarse particles or PM_{10}, composed predominantly of geological elements, were more strongly associated with total, respiratory, and cardiovascular mortality, than fine particles alone or PM_{10} with higher ratios of fine particles (i.e., mostly associated with combustion processes) (Bell et al., 2007; UNEP, 2007). Thus, based on current evidence, neither fine nor coarse particles can be considered harmless.

Although the risk of various adverse health effects increases with PM exposures, there is no indication or only weak evidence to suggest safe levels or a threshold below which adverse health outcomes would not be expected (Pope and Dockery, 2006; WHO, 2006). It is important to note that health outcomes have been found at concentrations that are not much higher than background levels, estimated for $PM_{2.5}$ at 3–5 $\mu g/m^3$ in the United States and Western European locations. Further, current evidence demonstrates that adverse effects occur in cities across the globe, including North and South America, Europe and Asia (WHO, 2006).

In spite of the existing epidemiological evidence documenting PM as the air pollutant with the most consistent and stronger associations with diverse adverse health outcomes, we are still far from understanding the mechanisms involved. Additionally, we do not have a clear picture regarding whether the reported adverse health effects should be fully attributed to some of the specific components of PM, to its composite mixture, or, moreover, whether PM is a proxy for the larger complex multi-air pollutants matrix.

The current evidence has been obtained mostly from criteria pollutants measurements. Nevertheless, it is clear that air pollution contains a wide range of substances, including "hazardous air pollutants" (HAPs). HAPs include metals, such as cadmium and mercury, and organics, such as benzene and polycyclic aromatic hydrocarbons (PAHs) (US-EPA 2006). Although selected HAPs have been monitored recently in some cities, it is not a common practice. Emerging literature indicates that air pollution components vary from city to city and, within a city, by region and season (Dominici et al., 2005). This raises the question of whether air pollution-related adverse health effects are the result of exposure to individual components or are they the result of complex interactions among pollutants. We need to strengthen capacities to collect data on multiple pollutants, as well as to develop methodologies to assess the effects of mixtures on the chronic multifactorial disease processes associated with air pollution exposures (e.g., atherosclerosis, cancer, etc.) (Pope and Dockery, 2006).

Uncertainties linked to findings emerging from epidemiological studies, such as difficulty in establishing causality, identification of thresholds, assessing biological plausibility, identification of mechanistic relationships, assessing effects of pollutants' mixtures, and so on, require complementary experimental research. Building the evidence necessary to understand and manage environmentally linked health problems could follow a path like the one presented in Figure 9.1. After identifying, characterizing, and measuring air pollutants in the atmosphere (air monitoring),

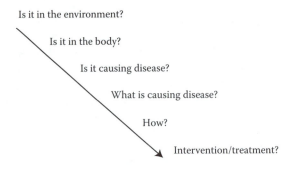

FIGURE 9.1 Schematic representation of the major steps of a theoretical pathway useful in the study of environmentally related health effects; these steps can be applied to air pollution.

estimating population exposures, and evaluating the relationship with potential health effects (epidemiology), there is an opportunity for experimental research focused on understanding biological plausibility, the role of relevant physiological components and the cellular mechanisms involved in the causation of adverse health effects (questions 3, 4, and 5 in the figure). Current understanding of the health effects related to air pollution has not arisen by following the linear path proposed in the figure. However, prior research has allowed the use of scientific evidence to formulate and design policy strategies and regulatory actions that, although far from complete, have been useful to mitigate the problem in various cities around the world (UNEP, 2007). For instance, in some megacities, maximum concentrations of particles and ozone have decreased significantly and the national ambient air quality standards are less frequently exceeded (Molina et al., 2004; Molina and Molina, 2004). This is the case for the Mexico City Metropolitan Area (MCMA) where improving fuel quality and introducing better vehicular emission control technologies (i.e., three-way catalytic converters), among other control measures, have helped improve air quality (Molina and Molina, 2002).

In the following sections, we will present the experimental evidence that has fed the path of knowledge, using "simplified" experimental settings in which cells and various types of PM have been used. Cell cultures of target cell types known to be relevant to organs affected by PM, and exposures to particles from various emissions sources and different nature (from surrogate PM of known composition to real urban particles), constitute a range of variables included in experiments described below. Three main biological response patterns have been observed after exposing cells *in vitro* to PM: cytotoxicity (necrosis/apoptosis), cytokine production, and genotoxicity. However, the precise role of these types of cellular responses in the health outcomes observed in humans still needs to be better understood.

Evidence collected by epidemiologists correlating PM exposures associated with increased risks of morbidity and mortality has found support in several experimental studies, in which the effects of PM have been tested *in vivo* and *in vitro* (Schwarze et al., 2006; Alfaro-Moreno et al., 2007b). Also, mathematical models have been used to predict deposition patterns of PM in the respiratory tract (Balashazy et al.,

2003). Additionally, statistical analysis with tools classically used in epidemiology has evaluated the association of various PM components with various biological effects (Becker et al., 2005a; Veranth et al., 2006; Rosas Perez et al., 2007). In the section below, we will discuss various aspects related to pulmonary and systemic effects associated with PM, together with the *in vivo* and *in vitro* evidence that has been collected so far.

9.2 PULMONARY EFFECTS OF PM

It is evident that the main impact of an inhaled particle is on the respiratory tract. Depending on aerodynamic size, particles can reach the upper airways (TSP), the lower airways (PM_{10}) or the alveoli ($PM_{2.5}$ and ultrafine PM) or may be translocated out of the respiratory tract and cause systemic damage (nanoparticles). Mathematical models, supported by animal studies, have shown that smaller particles can also be deposited in the airways (Balashazy et al., 2003), with a nonhomogeneous depositional pattern. Current evidence shows the existence of target hotspots, that is, high PM concentrations occurring at the bronchial and bronchiolar bifurcations (Balashazy et al., 2003) (Figure 9.2). This is of crucial importance for the analysis of the *in vitro* results, since PM concentrations at deposition hotspots could be similar to the ones used in *in vitro* analysis. Another aspect to consider is the difficulty of collecting representative PM samples for experimental studies that span relevant size ranges, and are representative of different sources. Therefore, some groups have used surrogate particles (e.g., residual oil fly ash—ROFA) to evaluate the role of size, physicochemical properties, or composition, while other groups have performed their experiments with collected real-world urban PM.

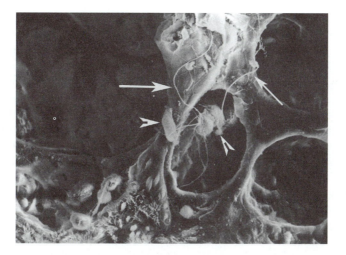

FIGURE 9.2 Scanning electron microscopy image of an airway bifurcation where particles preferentially deposit (arrows) after experimental inhalation in rats exposed to asbestos fibers. This correlates with particle deposition hot spots identified by modeling. Macrophages (arrowheads) migrate to deposited fibers several hours after exposure.

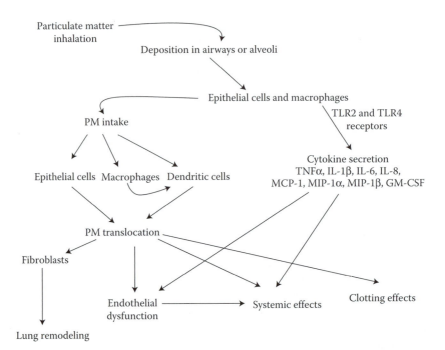

FIGURE 9.3 Possible chain of events related to PM exposure. Deposited PM in the airways or the alveoli can induce the production of cytokines that will activate other cells creating local or even a systemic effect. The translocation of the PM may explain effects induced by PM at distant sites.

Experiments have identified the cellular participants and the mechanisms involved in the response to PM exposure. The sequence of events is shown schematically in Figure 9.3. After PM is deposited in the lungs, particles will interact with macrophages or with epithelial cells. The presence of PM triggers cell responses that can be mediated by cellular receptors that, in turn, trigger a cascade of events within the cell that result in cell recruitment to remove deposited particles, the secretion of inflammatory factors (e.g., cytokines), or tissue remodeling (e.g., fibroblast/myofibroblasts recruitment/proliferation). If particles are not removed, they can be translocated beyond the epithelial barrier with the help of dendritic cells or by trans/interepithelial transport leading to effects on the interstitium or even beyond. They can reach the vascular system activating vessel endothelial cells or even distant organs. In the following sections, we will describe the role of molecules and cells that seem to have crucial roles in the biological response to PM exposure.

Among the many molecules present on the cell surface, there are two that have been discerned to play an important role in the response of epithelial cells and macrophages to PM exposure. The Toll-like receptors (TLRs) TLR-2 and TLR-4 belong to the TLR family, which comprises 11 members. TLR-4 is one of the most studied family members and it responds to the presence of lipopolysaccharides, molecules that are present in gram-negative bacteria. On the other hand, TLR-2 responds to lipoproteins and peptidoglycans, found on gram-positive bacteria (Tsan, 2006). The roles of these two receptors in response to PM exposure vary, macrophage response is related to the TLR-4 receptor, while epithelial cells respond via the TLR-2 (Becker et al., 2005b). When the TLR-2 receptor is blocked with an antibody on epithelial cells exposed to PM, the secretion of interleukin-8 (IL-8) is inhibited, but no inhibition is observed in the same cell line when the TLR-4 is inhibited directly or indirectly. In the case of alveolar macrophages, if the TLR-4 is inhibited, the secretion of interleukin-6 (IL-6) is inhibited. This shows that the response of two different cell types (macrophages and epithelial cells) to PM is related to different mechanisms with resulting differentiated outcomes (IL-6 versus IL-8 secretion). It is not clear if these responses can be attributed solely to the presence of lipopolysaccharides (endotoxins) or gram-negative bacteria in PM or are a result of the induction of other molecules, such as heat shock protein 70 (Hsp70), that are also activated after PM exposure.

Another group of substances that also play a crucial role in the response to PM exposure are cytokines. Cytokines are a group of peptides and proteins used by the cells as signaling molecules to communicate with each other. There are different cytokines related to specific tasks in the inflammatory response. They orchestrate the sequence of events that lead to cell recruitment (e.g., monocytes) via endothelial cell activation, necessary for further cell migration into the tissue, to mention a typical example (Male et al., 2006). Various cytokines participate in the cell response to PM. Tumor necrosis factor α (TNFα), IL-6, and IL-8 are three cytokines that have consistently been reported as important participants in the response to PM. Others, like interleukin-1β and interferon γ (IFNγ), play a crucial role at the immunological level. Growth factors, such as granulocyte colony-stimulating factor (G-CSF) and granulocyte macrophage colony-stimulating factor (GM-CSF), growth factor receptors, like platelet-derived growth factor receptor α (PDGF-Rα), and cell recruiting chemokines, such as monocyte chemotactic protein-1 (MCP-1), regulated upon activation, normal T-cell expressed, and secreted (RANTES), macrophage inflammatory protein (MIP)-1α, and MIP-1β, have been also described as important mediators of cell proliferation.

As previously mentioned, the first cytokines to be linked to PM exposure were TNFα, IL-6, and IL-8 (Monn and Becker, 1999; Alfaro-Moreno et al., 2002). TNFα is a molecule that triggers many different inflammatory-related responses, like the activation and recruitment of inflammatory cells and activation of endothelial cells. IL-6 is also a proinflammatory molecule, and elevations in TNFα levels induce the expression of IL-6. However, elevation of IL-6 downregulates TNFα; hence, these two molecules participate in a feedback cycle. Nevertheless, the secretion of IL-6 after exposure to PM can be independent of the secretion of TNFα. Exposing cells to PM_{10} collected from different MCMA locations, and therefore, associated with distinct emission sources and chemical composition, results in different patterns of inducing the expression of TNFα and IL-6. TNFα expression did not vary with PM location, while IL-6 showed up to 5-fold difference depending on PM location. These results indicate that IL-6 induction from PM_{10} exposure does not depend entirely on TNFα secretion and that differences in PM composition may play an important role in inducing biological effects (Alfaro-Moreno et al., 2002).

IL-8 is an important cytokine responsible for neutrophil recruitment. Most experimental studies report an upregulation of IL-8 levels after *in vivo* and *in vitro* exposures to PM (Monn and Becker, 1999). Nevertheless, recent reports show contradictory results, showing inhibition of IL-8 secretion by bronchial epithelial cells after exposure to different soil-derived PM samples, and decreased IL-8 plasma levels in children chronically exposed to urban air pollution (Calderon-Garcidueñas et al., 2003; Veranth et al., 2006). These results may indicate, as they did for IL-6 induction, that PM composition is relevant to the IL-8 response.

TNFα is mainly secreted by macrophages, although epithelial cells can also produce it. TNFα activates other cells to produce IL-6 and can also induce the secretion of IL-8, RANTES, growth regulated oncogene-alpha (GROα), GM-CSF and G-CSF. TNFα is a classical activator of endothelial cells, enhancing the expression of adhesion molecules and facilitating the migration of inflammatory cells. If the secretion of IL-8 is enhanced, neutrophilia increases. G-CSF and GM-CSF are cytokines that can induce systemic effects. TNFα is not the only factor that can induce the upregulation of other cytokines. The role of endotoxins is also very important. Endotoxins are capable of inducing the expression of MIP-1α and IL-1β, which are also, in turn, capable of inducing other cytokines (Figure 9.4). For instance, MIP-1α induces the secretion of TNFα, IL-1β, and IL-6, while IL-1β induces the secretion of G-CSF, GROα, IL-8, and MCP-1. The different mediators mentioned above have been related to *in vivo* and *in vitro* PM exposures; however, the sequence in which their secretion occurs and the relative participation of different cells in the regulation, amplification, and inhibition of each mediator are poorly understood.

The cascade of cytokine responses related to PM exposure is a very complex issue; different responses are observed when a single cell strain is exposed to PM. If we consider effects on different interacting cell types, the sociology of the cells becomes an important issue. It has been demonstrated that when epithelial cells are exposed to PM in the presence of macrophages or conditioned medium from previously exposed macrophages to PM, cytokine profile secretion is enhanced

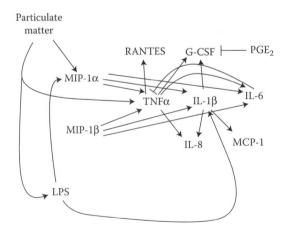

FIGURE 9.4 PM can induce a complex pattern of cytokine responses. PM or some specific components like endotoxin can induce the secretion of TNFα, MIP-1α, or IL-1β that stimulate other cytokines (i.e., RANTES, IL-6, IL-8, G-CSF, etc.).

(Ishii et al., 2004, 2005). Further work is necessary in this area to link the understanding of cellular effects with *in vivo* effects.

9.3 TRANSLOCATION OF PM

An important question related to the PM effects issue is how they induce effects beyond the respiratory tract or beyond the respiratory epithelial barrier. There are at least three hypotheses proposed to address this question.

First, the local induction of inflammatory factors by macrophages and epithelial cells in the lungs can induce activation of distant cells, such as endothelial cells. For example, TNFα can induce a dysfunctional phenotype of endothelial cells (Lum and Roebuck, 2001).

Second, the presence of soluble elements that compose PM may reach and activate distant cells. This has been partially demonstrated by us, when using the soluble fraction of PM to activate monocytes and endothelial cells (Alfaro-Moreno et al., 2007a). Also, other authors have reported the presence of soluble metals in distant organs (Kodavanti et al., 2008).

Third, PM can be translocated beyond the epithelial barrier and even cross the endothelial barrier reaching the bloodstream. Several studies have shown that PM translocation is possible, although the mechanisms are not fully understood.

With regard to PM translocation, Nemmar et al. (2002a) showed that humans exposed by inhalation to nanoparticles labeled with technetium-99 had a radioactive signal in the bladder and in the liver. These results are controversial, since another study did not observe particle translocation (Mills et al., 2006). This negative result could be related to particle clumping and/or the specific type of nanoparticles used by the authors, since further evidence using animal models showed the occurrence of PM translocation after intratracheal instillation (Nemmar et al., 2001).

There are two main hypotheses that attempt to explain how PM can translocate to other sites. Some evidence suggests that epithelial cells uptake particles and translocate them into lung capillaries (Kato et al., 2003). Recently, the role of dendritic cells in PM translocation has been shown *in vitro* (Blank et al., 2007). These authors demonstrate that dendritic cells are capable of crossing the epithelial barrier, phagocytizing the particles and translocating them to the other side of the barrier. These findings may be relevant for asthmatics and allergic individuals, who have increased populations of dendritic cells.

Other authors suggest that PM diffusion can occur. Geys et al. (2006) showed that the intercellular epithelial cell junctions (tight junctions) are lost temporally after adding PM. This evidence suggests that PM can translocate between the cells rather than through the cells.

Whatever be the translocation mechanism, once particles translocate into the interstitium they can contact fibroblasts, activating them and inducing airway remodeling. We have previously shown that PM upregulates the expression of PDGF-Rα (Bonner et al., 1998), making myofibroblasts more prone to migration and proliferation after stimulation with PDGF. These results suggest significant associations between PM exposure and increased risk of developing COPD (Schikowski et al.,

2005) or asthma attacks (Ingram and Bonner, 2006), diseases in which airway remodeling takes place.

Studies conducted with ultrafine PM show direct penetration of PM into the bloodstream (Brook et al., 2004). Small size PM, fine and ultrafine fractions, gain access to the bloodstream reaching various organs, including the brain. In dogs and humans exposed to high ambient air, PM neurodegenerative processes have been identified histologically (Peters et al., 2006).

9.4   EXTRAPULMONARY EFFECTS INDUCED BY PM

Chronic exposures to PM have been positively associated with adverse extrapulmonary health outcomes. For instance, the cohort study conducted by the American Cancer Society with over a 16 year follow-up reported that a $PM_{2.5}$ elevation of 10 µg/m^3 increased the risk of ischemic heart disease by a factor of 1.18 (95% CI: 1.14–1.23), and that the risk of mortality related to arrhythmia, heart failure, and cardiac arrest increased the risk ratio to 1.13 (95% CI: 1.05–1.21) (Pope et al., 2004). Similarly, the classic Harvard Six Cities study (Dockery et al., 1993), showed an adjusted mortality-rate ratio for the most polluted city compared with the least polluted one equal to 1.25 (95% CI: 1.08–1.47), with the largest impact of fine PM on cardiovascular-related deaths.

What are the mechanisms related to these health effects? *In vivo* and *in vitro* evidence has been collected in the last few years, and several possibilities have been proposed: (1) induction of endothelial dysfunction phenotype; (2) increase in clotting processes; (3) induction of changes in heart rate variability; and (4) induction of changes in blood density.

The induction of endothelial dysfunction phenotype by PM has been shown *in vitro* and *in vivo*. Endothelial cells exposed to PM show upregulation in the expression of E-selectin, P-selectin, intercellular adhesion molecule 1 (ICAM-1), platelet endothelial cell adhesion molecule (PECAM), and vascular endothelial cell adhesion molecule (VCAM) (Alfaro-Moreno et al., 2007a; Montiel-Davalos et al., 2007). The expression of these molecules is associated with increases in the adhesion of monocytes to the endothelial monolayer. *In vivo* studies by Nemmar et al. (2007) reported that P-selectin plays an important role in the inflammatory and prothrombotic effects induced after intratracheal instillation of carbon nanotubes in a mice model. Another *in vivo* study showed that after intratracheal instillation of ROFA in rats, systemic vascular dysfunction was observed, in addition to the local inflammation in the lung (Nurkiewicz et al., 2006).

Different groups have shown that clotting formation is related to exposure to PM. Nemmar et al. (2002b) showed that the intravenous injections of positively charged (amine-modified) polystyrene particles with 60 nm diameter induced an increase in the prothrombotic tendency measured in the femoral vein in a hamster model. This effect was partially related to platelet activation. They reported similar results when the particles were administrated by intratracheal instillation (Nemmar et al., 2002b). These results have been verified independently using an ear vein thrombosis model (Silva et al., 2005). Interestingly, the use of negatively charged polystyrene particles did not induce an increase in clotting processes, supporting the idea that the composition of PM plays an important role in the related effects.

Other studies have shown that the use of surrogate PM, such as diesel exhaust particles (DEPs), induces marked pulmonary inflammation, accompanied by enhanced venous and arterial thrombosis and platelet activation, 1 h after their deposition in the lungs (Nemmar et al., 2003b). These effects persist for 24 h after the instillation. The mechanism of these prothrombotic effects seems to be related to the histamine receptor (H1-receptor) (Nemmar et al., 2003a). Stabilization of mast cells prevented these effects, suggesting that mast cells may play an important role in the inflammatory and the prothrombotic effects (Nemmar et al., 2004). The cross talk between macrophages and neutrophils seems to be relevant. Neutrophil enzymes may be responsible for the priming of platelet activation and may contribute to the development of a thrombotic tendency (Nemmar et al., 2005). Recently, the role of IL-6 in the prothrombotic effect of PM has been investigated. When mice were exposed to PM_{10}, bleeding times were reduced, but when IL-6 knock out mice were exposed to the same particles, the animals' bleeding time was not reduced (Mutlu et al., 2007). Tissue factor is another molecule related to the control of clotting formation; it modulates the extrinsic pathway of the blood coagulation cascade. Karoly et al. (2007) exposed pulmonary artery endothelial cells of humans to UFPs and an upregulation in the expression of the tissue factor after 2 and 24 h of exposure was observed. Interestingly, soluble and insoluble fractions of the UFPs were capable of inducing similar effects.

Using an apoE$^{-/-}$ mice model (apolipoprotein E knock out animal that has been used as an atherosclerosis model), Sun et al. (2005) showed that long-term exposure to low concentrations of $PM_{2.5}$ altered the vasomotor tone and induced vascular inflammation and atherosclerosis.

Some studies have shown that PM is associated with increases in endothelin-1 levels *in vivo* and *in vitro* (Chauhan et al., 2005; Calderon-Garcidueñas et al., 2007). Endothelin-1 is responsible for induction of vasoconstriction, and has been correlated to pulmonary and cardiovascular conditions (Luscher and Barton, 2000). A population of MCMA children chronically exposed to urban pollution was compared with a population of children living in a town with similar geographical conditions but with much lower levels of air pollutants. The children from Mexico City showed increased endothelin-1 plasma levels, which were correlated to mean pulmonary arterial pressure (Calderon-Garcidueñas et al., 2007). All children included in the study were clinically healthy; a longer follow-up of these populations may help to better understand risk factors associated with pulmonary hypertension.

PM systemic effects may occur even if particles never reached the lungs. The impact of inhaled PM on the nervous system was shown by Elder et al. (2006). These authors investigated changes in the breathing pattern, heart rate, and heart rate variability and reported that UFP may reach the central nervous system through the olfactory neuronal pathway.

9.5 OXIDATIVE STRESS INDUCED BY PM

Oxidative stress has been proposed as the central mechanism that mediates PM-induced biological effects (Kunzli and Tager, 2005). Oxidative stress embraces a wide spectrum of conditions that modify cellular redox status. Oxidative stress can be defined as an imbalance between oxidant and antioxidant responses in the organism with enhanced oxidants producing an increased presence of reactive oxygen species (ROS) at an intracellular level. Oxidative stress has been identified as a causative factor of several conditions like asthma and COPD (Tao et al., 2003); genotoxicity and tumor promotion (Knaapen et al., 2004); cardiovascular alterations and hypertension (Peters, 2005); and, neurodegenerative disorders, such as Parkinson's and Alzheimer's disease (Qiang et al., 2004).

The main deleterious cellular responses modulated by ROS comprise DNA oxidative damage; oxidation of biological membranes; interference with the cell proliferation and growth regulatory processes through the activation of signaling pathways such as mitogen-activated protein (MAP) kinases or protein kinase C (PKC); and activation of transcription factors that trigger cell death signals.

As previously mentioned, PM is a complex mixture of organic and inorganic compounds that vary in size, shape, origin, and chemical composition. The mechanisms by which they exert cell damage are not fully understood, although recent experimental evidence from different research groups points to a key participation of ROS generated from redox-active transition metals, redox cycling quinoids, and PAHs contained in PM. Studies using electron paramagnetic resonance showed that ferrous ions, redox cycling of quinones and Fenton-type reactions induced by transition metals, contained in urban PM, participate in the generation of hydroxyl radicals and superoxide anions in the presence and absence of H_2O_2 (Valavanidis et al., 2005a). The generation of ROS after PM exposure can induce cellular oxidative stress and biological effects, such as necrosis, apoptosis, DNA damage, inflammation, cell proliferation, and mitochondrial damage. These biological effects can be directly induced by PM components stimulating intracellular sources of ROS, or indirectly from ROS produced by proinflammatory mediators released by PM-stimulated macrophages (Gonzalez-Flecha, 2004). The persistent exposure to PM can have important consequences on cellular homeostasis, causing direct toxic damage to cells that come in contact with them such as lung, myocardial or neuronal cells or inducing chronic inflammation that results in a general systemic inflammation and oxidative stress status (Kunzli and Tager, 2005; Pope and Dockery, 2006).

It is important to note that other pollutants, such as SO_2 and ozone, can also contribute to adverse health outcomes from exposure to air pollution. SO_2 is a colorless asphyxiating gas, formed during combustion of sulfur-containing fossil fuels, and it is more a reducing agent than an oxidizing one. It is not a free radical, since it does not contain unpaired electrons. However, it can react with water and oxidants to form sulfite, sulfate, and bisulfate ions that can generate ROS, as well as sulfite, sulfate, and bisulfate radicals, which in turn may cause lung damage and bronchoconstriction (Halliwell and Gutteridge, 2007). Ozone is a highly irritating colorless gas with a characteristic odor; although it is not a free radical, it is a highly oxidative pollutant. Low concentrations of O_3 (0.5 ppm) can cause lung damage in a few hours, by inducing inflammation, alveolar macrophage activation, and neutrophil recruitment. Ozone can oxidize proteins by attacking sulfhydrl groups (–SH), tyrosine, histidine, and methionine residues. Lipids also can be direct targets of O_3. Nasal human fluids are rich in urates, which react with O_3 and NO_2 favoring nasal irritation (Halliwell and Gutteridge, 2007).

Impaired function or absence of antioxidant cellular elements contributes to worsening the oxidative cell damage from the exposure to PM, SO_2, and O_3. Cellular exposure to oxidant agents, such as atmospheric pollutants, involves a coordinated cell response in which gene induction of antioxidant proteins is a key player. Some of these proteins are enzymes that directly or indirectly take part in the antioxidant response, allowing the cell to defend against ROS through:

1. Antioxidant enzymes such as superoxide dismutase (SOD), catalase (CAT), glutathione peroxidase (GPx), and glutathione-S-transferase (GST)
2. Stress proteins with enzymatic activity, such as heme-oxygenase (HO-1)
3. Nonenzymatic mechanisms regulated by intracellular molecules like glutathione (GSH), thioredoxin (TRx), vitamin E, vitamin C, and catechins (Klaunig and Kamendulis, 2004)

Koike et al. (2004) report the expression of HO-1, an enzyme regulated by transcription factors such as Nrf-2, NFκB, and AP-1, through the ARE pathway, after exposure to PM in rat alveolar epithelial cells. This suggests HO-1 can serve as a marker of oxidant response associated with PM exposure and induced lung injury (Jin and Choi, 2005).

Induction of oxidative damage by PM has been extensively studied in controlled exposures, but little evidence on open populations exposed to ambient PM and alterations of the antioxidant response are available.

Cytotoxicity, malondialdehyde formation (product of lipid peroxidation) and apoptotic cell death induction were related with a compensatory increased expression of antioxidants such as CAT, metallothionein, and the Nrf-2 transcription factor in rat lung epithelial cells exposed *in vitro* to PM (Choi et al., 2004). Studies in rats exposed to concentrated fine PM for 5 h also resulted in a reversible induction of lipid peroxidation and antioxidant enzymes in the heart and lung, suggesting an adaptive response to acute exposure (Gurgueira et al., 2002). In another study, single exposures to very high doses (7.5 and 37.5 mg/kg) of fine PM in rats also showed increased lipid peroxidation in organs like the heart, liver, lung, and testicles, but a significant reduction in SOD, CAT, and GPx activities in the lung, liver, kidneys, and brain. A depletion of GSH levels was observed. Although this systemic toxicity affecting the respiratory and cardiovascular systems was attributable to a pro-oxidant/antioxidant imbalance (Liu and Meng, 2005), it is important to further investigate alterations in the pro-oxidant/antioxidant balance using smaller doses closer to "real-world" exposures.

Several studies have shown that PM and/or specific PM chemical components participate in the generation of ROS and cellular oxidative damage. They have used antioxidant compounds like *N*-acetyl cysteine (NAC), SOD, and CAT and metallic scavengers like deferoxamine and intracellular signaling inhibitors (Dellinger et al., 2001; Zafarullah et al., 2003; Karlsson et al., 2005). Studies with combustion by-products or high metal-content PM indicate that there is oxidative stress in their induced biological effects. Rhoden et al. (2005) observed that the exposure of rats to concentrated PM increased cardiac oxidative stress and damage through autonomic signaling, and consequently, induced cardiac function alterations. *In vivo* studies show that the induction of an inflammatory response and lung oxidative damage in rats exposed by intratracheal instillation to carbon PM (mainly organic compound content) were diminished when a simultaneous exposure to NAC was performed (Pinho et al., 2005). The chemical composition of PM can influence the cellular response. A study using naked DNA from human epithelial cells showed that the metallic component of PM is an important ROS generator that contributes to oxidative damage (Prahalad et al., 2001).

Macrophages and phagocytes after *in vitro* and *in vivo* exposure to various kinds of particles like ROFA, DEPs, quartz particles, and ambient PM result in the generation of ROS through processes that involve the mitochondria. Mitochondrial mechanisms involved include respiration through the activation of NADPH enzymes, the respiratory burst and the activation of inducible nitric oxidase synthetase (iNOS). These processes translate into cytotoxicity, inflammation, DNA damage, and the stimulation of cellular proliferation as final cellular outputs involving oxidative stress (Knaapen et al., 2004).

9.6   CARCINOGENIC RESPONSES INDUCED BY PM

A striking epidemiologically observed health outcome related to air pollution exposure, specifically PM, is a consistent association with lung cancer mortality (Pope et al., 2002). The reported risk (8% increased risk of lung cancer mortality per 10 μg/m^3 increase in $PM_{2.5}$ concentrations) is relatively small compared to other known risk factors, like smoking. However, it is an important observation, since it is related to nonvoluntary exposures and adds another health outcome to the list of concerns linked to air pollution. This imposes new challenges for understanding the underlying mechanisms. Air pollution contains many potentially carcinogenic substances, including benzo [α] pyrene, benzene, 1,3-butadiene, dioxins, metals, oxidants, such as ozone and NO_2, as well as sulfur and other nitrogen oxides. Some of these may be particle components (Kyrtopoulos et al., 2001; Aust et al., 2002; Cohen, 2003; Sorensen et al., 2003; Knox, 2005), but still we know very little about their relative contribution to increased risk of lung cancer mortality.

In general, cancer occurrence is strongly related to environmental factors. According to the WHO, outdoor air pollution (using PM as the exposure variable) accounts for approximately 1.4% of the mortality the urban population worldwide; and the fraction of that attributable to trachea, bronchus, and lung cancer is 5% (WHO, 2007; Ostro, 2004).

WHO estimates are based on existing evidence relating outdoor air pollution to lung cancer mortality (Beeson et al., 1998; Pope et al., 2002; Knaapen et al., 2004; Pope and Dockery, 2006; Boffetta et al., 2007); however, methodological limitations prevent reaching final conclusions regarding causation (Ostro, 2004). Nevertheless, cumulative new evidence indicates the existence of significant associations between ambient PM exposures and other cancers, including childhood cancers (Knox, 2005), gastric cancer (Sjodahl et al., 2007), and breast cancer (Nie et al., 2007). In order to understand the associations between cancer mortality and air pollution exposures, we require information including the use of monitoring data from a more comprehensive group of pollutants (i.e., HAPs, in addition to criteria pollutants).

Among the routinely monitored air pollutants, PM is regarded as the main risk factor associated with increased lung cancer mortality risk, but precise mechanisms have not been identified. Theoretically, induction or inhibition of a number of metabolic enzymes (e.g., CYPs, GST, tyrosine kinase, etc.), disrupting homeostasis through oxidative stress, organelle dysfunction, ionic channel interference, signal transduction interference, alteration in hormone–receptor interactions, and interference with growth and differentiation may be among them. These processes may result in cytotoxicity, DNA damage, mutagenicity, and stimulation of proinflammatory cytokine production (Lee and Lee, 2006; Valko et al., 2006). Oxidative stress is an important mechanism proposed to promote a procarcinogenic environment in the cell. There is abundant evidence suggesting that ROS/reactive nitrogen species (RNS) mediate particle-induced genotoxicity and mutagenesis, but little information is available on the subsequent steps leading to neoplastic changes (Knaapen et al., 2004). Particles themselves, as well as particle-elicited events, such as activation of pathways of inflammation and proliferation, have been suggested as important players in particle-induced genotoxicity, mutagenesis, and carcinogenesis. Besides proinflammatory effects, particles may also induce genotoxicity given their ability to carry surface-adsorbed carcinogenic components into the lung. Each of these pathways can impact on the balance between genotoxicity and proliferation and protective mechanisms in which DNA repair and apoptosis are involved.

Most of these studies concentrate on understanding pathways involving organic compounds such as PAHs, dioxins (Matsumoto et al., 2007), or metals (Valavanidis et al., 2005b). However, we should keep in mind that PM is not a single component pollutant, but rather a mixture of pollutants from various sources, subject to continuous variation in composition and reactivity.

Two main modes of PM-related genotoxic lung cell damage can be considered: primary and secondary genotoxicity. Primary genotoxicity is defined as genetic damage elicited by particles in the absence of pulmonary inflammation. Secondary genotoxicity implies a pathway of genetic damage resulting from oxidative DNA attack by ROS/RNS, generated during particle-elicited inflammation. Inflammation has long been associated with the development of cancer elsewhere (Rakoff-Nahoum, 2006).

Conceptually, primary genotoxicity might operate via various mechanisms, such as the action of ROS (e.g., as generated from reactive particle surfaces), or DNA-adduct formation by reactive metabolites of particle-associated organic compounds (e.g., PAHs). Currently, scientific literature merely indicates that the tumorigenesis of poorly soluble particles such as TiO_2, black carbon, and DEPs involves a mechanism of secondary genotoxicity (Schins and Knaapen, 2007).

Cellular response to genotoxic stress may depend on the cell type being insulted, toxicant concentration, and duration of exposure. Although mutagenic mechanisms are known to play a role in many types of cancer, assessment of air pollutants mutagenicity must be conducted and analyzed carefully, due to the dynamic interactions existing in such complex mixture of contaminants (Claxton and Woodall, 2007). Rodent carcinogenicity models have confirmed the epidemiological evidence that particulate air pollutants are often carcinogenic (Claxton and Woodall, 2007; Yauk et al., 2008). Differences in the response of various strains of inbred mice indicate that the genetic background of the individual influences the likelihood of the tumorigenic response. Most of the cancers associated with air pollutants are probably due to multiple genotoxic agents present in a complex mixture.

Genotoxic cell damage requires compensatory cellular activities to maintain DNA integrity and subsequently prevent cell dysfunction and uncontrolled growth and proliferation. Cell cycle control, DNA repair, and apoptosis are among those involved in the survival of the cell and maintenance of the cell in a nonneoplastic satisfactory fate. The p53 tumor suppressor protein plays a central role in maintaining genomic integrity as part of the DNA damage-control pathways. p53 represents a major marker for cell cycle regulation and apoptosis, after PM experimental exposure. Cells exposed to PM upregulate p53 (García-Cuellar et al., 2002) and p53 upregulation correlates with the induction of apoptosis (Soberanes et al., 2006). More studies are still necessary to understand the role of this protein in PM-induced DNA damage and carcinogenic pathways.

Besides mutagenesis, epigenetic modifications provide a plausible link between the environment and gene expression alterations that might lead to pathological phenotypes. Epigenetic changes confer a selective advantage on precancerous cells that may already harbor other lesions such as mutations, DNA repair defects, and/or aberrant cell cycle control. An increasing body of evidence from animal studies demonstrates that the heritable environmentally induced epigenetic modifications underlie reversible trans-generational phenotypic alterations (Jirtle and Skinner, 2007; Yauk et al., 2008). This epigenetic mechanism represents a candidate to explain how early life exposures to traffic emissions/air pollution influence the risk of developing cancer later in adult life, for example, breast cancer (Nie et al., 2007). Information on epigenetic alterations promises to serve, in conjunction with knowledge on DNA sequence alterations, as an important piece of evidence necessary to understand the biological responses to environmental exposures (Sutherland and Costa, 2003; Wade and Archer, 2006). A major difference between epigenetic and genetic outcomes is that while the DNA sequence is static, the epigenome is a dynamic entity that changes with cell type, stage of cell cycle, responses to biological signaling systems, and to environmental changes. Deciphering how the epigenome responds to environmental exposures and how it could aid in predicting risk of disease holds great promise and will undoubtedly prove an important complement to mutational analyses.

Epigenetic modifications do no alter gene sequence, but rather gene expression, via packing and unpacking the DNA. Two molecular mechanisms that mediate epigenetic phenomena are DNA methylation and post-translational histone protein modifications such as acetylation, methylation, ubiquitination, sumoylation, and phosphorylation (Jaenisch and Bird, 2003). Changes in DNA methylation status are frequently associated with carcinogenesis. There are several possible mechanisms by which DNA methylation may participate in carcinogenesis. Hypermethylation of promoter regions may contribute to carcinogenesis by silencing tumor suppressor genes, whereas hypomethylation in oncogene promoters may lead to their overexpression. Hypomethylation may also lead to increased mutation rates (Chen et al., 1998).

Examples of environmental toxicants with epigenetic effects known to be present in PM (HEI, 2002; Rosas Perez et al., 2007) include nickel (Ni) and arsenic (As). Ni is a potent human and animal carcinogen. Despite its well-known carcinogenic properties, Ni is a nonmutagenic or weakly mutagenic agent; instead, this metal may act as an epigenetic carcinogen by altering the expression of some genes via DNA methylation and/or histone acetylation (Sutherland and Costa, 2003). Arsenic is also a human carcinogen, although the mechanisms involved are not completely understood. Transformation appeared to be related to global DNA hypomethylation and proto-oncogene hypomethylation is a possible contributor to arsenic carcinogenesis (Sutherland and Costa, 2003).

In spite of the epidemiological evidence relating air pollution with an increased risk of cancer mortality, experimental evidence supporting its carcinogenic effect is sparse. A range of experimental studies suggest PM-bound metals are responsible for genotoxic effects. However, the identification of the genotoxic contributions of specific metals contained in PM is difficult to establish (Osornio-Vargas et al., 2003). Zinc, copper, vanadium, iron, and nickel could be more relevant than other metals, owing to their capacity to promote ROS through Fenton chemistry. Soluble organic compounds appear to be strongly implicated in PM-induced cancer, but data from epidemiological and experimental studies are not yet sufficient to prove this point.

9.7 CONCLUSIONS, CONTROVERSIES, AND FUTURE DIRECTIONS

It is clear that *in vitro* experimentation has made important contributions to the understanding of PM-related health impacts, providing evidence for biological plausibility and the understanding of some of the underlying cellular mechanisms involved in the disease processes. We know that PM can trigger inflammatory responses, alter DNA or cause irreparable cellular damage resulting in cell death (Schwarze et al., 2006; Rosas Perez et al., 2007). These responses can occur through direct effects on cells or as a consequence of secondary reactions triggered by particles, including inflammation processes. Cell activation via surface receptors represents an example of the former type of mechanism (Bonner et al., 1998), and the second generation of ROS exemplifies reactions triggered by PM (Valavanidis et al., 2005a). The latter is the central piece of oxidative stress: the current working hypothesis to explain PM-related biological and adverse health outcomes.

Oxidative stress has been linked to exposures of specific components of the complex PM mixture, such as metals and some organics. ROS generation can also result from the inflammatory processes or from intracellular reactions centered in the mitochondria (Soberanes et al., 2006).

In spite of significant gains, there are uncertainties related to sampling methods; exposure–dose relationship in ambient environments; more precise interpolation of results from cellular studies to humans; the role of individual versus mixtures of PM components; the effects of particle properties on specific adverse health outcomes; and the relationship between PM and the more complex mix of polluted air.

Methodological uncertainties also exist in the sampling of urban PM for toxicological analysis. For example, how well does the collected sample represent the original PM suspended in ambient air or how efficiently can PM be recovered from the sampling substrate, which may be membranes or water. Air quality standards deal with PM size and mass fractions; thus, the majority of sampling methods have been designed for gravimetric analysis and not particle recovery. Currently, there are no standardized methods for the collection of PM for experimentation purposes. Methods should be designed to address issues such as PM mass and representativeness of sample size and composition. Interestingly, observed PM biological effects seem consistent, independently of the methods used to collect this pollutant. However, studying and weighing the relative participation of PM components in observed biological effects require understanding the uncertainties introduced by differences in collection methods and controlling analyses for sampling differences used by various research groups.

How PM concentrations used for *in vitro* studies translated into real human open population exposures is an unanswered question. Experimental studies have observed cellular effects induced with concentrations as low as 2.5 $\mu g/cm^2$, but optimal acute effects have been consistently reported around 40 $\mu g/cm^2$. Although some may consider this as a very high concentration not likely to be relevant for open general human population exposures, results from modeling and animal studies offer a different angle to analyzing this problem. Modeling efforts (Balashazy et al., 2003) and animal studies (Brody et al., 1981) have shown that particles deposit on hot spots at airway bifurcations, producing deposition enhancement factors of 100. This implies that a person would need to breathe in an approximate mass of 1000 mg of PM to reach a concentration of 40 $\mu g/cm^2$ at hot spots on airway bifurcations, given a human airways surface area of 2471 cm^2 (Mercer et al., 1994). Breathing such high PM masses (around 1000 mg) is not far fetched, since a person could breathe it at a rate of 30 L/min of air with average PM concentrations of 50 $\mu g/m^3$. Urban populations may be frequently exposed to these PM levels, as indicated by fixed site monitoring data from some large metropolitan areas, like Mexico City, where the 98th percentile of 24-h means may exceed 120 $\mu g/m^3$. Also, this level is consistent with the recently published WHO 24-h PM_{10} mean air quality guideline (SEMARNAT-2007, 2007). Although more research is needed in this regard, available information indicates that experimentally relevant concentrations are not that far from PM concentrations observed in urban settings affecting open human populations.

Linking experimental studies with health outcomes related to PM exposures in urban populations requires much additional insight. Proinflammatory cell effects may well have direct relevance to human health conditions like asthma, chronic bronchitis, and even some cardiovascular diseases. However, the relevance of cell death by necrosis/apoptosis on adverse health outcomes is not fully understood. Impaired defense/repair mechanisms could be invoked to explain outcomes like impaired lung growth, synergisms with infectious agents complicating acute respiratory adverse conditions, or enhanced susceptibility to microbial infections.

For human cancer occurrence, experiments have provided evidence of PM's capacity to damage DNA. DNA degradation or DNA adduct formation are the most commonly reported *in vitro* genotoxic effects; however, a better understanding of the mechanisms involved is still needed. Effects on DNA repair machinery, interaction with gene promoters or epigenetic regulation are examples of areas that have been sparsely studied and require more research. Characterizing the expression profiles of epigenetically labile genes that are susceptible to environmental deregulation will ultimately identify epigenetic biomarkers of environmental exposures and disease. These epigenetic biomarkers will hopefully allow for the identification of individuals with a propensity for adult onset of disease after exposure to air pollutants. Understanding how the environment exposures influence human health and disease will ultimately require a comprehensive knowledge of the human epigenome.

So far, understanding of the effects of various PM components has focused on the identification of specific PM toxics producing specific cellular effects. Orthodox experimental paradigms focus on toxicological evaluations on a one-to-one basis for toxicant effects. Metals and organics have been identified as major contributors to oxidative stress via ROS generation. However, there is indication that cellular responses result from interactions among PM components that lead to complex patterns of cellular responses. Producing and interpreting results obtained with PM-related mixtures represent an intellectual and methodological challenge. In this respect, outstanding new research has been conducted using concentrated ambient particles (CAPs) (Ghio and Huang, 2004). The use of CAPs allows the possibility of investigating PM component interactions and interactions between PM and pollutant gases. Also, recent experimental work features applied statistical analyses commonly used by epidemiologists to try to disentangle the cellular effects of the various PM components (Becker et al., 2005a; Veranth et al., 2006; Rosas Perez et al., 2007). These publications have been able to link PM components or PM originated from specific emission sources to various patterns of cellular responses. Although these results are statistical approximations, they provide clues to generate a hypothesis relating PM composition with differential cellular responses that may produce specific adverse human health outcomes. This is an example where toxicological evidence feeds and is complementary to associations observed by epidemiological studies.

9.8   CONCLUDING POINTS

Environmental exposures are involved in the causation or aggravation of several commonly occurring human diseases. Urban air pollution contributes to the global burden of disease, as shown by epidemiological and clinical studies.

Adverse health outcomes that have been linked to PM exposures include cardiopulmonary and cardiovascular morbidity and mortality, as well as extrapulmonary diseases, linked to systemic inflammation.

Findings from epidemiological studies show that PM contributes to the majority of the associated adverse health outcomes.

Epidemiological observations provide the framework for assessing the biological plausibility through experimental scientific work.

Experimental evidence indicates that respiratory and cardiovascular system toxicity can be attributed to proinflammatory conditions, oxidative stress (pro-oxidant/antioxidant imbalance), and neural stimulation derived from PM exposure.

Experimentation with PM indicates that exposed cells can be activated or killed.

The significance of the observed cellular effects on specific diseases requires further study.

Cells respond to various PM properties including: concentration, size, shape, and composition. Therefore, health response is linked to emission source.

Research to identify PM characteristics that determine toxicity will be very relevant for air quality management, since results will guide us to the most important sources for control.

Understanding carcinogenic effects induced by PM is still at its early stages.

In spite of the existing uncertainties (precise evaluation of human exposures, sample collection, mechanisms of damage, etc.), the relation between PM air pollution and human health effects has been established; available information has allowed for important advancement in regulatory actions that have resulted in the improvement of air quality at various locations worldwide.

Uncertainties include the role of PM as the main determinant of adverse health effects or as a proxy for some, or for complex mixtures of, air pollutants.

Future research on PM should consider the following:

   i.   Using samples from ambient air at relevant doses, relative to open population exposures.
   ii. Research oriented toward better understanding the potential effects of cumulative exposures (i.e., the lifetime course of the effects) and exposures occurring within susceptibility time windows in the development and growth of children.
   iii. Development of methodologies suitable for assessing the toxicological and adverse health outcomes of exposures to multipollutants, including the identification of the mechanisms and outcomes associated with different PM components, and interactions among them.

ACKNOWLEDGMENTS

Some of the original work discussed in this chapter was supported by: CONACyT (43138-M and 52830), CAM-MIT, and MILAGRO-CAM. E. Alfaro-Moreno was supported by the Belgian Science Policy (2007–2008) and the European Respiratory Society (Scholarship 80). C. García-Cuellar was supported by a "Parker B. Francis fellowship" from the Francis Fellowship Program. All authors contributed equally in writing the present chapter.

REFERENCES

Alfaro-Moreno, E., Martinez, L., García-Cuellar, C., et al. 2002. Biologic effects induced *in vitro* by PM10 from three different zones of Mexico City. *Environ. Health Perspect.* 110, 715–720.

Alfaro-Moreno, E., Lopez-Marure, R., Montiel-Davalos, A., et al. 2007a. E-Selectin expression in human endothelial cells exposed to PM10: The role of endotoxin and insoluble fraction. *Environ. Res.* 103, 221–228.

Alfaro-Moreno, E., Nawrot, T.S., Nemmar, A., et al. 2007b. Particulate matter in the environment: Pulmonary and cardiovascular effects. *Curr. Opin. Pulm. Med.* 13, 98–106.

Aust, A.E., Ball, J.C., Hu, A.A., et al. 2002. Particle characteristics responsible for effects on human lung epithelial cells. *Res. Rep. Health Eff. Inst.* 110, 1–65; discussion 67–76.

Balashazy, I., Hofmann, W., and Heistracher, T. 2003. Local particle deposition patterns may play a key role in the development of lung cancer. *J. Appl. Physiol.* 94, 1719–1725.

Becker, S., Dailey, L.A., Soukup, J.M., et al. 2005a. Seasonal variations in air pollution particle-induced inflammatory mediator release and oxidative stress. *Environ. Health Perspect.* 113, 1032–1038.

Becker, S., Dailey, L., Soukup, J.M., et al. 2005b. TLR-2 is involved in airway epithelial cell response to air pollution particles. *Toxicol. Appl. Pharmacol.* 203, 45–52.

Beeson, W.L., Abbey, D.E. and Knutsen, S.F. 1998. Long-term concentrations of ambient air pollutants and incident lung cancer in California adults: Results from the AHSMOG study. Adventist Health Study on Smog. *Environ. Health Perspect.* 106, 813–823.

Bell, M.L., Ebisu, K., and Belanger, K. 2007. Ambient air pollution and low birth weight in Connecticut and Massachusetts. *Environ. Health Perspect.* 115, 1118–1124.

Blank, F., Rothen-Rutishauser, B., and Gehr, P. 2007. Dendritic cells and macrophages form a transepithelial network against foreign particulate antigens. *Am. J. Respir. Cell Mol. Biol.* 36, 669–677.

Boffetta, P., McLaughlin, J.K., la Vecchia, C., et al. 2007. "Environment" in cancer causation and aetiological fraction: Limitations and ambiguities. *Carcinogenesis* 28, 913–915.

Bonner, J.C., Rice, A.B., Lindroos, P.M., et al. 1998. Induction of the lung myofibroblast PDGF receptor system by urban ambient particles from Mexico City. *Am. J. Respir. Cell Mol. Biol.* 19, 672–680.

Brody, A.R., Hill, L.H., Adkins, B.J., et al. 1981. Chrysotile asbestos inhalation in rats: Deposition pattern and reaction of alveolar epithelium and pulmonary macrophages. *Am. Rev. Respir. Dis.* 123, 670–679.

Brook, R.D., Franklin, B., Cascio, W., et al. 2004. Air pollution and cardiovascular disease: A statement for healthcare professionals from the Expert Panel on Population and Prevention Science of the American Heart Association. *Circulation* 109, 2655–2671.

Calderon-Garcidueñas, L., Mora-Tiscareno, A., Fordham, L.A., et al. 2003. Respiratory damage in children exposed to urban pollution. *Pediatr. Pulmonol.* 36, 148–161.

Calderon-Garcidueñas, L., Vincent, R., Mora-Tiscareno, A., et al. 2007. Elevated plasma endothelin-1 and pulmonary arterial pressure in children exposed to air pollution. *Environ. Health Perspect.* 115, 1248–1253.

Chauhan, V., Breznan, D., Thomson, E., et al. 2005. Effects of ambient air particles on the endothelin system in human pulmonary epithelial cells (A549). *Cell Biol. Toxicol.* 21, 191–205.

Chen, R.Z., Pettersson, U., Beard, C., et al. 1998. DNA hypomethylation leads to elevated mutation rates. *Nature* 395, 89–93.

Choi, J.H., Kim, J.S., Kim, Y.C., et al. 2004. Comparative study of PM2.5- and PM10-induced oxidative stress in rat lung epithelial cells. *J. Vet. Sci.* 5, 11–18.

Claxton, L.D. and Woodall, G.M.J. 2007. A review of the mutagenicity and rodent carcinogenicity of ambient air. *Mutat. Res.* 636, 36–94.

Cohen, A.J. 2003. Air pollution and lung cancer: What more do we need to know? *Thorax* 58, 1010–1012.

Dellinger, B., Pryor, W.A., Cueto, R., et al. 2001. Role of free radicals in the toxicity of airborne fine particulate matter. *Chem. Res. Toxicol.* 14, 1371–1377.

Dockery, D.W., Pope, C.A.R., Xu, X., et al. 1993. An association between air pollution and mortality in six U.S. cities. *N. Engl. J. Med.* 329, 1753–1759.

Dominici, F., McDermott, A., Daniels, M., et al. 2005. Revised analyses of the national morbidity, mortality, and air pollution study: Mortality among residents of 90 cities. *J. Toxicol. Environ. Health A* 68, 1071–1092.

Dominici, F., Peng, R.D., Zeger, S.L., et al. 2007. Particulate air pollution and mortality in the United States: Did the risks change from 1987 to 2000? *Am. J. Epidemiol.* 166, 880–888.

Elder, A., Gelein, R., Silva, V., et al. 2006. Translocation of inhaled ultrafine manganese oxide particles to the central nervous system. *Environ. Health Perspect.* 114, 1172–1178.

García-Cuellar, C., Alfaro-Moreno, E., Torres-Flores, V., et al. 2002. Differential expression of p53 and PARP in mouse fibroblast (L929) exposed to urban air particles (PM10) from different zones of Mexico City. *Toxicol. Lett.* 135, s140.

Geys, J., Coenegrachts, L., Vercammen, J., et al. 2006. *In vitro* study of the pulmonary translocation of nanoparticles: A preliminary study. *Toxicol. Lett.* 160, 218–226.

Ghio, A.J. and Huang, Y.C. 2004. Exposure to concentrated ambient particles (CAPs): A review. *Inhal. Toxicol.* 16, 53–59.

Gonzalez-Flecha, B. 2004. Oxidant mechanisms in response to ambient air particles. *Mol. Aspects Med.* 25,169–182.

Gurgueira, S.A., Lawrence, J., Coull, B., et al. 2002. Rapid increases in the steady-state concentration of reactive oxygen species in the lungs and heart after particulate air pollution inhalation. *Environ. Health Perspect.* 110, 749–755.

Halliwell, B. and Gutteridge, J.M.C. (Eds). 2007. *Free Radicals in Biology and Medicine*, 4th edition, ISBN13: 9780198568698; ISBN10: 019856869X; May 2007.Oxford University Press, New York, p. 105.

HEI. 2002. Understanding the health effects of components of the particulate matter mix: Progress and next step. Insight from HEI's Research Programs. HEI Perspectives, http://pubs.healtheffects.org/getfile.php?u = 244

Hinds, W.C., Liu, W.C., and Froines, J.R. 1985. Particle bounce in a personal cascade impactor: A field evaluation. *Am. Ind. Hyg. Assoc. J.* 46, 517–523.

Ingram, J.L. and Bonner, J.C. 2006. EGF and PDGF receptor tyrosine kinases as therapeutic targets for chronic lung diseases. *Curr. Mol. Med.* 6, 409–421.

Ishii, H., Fujii, T., Hogg, J.C., et al. 2004. Contribution of IL-1 beta and TNF-alpha to the initiation of the peripheral lung response to atmospheric particulates (PM10). *Am. J. Physiol. Lung Cell. Mol. Physiol.* 287, L176–L183.

Ishii, H., Hayashi, S., Hogg, J.C., et al. 2005. Alveolar macrophage–epithelial cell interaction following exposure to atmospheric particles induces the release of mediators involved in monocyte mobilization and recruitment. *Respir. Res.* 6, 87.

Jaenisch, R. and Bird, A. 2003. Epigenetic regulation of gene expression: How the genome integrates intrinsic and environmental signals. *Nat. Genet.* 33(Suppl.), 245–254.

Jin, Y. and Choi, A.M. 2005. Cytoprotection of heme oxygenase-1/carbon monoxide in lung injury. *Proc. Am. Thorac. Soc.* 2, 232–235.

Jirtle, R.L. and Skinner, M.K. 2007. Environmental epigenomics and disease susceptibility. *Nat. Rev. Genet.* 8, 253–262.

Karlsson, H.L., Nilsson, L., and Moller, L. 2005. Subway particles are more genotoxic than street particles and induce oxidative stress in cultured human lung cells. *Chem. Res. Toxicol.* 18, 19–23.

Karoly, E.D., Li, Z., Dailey, L.A., et al. 2007. Up-regulation of tissue factor in human pulmonary artery endothelial cells after ultrafine particle exposure. *Environ. Health Perspect.* 115, 535–540.

Kato, T., Yashiro, T., Murata, Y., et al. 2003. Evidence that exogenous substances can be phagocytized by alveolar epithelial cells and transported into blood capillaries. *Cell Tissue Res.* 311, 47–51.

Klaunig, J.E. and Kamendulis, L.M. 2004. The role of oxidative stress in carcinogenesis. *Annu. Rev. Pharmacol. Toxicol.* 44, 239–267.

Knaapen, A.M., Borm, P.J., Albrecht, C., et al. 2004. Inhaled particles and lung cancer. Part A: Mechanisms. *Int. J. Cancer* 109, 799–809.

Knox, E.G. 2005. Childhood cancers and atmospheric carcinogens. *J. Epidemiol. Community Health* 59, 101–105.

Kodavanti, U.P., Schladweiler, M.C., Gilmour, P.S., et al.2008. The role of particulate matter-associated zinc in cardiac injury in rats. *Environ. Health Perspect.* 116, 13–20.

Koike, E., Hirano, S., Furuyama, A., et al. 2004. cDNA microarray analysis of rat alveolar epithelial cells following exposure to organic extract of diesel exhaust particles. *Toxicol. Appl. Pharmacol.* 201, 178–185.

Kunzli, N. and Tager, I.B. 2005. Air pollution: From lung to heart. *Swiss Med. Wkly* 13, 697–702.

Kyrtopoulos, S.A., Georgiadis, P., Autrup, H., et al. 2001. Biomarkers of genotoxity of urban air pollution. Overview and descriptive data from a molecular epidemiology study on populations exposed to moderate-to-low levels of polycyclic aromatic hydrocarbons: The AULIS project. *Mutat. Res.* 496, 207–228.

Lee, K.W. and Lee, H.J. 2006. Biphasic effects of dietary antioxidants on oxidative stress-mediated carcinogenesis. *Mech. Ageing Dev.* 127, 424–431.

Liu, X. and Meng, Z. 2005. Effects of airborne fine particulate matter on antioxidant capacity and lipid peroxidation in multiple organs of rats. *Inhal. Toxicol.* 17, 467–473.

Lum, H. and Roebuck, K.A. 2001. Oxidant stress and endothelial cell dysfunction. *Am. J. Physiol. Cell Physiol.* 280, C719–C741.

Luscher, T.F. and Barton, M. 2000. Endothelins and endothelin receptor antagonists: Therapeutic considerations for a novel class of cardiovascular drugs. *Circulation* 102, 2434–2440.

Male, D., Brostoff, J., Roth, D., et al. 2006. *Immunology*, Elsevier, Amsterdam, The Netherlands, pp. 6–7.

Matsumoto, Y., Ide, F., Kishi, R., et al. 2007. Aryl hydrocarbon receptor plays a significant role in mediating airborne particulate-induced carcinogenesis in mice. *Environ. Sci. Technol.* 41, 3775–3780.

Mercer, R.R., Russell, M.L., Roggli, V.L., et al. 1994. Cell number and distribution in human and rat airways. *Am. J. Respir. Cell Mol. Biol.* 10, 613–624.

Mills, N.L., Amin, N., Robinson, S.D., et al. 2006. Do inhaled carbon nanoparticles translocate directly into the circulation in humans? *Am. J. Respir. Crit. Care Med.* 173, 426–431.

Molina, L.T. and Molina, M.J. 2002. *Air Quality in the Mexico Megacity: An Integrated Assessment*. Kluwer Academic Publishers, Dordrecht, The Netherlands, 384pp.

Molina, M.J. and Molina, L.T. 2004. 2004 Critical review: Megacities and atmospheric pollution. *J. Air Waste Manag. Assoc.* 54, 644–680.

Molina, L.T., Molina, M.J., Slott, R., et al. 2004. 2004 Critical review supplement: Air quality in selected megacities. *J. Air Waste Manag. Assoc.* http://www.awma.org

Monn, C. and Becker, S. 1999. Cytotoxicity and induction of proinflammatory cytokines from human monocytes exposed to fine (PM2.5) and coarse particles (PM$_{10}$-2.5) in outdoor and indoor air. *Toxicol. Appl. Pharmacol.* 155, 245–252.

Montiel-Davalos, A., Alfaro-Moreno, E., Lopez-Marure, R. 2007. PM$_{2.5}$ and PM$_{10}$ induce the expression of adhesion molecules and the adhesion of monocytic cells to human umbilical vein endothelial cells. *Inhal. Toxicol.* 19 (Suppl. 1), 91–98.

Mutlu, G.M., Green, D., Bellmeyer, A., et al. 2007. Ambient particulate matter accelerates coagulation via an IL-6-dependent pathway. *J. Clin. Invest.* 117, 2952–2961.

Nemmar, A., Vanbilloen, H., Hoylaerts, M.F., et al. 2001. Passage of intratracheally instilled ultrafine particles from the lung into the systemic circulation in hamster. *Am. J. Respir. Crit. Care Med.* 164, 1665–1668.

Nemmar, A., Hoet, P.H., Vanquickenborne, B., et al. 2002a. Passage of inhaled particles into the blood circulation in humans. *Circulation* 105, 411–414.

Nemmar, A., Hoylaerts, M.F., Hoet, P.H., et al. 2002b. Ultrafine particles affect experimental thrombosis in an *in vivo* hamster model. *Am. J. Respir. Crit. Care Med.* 166, 998–1004.

Nemmar, A., Nemery, B., Hoet, P.H., et al. 2003a. Pulmonary inflammation and thromboge-nicity caused by diesel particles in hamsters: Role of histamine. *Am. J. Respir. Crit. Care Med.* 168, 1366–1372.

Nemmar A., Hoet, P.H., Dinsdale, D., et al. 2003b. Diesel exhaust particles in lung acutely enhance experimental peripheral thrombosis. *Circulation* 107, 1202–1208.

Nemmar, A., Hoet, P.H., Vermylen, J., et al. 2004. Pharmacological stabilization of mast cells abrogates late thrombotic events induced by diesel exhaust particles in hamsters. *Circulation* 110, 1670–1677.

Nemmar, A., Nemery, B., Hoet, P.H., et al. 2005. Silica particles enhance peripheral thrombo-sis: Key role of lung macrophage-neutrophil cross-talk. *Am. J. Respir. Crit. Care Med.* 171, 872–879.

Nemmar, A., Hoet, P.H., Vandervoort, P., et al. 2007. Enhanced peripheral thrombogenicity after lung inflammation is mediated by platelet-leukocyte activation: Role of P-selectin. *J. Thromb. Haemost.* 5, 1217–1226.

Nie, J., Beyea, J., Bonner, M.R., et al. 2007. Exposure to traffic emissions throughout life and risk of breast cancer: The western New York exposures and breast cancer (WEB) study. *Cancer Causes Control* 18, 947–955.

Nurkiewicz, T.R., Porter, D.W., Barger, M., et al. 2006. Systemic microvascular dysfunction and inflammation after pulmonary particulate matter exposure. *Environ. Health Perspect.* 114, 412–419.

Osornio-Vargas, A.R., Bonner, J.C., Alfaro-Moreno, E., et al. 2003. Proinflammatory and cytotoxic effects of Mexico City air pollution particulate matter *in vitro* are dependent on particle size and composition. *Environ. Health Perspect.* 111, 1289–1293.

Ostro, B. 2004. *Outdoor Air Pollution. Assessing the Environmental Burden of Disease at National and Local Levels*, Annette Prüss-Üstün Diarmid Campbell-Lendrum, Carlos Corvalán, Alistair Woodward (eds). Environmental Burden of Disease Series, No. 5, http://www.who.int/quantifying_ehimpacts/publications/ebd5/en/; World Health Organization, Geneva, p. 62.

Pedersen, M., Vinzents, P., Petersen, J.H., et al. 2006. Cytogenetic effects in children and mothers exposed to air pollution assessed by the frequency of micronuclei and fluores-cence *in situ* hybridization (FISH): A family pilot study in the Czech Republic. *Mutat. Res.* 608, 112–120.

Peters, A. 2005. Particulate matter and heart disease: Evidence from epidemiological studies. *Toxicol. Appl. Pharmacol.* 207(Suppl. 2), 477–482.

Peters, A., Veronesi, B., Calderon-Garcidueñas, L., et al. 2006. Translocation and potential neu-rological effects of fine and ultrafine particles a critical update. *Part Fibre Toxicol.* 3, 13.

Pinho, R.A., Silveira, P.C., Silva, L.A., et al. 2005. *N*-acetylcysteine and deferoxamine reduce pulmonary oxidative stress and inflammation in rats after coal dust exposure. *Environ. Res.* 99, 355–360.

Pope, C.A.R., Burnett, R.T., Thun, M.J., et al. 2002. Lung cancer, cardiopulmonary mortality, and long-term exposure to fine particulate air pollution. *JAMA* 287, 1132–1141.

Pope, C.A.R., Burnett, R.T., Thurston, G.D., et al. 2004. Cardiovascular mortality and long-term exposure to particulate air pollution: Epidemiological evidence of general pathophysiological pathways of disease. *Circulation* 109, 71–77.

Pope C.A.R. and Dockery D.W. 2006. Health effects of fine particulate air pollution: Lines that connect. *J. Air Waste Manag. Assoc.* 56, 709–742.

Prahalad, A.K., Inmon, J., Dailey, L.A., et al. 2001. Air pollution particles mediated oxidative DNA base damage in a cell free system and in human airway epithelial cells in relation to particulate metal content and bioreactivity. *Chem. Res. Toxicol.* 14, 879–887.

Qiang, W., Cahill, J.M., Liu, J., et al. 2004. Activation of transcription factor Nrf-2 and its downstream targets in response to moloney murine leukemia virus ts1-induced thiol depletion and oxidative stress in astrocytes. *J. Virol.* 78, 11926–11938.

Rakoff-Nahoum, S. 2006. Why cancer and inflammation? *Yale J. Biol. Med.* 79, 123–130.

Rhoden, C.R., Wellenius, G.A., Ghelfi, E., et al. 2005. PM-induced cardiac oxidative stress and dysfunction are mediated by autonomic stimulation. *Biochim. Biophys. Acta* 1725, 305–313.

Rosas Perez, I., Serrano, J., Alfaro-Moreno, E., et al. 2007. Relations between PM10 composition and cell toxicity: A multivariate and graphical approach. *Chemosphere* 67, 1218–1228.

Samet, J. and Krewski, D. 2007. Health effects associated with exposure to ambient air pollution. *J. Toxicol. Environ. Health A* 70, 227–242.

Schikowski, T., Sugiri, D., Ranft, U., et al. 2005. Long-term air pollution exposure and living close to busy roads are associated with COPD in women. *Respir. Res.* 6, 152.

Schins, R.P. and Knaapen, A.M. 2007. Genotoxicity of poorly soluble particles. *Inhal. Toxicol.* 19(Suppl. 1), 189–198.

Schwarze, P.E., Ovrevik, J., Lag, M., et al. 2006. Particulate matter properties and health effects: Consistency of epidemiological and toxicological studies. *Hum. Exp. Toxicol.* 25, 559–579.

SEMARNAT-2007. 2007. Tercer almanaque de datos y tendencias de la calidad del aíre en nueve ciudades mexicanas, Instituto Nacional de Ecología, México, D.F.

Silva, V.M., Corson, N., Elder, A., et al. 2005. The rat ear vein model for investigating *in vivo* thrombogenicity of ultrafine particles (UFP). *Toxicol. Sci.* 85, 983–989.

Sjodahl, K., Jansson, C., Bergdahl, I.A., et al. 2007. Airborne exposures and risk of gastric cancer: A prospective cohort study. *Int. J. Cancer* 120, 2013–2018.

Soberanes, S., Panduri, V., Mutlu, G.M., et al. 2006. p53 mediates particulate matter-induced alveolar epithelial cell mitochondria-regulated apoptosis. *Am. J. Respir. Crit. Care Med.* 174, 1229–1238.

Sorensen, M., Autrup, H., Moller, P., et al. 2003. Linking exposure to environmental pollutants with biological effects. *Mutat. Res.* 544, 255–271.

Spengler, J.D. and Wilson, R. 1996. *Particles in Our Air: Concentrations and Health Effects.* Harvard University Press, Cambridge, MA.

Sun, Q., Wang, A., Jin, X., et al. 2005. Long-term air pollution exposure and acceleration of atherosclerosis and vascular inflammation in an animal model. *JAMA* 294, 3003–3010.

Sutherland, J.E. and Costa, M. 2003. Epigenetics and the environment. *Ann. N. Y. Acad. Sci.* 983, 151–160.

Tao, F., Gonzalez-Flecha, B., and Kobzik, L. 2003. Reactive oxygen species in pulmonary inflammation by ambient particulates. *Free Radic. Biol. Med.* 35, 327–340.

Tsan, M.F. 2006. Toll-like receptors, inflammation and cancer. *Semin. Cancer Biol.* 16, 32–37.

UNEP. 2007. Global Environment Outlook: Environment for development (GEO-4), http://www.unep.org/geo/geo4/report/02 Atmosphere.pdf Chapter 2: Atmosphere, pp. 39–80.

US-EPA. 2000. National air toxics program: The integrated urban strategy. Report to Congress http://www.epa.gov/ttn/oarpg/t3/reports/urbanrtc.zip

US-EPA. 2006. What are the six common air pollutions? http://www.epa.gov/air/urbanair/6poll.html

Valavanidis, A., Fiotakis, K., Bakeas, E., et al. 2005a. Electron paramagnetic resonance study of the generation of reactive oxygen species catalysed by transition metals and quinoid redox cycling by inhalable ambient particulate matter. *Redox Rep.* 10, 37–51.

Valavanidis, A., Vlahoyianni, T., and Fiotakis, K. 2005b. Comparative study of the formation of oxidative damage marker 8-hydroxy-2'-deoxyguanosine (8-OHdG) adduct from the nucleoside 2'-deoxyguanosine by transition metals and suspensions of particulate matter in relation to metal content and redox reactivity. *Free Radic. Res.* 39, 1071–1081.

Valko, M., Rhodes, C.J., Moncol, J., et al. 2006. Free radicals, metals and antioxidants in oxidative stress-induced cancer. *Chem. Biol. Interact.* 160, 1–40.

Veranth J.M., Moss, T.A., Chow, J.C., et al. 2006. Correlation of *in vitro* cytokine responses with the chemical composition of soil-derived particulate matter. *Environ. Health Perspect.* 114, 341–349.

Wade, P.A. and Archer, T.K. 2006. Epigenetics: Environmental instructions for the genome. *Environ. Health Perspect.* 114, A140–A141.

WHO. 2006. *WHO Air Quality Guidelines. Global Update 2005.* WHO Regional Office for Europe. Publications, Copenhagen, Denmark.

WHO. 2007. Environmental burden of disease: Country profiles. http://www.who.int/quantifying_ehimpacts/national/countryprofile/intro/en/index.html

Yauk, C., Polyzos, A., Rowan-Carroll, A., et al. 2008. Germ-line mutations, DNA damage, and global hypermethylation in mice exposed to particulate air pollution in an urban/industrial location. *Proc. Natl Acad. Sci. USA* 105, 605–610.

Zafarullah, M., Li, W.Q., Sylvester, J., et al. 2003. Molecular mechanisms of *N*-acetylcysteine actions. *Cell Mol. Life Sci.* 60, 6–20.

Section III

Health Risk Assessment
and Management

10 Emission of Airborne Particulate Matter in Indoor Environments

Exposure and Risk Assessment

Rajasekhar Balasubramanian, See Siao Wei, and Sathrugnan Karthikeyan

CONTENTS

10.1 INTRODUCTION

Air pollution is a major global environmental problem, particularly in urban areas where pollution levels are of great concern. It is responsible for a variety of environmental and health issues depending on the type of contaminants that are present in

TABLE 10.1
Classification of PM

d_a (µm)	Particle Terminology
$d_a > 10$	Suspended particulate matter (SPM) or total suspended particles (TSP)
$d_a \leq 10$	PM_{10} or inhalable fraction
$d_a \leq 2.5$	$PM_{2.5}$ or respirable fraction
$d_a > 10$	Supercoarse particles
$2.5 < d_a \leq 10$	Coarse fraction in respirable SPM
$0.1 < d_a \leq 2.5$	Fine fraction in respirable SPM
$0.05 < d_a \leq 0.1$	Ultrafine particles (UFPs)
$d_a \leq 0.05$	Nanoparticles

the atmosphere. The six most common (or "criteria") pollutants, for which national ambient air quality standards are set in many countries, are carbon monoxide (CO), nitrogen dioxide (NO_2), sulfur dioxide (SO_2), ozone (O_3), lead (Pb), and airborne particulate matter (PM). Over the past few decades, special attention has been paid to PM, which is a complex mixture of physically and chemically diverse substances that exist as solid particles or liquid droplets over a wide range of sizes. Airborne particles are classified into different categories based on aerodynamic diameters (d_a) as shown in Table 10.1.

PM typically ranges from 0.001 to 100 µm, but special focus is currently paid to the respirable fraction, that is, $PM_{2.5}$, as such particles can cause damage to human health. Due to their smaller sizes and aerodynamic behaviors, these particles can travel deeply into the lung and be trapped for a long period of time or even permanently. Both experimentally determined regional deposition data and mathematically derived deposition fraction in human lung models demonstrated that smaller particles such as fine particles, UFPs, and nanoparticles deposit more efficiently in the tracheobronchial and alveolar regions, whereas coarse fraction of respirable suspended PM (SPM) is more likely to be confined to the laryngeal and nasal regions (Heyder et al., 1986; Martonen et al., 2002). Given that the kinetics of the particle clearance processes is much slower in the lower respiratory tract (Miller, 1999), smaller particles containing toxic compounds have greater potential to affect human health adversely than larger particles with similar chemical composition. Some particles may even find their way into the bloodstream and eventually end up in other major organs such as liver, kidney, heart, and brain (Takenaka et al., 2001; Oberdorster et al., 2004), where similar inflammatory reactions can be expected.

As a result of the entry of smaller particles into the lungs, there exist a higher risk of acute (short-term) and chronic (long-term) health effects with an increase in levels of $PM_{2.5}$ than with an increase in levels of PM_{10}. While Samet et al. (2000) found a 0.5%, 1%, and 2% rise in mortality, cardiovascular, and respiratory admissions, respectively, per 10 µg m^{-3} increase in PM_{10}, the risk factors are even more considerable for $PM_{2.5}$, which has a 4%, 6%, and 8% increase, respectively, in all-cause, cardiopulmonary, and lung cancer mortalities for every 10 µg m^{-3} increase in $PM_{2.5}$

(Pope et al., 2002). The same conclusion is also drawn from a number of other recent studies (e.g., Burnett et al., 1999; Simpson et al., 2000). These epidemiological studies offer an estimate of the various health risks faced by the exposed population. However, they are often too time-consuming and costly and can be confounded by a number of other parameters, such as the presence of other gaseous pollutants, age, gender, weather, demographics, and socioeconomic status (Levy et al., 2000; Valberg, 2004). A simple quantitative approach to health risk assessment based on available toxicological data is therefore preferred as an alternative measure. Figure 10.1 illustrates the conceptual framework and the four key steps (labeled 1–4 in Figure 10.1) involved in the determination of potential health effects of exposure to particulate air pollution and, whenever possible, mitigation or remediation of the problems by means of effective control strategies (United States Department of Energy, 1999).

The initial step in any risk assessment process is data collection and evaluation. There are two possible approaches to obtain data sets that are relevant so that the ultimate aim of quantifying the risks faced by a targeted population can be met. The first approach is to measure the emission rates of a variety of particulate pollutants of health concern at the point of release (labeled 1a in Figure 10.1) and then predict the actual inhaled concentration of an exposed individual some distances away by means of air dispersion modeling. Several air quality models, which employ mathematical techniques to simulate the changes in concentration due to physical and chemical processes in the atmosphere, have been developed for this purpose. The Support Center for Regulatory Atmospheric Modeling (SCRAM) under the Air Quality Modeling Group (AQMG) at the United States Environmental Protection Agency

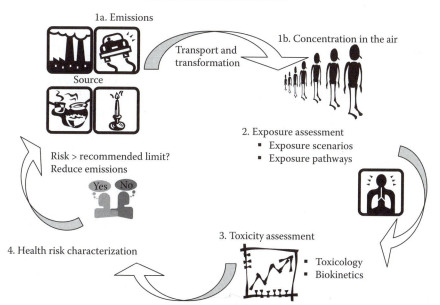

FIGURE 10.1 Conceptual framework and guidelines of health risk assessment.

(U.S. EPA) provides the source codes and associated documentation and guidance for a number of commonly used models. The second approach is to monitor the particulate pollutants directly at the point of exposure (labeled 1b in Figure 10.1).

In the first approach, the sources of interest have to be first identified. Particles can originate from both natural and anthropogenic (man-made) sources and are directly emitted into the atmosphere (primary particles) or formed from reaction of atmospheric gases (secondary particles). Since secondary emissions arise indirectly from primary emissions, control of particle emissions usually deals with the reduction of primary particle emissions. Biogenic, microbial, crustal, marine, and volcanic aerosols are all examples of primary particles from natural sources, while primary anthropogenic sources are mainly combustion processes burning coal, oil, natural gas, biomass, and other fuels. In general, particles generated by combustion sources are smaller in size within the fine range (Lighty et al., 2000) and are therefore of greater concern from a health standpoint. One of the main outdoor sources of primary $PM_{2.5}$ is internal combustion engines in petroleum- and diesel-fueled automobiles (Abu-Allaban et al., 2002; Lee et al., 2003; Vallius et al., 2005). On the other hand, smoking and cooking are two important sources influencing the indoor $PM_{2.5}$ concentration significantly (Wallace, 1996; Chao and Cheng, 2002). Tucker (2000) provides some estimates of the total emission rate of a few indoor and outdoor sources.

Once a particular source is selected, controlled laboratory experiments are usually carried out to determine their emission factors, as discussed by Zhang and Morawska (2002). Among the important combustion sources examined are gasoline and diesel engines (Marr et al., 1999; Kelly et al., 2003), biomass burning (Hays et al., 2005), environmental tobacco smoke (Klepeis et al., 2003), and incense burning (Lung and Hu, 2003). The repository, SPECIATE, maintained by the U.S. EPA contains the PM speciated profiles for an assortment of sources. Overall, this method helps to gain a better understanding of the amount and type of particles emitted from a single source. However, people are usually exposed to a mix of particulate pollutants from many different sources and thus the second approach may be desired. Other than monitoring ambient air (Marcazzan et al., 2001; Balasubramanian et al., 2003; Sharma and Maloo, 2005), indoor air quality has recently been widely studied as it is recognized that most people spend more than 80% of their time indoors (Morawska et al., 2003; Liu et al., 2004), in particular in occupational settings where people are exposed to high levels of particulates (Richardson, 2003; Bakoglu et al., 2004).

In summary, the first step directly measures or indirectly estimates the concentration of the harmful particulate contaminants to which people are exposed in either outdoor or indoor environments. However, as individuals are not in constant contact with the same environment throughout the day, other factors such as the magnitude, frequency, duration, and route of exposure (inhalation, ingestion, and skin absorption) have to be considered too. For airborne pollutants, the main pathway by which they come into contact with people is through inhalation. Such exposure assessment is taken into account in the second step of the risk assessment process. Once the appropriate exposure routes and scenarios have been identified, the inhaled dose can then be calculated as the amount of contaminant that is in contact with the body at an exchange boundary per unit body weight per unit time or the amount of contaminant that is absorbed by the body per unit body weight per unit time.

The third step is toxicity assessment, which estimates the relationship between the extent of exposure and the increased probability and/or severity of the adverse effects. Toxicity profiles for selected carcinogens, noncarcinogenic chemicals, and radionuclides can be obtained from the Risk Assessment Information System (RAIS) that collates all relevant information presented in the U.S. EPA's Integrated Risk Information System (IRIS) and the Health Effects Assessment Summary Tables (HEAST), and other published works. The last step, risk characterization, incorporates all the previous steps to calculate the risk resulting from exposure to chemicals for an exposed individual.

10.2   OBJECTIVES

In the present study conducted in Singapore, the goal was to monitor cooking emissions and their associated health risks in indoor environments. Cooking using liquefied petroleum gas (LPG) is considered to be one of the important combustion sources of aerosols indoors as it releases a significant amount of $PM_{2.5}$ in residential homes and commercial buildings with nonsmokers, and has been linked to lung cancer and respiratory diseases (Jarvis et al., 1996; Ko et al., 2000). A comprehensive study of the characterization and risk assessment of $PM_{2.5}$ of cooking origin is especially important for Singapore since it is popularly known as a food paradise and has more than 21,400 food stalls in food courts, private eating establishments, and hawker centers as of 2004. Being a multiracial country, most food or eating places around Singapore feature a magnificent showcase of diverse local and international food. Hence, a multitude of cooking methods in different types of stalls are used, which need to be further investigated from the risk assessment standpoint.

The objectives of this study were to use a combination of controlled experiments and real-world measurements to evaluate the potential health risks faced by the chefs. To demonstrate the importance of both physical and chemical properties of particles in determining risks, this chapter presents both the physical characteristics (derived from controlled experiments) and the chemical properties (derived from real-world studies), and their relative potential risks in the respective sections given below.

10.3   CASE STUDY 1 (CONTROLLED EXPERIMENTS)

10.3.1   SAMPLING SITE

This study was carried out during March 2005 in the kitchen of a residential house in Singapore. Figure 10.2 shows the layout of the tiled kitchen (~21 m³), which has two-burner gas stoves connected to the local town gas system. Only one of the two burners was used during the cooking experiments, and the instrument was placed on an elevated platform with its sample port facing the used burner. The sampling point was fixed at ~1.5 m above the ground, that is, ~0.5 m above the gas stove,

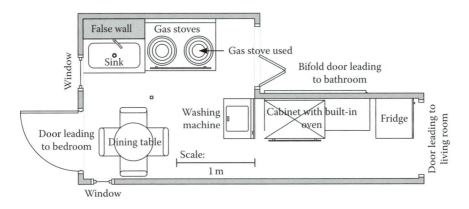

FIGURE 10.2 Layout of the kitchen, (Reproduced from See, S.W. and Balasubramanian, R. 2008. *Atmospheric Environment* 42(39), 8852–8862. With permission.)

to simulate the human breathing zone. During the aerosol sampling, no particle sources existed in the kitchen except for cooking. All the doors and windows in the kitchen were closed so as to prevent infiltration of particles from outdoor sources. During the course of experiments, no individuals were present in the house except for the investigator, and no human activities other than cooking took place in the house.

10.3.2 SAMPLE COLLECTION

The particle sizing and monitoring instrument used in this work was a TSI Model 3034 Scanning Mobility Particle Sizer (SMPS; TSI Incorporated, MN, USA). The SMPS measures the number of particles at an inlet flow rate of 1 L min^{-1} in the range of 10–500 nm over 54 channels and determines the total particle number concentration up to 10^7 cm^{-3}. The sampling time was set to 3 min, which corresponded to one scan per sample.

Five commonly used cooking methods were investigated in this study in terms of their association with indoor particulate air pollution in the kitchen: steaming (to cook over boiling water), boiling (to cook in a liquid heated to, or past its boiling point), pan-frying (to fry in a small amount of oil), stir-frying (to fry quickly in a small amount of oil over high heat while stirring continuously), and deep-frying (to fry by immersing in hot oil). In all cases, a pack (150 g) of plain tofu (soybean curd), bought from a local supermarket, was cut into 10 circular pieces measuring ~3.0 cm (diameter) by ~1.5 cm thickness and cooked in the same Chinese wok generating a steady heat of 3.0 kW for a sufficiently long time to obtain a representative number of samples per cooking method each time. The experimental conditions of different cooking methods can be found in Table 10.2. Between each cooking experiment, the wok was washed thoroughly. The next experiment commenced when the particles' levels returned to their baseline level; this level refers to the particle number concentration of all sizes within ±10% of the background concentration measured in the kitchen when there was no cooking.

TABLE 10.2

Experimental Conditions for Different Cooking Methods

Method	Properties of Water/ Oil Used	Volume of Water/ Oil Used (cm³)	Maximum Temperature Reached by Water/Oil (°C)
Steaming	Tap water	1000	100
Boiling	Specific gravity = 1.0	1000	100
	Boiling point = 100°C		
Pan-frying	Corn oil	15	190
Stir-frying	Specific gravity = 0.9	15	190
Deep-frying	Boiling point = 245°C	1000	190
	Smoking point = 235°C		

Source: Reproduced from See, S.W. and Balasubramanian, R. 2008. *Atmospheric Environment* 42(39), 8852–8862. With permission.

10.3.3 RESULTS AND DISCUSSION (CASE STUDY 1)

10.3.3.1 Number Concentration of Particles

The number concentrations of nanoparticles (10–50 nm), UFPs (50–100 nm), and accumulation mode particles (100–500 nm) are given in Table 10.3 for background (with no cooking) and the five different cooking methods, and their respective percentages. The size distributions of the particle concentration are illustrated in Figure 10.3 for different weight units. The horizontal axis represents the particle size (D_p) on a logarithmic scale and the vertical axis represents the normalized number concentration ($dN/d \log D_p$), which allows the size distribution to be compared regardless of the channel resolution. The number of modes and the modal diameters can be determined by visual inspection of the individual figures.

TABLE 10.3

Number Concentration of Size-Differentiated Particles Associated with Different Cooking Methods

D_p (µm)	Particle Number Concentration (10⁴ cm⁻³)					
	Background	Steaming	Boiling	Stir-Frying	Pan-Frying	Deep-Frying
0.01–0.05	1.7 ± 0.6	3.0 ± 0.6	4.3 ± 2.9	6.4 ± 3.3	8.6 ± 4.0	53.8 ± 26.5
	(68 ± 7%)	(55 ± 6%)	(62 ± 11%)	(69 ± 3%)	(78 ± 3%)	(90 ± 7%)
0.05–0.1	0.5 ± 0.2	1.4 ± 0.4	1.5 ± 0.4	2.1 ± 1.9	1.4 ± 0.4	4.2 ± 1.8
	(19 ± 4%)	(27 ± 4%)	(22 ± 7%)	(23 ± 6%)	(12 ± 2%)	(7 ± 4%)
0.1–0.5	0.3 ± 0.1	1.0 ± 0.3	1.1 ± 0.4	0.8 ± 0.1	1.0 ± 0.4	1.4 ± 0.5
	(13 ± 3%)	(18 ± 3%)	(16 ± 4%)	(8 ± 3%)	(9 ± 1%)	(2 ± 3%)
Total	2.5 ± 0.7	5.4 ± 1.2	6.9 ± 3.5	9.3 ± 5.3	11.0 ± 4.8	59.5 ± 27.3

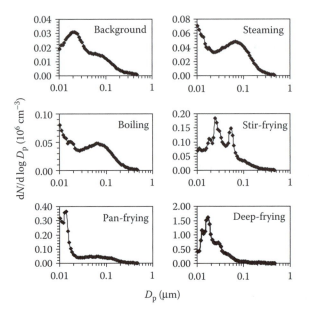

FIGURE 10.3 Size distribution of the number concentration associated with different cooking methods.

From the mean number concentrations, it can be seen that gas cooking emitted a substantial amount of particles. A 2–24-fold increase in the total counts over the background level was observed during gas cooking, especially during oil-based cooking methods such as pan-frying, stir-frying, and deep-frying, compared to water-based cooking methods such as steaming and boiling. This trend in the particle number concentrations can be attributed to high temperature heating of cooking oil (fatty acids), which presumably generated more particles than the boiling of water. In addition, water, with a boiling point of 100°C, is much more volatile than corn oil, which has a boiling point of 245°C; therefore, under high temperature in the kitchen, water droplets most likely exist in the gaseous phase than in the particulate phase, while less volatile oil droplets tend to remain as particles.

As can be seen from the table, deep-frying produced the highest number of particles, followed by stir-frying, pan-frying, boiling, and finally steaming. As the same food was cooked under the same heat setting on the stove, the much higher number concentration of particles during deep-frying compared to stir-frying and pan-frying can be attributed to the larger quantity of cooking oil used and not the temperature of the oil (Siegmann and Sattler, 1996) since the temperature was stabilized at ~190°C before tofu was added, which in turn was kept at room temperature to reduce the drop in temperature. Both stir-frying and pan-frying methods used the same amount of oil. As a result, the resulting particle counts were comparable between these two methods. However, the number concentration during stir-frying was found to be slightly lower since the turbulence created by continuous stirring action could have caused a more effective dispersion of particles. Moreover, the temperature of the oil was observed to decrease slightly by ~5°C during stirring, which might explain the lower concentration. On the other hand, contrary to the conclusions reached by Siegmann and Sattler (1996) who found the peak diameter to increase with temperature, the modal diameter did not decrease in the case of stir-frying in spite of the lower temperature. This could be a result of the different techniques employed in the two studies; there was no stirring process in the other study. Steaming also released lower particle counts than boiling as the wok was covered during steaming and thus particles could have been trapped inside.

This association between particle counts and different cooking methods appeared to be similar to the observations reported in the literature previously. For example, Li et al. (1993) reported that the total number mean concentrations of particles in the size range of 17–886 nm were 1.9×10^5, 3.3×10^5, and 4.0×10^5 cm^{-3} during the cooking of vegetable soup (considered to be boiling according to this study), scrambling eggs (stir-frying), and frying chicken (deep-frying), respectively. Dennekamp et al. (2001) reported a peak UFP concentration of 1.1×10^5 cm^{-3} when boiling water, 1.4×10^5 cm^{-3} when stir-frying vegetables, and 5.9×10^5 cm^{-3} when pan-frying bacon.

The findings from this study further reveal that cooking with oil (stir-frying, pan-frying, and deep-frying) liberated more harmful submicron-sized particles than cooking with water (steaming and boiling). Nanoparticles accounted for 69% (stir-frying) to 90% (deep-frying) of all particles during oil-based cooking as compared to 55% during steaming and 62% during boiling. On the other hand, water-based cooking produced more ultrafine and accumulation mode particles presumably due to the higher humidity in the kitchen. The water vapor generated, while boiling water, could have condensed on preexisting nanoparticles to form larger particles in the ultrafine and accumulation mode. This becomes apparent by the shift in the modal diameter over the course of steaming and boiling experiments as illustrated by Figure 10.4. Three spectra representing the size distribution at the start, the middle, and the end of the cooking experiment are shown. It can be seen that the peak shifted toward the right as the cooking progressed. The number of particles in the nano range decreased, while that in the ultrafine and accumulation mode increased. Such a trend was not observed during the three different frying operations.

During steaming and boiling, the size distributions of the number concentration were clearly bimodal with a dominant peak <10 nm, a smaller peak at 70–80 nm, and a local minimum at ~25 nm. Pan-frying also showed a somewhat similar profile although the mode in the ultrafine range was not so obvious. This is in good agreement with the study conducted by Wallace et al. (2004) despite the fact that the majority of the cooking episodes actually involved frying. The latter study revealed a bimodal distribution with a peak at or below 10 nm and a second peak at ~60 nm with an intervening minimum occurring at ~16 nm from 24 cooking episodes. However, it was found that stir-frying and deep-frying methods showed more complex and jagged profiles with at least three peaks, although all the modal diameters were confined to the ultrafine and nano ranges.

Dennekamp et al. (2001) noted a larger particle size at peak concentration during pan-frying bacon (69 nm) and stir-frying (41 nm) than during boiling water (17 nm). Wallace et al. (2004) observed a shift toward larger particle sizes during elaborate dinner cooking in comparison to simpler breakfast cooking (mainly boiling water and using an electric toaster). These observations are consistent with the findings

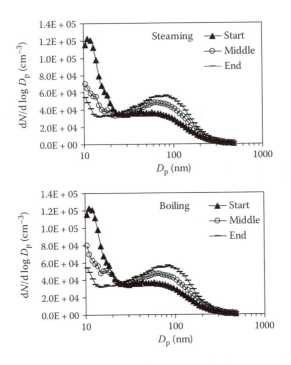

FIGURE 10.4 Representative size distributions of particle number concentrations at different times during steaming and boiling.

reported in this chapter. Steaming and boiling produced a peak at <10 nm, whereas the modal diameter was found to be 24.6 nm during stir-frying, 13.8 nm during pan-frying, and 17.2 nm during deep-frying.

10.3.3.2 Health Risk Assessment

One of the most important parameters in the evaluation of potential health impacts based solely on physical characteristics is the amount of particles that can be deposited in the lung, which is mainly governed by D_p (nm). For particles of aerodynamic diameter, 10–500 nm, almost all particles would be deposited in the alveolar region (Heyder et al., 1986). A function was derived using the mean experimental values of the deposition fraction of 5–700 nm particles in three healthy human subjects during nasal breathing reported by Heyder et al. (1986). FindGraph (UNIPHIZ Lab) software was employed for curve fitting and the best-fit function was found to be as per the following equations:

$$f_j = 0.8550 - \frac{2.991 D_{p_j}}{178.1 + D_{p_j}} + \frac{3.112 D_{p_j}}{525.7 + D_{p_j}} \quad \text{(Standard error} = 0.02102), \quad (10.1)$$

$$N = \sum_{\text{all } j} N_j = \sum_{\text{all } j} C_j \times f_j, \quad (10.2)$$

TABLE 10.4
Particle Deposition in the Lung

	Background	Steaming	Boiling	Stir-Frying	Pan-Frying	Deep-Frying
N (10^4 cm^{-3})	1.3	2.7	3.6	5.5	6.7	37.6
f	0.54	0.49	0.53	0.55	0.61	0.63

$$f = \frac{N}{C},$$

(10.3)

where f_j and f denote the deposition fraction of particles of size j and that of all particles from 10 to 500 nm, respectively. C_j (cm^{-3}) and C (cm^{-3}) are number concentration of particles of size j and that of all particles from 10 to 500 nm, respectively. N_j (cm^{-3}) and N (cm^{-3}) represent the number of particles of size j and that of all particles from 10 to 500 nm that can be deposited in the lung per cm^3 of air breathed in, respectively. The above equations assume that all particles contained in cooking emissions are breathed in.

As evaluated from Equations 10.1 through 10.3, almost half or more than half of the particles are deposited in the lung (see Table 10.4). Such a large number of deposited particles can result in particle overload, which will in turn lead to acute inflammation and impairment of alveolar macrophage-mediated lung clearance. This vicious cycle is likely to continue as more particles are accumulated due to retarded particle clearance, which could eventually result in lung tumor following a sequence of pulmonary events (Oberdorster et al., 1995). Von Klot et al. (2005) found that cardiac readmissions increased with the number concentration at a rate ratio of 1.026 per 10^4 cm^{-3}. Therefore, exposure to cooking emissions is of health concern and should be reduced as much as possible. If unavoidable, cooking should take place in a well-ventilated setting so as to decrease the number concentration and lessen the potential health impacts.

10.4 CASE STUDY 2 (REAL-WORLD ENVIRONMENT)

10.4.1 DESCRIPTION OF SAMPLING SITES

Three food stalls (one Chinese, one Malay, and one Indian) in one of the canteens within the National University of Singapore (NUS) Kent Ridge campus were chosen as the sampling sites. These commercial food stalls are located in the same food court, and are not air-conditioned, that is, no mechanical ventilation is employed. The stalls are small, and have an identical layout with a floor area of approximately 6–8 m^2. As a result, they experience almost the same level of natural ventilation and also have similar air dispersion conditions. Although the three kitchens are situated in the same food court, the indoor air in each stall is unlikely to be influenced by pollution from other stalls as they are physically separated.

All three food stalls are naturally ventilated through the front counter and the back door during their operating hours (also known as the cooking hours) from 07:30

to 19:30 and are completely closed otherwise. There are four LPG stoves below an exhaust fume extractor on the right-hand side and the sampler was placed on the opposite side of the stoves, 1.5 m above the ground to simulate the breathing zone. Air particulate sampling was carried out during both cooking (07:30 to 19:30) and noncooking hours (19:30 to 07:30 on the following day) under identical sampling conditions to assess the contribution of cooking activities to the $PM_{2.5}$ and polycyclic aromatic hydrocarbons (PAHs) concentrations measured indoors, and the corresponding health risk posed by the inhalation of cooking emissions.

10.4.2 SAMPLE COLLECTION

Air particulate sampling was carried out under ambient conditions at these three locations over a period of two weeks, from March 13, 2005 to March 26, 2005. A MiniVol portable air sampler (Airmetrics, OR, USA) was calibrated with a Gilibrator-2 standard flow Primary Air Flow Calibrator (Gilian Instrument Co., NJ, USA) before and after sampling. This sampler was used in each of the kitchens to collect $PM_{2.5}$ by drawing the ambient air at a flow rate of ~0.005 m^3 min^{-1} through size-selective inlets (PM_{10} and $PM_{2.5}$ impactors) and subsequently through a 47 mm quartz microfilter-A (QMA) precombusted quartz filter (Whatman International Ltd, Kent, UK).

Before air sampling, clean quartz filters were maintained in a dry box at a constant temperature of 25°C and constant humidity of 35% for at least 24 h before weighing with a microbalance (readability to 1 µg; Sartorius AG, Goettingen, Germany) just before use. After sampling, the exposed filters were stored in sterile petri dishes (Gelman Sciences Inc., MI, USA) and again placed in the dry box for at least 24 h before final weight measurements, after which they were stored in the refrigerator at 4°C until extraction and chemical analyses. The mass concentration of $PM_{2.5}$ was calculated as Equation 10.4:

$$PM_{2.5} = \frac{\text{amount of particles collected onto the filter (µg)}}{\text{flow rate of air (m}^3\text{ min}^{-1}) \times \text{sampling time (min)}} \text{ µg m}^{-3}. \quad (10.4)$$

As the flow rate of the MiniVol sampler was calibrated at room temperature (24°C) and atmospheric pressure (760 torr, or 101,325 Pa), the actual flow rate of air was adjusted for different temperatures and pressures that existed at the sampling sites using the ideal gas law.

10.4.3 EXTRACTION AND CHEMICAL ANALYSES

The quartz filters were extracted and analyzed for PAHs following the procedures described by Karthikeyan et al. (2006). Briefly, the filters were treated with 20 mL of 1:1 (v/v) reagent grade acetone:hexane in an MLS1200 MEGA closed vessel microwave digestion system (Milestone srl, Sorisole (BG), Italy) for 20 min at 150 W microwave irradiation. The extracts were concentrated to 3 mL using a rotary evaporator and then to near dryness with a gentle stream of nitrogen under low temperature at ~20°C and redissolved in 1 mL of the extraction solvent for PAHs analysis. The evaporative loss of low-molecular-weight PAHs was minimized by carrying out the drying process carefully at low temperature using a small steady flow of nitrogen.

Acenaphthene-d_{10} was used as a surrogate, and its recovery efficiency evaluated, which was in the range of $90 \pm 10\%$.

Sixteen PAHs that are regarded as priority pollutants by the USEPA were analyzed, namely naphthalene (Nap), acenaphthene (Ace), acenaphthylene (Acy), fluorene (Flu), phenanthrene (Phe), anthracene (Ant), fluoranthene (Flt), pyrene (Pyr), benz[a]anthracene (BaA), chrysene (Chr), benzo[b]fluoranthene (BbF), benzo[k]fluoranthene (BkF), benzo[a]pyrene (BaP), indeno[1,2,3,cd]pyrene (Ind), dibenz[a,h] anthracene (DBA), and benzo[g,h,i]perylene (BPe).

A Hewlett Packard 6890 series gas chromatography-mass spectrometry (GC-MS) (Agilent Technologies, CA, USA) fitted with a DB-5MS 5% phenyl-methylpolysiloxane 30 m long $\times$ 0.2 mm internal diameter $\times$ 0.25 µm film thickness capillary column (J&W Scientific, CA, USA) was used. The GC-MS was operated under the following conditions: splitless injection of 2 µL, split opening after 30 s and injector temperature at 280°C; the oven temperature program was 50°C (hold 2 min); 50–200°C at 10°C min^{-1} (hold 1 min); 200–300°C at 5°C min^{-1} (hold 8 min). The detector was run in electron impact mode with an electron energy of 70 eV and an ion source temperature of 230°C. Helium at a constant flow rate of 0.8 mL min^{-1} was used as a carrier gas. PAHs were monitored using selected ion monitoring mode (SIM). In order to get maximum sensitivity, the 16 ions were divided into groups (seven intervals of retention time), and the detector monitors only the ions programmed for each group. The identification of individual PAHs was based on the comparison of retention times (chromatographic column) and mass spectra (mass detector) of PAHs in aerosol samples with those of PAH standards (full scan mode).

Prior to sample analyses, the GC-MS was calibrated with three different concentrations (200, 500, and 1000 times dilution) of an EPA 610 Polynuclear Aromatic Hydrocarbons Mix containing the 16 PAHs (Supleco, 100 ppm for Phe, Ant, Pyr, BaA, Chr, BkF, BaP, Ind; 200 ppm for Flu, Flt, BbF, DBA, Bpe; 1000 ppm for Nap and Acy; 2000 ppm for Ace). In addition, the recoveries of PAHs were determined by processing four sets of SRM 1649a urban dust (National Institute of Standards and Technology, MD, USA) in the same manner as the samples and compared against the certified values. The retention times, major ions (m/z), regression coefficient for the calibration and the recoveries are listed in Table 10.5. Information on the limit of detection [LOD, 0.3×10^{-3} ppm (BaP) to 8.81×10^{-3} ppm (Flt)] and the limit of quantification [LOQ, 0.59×10^{-3} ppm (BaP) to 17.63×10^{-3} ppm (Flt)] is given in one of our recent publications (Karthikeyan et al., 2006).

10.4.4   RESULTS AND DISCUSSION (CASE STUDY 2)

10.4.4.1   Mass Concentrations of PM$_{2.5}$ and PAHs

The average mass concentrations of PM$_{2.5}$ (reported in µg m^{-3}) and PAHs (reported in ng m^{-3}) during cooking and noncooking hours are given in Table 10.6 in terms of their mean values with the corresponding standard deviation. The background concentrations measured for all the three stalls during noncooking hours are quite similar, and the average concentrations are presented in the table. As can be seen from the table, both the mass concentrations of PM$_{2.5}$ and PAHs and the percentage of PAHs in PM$_{2.5}$ were highest at the Malay stall (PM$_{2.5}$: 245.3 µg m^{-3}, PAHs: 609.0 ng m^{-3},

TABLE 10.5

Regression Coefficients and Recoveries of PAHs

PAHs	Retention Time (min)	Major Ion (m/z)	r^2	Measured Concentration (ppm)	Certified Concentration (ppm)	Recovery (%)
Nap	10.77	128	0.997	10.60 ± 3.80	—	—
Ace	14.50	152	0.994	0.42 ± 0.03	—	—
Acy	14.93	154	0.998	0.36 ± 0.03	—	—
Flu	16.16	166	0.999	0.56 ± 0.04	—	—
Phe	18.64	178	0.999	5.16 ± 0.40	4.14 ± 0.37	124.6 ± 9.7
Ant	18.81	178	1.000	0.54 ± 0.01	0.43 ± 0.09	125.3 ± 3.0
Flt	22.73	202	1.000	6.44 ± 0.46	6.45 ± 0.18	99.8 ± 7.2
Pyr	23.57	202	1.000	5.56 ± 0.43	5.29 ± 0.25	105.1 ± 8.1
BaA	28.66	228	0.999	2.09 ± 0.16	2.21 ± 0.07	94.5 ± 7.1
Chr	28.81	228	0.997	4.02 ± 0.28	3.05 ± 0.06	131.7 ± 9.3
BbF	33.16	252	0.995	6.81 ± 0.47	6.45 ± 0.64	105.6 ± 7.2
BkF	33.27	252	0.990	1.63 ± 0.13	1.91 ± 0.03	85.3 ± 6.8
BaP	34.40	252	0.995	2.37 ± 0.18	2.51 ± 0.09	94.6 ± 7.3
Ind	38.44	276	0.999	3.37 ± 0.28	3.18 ± 0.72	106.0 ± 8.7
DBA	38.61	278	0.999	0.31 ± 0.07	0.29 ± 0.02	107.0 ± 24.0
Bpe	39.35	276	0.990	3.59 ± 0.27	4.01 ± 0.91	89.5 ± 6.7

Source: Reproduced from See, S.W. and Balasubramanian, R. 2006. *Journal of Environmental Monitoring* 8, 369–376. With permission.

PAHs in $PM_{2.5}$: 0.25%), followed by the Chinese stall (201.8 µg m^{-3}, 141.0 ng m^{-3}, 0.07%), and then the Indian stall (186.9 µg m^{-3}, 37.9 ng m^{-3}, 0.02%). While the level of $PM_{2.5}$ increased by factors of 8.3, 6.9, and 6.4, that of PAHs was enhanced by factors of 14.3, 61.6, and 3.8 compared to the background levels at the Chinese, Malay, and Indian stalls, respectively. This difference among the three food stalls in terms of the particulate pollution levels may be explained by the different cooking activities that took place in the individual stalls.

It has been documented that the major factors contributing to the amount and type of pollutants released from food cooking include the type of fuel (Oanh et al., 1999), oil (Wu et al., 1998), food (McDonald et al., 2003), and cooking methods (Lee et al., 2001) employed during the operation. Dispersion of cooking emissions within each stall is also likely to affect the particulate concentrations. However, all the three food stalls have identical layouts, and are naturally ventilated with almost equal air change rates. Therefore, the differences in the concentrations of $PM_{2.5}$ and PAHs between the stalls are taken to be independent of dispersion conditions. In addition, as pointed out earlier, all the three stalls use only LPG as fuel and vegetable oil, for cooking. The three ethnic food stalls cooked a variety of vegetables, meat, and fish, but in different amounts. Hence, the variables considered here are the relative quantity of food cooked, the relative amount of time spent on cooking, and the cooking methods used.

TABLE 10.6

Average Mass Concentrations of $PM_{2.5}$ and PAHs

		Chinese Stall	Malay Stall	Indian Stall	Background
$PM_{2.5}$	/µg m^{-3}	201.8 ± 140.5	245.3 ± 77.1	186.9 ± 43.6	29.4 ± 7.6
Nap	/ng m^{-3}	1.9 ± 3.8	2.8 ± 5.1	3.9 ± 3.2	0.7 ± 1.0
Ace	/ng m^{-3}	1.0 ± 0.8	3.1 ± 2.8	1.1 ± 0.3	0.4 ± 0.1
Acy	/ng m^{-3}	2.4 ± 2.0	5.6 ± 5.0	2.7 ± 0.8	1.8 ± 1.1
Flu	/ng m^{-3}	3.8 ± 2.4	9.2 ± 9.3	3.9 ± 1.1	1.1 ± 0.3
Phe	/ng m^{-3}	11.5 ± 8.7	15.7 ± 10.5	9.5 ± 4.3	2.2 ± 0.4
Ant	/ng m^{-3}	3.0 ± 1.3	6.1 ± 4.9	2.6 ± 1.2	0.5 ± 0.1
Flt	/ng m^{-3}	6.9 ± 9.9	30.7 ± 47.3	1.6 ± 0.7	0.3 ± 0.0
Pyr	/ng m^{-3}	10.9 ± 14.5	18.1 ± 27.3	2.9 ± 1.1	0.6 ± 0.0
BaA	/ng m^{-3}	3.8 ± 5.1	23.1 ± 24.2	1.0 ± 0.5	0.2 ± 0.0
Chr	/ng m^{-3}	5.8 ± 8.1	48.7 ± 50.7	1.0 ± 0.5	0.1 ± 0.1
BbF	/ng m^{-3}	21.8 ± 34.8	122.4 ± 125.9	1.9 ± 1.4	0.4 ± 0.2
BkF	/ng m^{-3}	3.7 ± 5.9	23.1 ± 27.4	0.5 ± 0.5	0.1 ± 0.1
BaP	/ng m^{-3}	5.6 ± 7.6	16.0 ± 20.5	0.9 ± 0.6	0.3 ± 0.2
Ind	/ng m^{-3}	24.4 ± 41.9	105.9 ± 143.4	1.3 ± 1.0	0.4 ± 0.3
DBA	/ng m^{-3}	2.7 ± 4.3	8.3 ± 11.1	1.1 ± 1.4	0.1 ± 0.0
BPe	/ng m^{-3}	31.9 ± 52.9	170.1 ± 239.1	2.1 ± 1.5	0.6 ± 0.4

Source: Reproduced from See, S.W. and Balasubramanian, R. 2006. *Journal of Environmental Monitoring* 8, 369–376. With permission.

The quantity of food cooked and the total time spent on cooking on each day were estimated by the respective chefs. Food was cooked using one or more of the following five cooking methods: deep-frying (to fry by immersing in hot oil), stir-frying (to fry quickly in a small amount of oil at high heat while stirring continuously), pan-frying (to fry in a small amount of oil), boiling (to cook in a hot liquid kept just below its boiling point), and steaming (to cook over boiling water).

The Malay stall was found to be the most polluted. However, relatively less food was cooked in this stall on a daily basis compared to the other two stalls (~30 kg compared to ~45 kg at the Chinese stall and ~40 kg at the Indian stall). The time spent on cooking was about 10 h while it was about 8 h at the Chinese stall and about 20 h at the Indian stall; this estimation was done based on the number of gas stoves used for cooking. It therefore appears that the higher mass concentrations of $PM_{2.5}$ and PAHs at the Malay stall as compared to the other two stalls are associated with the cooking method used. Specifically, deep-frying is the preferred cooking method at the Malay stall as it offers a number of deep-fried snacks such as fried bread, banana cakes, bananas, and curry puffs besides rice and side dishes. On the other hand, the most common cooking method at the Chinese stall was stir-frying where the ingredients are raw or are partially cooked by pan-frying, and the process itself takes only a few minutes. At the Indian stall, the cooking method is boiling, as Indian curry is a very popular dish at the stall, and this recipe requires boiling until the ingredients are tender.

A comparison of the different cooking methods used at the three ethnic food stalls implies that deep-frying generates more $PM_{2.5}$ and PAHs than any other cooking method, which could be due to the higher temperature maintained during cooking and the larger amount of oil used in deep-frying. Acrylamide, a cancer-causing chemical produced when starchy foods such as potatoes are fried at high temperatures, has been reported to increase in concentration with temperature (McDonald et al., 2003). Likewise, both $PM_{2.5}$ and airborne PAHs could follow the same trend when high-temperature cooking is used. This postulation is supported by the higher mass concentrations of $PM_{2.5}$ and PAHs and percentage of PAHs in $PM_{2.5}$ measured at the Chinese stall than at the Indian stall since the stir-frying cooked method involves a higher temperature and uses more oil than boiling. In addition, the larger quantity of food cooked at the Chinese stall could also contribute to the higher level of particulate pollution. The results obtained here is consistent with what was found in the controlled experiments.

10.4.4.2 Health Risk Assessment

To obtain a better estimate of the health risks associated with gas cooking at the different stalls, the mass concentrations of $PM_{2.5}$ and PAHs were compared to the present regulatory standards that were established to protect public or workers' health with an adequate safety margin. In 1997, the USEPA established $PM_{2.5}$ National Ambient Air Quality Standard (NAAQS) at 15 µg m^{-3} for the annual standard (three year average of the annual arithmetic mean concentrations) and 65 µg m^{-3} (three year average of the 98th percentile of 24 h concentrations) for the 24 h standard; other $PM_{2.5}$ standards also exist for different countries. However, similar indoor air quality standards or guidelines do not currently exist for $PM_{2.5}$. The $PM_{2.5}$ levels measured at the food stalls over a 24-h period were 115.6, 137.4, and 108.1 µg m^{-3} at the Chinese, Malay, and Indian stalls, respectively, which are far in excess of the 24 h NAAQS and are thus of health concern.

The intercomparison of $PM_{2.5}$ and PAHs mass concentrations with the regulatory standards or exposure limits is just a qualitative estimate. A better understanding of the associated health risks at the food stalls can be achieved by calculating the total excess lifetime cancer risk ($ELCR_{total}$) according to Equation 10.5:

$$ELCR_{total} = \sum_{all\ i} C_i \times \left[\frac{EF \times ED \times IR_{air}}{BW \times AT} \right] \times SF_i \qquad (10.5)$$

where definitions, units, and default values of parameters used in Equation 10.5 are given in Table 10.7. And, Table 10.8 states toxicity profiles of different PAHs as per the USEPA classification. According to the USEPA, a one in a million chance of an additional human cancer over a 70 year lifetime is the level of risk considered acceptable or inconsequential, whereas a lifetime risk of one in thousand or greater is considered serious and is a high priority for attention.

$ELCR_{total}$ was estimated to be 4.86×10^{-6}, 1.81×10^{-5}, and 8.73×10^{-7} at the Chinese, Malay, and Indian stalls, respectively. The first two values are higher than the recommended acceptable limit of 10^{-6} for ELCR. These calculations suggest that the chefs and other workers in the Chinese and Malay kitchens and possibly the clients visiting the food stalls are exposed to an exceedingly large amount of fine particles containing carcinogenic PAHs. Thus, the human exposure to cooking emissions in the food stalls is of serious health concern.

TABLE 10.7
Health Risk Assessment Parameters

Abbreviation	Definition	Unit	Default Value
Ci	Concentration of PAH i	mg m^{-3}	NA
EF	Exposure frequency	day/yr	350
ED	Exposure duration	yr	30
IR$_{air}$	Total inhalation rate	m^3/day	20
BW	Adult body weight	kg	70
AT	Averaging time	yr × day/yr	70 × 365
SF$_i$	Inhalation slope factor	(mg/kg day)$^{-1}$	See Table 10.8

TABLE 10.8
Toxicity Profiles of PAHs

PAHs	USEPA Classification*	IARC Classification[a]	SF (mg kg^{-1} d^{-1})$^{-1}$
Nap	C	2B	1.2×10^{-1}
Ace	—	—	
Acy	D	—	
Flu	D	3	
Phe	D	3	
Ant	D	3	
Flt	D	3	
Pyr	D	3	
BaA	B2	2A	3.1×10^{-1}
Chr	B2	3	3.1×10^{-3}
BbF	B2	2B	3.1×10^{-1}
BkF	B2	2B	3.1×10^{-2}
BaP	B2	2A	3.1×10^{0}
Ind	B2	2B	3.1×10^{-1}
DBA	B2	2A	3.1×10^{0}
BPe	D	3	

Source: Reproduced from See, S.W. and Balasubramanian, R. 2006. *Journal of Environmental Monitoring* 8, 369–376. With permission.

[a] USEPA Class B2 and IARC Class 2A: probable human carcinogens; USEPA Class C and IARC Class 2B: possible human carcinogens; USEPA Class D and IARC Class 3: not classifiable as to human carcinogenicity.

10.5 CONCLUSIONS

Cooking on a gas stove generated a significant amount of fine particles, UFPs, and nanoparticles. In the first part of this study, it is found that the total number concentration of 10–500 nm particles increased by 2–24 times over the background levels, with more than 80% being in the nanometer size range. There were some variations in the particle concentration and size distribution when different cooking methods were used to cook the same quantity and type of food. It was observed that the deep-frying method emitted the most number of particles and the highest portion of nanoparticles, followed by pan-frying, stir-frying, boiling, and steaming. This trend between the particle number concentrations and different cooking methods implies that cooking with oil causes more particulate emissions than that with water. A larger proportion of ultrafine and accumulation mode particles was observed during water-based cooking, which is thought to be due to the hygroscopic growth of freshly emitted particles in the presence of high humidity.

The same observations were seen in the real-world kitchens. The Malay stall where food was mainly deep-fried showed the highest concentration of $PM_{2.5}$ and PAHs, followed by the Chinese stall (stir-frying) and the Indian stall (boiling). The quantity of food cooked could also contribute to the increase in the level of particulate pollution in kitchens. The equivalent mass concentrations of BaP were calculated for all the three food stalls, and they are higher than the European limit ($0.7–1.3$ ng m^{-3}). The health risk assessment indicated that the occupational exposure to cooking emissions in the three commercial kitchens is of serious health concern. Effective protective measures should therefore be undertaken to reduce cooking emissions and/or minimize human exposure to such emissions, especially when a large quantity of food is cooked using deep-frying or stir-frying.

ACKNOWLEDGMENTS

This chapter is based on a study funded by the NUS ARF through Grant No. RP-279-000-142-112. We are grateful to the National University of Singapore for the financial support provided to carry out this project.

REFERENCES

Abu-Allaban, M., Gertler, A.W., and Lowenthal, D.H. 2002. A preliminary apportionment of the sources of ambient PM_{10}, $PM_{2.5}$, and VOCs in Cairo. *Atmospheric Environment* 36, 5549–5557.

Bakoglu, M., Karademir, A., and Ayberk, S. 2004. An evaluation of the occupational health risks to workers in a hazardous waste incinerator. *Journal of Occupational Health* 46, 156–164.

Balasubramanian, R., Qian, W.B., Decesari, S., Facchini, M.C., and Fuzzi, S. 2003. Comprehensive characterization of $PM_{2.5}$ aerosols in Singapore. *Journal of Geophysical Research—Atmospheres* 108, Article No. 4523.

Burnett, R.T., Smith-Doiron, M., Stieb, D., Cakmak, S., and Brook, J.R. 1999. Effects of particulate and gaseous air pollution on cardiorespiratory hospitalizations. *Archives of Environmental Health* 54, 130–139.

Chao, C.Y. and Cheng, E.C. 2002. Source apportionment of indoor $PM_{2.5}$ and PM_{10} in homes. *Indoor and Built Environment* 11, 27–37.

Dennekamp, M., Howarth, S., Dick, C.A., Cherrie, J.W., Donaldson, K., and Seaton, A. 2001. Ultrafine particles and nitrogen oxides generated by gas and electric cooking. *Occupational and Environmental Medicine* 58, 511–516.

Hays, M.D., Fine, P.M., Geron, C.D., Kleeman, M.J., and Gullett, B.K. 2005. Open burning of agricultural biomass: Physical and chemical properties of particle-phase emissions. *Atmospheric Environment* 39, 6747–6764.

Heyder, J., Gebhart, J., Rudolf, G., Schiller, C.F., and Stahlhofen, W. 1986. Deposition of particles in the human respiratory tract in the size range 0.005–15 μm. *Journal of Aerosol Science* 17, 811–825.

Jarvis, D., Chinn, S., Luczynska, C., and Burney, P. 1996. Association of respiratory symptoms and lung function in young adults with use of domestic gas appliances. *Lancet* 347, 426–431.

Karthikeyan, S., Balasubramanian, R., and See, S.W. 2006. Optimization and validation of a low temperature microwave-assisted extraction method for analysis of polycyclic aromatic hydrocarbons in airborne particulate matter. *Talanta* 69, 79–86.

Kelly, K.E., Wagner, D.A., Lighty, J.S., et al. 2003. Characterization of exhaust particles from military vehicles fuelled with diesel, gasoline, and JP-8. *Journal of the Air and Waste Management Association* 53, 273–282.

Klepeis, N.E., Apte, M.G., Gundel, L.A., Sextro, R.G., and Nazaroff, W.W. 2003. Determining size-specific emission factors for environmental tobacco smoke particles. *Aerosol Science and Technology* 37, 780–790.

Ko, Y.C., Cheng, L.S., Lee, C.H., et al. 2000. Chinese food cooking and lung cancer in women non-smokers. *American Journal of Epidemiology* 151, 140–147.

Lee, P.K.H., Brook, J.R., Dabek-Zlotorzynska, E., and Mabury, S.A. 2003. Identification of the major sources contributing to $PM_{2.5}$ observed in Toronto. *Environmental Science and Technology* 37, 4831–4840.

Lee, S.C., Li, W.M., and Chan, L.Y. 2001. Indoor air quality at restaurants with different styles of cooking in metropolitan Hong Kong. *Science of the Total Environment* 279, 181–193.

Levy, J.I., Hammitt, J.K., and Spengler, J.D. 2000. Estimating the mortality impacts of particulate matter: What can be learned from between-study variability? *Environmental Health Perspectives* 108, 109–117.

Li, C.S., Lin, W.H., and Jenq, F.T. 1993. Size distributions of submicrometer aerosols from cooking. *Environment International* 19, 147–154.

Lighty, J.S., Veranth, J.M., and Sarofim, A.F. 2000. Combustion aerosols: Factors governing their size and composition and implications to human health. *Journal of the Air and Waste Management Association* 50, 1565–1618.

Liu, Y.S., Chen, R., Shen, X.X., and Mao, X.L. 2004. Wintertime indoor air levels of PM_{10}, $PM_{2.5}$ and PM_1 at public places and their contributions to TSP. *Environment International* 30, 189–197.

Lung, S.C.C. and Hu, S.C. 2003. Generation rates and emission factors of particulate matter and particle-bound polycyclic aromatic hydrocarbons of incense sticks. *Chemosphere* 50, 673–679.

Marcazzan, G.M., Vaccaro, S., Valli, G., and Vecchi, R. 2001. Characterisation of PM_{10} and $PM_{2.5}$ particulate matter in the ambient air of Milan (Italy). *Atmospheric Environment* 35, 4639–4650.

Marr, L.C., Kirchstetter, T.W., Harley, R.A., Miguel, A.H., Hering, S.V., and Hammond, S.K. 1999. Characterization of polycyclic aromatic hydrocarbons in motor vehicle fuels and exhaust emissions. *Environmental Science and Technology* 33, 3091–3099.

Martonen, T.B., Zhang, Z., Yue, G., and Musante, C.J. 2002. 3-D Particle transport within the human upper respiratory tract. *Journal of Aerosol Science* 33, 1095–1110.

McDonald, J.D., Zielinska, B., Fujita, E.M., Sagebiel, J.C., Chow, J.C., and Watson, J.G. 2003. Emissions from charbroiling and grilling of chicken and beef. *Journal of the Air and Waste Management Association* 53, 185–194.

Miller, F.J. 1999. Dosimetry of particles: Critical factors having risk assessment implications. *Inhalation Toxicology* 12, 389–395.

Morawska, L., He, C.R., Hitchins, J., Mengersen, K., and Gilbert, D. 2003. Characteristics of particle number and mass concentrations in residential houses in Brisbane, Australia. *Atmospheric Environment* 37, 4195–4203.

Oanh, N.T.K., Reutergårdh, L.B., and Dung, N.T. 1999. Emission of polycyclic aromatic hydrocarbons and particulate matter from domestic combustion of selected fuels. *Environmental Science and Technology* 33, 2703–2709.

Oberdorster, G. 1995. Lung particle overload—implications for occupational exposures to particles. *Regulatory Toxicology and Pharmacology* 21, 123–135.

Oberdorster, G., Sharp, Z., Atudorei, V., et al. 2004. Translocation of inhaled ultrafine particles to the brain. *Inhalation Toxicology* 16, 437–445.

Pope, C.A., Burnett, R.T., Thun, M.J., Calle, E.E., Krewski, D., Ito, K., and Thurston, G.D. 2002. Lung cancer, cardiopulmonary mortality, and long-term exposure to fine particulate air pollution source. *JAMA—Journal of the American Medical Association* 287, 1132–1141.

Richardson, G.M. 2003. Inhalation of mercury-contaminated particulate matter by dentists: An overlooked occupational risk. *Human and Ecological Risk Assessment* 9, 1519–1531.

Samet, J.M., Dominici, F., Curriero, F.C., Coursac, I., and Zeger, S.L. 2000. Fine particulate air pollution and mortality in 20 US Cities, 1987–1994. *New England Journal of Medicine* 343, 1742–1749.

See, S.W. and Balasubramanian, R. 2006. Health risk assessment of occupational exposure to particulate-phase polycyclic aromatic hydrocarbons associated with Chinese, Malay and Indian cooking. *Journal of Environmental Monitoring* 8, 369–376.

See, S.W. and Balasubramanian, R. 2008. Chemical characteristics of fine particles emitted from different gas cooking methods. *Atmospheric Environment* 42(39), 8852-8862.

Sharma, M. and Maloo, S. 2005. Assessment of ambient air PM_{10} and $PM_{2.5}$ and characterization of PM_{10} in the city of Kanpur, India. *Atmospheric Environment* 39, 6015–6026.

Siegmann, K. and Sattler, K. 1996. Aerosol from hot cooking oil, a possible health hazard. *Journal of Aerosol Science* 27, S493–S494.

Simpson, R., Denison, L., Petroeschevsky, A., Thalib, L., and Williams, G. 2000. Effects of ambient particle pollution on daily mortality in Melbourne, 1991–1996. *Journal of Exposure Analysis and Environmental Epidemiology* 10, 488–496.

Takenaka, S., Karg, E., Roth, C., et al. 2001. Pulmonary and systemic distribution of inhaled ultrafine silver particles in rats. *Environmental Health Perspectives* 109, 547–551.

Tucker, W.G. 2000. An overview of $PM_{2.5}$ sources and control strategies. *Fuel Processing Technology* 65, 379–392.

United States Department of Energy. 1999. Guidance for Conducting Risk Assessments and Related Risk Activities for the DOE-ORO Environmental Management Program (BJC/OR-271). United States Department of Energy, Office of Environmental Management.

Valberg, P.A. 2004. Is PM more toxic than the sum of its parts? Risk-assessment toxicity factors vs. PM-mortality "effect functions". *Inhalation Toxicology* 16, 19–29.

Vallius, M., Janssen, N.A.H., Heinrich, J., et al. 2005. Sources and elemental composition of ambient $PM_{2.5}$ in three European cities. *Science of the Total Environment* 337, 147–162.

von Klot, S., Peters, A., Aalto, P., et al. 2005. Ambient air pollution is associated with increased risk of hospital cardiac readmissions of myocardial infarction survivors in five European cities. *Circulation* 112, 3073–3079.

Wallace, L. 1996. Indoor particles: A review. *Journal of the Air and Waste Management Association* 46, 98–126.

Wallace, L.A., Emmerich, S.J., and Howard-Reed, C. 2004. Source strengths of ultrafine and fine particles due to cooking with a gas stove. *Environmental Science and Technology* 38, 2304–2311.

Wu, P.F., Chiang, T.A., Wang, L.F., Chang, C.S., and Ko, Y.C. 1998. Nitro-polycyclic aromatic hydrocarbon contents of fumes from heated cooking oils and prevention of mutagenicity by catechin. *Mutation Research* 403, 29–34.

Zhang, J.F. and Morawska, L. 2002. Combustion sources of particles: 2. Emission factors and measurement methods. *Chemosphere* 49, 1059–1074.

11 Estimation of Health Impacts due to PM$_{10}$ in Major Indian Cities

Prabhakar Nema and Sanjeev K. Goyal

CONTENTS

11.1 INTRODUCTION

Exposure to air pollution causes a number of health problems. Air pollutants that are inhaled have serious impacts on human health affecting the lungs and the respiratory system. Different people are affected by air pollution in different ways. Poor people, undernourished people, very young and very old, and people with preexisting respiratory disease and other health problems are more at risk. The effects of air pollution on health are very complex and the pollutants can have synergistic effects when people are exposed to a number of pollutants present in the atmosphere.

PM$_{10}$ are the suspended particulates having a diameter of less than 10 μm and are small enough to enter the respiratory tract and pulmonary system of human being. Anthropogenic sources of PM$_{10}$ include fuel combustion in industries and homes, traffic/road dust, construction activities, solid waste disposal, waste incinerators, refuse/biomass burning, and so on. Naturally occurring processes, such as dust storms, volcanic eruptions, forest fires, soil erosion, and so on, also contribute significantly to the particulate pollution.

WHO (2004) presented a systematic review of health aspects of air pollution in Europe, which included the findings of various epidemiological studies. It was reported that the most severe effects in terms of the overall health burden include a significant reduction in life expectancy of the average population by a year or more, which is linked to the long-term exposure to high levels of air pollution with particulate matter (PM).

The most significant health effects of air pollution in Asian cities are associated with exposure to PM: premature deaths from heart and lung diseases, chronic bronchitis, asthma, and other forms of respiratory illness (ESMAP, 2004). Health aspects of air pollution, particularly with respect to PM, have been presented by WHO (2003), as given in Table 11.1.

The WHO estimates that every year 800,000 people die prematurely worldwide as a result of lung cancer, cardiovascular diseases, and respiratory diseases caused by outdoor pollution. Approximately 150,000 of these deaths are estimated to occur in South Asia alone (World Bank, 2003a). Between 100 and 150 million people around the globe have asthma and the figure is increasing. Over 180,000 people die worldwide each year due to asthma and India has an estimated 15–20 million asthmatics (Anon, 2004).

A huge amount is spent on treatment of diseases caused by air pollution. The US spends more than US$ 6 billion on asthma, and Britain spends about US$ 1.8 billion on health care for asthma and the man-days lost due to illness. In Australia, annual direct and indirect medical costs associated with asthma reach almost US$ 460 million. In Malaysia, the average monthly cost for treating a child is US$ 15.56. India is also spending a large amount on treatment of asthma, which is mainly caused by particulate pollution. Asthma cannot be cured, but by reducing the level of particulate concentration in air, the number of cases of asthma can surely be brought down (Anon, 2004).

PM_{10} can have different chemical composition and associated health impacts. Because of a variety of anthropogenic activities concentrated in and around urban areas, the levels there are usually far higher than in the rural areas. A prerequisite for their control/mitigation would be to identify potential sources and the population groups prone to PM_{10} exposure. The following sections describe the air quality standards for PM_{10}, observed levels of PM_{10} in major Indian cities along with the estimated cases of mortality and morbidity in these cities, which are classified into metro, industrial, fast developing (e.g., Information Technology oriented), and other cities.

TABLE 11.1
Short-Term and Long-Term Health Effects of PM

Effects Related to Short-Term Exposure	Effects Related to Long-Term Exposure
• Lung inflammatory reactions	• Increase in lower respiratory symptoms
• Respiratory symptoms	• Reduction in lung function in children
• Adverse effects on the cardiovascular system	• Increase in chronic obstructive pulmonary disease
• Increase in medication usage	• Reduction in lung function in adults
• Increase in hospital admissions	• Reduction in life expectancy, owing mainly to cardiopulmonary mortality and probably to lung cancer
• Increase in mortality	• Increase in mortality

11.2 STANDARDS FOR PM$_{10}$ AND ITS STATUS IN MAJOR INDIAN CITIES

The Central Pollution Control Board (CPCB), the Apex regulatory agency in India, has stipulated National Ambient Air Quality Standards (NAAQS), as presented in Table 11.2. The WHO guidelines are also given in Table 11.2 for comparison, which are independent of land-use category.

Monitoring for PM$_{10}$ was started in the early 1990s at 30 monitoring stations operated by the National Environmental Engineering Research Institute (NEERI), and was extended later to other stations in the late 1990s under National Air Monitoring Programme of CPCB. PM$_{10}$ data with CPCB (www.cpcb.nic.in) have been analyzed and the annual mean levels of PM$_{10}$ in the residential areas of major Indian cities during 2001, 2002, and 2003 are presented in Figure 11.1 along with the exceedance factors (EFs). EF is defined as the ratio of pollutant concentration to the Standard for that pollutant. The descriptor categories used to describe the extent of pollution level are critical (EF: >1.5), high (EF: 1.0–1.5), moderate (EF: 0.5–1.0), and low (EF: <0.5).

The analysis indicates that, in general, PM$_{10}$ levels exceeded the CPCB standard of 60 μg/m^3 in all the cities during 2001–2003 with Ahmedabad and Kanpur showing the highest concentrations. The WHO guideline of 40 μg/m^3 was also exceeded by a factor of 2–5 in these cities. A decreasing trend in PM$_{10}$ concentrations was seen in Ahmedabad, Bhopal, Kanpur, Kochi, Nagpur, and Surat, whereas increasing trends were observed in Bangalore, Delhi, Jaipur, and Pune. Such variations could not be explained in definitive terms and are attributable to various unorganized/unidentified local area source activities. The study of various pollution sources present in each city together with prevailing meteorological conditions over the years should explain these variations.

It is not only the ambient air quality in the cities but also the indoor air quality in the rural and the urban areas that is causing concern. Studies conducted show that indoor environments sometimes can have levels of pollutants that are actually higher than levels found outdoor. Pollution in the indoor environment can increase the risk of illness because (1) the pollutants are released in proximity to the building occupants; (2) inadequate ventilation can increase their levels by not carrying indoor air

---

TABLE 11.2
National Ambient Air Quality Standards and WHO Guidelines for PM$_{10}$

	Concentration (μg/m^3)			
Averaging Time	Industrial Area	Residential Area	Sensitive Area	WHO Guidelines (μg/m^3)
Annual	120	60	50	40
24 h	150	100	75	70

Note: PM$_{10}$ is measured by the high-volume sampling (Gravimetric) method (CPCB, 1994).

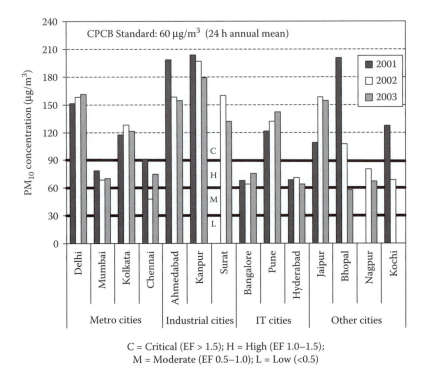

C = Critical (EF > 1.5); H = High (EF 1.0–1.5);
M = Moderate (EF 0.5–1.0); L = Low (<0.5)

FIGURE 11.1   PM$_{10}$ levels in major Indian cities: 2001–2003.

pollutants out of the home; and (3) long exposure time as people spend as much as 80–90% of their time indoor (Mukhi and Khare, 2006).

In the developing countries, a fairly large portion of the population is dependent on biomass for their energy requirements. These include wood, charcoal, agricultural residue, and animal waste. Open fuel burning for cooking and heating are common in the households both in rural and urban areas. In such households, the children and women are most likely to be affected, as they are the group that spends more time indoor. PM is the main pollutant of concern, and is reported to be causing maximum number of deaths in rural areas of India. Many of the deaths are due to acute respiratory infections in children; others are due to cardiovascular diseases, lung cancer, and chronic respiratory diseases in adults. If emissions are high and ventilation is poor, household use of coal and biomass can severely affect the indoor air quality.

Smith (2000) has evaluated the health effects of indoor air pollution with special reference to India. He examined the existing epidemiological studies and estimated the total burden of disease (mortality and morbidity) attributable to the use of solid fuels in adult women and young children, who jointly receive the highest exposures because of their household roles. He conservatively estimated that some 400–550 thousand premature deaths can be attributed annually to the use of biomass fuels in these population groups.

11.3 ESTIMATION OF HEALTH IMPACTS DUE TO PM_{10}

Among all air pollutants, PM causes the most numerous and serious effects on human health because it contains a broad range of diverse toxic substances, and PM_{10} may be considered to be a reliable indicator of the impact of global air pollution. The major constituents of PM are water-soluble inorganic ions [NH_4^+, NO_3^-, SO_4^{2-}, Na^+, K^+, Mg^{2+}, Cl^- (sum of the concentrations of NH_4^+, NO_3^- and SO_4^{2-} represents the secondary inorganic particle fraction)], elemental and organic carbon (EC and OC), and trace elements (Al, As, Ba, Br, Ca, Ce, Cd, Cu, Fe, Ga, K, La, Mg, Mn, Mo, Nd, Ni, Pb, Rh, Sb, Se, Tl, V, Y). The chemical composition of PM_{10} depends on the sources of origin and the activities influencing the sampling site. A study conducted at three sites representing suburban area, Kerbside location, and rural area in Switzerland corroborates this fact as shown in Table 11.3.

11.3.1 METHODOLOGIES

Epidemiological studies are required to determine potential relationships between a variety of environmental factors and human diseases. They are characterized by statistical analysis of the data collected on the health status of the individuals, pollutant exposures, and potential confounding factors. Such studies often provide evidence of possible causal relationships between pollutant exposures and observed or reported health effects. In general, epidemiological studies become more important as the risk attributable to atmospheric pollutants becomes smaller and the duration of exposure required to produce effects becomes longer. Such studies have been particularly useful in identifying acute effects of elevated short-term pollutant exposures. These effects may include pulmonary function changes, asthmatic attacks, and increased mortality (Godish, 2004).

In the absence of specific epidemiological studies conducted in the city/country of interest, the experience of other countries where such studies have been done can be utilized through meta-analysis approach. Meta-analysis allows deriving "best estimates" of the relationship between air pollution and health endpoints, which can be applied with greater confidence to other situations/cities/countries than an estimate

---

TABLE 11.3
Chemical Composition (%) of PM_{10} at Sub-Urban, Kerbside, and Rural Areas

Constituents	Sub-Urban	Kerbside	Rural
Organic material	20	23	17
Minerals and trace metals	13	22	17
Secondary inorganic aerosols	36	20	35
Elemental carbon	8	14	6
Unidentified	23	21	25

Source: Gehrig, R., et al. 2001. Contributions of road traffic to ambient PM_{10} and $PM_{2.5}$ concentrations, STRC 1st Swiss Transport Research Conference, Monte Verità/Ascona, March 1–3, pp. 5–8, http://www.strc.ch/conferences/2001/gehrig.pdf

from an individual study. Ostro (1994), Pope and Dockery (1994), and Schwartz (1994) have carried out meta-analysis reviews.

Lvovsky et al. (2000) reviewed the impact of exposure to PM with special reference to India and discussed the issues of economic valuation of sickness and premature deaths due to outdoor air pollution. World Bank (2003b) has also described the methodology for the valuation of health impacts due to premature deaths and illness from air pollution, and has shown how the monetized value of health benefits from reducing air pollution can be useful to policy-makers for implementation of air pollution control programs.

The impacts of health are divided into mortality and morbidity, which are derived from air quality data using dose–effect relationships. In principle, such relations are found by statistical comparison of death rates and morbidity in areas with different air quality. Appropriate dose–effect relationships have been estimated by Ostro (1994) for US cities, and Cropper et al. (1997) and Lvovsky et al. (2000) for India, based on the meta-analysis approach.

The pooled estimate of 0.84% change in all-cause mortality relative to a 10 $\mu g/m^3$ change in PM_{10} concentration was derived from a number of acute exposures studied in an earlier World Bank review (Lvovsky et al., 2000). Similar to the estimates of mortality, concentration–response functions were also derived for morbidity impacts such as respiratory hospital admissions, chronic bronchitis, cough, asthma, and so on, as summarized in Table 11.4.

It is emphasized that these estimates implicitly incorporate the baseline incidences of the morbidity endpoints (such as chronic bronchitis) in the locations where the studies were conducted, and as a result they are likely to underestimate the morbidity impacts for Indian cities (since the baseline incidence in Indian cities can be expected to be higher than in cities in industrialized countries where most of the studies have been conducted).

TABLE 11.4

Concentration–Response Function Slope per 1 $\mu g/m^3$ Change in Mean Annual Level of PM_{10}

Health Effects	Units	Slope Function
Mortality	Percentage change	0.084
Morbidity		
Chronic bronchitis (CrBr)	Per 100,000 adults	6.12
Respiratory hospital admissions (RHA)	Per 100,000 population	1.2
Asthma attack (A)	Per 100,000 asthmatics	3260
Emergency room visits (ERV)	Per 100,000 population	23.54
Restricted day activities (RDA)	Per 100,000 adults	5700
Lower respiratory illness (LRI)	Per 100,000 children	169
Respiratory symptoms days (RSD)	Per 100,000 adults	18,300

Source: Cropper, M., et al. 1997. The health effects of air pollution in Delhi, India., Policy Research Working Paper 1860, DECRG, World Bank, Washington DC.

Excess mortality/morbidity due to PM_{10} can be estimated as

Excess mortality/morbidity = Slope function × (PM_{10} levels–standard) × PE,

where PE is population exposed to a specific PM_{10} concentration. PM_{10} level is annual average concentration of PM_{10} in residential/mixed use activity area and the standard is the corresponding statutory limit for PM_{10} (60 μg/m³).

In the case of mortality, the slope function refers to percentage change in the mortality and hence it is multiplied by the crude mortality rate (CMR) to get the absolute number of cases. Similarly, the other slope functions are multiplied by the adult or children fraction of the population as appropriate. In the case of asthma attacks, the slope function is multiplied by the fraction of asthmatics in the population.

11.3.2 ESTIMATED HEALTH IMPACTS FOR MAJOR INDIAN CITIES

Using the above equation, mortality and morbidity were estimated for selected 14 major cities of India having population more than a million. These cities have been classified as follows:

- Delhi, Mumbai, Kolkata, Chennai—metro cities
- Ahmedabad, Kanpur, Surat—industrial cities
- Bangalore, Pune, Hyderabad—IT cities
- Jaipur, Bhopal, Nagpur, Kochi—other cities

All the data have been used for the year 2001 (reference year). Total city population along with the annual average concentration levels of PM_{10} in the residential areas in these cities is presented in Figure 11.2.

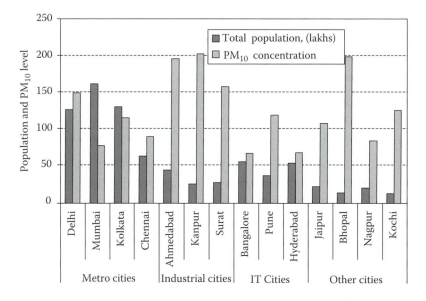

FIGURE 11.2   Total population and observed PM_{10} concentration in major Indian cities.

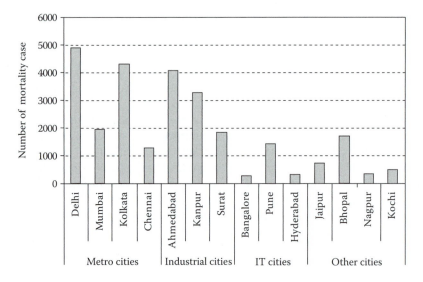

FIGURE 11.3   Estimated number of mortality cases in major Indian cities.

Population of these cities varies from 13.55 lakhs for Kochi to 163.68 lakhs for Mumbai. PM_{10} levels are recorded highest in Bhopal (201 $\mu g/m^3$) and lowest in Bangalore (68 $\mu g/m^3$).

All the cities except Mumbai and Kolkata have shown higher PM_{10} level per lakh population. The ratio of PM_{10} level ($\mu g/m^3$) to population (per lakh) is more in industrial and other cities as compared to metro and IT cities. The estimated number of cases of mortality is depicted in Figure 11.3.

Among the metro cities, the highest number of mortality cases is observed in Delhi (4889), followed by Kolkata (4303), Mumbai (1959), and Chennai (1272). Among the industrial cities, Ahmedabad ranks first, followed by Kanpur and Surat. Among the IT cities, Pune is found to have the highest number of estimated mortality cases followed by Hyderabad and Bangalore.

Among the other cities category, Bhopal ranks first followed by Jaipur and Kochi, and Nagpur is at the bottom. Out of all the 14 cities, Delhi and Bangalore are at the top and bottom levels, respectively.

Morbidity impacts have been estimated in terms of number of cases of chronic bronchitis, respiratory hospital admissions, asthma attacks, lower respiratory illness, emergency room visits, restricted day activities, and RSD. The number of the morbidity cases in the 14 cities classified under four categories is shown in Figures 11.4 through 11.7. The morbidity pattern in different cities is observed to be similar to that of mortality.

11.3.3   ECONOMIC VALUATION OF THE HEALTH IMPACTS

World Bank (2003b) has done economic valuation of health impacts based on the approach of Willingness-to-Pay (WTP) to reduce the risk of getting sick or dying prematurely. In valuing the mortality cases, this approach uses Value of Statistical

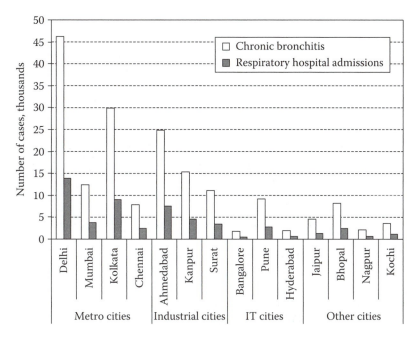

FIGURE 11.4 Estimated number of cases of chronic bronchitis and respiratory hospital admissions in major Indian cities.

Life (VOSL), which is not the same as valuing actual life. It involves placing a value on reduction in overall risk that people face. The VOSL is normally adjusted for the difference between the average number of years lost by people dying due to air pollution and the average number of years lost by people surveyed by the labor market studies, on which the majority of VOSL estimates are based. Based on the WTP

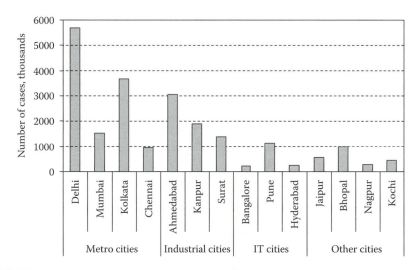

FIGURE 11.5 Estimated number of cases of asthma attacks in major Indian cities.

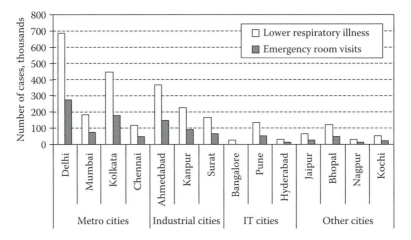

FIGURE 11.6   Estimated number of cases of lower respiratory illness and emergency room visits in major Indian cities.

approach, Lvovsky et al. (2000) have estimated total costs associated with mortality and morbidity cases for Delhi, as presented in Table 11.5.

11.3.4   UNCERTAINTY IN HEALTH IMPACT ASSESSMENT

It is important to note that the estimated cases of mortality and morbidity largely depend on the dose–response relationship, which is primarily developed in the United States and such relationships under Indian conditions are not available. This can therefore result in significant variations in the health impact estimates.

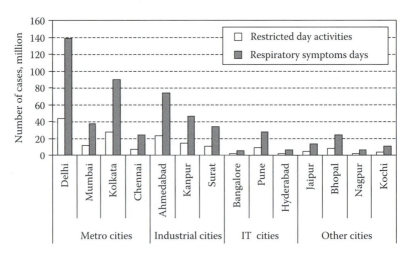

FIGURE 11.7   Estimated number of cases of restricted day activities and RSD in major Indian cities.

TABLE 11.5
Health Effect Cost per Case for Delhi (2003)

Health Effect	Cost (Rs./per Case)
Mortality	15,72,419
Morbidity	
CrBr	189,273
RHA	4101
A	61
ERV	122
RDA	51
LRI	43
RSD	43

However, in the absence of real-time data, the meta-analysis approach is widely accepted by Scientists the world over. Pope and Dockery (2006) analyzed the data from various research studies carried out on different health aspects having different time scales (short term and long term) in a single city and multiple cities. The analysis showed reasonable consistent associations between cardiopulmonary mortality and daily changes in PM levels.

Another important aspect is the measured annual mean concentration of PM$_{10}$. Measurement of representative PM$_{10}$ exposure level in a city should be ascertained. Further, the PM$_{10}$ concentration levels may vary significantly from one year to another (as shown in Figure 11.1). Measured PM$_{10}$ levels may be lower or higher in the successive years, thus affecting the health impact estimates. Percent variation in PM$_{10}$ concentration levels in 2002 and 2003 as compared to the previous year is presented in Figure 11.8, which shows large variations particularly in the case of Chennai, Jaipur, Bhopal, and Kochi. The associated effects on health impact estimates are presented in Figure 11.9. A perusal of these figures indicates that one needs to apply his discretion while using this information. Uncertainty in measurement of PM can vary up to 50% (Goyal, 2003).

11.4   SUMMARY

Health effects of air pollution with particular reference to PM$_{10}$ (recognized as the major cause of health problems in urban areas) have been discussed. Concentration levels of PM$_{10}$ in major urban centers in India during 2001, 2002, and 2003 are presented. Analysis of data indicates that in most of the cities, PM$_{10}$ levels exceed 2–3 times the permissible limit of 60 µg/m^3. Further, mortality and morbidity associated with PM$_{10}$ have been estimated and presented for 14 major cities in India. Based on PM$_{10}$ concentration levels of 2001, Delhi is found to be worst affected and Bangalore is found to be least affected in terms of health impacts. Uncertainty associated with the estimation of health impacts highlights that PM$_{10}$ levels vary considerably over time and hence involve large variations in estimates. Correct measurement of pollution levels and choosing their representative values are important inputs to obtain

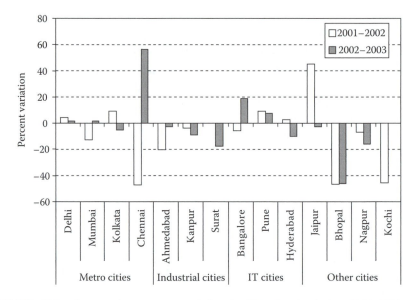

FIGURE 11.8 Comparative percent variation in PM_{10} levels during 2001–2002 and 2002–2003.

better estimates. Therefore, a major step toward mitigating/minimizing adverse health impacts of air pollution would be to collect systematic and reliable air quality data, which in turn would facilitate delineating appropriate strategies/action plans for air quality management.

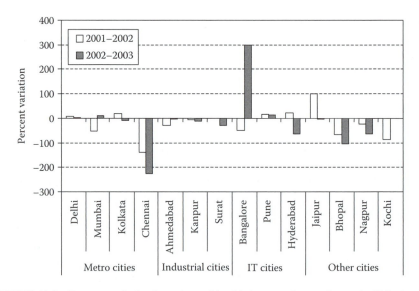

FIGURE 11.9 Percent variation in estimated health impacts due to change in PM_{10} levels during 2001–2002 and 2002–2003.

Future research should focus on

- Epidemiological studies to develop the dose–response relationship between pollutant concentration and the resultant health impact under Indian conditions.
- Measurement of representative pollutant exposure concentration levels in ambient air as well as in indoor air in urban and rural areas.
- Evaluation of PM$_{10}$ control strategies vis-à-vis impact on health through epidemiological studies. Development of cost-effective abatement policies depends heavily upon: (1) a complete knowledge of the source apportionment of airborne PM; (2) knowledge of how new emission controls will have an impact upon emissions of primary pollutants and hence upon airborne concentrations; and (3) reliable source–receptor relationship for secondary pollutants, which also form PM (Harrison et al., 2008). Maenhaut (2008) emphasized the need for continued and expended aerosol chemical composition monitoring and research on a global scale.
- Further, Pope and Dockery (2006), after reviewing a number of health impact studies, emphasized the need for research to understand (1) the most susceptible category of people; (2) the impacts of PM exposure on infant mortality and various birth outcomes, including fetal growth, premature birth, intra-uterine mortality, and birth defects; (3) the effect of ambient PM on lung cancer risk; and (4) the role of various characteristics and constituents of PM, and what is the relative importance of various sources and related co-pollutants.

REFERENCES

Anon. 2004. A disease on the rise. *Down to Earth* March 15, 33–34, www.downtoearth.org.in
CPCB. 1994. CPCB—NAAQS Notification, April 11, 1994.
Cropper, M., Simon, N., Alberini, A., and Sharma, P.K. 1997. The health effects of air pollution in Delhi, India, Policy Research Working Paper 1860, DECRG, World Bank, Washington DC.
ESMAP. 2004. Toward cleaner urban air in South Asia: Tackling transport pollution, understanding sources, UNDP/World Bank Report, 281/04.
Gehrig, R., Hüglin, C., and Hofer, P. 2001. Contributions of road traffic to ambient PM$_{10}$ and PM$_{2.5}$ concentrations, STRC 1st Swiss Transport Research Conference, Monte Verità/Ascona, March 1–3, pp. 5–8, http://www.strc.ch/conferences/2001/gehrig.pdf
Godish, T. 2004. *Chapter on Health Effects in Air Quality*, 4th Edition, Lewis Publishers, CRC Press LLC, USA, pp. 141–182.
Goyal, S.K. 2003. Estimation of error in flow measurement during ambient particulate matter monitoring, *Indian Journal of Environmental Protection* 23(8), 850–856.
Harrison, R.M., Stedman, J., and Derwent, D. 2008. New direction: Why are PM$_{10}$ concentrations in Europe not falling? *Atmospheric Environment* 42, 603–606.
Lvovsky, K., Hughes, G., Maddison, D., Ostro, B., and Pearce, D. 2000. Environmental costs of fossil fuels: A rapid assessment methodology with application to six cities, World Bank, Washington, DC, http://www.cleanairnet.org/cai/1403/articles_34285_env_cost_fossil_fuels.pdf

Maenhaut, W. 2008. New direction: Future needs for global monitoring and research of aerosol composition, *Atmospheric Environment* 42, 1070–1072.

Mukhi, N. and Khare, M. 2006. Indoor air pollution: Part I, *Journal of the IPHE* 2006–2007, 2, 34–38.

Ostro, B. 1994. Estimating the health effects of air pollution: A methodology with application to Jakarta, Working paper 1301, World Bank, Washington, DC, pp. 33–39.

Pope, C.A. and Dockery, D.W. 1994. Acute respiratory effects of particulate air pollution, *Annual Review of Public Health* 15, 107–132.

Pope, C.A. and Dockery, D.W. 2006. Health effects of fine particulate air pollution: Lines that connect, *Journal of the Air and Waste Management Association* 56, 709–742.

Schwartz, J. 1994. Air pollution and daily mortality: A review and meta-analysis, *Environmental Research* 64, 36–52.

Smith, K.R. 2000. National burden of disease in India from indoor air pollution. *Proceedings of the National Academy of Sciences, USA* 97(24), 13286–13293, http://www.pnas.org/content/97/24/13286.full.pdf

WHO. 2003. Health aspects of air pollution with particulate matter, ozone and nitrogen dioxide, *Report of a WHO Working Group*, Bonn, Germany, January, http://www.euro.who.int/ document/ e79097.pdf

WHO. 2004. Health aspects of air pollution—results from the WHO Project "Systematic review of health aspects of air pollution in Europe," http://www.who.euro.who.int/air/publications/20030616_2

World Bank. 2003a. Health impacts of outdoor air pollution, in urban air pollution—South Asia urban air quality management Briefing Note, No. 11, February, pp. 1–4.

World Bank. 2003b. Economic valuation of the health benefits of reduction in air pollution, in Urban air pollution—South Asia urban air quality management Briefing Note, No. 12, February, pp. 1–4.

12 Health Risk Assessment and Management for Air Toxics in Indian Environment

Manju Mohan and Bhola R. Gurjar

CONTENTS

12.1 INTRODUCTION

A number of accidents worldwide such as those at Flixborough (1974), Seveso (1976), Bhopal (1984), Pasadena (1989), and so on have led to growing concern about the potential hazards and risks involved in chemical process industries (CPIs; Kim et al., 1995; Khan and Abbasi, 1998). Such industrial accidents may cause serious injury to people and/or severe damage to infrastructures within and beyond the immediate vicinity of the workplace. Top 10 industrial disasters based on fatality estimates, which occurred between 1945 and 1990, are shown in Table 12.1 (UNEP, 2000).

In addition to accidental releases of extremely hazardous chemicals, the continuous release of toxic pollutants from major industrial facilities and other anthropogenic activities may also cause adverse effects on human health and the environment. There are various guidelines, acts, laws, and regulations to reduce and control the

TABLE 12.1

Top 10 Industrial Disasters Based on Fatality Estimates (1945–1990)

Sl. No.	Year	Location	Type/Agent	Deaths[a]
1.	1984	Bhopal, India	Toxic vapor/methyl isocyanate	2750–3849
2.	1982	Salang Pass, Afghanistan	Toxic vapor/carbon monoxide	1550–2700
3.	1956	Cali, Colombia	Explosion/ammunitions	1200
4.	1958	Kyshtym, USSR	Radioactive leak	1118[b]
5.	1947	Texas City, TX	Explosion/ammonium nitrate	576
6.	1989	Acha Ufa, USSR	Explosion/natural gas	500–574
7.	1984	Cubato, Brazil	Explosion/gasoline	508
8.	1984	St. Juan Ixhautepec, Mexico	Explosion/natural gas	478–503
9	1983	Nile River, Egypt	Explosion/natural gas	317
10.	1986	Chernobyl, USSR	Explosion/radioactivity	31–300[b]

Source: Adapted from Gurjar, B.R. 1999. Environmental Risk Analysis for Industrial Siting, Planning and Management. Ph.D. Thesis, Indian Institute of Technology, New Delhi.

[a] Estimates vary depending on the source(s) used; therefore ranges are provided where there are differences in the total.

[b] Total number of deaths are hard to gauge since the reported fatality figures only reflect immediate deaths, not the longer-term deaths associated with radioactive exposure.

risks in India. These aid in preparing and implementing the emergency response plans to respond in a catastrophic situation. For example, Manufacture, Storage, and Import of Hazardous Chemicals (MSIHC) Rules, 1989 (later amended in 1994 and 1999), Hazardous Waste (Management and Handling), that is, HW (M & H) Rules 1989 (later amended in 1997 and 1999), and Public Liability Insurance (PLI) Act, 1991, and so on are among those used for regulatory applications. Despite so many rules and acts, the hazardous chemicals continue to be handled in an unsafe and environmentally unsound manner. This is reflected in the many catastrophic accidents that occurred in past few decades such as at Bhopal (1984), Panipat (1993), Mumbai (1995), and Visakhapatnam (1997) as shown in Table 12.2.

Thus, there is also a substantial concern about the accidental release of hazardous materials as a result of growing awareness of the scale of tragedy that may accompany activities involving such materials. The development of appropriate regulatory measures requires the acceptable balance between economic benefit and potential harm accompanying such activities, which emphasizes a need for quantitative risk assessment (QRA) of the accidental releases of hazardous materials into the environment.

In scientific terms, risk is expressed as a combination of two factors: (i) the probability that an adverse event will occur and (ii) the consequences of the adverse event. More precisely, the risk is a combination of expected frequency and the consequence of a single accident or a group of accidents. This type of risk is associated with an emergency situation comprising an acute hazard (i.e., the potential for an injury or damage to occur as a result of an instantaneous or short-duration exposure) caused by an episodic event (i.e., an unplanned event of limited duration, usually associated

TABLE 12.2
Major Accidents that Occurred in India during the Last Two Decades

Year	Date/Month	Location	Origin of Accident	Products Involved	Number of Deaths	Injured
2008	20.04	Kanpur	Petrochemical factory fire[a]	Inflammable matter	1	40
2008	04.04	Ankleshwar, Gujarat	Chemical factory fire[b]	Solid/hazardous wastes	—	—
2003	11.06	Mohali	Pharmaceutical plant fire[c]	Cloud of thick smoke	—	30
1997	14.09	Visakhapatnam	Refinery fire		34	31
1997	21.01	Bhopal	Leakage (transport accident)	Ammonia		400
1995	12.03	Madras	Transport accident	Fuel	~100	23
1994	13.11	New Delhi	Fire at a chemical store	Toxic cloud (chemicals)		500
1994	4.1	Madhya Pradesh	Explosion (storage)	Fire crackers	30	100
1994	January	Thane District	Transport accident	Chlorine gas	4	298
1992	25.1	Tharia	Explosion, fire	Fireworks	>25	100
1992	29.04	New Delhi	Explosion (warehouse)	Chemicals	43	20
1991	December	Calcutta	Leakage from a pipeline	Chlorine		200
1991	November	Medran	Transport accident (leakage)	Inflammable liquid	93	25
1991	January	Ludhiana	Market	Fireworks	>40	
1991	January	New Bombay	Transport accident	Ammonia gas	1	150
1991	12.07	Meenampalti	Explosion (firework factory)	Fireworks	38	
1990	5.11	Nagothane	Leakage	Ethane and propane	32	22
1990	July	Lucknow	Leakage in an ice factory	Ammonia gas		200
1990	16.04	Near Patna	Leakage, transport accident	Gas	100	100
1990	15.04	Basti	Food poisoning	Sulplios	150	>150

Source of other data: Gurjar, B.R. and Mohan, M. 2002. *Risk: Health, Safety and Environment (USA)* 13(1/2), 1–30.

[a] *Source:* http://www.topnews.in/fire-kanpur-guts-chemical-factory-237006

[b] *Source:* http://indianewslive.com/?p=248

[c] *Source:* http://www.tribuneindia.com/2004/20041031/cth1.htm

with an accident). However, there is another type of risk known as chronic risk due to long-term exposure for months, years, or decades to toxic chemicals that manifest as chronic health problems. Risk from chronic exposure arises from activities associated with the production and use of food, energy, industrial and consumer goods, and from the waste produced through daily living. This chapter presents in detail an account of the following studies encompassing both acute and chronic risks.

1. *Acute risk:* IITD Heavy Gas Model for dense gas dispersion of airborne toxic materials for estimating individual and societal risk and for the estimation of vulnerable zones and user-oriented nomograms (Singh et al., 1991; Mohan et al., 1994, 1995)
2. *Chronic risk:* Potential health risks related to carcinogens in the atmospheric environment in India for cadmium, chromium, and nickel (Gurjar et al., 1996; Gurjar and Mohan, 2003a)
3. *Integrated risk analysis using both acute and chronic risk:* Case studies (Gurjar and Mohan, 2003b)

12.2   ACUTE RISK

A large number of chemical disasters in the past emphasize that the accidental release of hazardous materials is a grave problem. This is especially the case when the release leads to the quasi-instantaneous formation of a large cloud. In many situations, such a cloud will have a density that is greater than the ambient air. The ground-hugging tendency of dense toxic cloud makes a large part of population vulnerable to the exposure of hazardous material. Hence, it is important to develop appropriate models for heavy gas dispersion or negatively buoyant cloud. In this context, it is worth mentioning that appropriate heavy gas models (IIT Heavy Gas Model I and II) have been developed for this purpose and verified against field trials. Mohan et al. (1994, 1995) describe in detail the basics and application of the IITD Heavy Gas Model for dense gas dispersion of airborne toxic materials for estimating individual and societal risk and for the estimation of vulnerable zones and user-oriented nomograms, which is given in brief as below.

12.2.1   Model Formulation and Validation

IIT Heavy Gas Model is a numerical box model where governing equations take into account the relevant physical processes such as gravitational slumping, entrainment of air, cloud heating, and so on (Singh et al., 1991; Mohan et al., 1995). Numerical solution of the equations evaluates the cloud characteristics, namely, radius (length in the case of continuous release), height, density, temperature, and amount of entrained air at each time step. Particular emphasis has been placed on model validation by comparing their performance against relevant field trial data (Thorney Island, Burro Series, and Maplin Sands Trials) as well as with other models. On the basis of statistical evaluation, a good performance of the model has been established. The performance of the IIT Heavy Gas Model is close to the model showing the best performance among 11–14 other models developed in various countries. A detailed account of model formulation and validation is included in Mohan et al. (1995). Using the IIT Heavy Gas Model, the safe distance/vulnerable zones can be easily estimated for different meteorological and release conditions for the storage of various hazardous chemicals.

12.2.2 ESTIMATION OF VULNERABLE ZONES

A special feature of the IIT Heavy Gas model is to determine various zones in terms of concentration isopleths corresponding to IDLH (immediately dangerous to life and health), STEL (short-term exposure limit) and TLV (threshold limit value) on the city map for a given release and meteorological situation, as illustrated in Figure 14 of Singh et al. (1991). Thus, areas of potential impact could be easily identified and appropriate emergency preparedness measures could be taken in case of an industrial accident. Here, concentration isopleths are provided for 65 tonnes storage of chlorine at Shriram Foods and Fertilizers Industries (SFFI) in Delhi. The site and storage quantity are only for the demonstration purpose as it existed in the past. However, these do not confirm to the existing scenario. The meteorological scenario chosen for plotting the isopleths corresponds to stable atmospheric conditions with low wind speed and it represents the case for which the maximum value of IDLH distance is obtained. Based on climatological data on wind frequencies, we have considered North-Westerly wind direction to plot the isopleths. Therefore, the areas of potential impact lie in the South-Easterly direction of the site.

12.2.3 USER-ORIENTED RESULTS

Based on this model, certain user-oriented results have been provided that could be used by a layman for control and management in case of an emergency without really working on the complex mathematical simulation procedure on a computing machine. Figure 15 of the above-referred publication (Singh et al., 1991) provides a nomogram for different storage quantities of chlorine under various meteorological situations that may be prevalent during the entire year and the same are represented by scenario numbers. Here, the term "nomogram" refers to a graph from which safe distances are determined due to the accidental release of various quantities of a toxic material under different meteorological scenarios that could be present in any given year. The meteorological conditions at the time of the accident could be easily correlated with the scenario number by knowing the time of the accident, cloud cover, and observable wind effects in the surroundings.

The details about the meteorological classification are given in Table 3 of the previously cited publication and Mohan et al. (1994). Thus, by knowing the meteorological scenario and the quantity of the toxic material released during an accident, the safe distance could be easily ascertained. This study has been performed for eight toxic chemicals (chlorine, ammonia, ethylene oxide, hydrogen fluoride, carbon disulfide, oleum, phosgene, and hydrogen cyanide) at the behest of the Ministry of Environment and Forests, Government of India. Here, emphasis is placed on the presentation of the model output in a user-friendly manner such that, in case of an emergency, a nontechnical person at the site can also use these. In this regard, the modeling approach to estimate the safe or threshold planning quantities (TPQ) by the industries using dispersion models is elaborated by Mohan and Gurjar (1995, 2004).

12.3 CHRONIC RISK

Potential health risks related to carcinogens in the atmospheric environment in India for cadmium, chromium, and nickel have been estimated and published by authors in the following publications: Gurjar et al. (1996) and Gurjar and Mohan (2003a). Urbanization, industrialization, increased vehicular traffic, and use of fertilizers and pesticides in agriculture have resulted in increased contamination of the atmospheric environment by chemical pollutants.

India has developed many industries both to meet its own demands and for export purposes. The development of metallurgical engineering, heavy engineering, and various chemical industries in India has created new and complex potential health risks both for workers and for the community at large through exposure to toxic contaminants in the air environment. Consequently, quantitative analysis of human health risks has become increasingly important as a means of not only judging the degree of risk associated with chemical pollutants but also selecting control strategies that can reduce these risks to an acceptable level. This study assesses the individual and societal risks due to three carcinogenic metals, namely cadmium (Cd), chromium (Cr), and nickel (Ni), present in the atmospheric environment of different states of India. The ambient air interim guidelines for these carcinogenic metals were also derived at the risk level of 10^{-5}. The typical rate of occurrence of cancer is one in four; thus, in a group of 100,000 persons, 25,000 would be likely to develop a tumor during their lifetime. If the same 100,000 persons breathe air contaminated with 5×10^{-3} $\mu g/m^3$ (ambient air interim guideline concentration) cadmium daily for their lifetime, no more than one additional person (i.e., 25,001) would get cancer (Sidhu, 1987).

Table 12.3 shows the estimated individual risk, while Table 12.4 shows societal risk for cadmium, chromium, and nickel contamination in various states of India. The inference from these tables is briefly described here.

12.3.1 CADMIUM

It appears that cancer risk to humans from exposure to cadmium concentrations in ambient air in various states of India is very minimal, as shown in Tables 12.3 and 12.4. In other words, the low concentration of cadmium in ambient air in different states of India may not lead to increased development of cancer in humans. The maximum individual carcinogenic risk is found for Uttar Pradesh, West Bengal, and Rajasthan in equal measures.

12.3.2 CHROMIUM

The highest individual risk estimate attributable to human exposure to chromium was estimated for the residents of Chandigarh (Union Territory) followed by the risk to those living in the states of Orissa, Bihar, Haryana, Punjab, Uttar Pradesh, West Bengal,

TABLE 12.3

Incremental Individual Cancer Risk from Cadmium, Chromium, and Nickel Contamination in Certain Atmospheric Environments in India (Conservative Estimates for Mixed Population)

State	Incremental Individual Cancer Risk (10^{-5})		
	Cadmium	Chromium[a]	Nickel
Andhra Pradesh	2	30	2
Bihar	2	125	6
Chandigarh (UT)	1	195	7
Gujarat	2	48	1
Haryana	1	85	4
Himachal Pradesh	1	60	2
Karnataka	0	16	1
Kerala	1	17	1
Orissa	1	134	2
Punjab	1	77	3
Rajasthan	3	15	2
Tamil Nadu	1	16	1
Uttar Pradesh	3	67	6
West Bengal	3	63	2

Source: From Gurjar, B.R. and Mohan, M. 2003a. *Journal of Environmental Monitoring and Assessment (Kluwer Academic Publishers)* 82, 203–223. With permission.

Note: Input data: Inhalation rate = 0.6 m³/h, exposure time = 24 h/day; exposure frequency = 350 days/year; exposure duration = 60 years, absorption fraction = 1; body weight = 60 kg, averaging time period = 60 × 365 days; potency factors = 6.1 (Cd), 41 (Cr), and 1.19 (Ni).

[a]   Actual chromium risk will be less because risk estimates included total chromium, whereas only hexavalent chromium is carcinogenic in nature.

Himachal Pradesh, and Gujarat. The Union Territory of Chandigarh was a rural agricultural land prior to 1950, and therefore it is very surprising that the ambient concentration of chromium is high at this location. It may be worth investigating whether the industrial plants, particularly those producing cement located only a few miles east of Chandigarh, are releasing hexavalent chromium to the atmospheric environment. The cement-producing plants are important potential sources of atmospheric chromium. There was a huge cement plant located about 3 km east of Chandigarh (UT) during the period when these data were collected for Ministry of Environment and Forests (MOEF), Government of India (GOI). A large number of small-scale industries related to the manufacture of fasteners and doing the job of electroplating may also be additional sources contributing to the high chromium content in the atmosphere of Chandigarh. Chronic exposure (30–40 years) via inhalation of elevated levels of hexavalent chromium increases the likelihood of development of lung cancer in humans.

TABLE 12.4

Incremental Societal Cancer Risk from Cadmium, Chromium, and Nickel Contamination in Certain Atmospheric Environments in India

State	Population[a] (million)	Incremental Societal Cancer Risk[b]		
		Cadmium	Chromium[c]	Nickel
Andhra Pradesh	66.30	1326	19,890	1326
Bihar	86.34	1727	107,925	5180
Chandigarh (UT)	0.64	6	1248	45
Gujarat	41.17	823	19,762	412
Haryana	16.32	163	13,872	653
Himachal Pradesh	5.11	51	3066	102
Karnataka	44.82	63	7171	448
Kerala	29.01	290	4932	290
Orissa	31.51	315	42,223	630
Punjab	20.19	202	15,546	606
Rajasthan	43.88	1316	6582	878
Tamil Nadu	55.64	556	8902	556
Uttar Pradesh	138.76	4163	92,969	8326
West Bengal	67.98	2039	52,827	1360

Source: From Gurjar, B.R. and Mohan, M. 2003a. *Journal of Environmental Monitoring and Assessment (Kluwer Academic Publishers)* 82, 203–223. With permission.

[a]   Mahendra (1991).

[b]   Incremental societal cancer risk = incremental individual cancer risk (from Table 12.1) × population.

[c]   Actual chromium risk will be less because risk estimates included total chromium, whereas only hexavalent chromium is carcinogenic in nature.

12.3.3   NICKEL

It appears that the highest individual risk estimate attributable to human exposure to nickel is for the residents of Chandigarh (Union Territory), followed by those living in Uttar Pradesh, Bihar, Haryana, and Punjab (Table 12.3). It may be worth examining whether the high levels of nickel in the ambient air and the increased risk estimates of lung and nasal cancers attributable to exposure to nickel in some northern states of India (Union Territory of Chandigarh, Uttar Pradesh, Bihar, Haryana, and Punjab) are associated with the burning of coal. Burning of coal is one of the major sources of release of nickel to the atmospheric environment (ATSDR, 1993).

It appears that the estimated cancer risk attributable to chromium contamination of ambient air is relatively higher than that attributable to nickel (Table 12.3). However, the true carcinogenic risk from chromium, as shown in Table 12.3, is less, because estimates in Table 12.3 include total chromium while only the hexavalent chromium is carcinogenic in nature (USEPA, 1995).

Smoking contributes significantly to cadmium intake. Depending on the brand of cigarette and the number of cigarettes smoked, daily intake of cadmium may vary from 1 to 6 µg/day (Pandya, 1978). When 10% of the intake of cadmium is assumed to be absorbed in the lungs, it may cause about 3–19 times more risk to the smoker than the maximum risk that exists in Uttar Pradesh state (Gurjar and Mohan, 1996).

The societal risk is highest in Bihar for chromium and in Uttar Pradesh for nickel and cadmium (Table 12.4). This is understandable because, with respect to the total population, Uttar Pradesh and Bihar are the top two states among the 14 states shown in this table. The societal risk and population (total and density) are important factors for decision-making processes for future industrial growth and location (Gurjar and Mohan, 2003a).

It is impossible to develop precise risk estimates due to various uncertainties included both in available data and in the models that are used to calculate potency factors and effective concentration (Fiksel, 1985). However, this type of average estimation of risk levels can be useful for planning purposes. The method adopted here is helpful in estimating carcinogenic risk attributable to carcinogens present in the atmospheric environment. However, there are some limitations as well. The assumption of average population, for example, implicitly assumes a homogeneous population distribution with identical background factors such as sensitive populations, age, and sex. These limitations narrow the scope of this method, which should be kept in mind when using this method in planning and decision-making activities.

12.4 INTEGRATED RISK ANALYSIS USING BOTH ACUTE AND CHRONIC RISK: CASE STUDIES

The general practice to study various types of risk in isolation from each other could give erroneous conclusions. This is because, in real-life situations, a person or community may encounter the different combinations of environmental risks posed by the surroundings. Keeping this in mind, an integrated approach for environmental risk analysis (ERA) has been proposed to explicitly consider and define a comprehensive tool, specifically for toxic risks, that is broadly reflective of real-life risk situations (Gurjar and Mohan, 2003b).

To begin with, individual risk factors (IRF) and geosocietal risk factors (GSRF) have been estimated for two Indian industries that may cause acute risk due to an accidental release of chlorine in the atmosphere. Further, background risk factors (BRF) have been determined by converting the chances of extra cancer cases into mortality per year due to the presence of carcinogenic toxic elements present in the environment. The cumulative individual risk factors (CIRF) and cumulative geosocietal risk factors (CGSRF) have been calculated as the sum total of risks posed by acute and chronic toxic exposures for individual and societal risk, respectively (Gurjar and Mohan, 2003b).

The methodology is demonstrated with two industries located in Haryana state of India. These are (i) M/s Ballarpur Industries Ltd, Yamuna Nagar, and (ii) M/s Advanced Chemicals, Bahadurgarh (Haryana). The detailed methodology is covered in Gurjar and Mohan (2003b). The risk contours for the above two industries are shown in Figures 12.1 and 12.2, respectively.

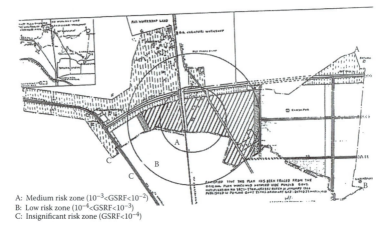

A: Medium risk zone (10^{-3}<GSRF<10^{-2})
B: Low risk zone (10^{-4}<GSRF<10^{-3})
C: Insignificant risk zone (GSRF<10^{-4})

FIGURE 12.1 Risk contours pertaining to a hypothetical catastrophic release of chlorine from M/s Ballarpur Industries Limited, Yamuna Nagar (Haryana). Scale: 1 cm = 0.754 km. (From Gurjar, B.R. and Mohan, M. 2003b. *Journal of Hazardous Materials (Elsevier)* 103(1–2), 25–30. With permission.)

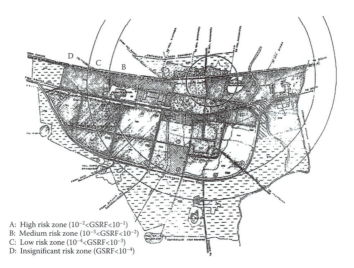

A: High risk zone (10^{-2}<GSRF<10^{-1})
B: Medium risk zone (10^{-3}<GSRF<10^{-2})
C: Low risk zone (10^{-4}<GSRF<10^{-3})
D: Insignificant risk zone (GSRF<10^{-4})

FIGURE 12.2 Risk contours pertaining to hypothetical catastrophic release of chlorine from M/s Advanced Chemicals, Bahadurgarh (Haryana). Scale: 1 cm = 0.445 km. (From Gurjar, B.R. and Mohan, M. 2003b. *Journal of Hazardous Materials (Elsevier)* 103(1–2), 25–30. With permission.)

12.5 CURRENT STATUS AND LIMITATIONS OF RISK ASSESSMENT TECHNIQUES

It is a worldwide experience that QRA is a valuable tool to improve the safety and efficiency level in CPIs. However, it is also a fact that this is a continuously evolving science, needing further refining. That is why the views on the potential uses of risk analysis differ. For example, most experts and policy-makers agree that risk analysis is a valuable tool to inform decisions, but they disagree about the extent to which risk estimates are biased and should be allowed to influence public policies to protect health and the environment. Often, it is agreed that it should be used to target and address the worst risks to health and the environment first, to achieve risk reduction in more cost-effective and flexible ways that minimize overall economic impacts, and to ensure that risk reduction achieved by regulations is worth the cost. Critics also charge that quantitative methods cannot assess very long-term or newly discovered threats. They also believe that quantitative cost–benefit analyses undervalue environmental and health benefits, exaggerate costs, and focus on relatively widespread but individually small costs and risks rather than on much larger costs and risks to smaller (and often more vulnerable) groups.

The crucial parts of a QRA come before and after the actual risk analysis, that is, making the correct initial assumptions and then interpreting results. An assumption for one case may not be appropriate for another and in case it is used, it may give highly debatable results. In a study conducted in 1988, for example, 11 teams used QRA on a small ammonia plant and their results for one hazard varied from 1 in 400 to 1 in 10 million (Kim et al., 1995). Further, it has been observed that descriptions of the likelihood of adverse effects may range from qualitative judgments to quantitative probabilities. Although risk assessments may include quantitative risk estimates, quantification of risks may not always be possible. Thus it is better to convey conclusions (and associated uncertainties) qualitatively than to ignore them because they are not easily understood or estimated.

Another problem is that the models drastically simplify what happens in real nature. This is the reason that for the same set of data, different models are liable to give highly varied results depending on the basic premises and assumptions used in the development of models (Smith et al., 2000). This makes it difficult to choose one model and reject the others. Another drawback of QRA is the need for accident and equipment failure data, which become scarcer, the safer plants become. Nevertheless, trends can be seen. One common cause of failure is "correlated failure," in which backing up one piece of equipment is assumed to increase safety. In an example of "external" correlated failure, an explosion would disable two generators located next to each other. An example of internal correlated failure would be when environmental factors damage the Teflon seals in two pumps of the same type, and a pressure surge takes both of them out. Human error is also becoming a more prominent factor in failures, as the trend toward automated equipment continues.

Furthermore, the quality of risk analysis depends on the adequacy of data and the validity of the method. For environmental hazards and most health and ecological effects, there are few data and methods are controversial. As a result, there is a growing perception that the risk analysis has not done a very good job predicting the ecological and health effects of many new technologies (Schierow and Linda-Jo, 2001).

Yet, despite various limitations of QRA as shown in Table 12.5 (CMA, 1987) and differences in attitude toward risk analysis, ERA is becoming more important globally. Risk-based decisions, whatever the context, seem to be the soundest guides for

TABLE 12.5
Classical Limitations of QRA

Five of the Most Global Limitations of QRA	
Issue	Description
Completeness	There can never be a guarantee that all accident, situations, causes, and effects have been considered
Model validity	Probabilistic failure models cannot be verified. Physical phenomena are observed in experiments and used in model correlations, but models are, at best, approximations of specific accident conditions
Accuracy/uncertainty	The lack of specific data on component failure characteristics, chemical and physical properties, and phenomena severely limits accuracy and can produce large uncertainties
Reproducibility	Various aspects of QRA are highly subjective; thus the results are very sensitive to the analyst's assumptions. Using identical data for a problem, models may generate widely varying answers when analyzed by different experts
Inscrutability	The inherent nature of QRA makes the results difficult to understand and use

Source: Adapted from Gurjar, B.R. 1999. Environmental Risk Analysis for Industrial Siting, Planning and Management. Ph.D. Thesis, Indian Institute of Technology, New Delhi.

ensuring adequate human health and environmental protection, while avoiding costly and unnecessarily stringent control on chemical exposures. It is expected that the use of risk analysis will increase in the future because of its versatile applications in cost-effective management of CPIs in particular and to ensure a safety and healthy environment to the public in general.

12.6   CONCLUSIONS

Use of dispersion modeling is demonstrated for the estimation of acute risk for estimating vulnerable zones. Nomograms from dispersion modeling techniques can be used for evacuation purposes at the time of an accidental toxic release from an industry.

Chronic risk estimates for certain air toxics, estimated from simple dose–response models for different states of India, are shown and their usefulness is discussed. Finally, the integrated risk assessment approach is demonstrated with case studies for Indian industries with possible adaptation for industrial siting and planning. Lastly, it is recommended to use with caution the dispersion models and QRA techniques with due importance to assumptions and limitations implied therein.

REFERENCES

ATSDR. 1993. Toxicological profile for cadmium. U.S. Department of Health and Human Services, Public Health Services. Agency for Toxic Substances and Disease Registry (ATSDR), Atlanta, GA.

CMA. 1987. *Evaluating Process Safety in the Chemical Industry: A Manager's Guide to Quantitative Risk Assessment (QRA).* Chemical Manufactures Association (CMA), UK.

Fiksel, J. 1985. Quantitative risk analysis for toxic chemicals in the environment. *Journal of Hazardous Material* 10, 227–240.

Gurjar, B.R. and Mohan, M. 2002. Environmental risk analysis: Problems and Perspectives in different countries. *Risk: Health, Safety and Environment (USA)* 13(1/2), 1–30.

Gurjar, B.R. and Mohan, M. 2003a. Potential health risks due to toxic contamination in the ambient environment of certain Indian states. *Journal of Environmental Monitoring and Assessment (Kluwer Academic Publishers)* 82, 203–223.

Gurjar, B.R. and Mohan, M. 2003b. Integrated risk analysis for acute and chronic exposure to toxic chemicals: A case study. *Journal of Hazardous Materials (Elsevier)* 103(1–2), 25–30.

Gurjar, B.R., Mohan, M., and Sidhu, K.S. 1996. Potential Health risks related to carcinogens in the atmospheric environment in India. *Regulatory Toxicology and Pharmacology* 24, 141–148.

Hanna, S.R., Strimaitis, D.G., and Chang, J.G. 1991. *Hazard Response Modeling Uncertainty (A Quantitative Method). Evaluation of Commonly Used Hazardous Gas Dispersion Models*, Vol. II. Report prepared by Sigma Research Corporation for the Air Force and the American Petroleum Institute.

Khan, F.I. and Abbasi, S.A. 1998. *Risk Assessment in Chemical Process Industries: Advanced Techniques.* Discovery Publishing House, New Delhi.

Kim, I., Moore, S., and Ondrey, G. 1995. Risk and the chemical process industry. *Chemical and Engineering* February, 26–29.

Mahendra, K.P. 1991. *India's Population: Heading Towards a Billion.* B.R. Publishing, Delhi.

Mohan, M. and Gurjar, B.R. 1995. Estimation of threshold planning quantities of extremely hazardous chemicals based on simple technical models, Institution of Engineers. (India). *Journal of Environmental Engineering*, 76, 17–21.

Mohan, M. and Gurnard, B.R. 2004. Risk based IIT-TPQ model to establish threshold planning quantities of hazardous substances. *Journal of the Air and Waste Management Association (USA)* 54, 495–503.

Mohan, M., Panwar, T.S., and Singh, M.P. 1995. Development of dense gas dispersion model for emergency preparedness, *Atmospheric Environment, Part B: Urban Atmosphere* 29(16), 2075–2087.

Mohan, M., Singh, M.P., and Chopra, H.V.K. 1994. Dense gas dispersion model for the assessment of Industrial hazards: A review. *Journal 'Mathematics Student'* 1–4, 33–73.

Pandya, C.B. 1978. Evaluation of occupational exposure to trace metals in small and medium scale industries. PhD Thesis, Gujarat University, Ahmedabad, India.

Schierow, Linda-Jo. 2001. IB94036: The role of risk analysis and risk management in environmental protection. Resources, Science, and Industry Division, The National Council for Science and Environment, 1725 K Street, NW, Suite 212, Washington, DC, 20006 (202) 530-5810, March 14, 2001.

Sidhu, K.S. 1987. *Health Risk Assessment. P-Dioxane.* Michigan Department of Public Health, Lansing, MI (September 27).

Singh, M.P., Mohan, M., Panwar, T.S., and Chopra, H.V.K. 1991. Atmospheric dispersion of dense materials for estimating vulnerable zones due to accidental release of toxic chemicals. *Risk Analysis* 11(3), 425–440.

Smith, K.R., Zhang, J., Uma, R., Kishore, V.V.N., Joshi, V., and Khalil, M.A.K. 2000. Greenhouse implications of household fuels: An analysis for India. *Annual Review of Energy and Environment* 25, 741–763.

UNEP. 2000. Issues for the 21st Century, http://www.unep.org/geo2000/english/0223.htm

USEPA. 1995. Integrated Risk Information Systems (IRIS). Online Computer Database. *File Chromium. United States Environmental Protection Agency (USEPA)*, Cincinnati, OH.

Section IV

Air Quality Management:
Techniques and Policy Aspects

13 The Economics of Air Pollution
Theories, Valuation Methods, and Policy Aspects

Dilek Demirbas

CONTENTS

13.1 INTRODUCTION: ENVIRONMENTAL DEGRADATION, HEALTH EFFECTS OF AIR POLLUTION, AND NEOCLASSICAL ENVIRONMENTAL ECONOMICS

For the last 200 years, largely through human activity, unprecedented environmental challenges and irreversible mass extinctions have been occurring on the Earth. Owing to this significant damage, some estimates suggest that not only are more than 100 species a day becoming extinct, but also our natural resources that sustain life on the planet—air, water, and soil—are becoming polluted or depleted on an alarming scale together with exponentially increasing human population growth (Desjardins, 2001). This means that as the prospects for continued degradation and depletion of natural resources multiply as a result of this population growth, natural resources on the planet such as clean water, clean air, and clean soil will become a luxury and resources for the survival of future generations will become scarce.

According to Schumacher (1973),

> The modern man does not experience himself as a part of nature but as an outside force destined to dominate and conquer it. He even talks of a battle with nature, forgetting that, if he won the battle he would find himself on the losing side. One reason overlooking this vital fact is that we are estranged from reality and inclined to treat as valueless everything that we have not made ourselves.... [W]e forgot the fact that what we use to make some of the capital which today helps us to produce is a small part of the total capital we are using. Far larger is the capital provided by nature and not by man. This larger part is now being used up at an alarming rate (p. 11).

It is true that human beings are greedy and treat nature as a renewable source. It is also frightening to know the extent to which the illusion of having solved the problem of production is still on the agenda. If we continue to ignore the difference between income and capital for nature, mistakenly treating nature as an income item rather than as a capital item, we will reach a stage where the adverse effects of damages to nature will not be reversible and we will continue to destroy our planet and possibly even cause our own extinction.

To bring to an end these arguments, over the last half century developed countries have begun to reverse health effects and to reduce the cost of environmental pollution in urban cities. For this reason, Environmental Economics has emerged as a subfield of economics to deal with environmental issues using standard methods of Neoclassical Economics and to undertake theoretical and empirical studies of the economic effects of national or local environmental policies around the world. As a result, issues such as the costs and benefits of pollution, alternative environmental policies to deal with air pollution, water quality, toxic substances, solid waste, and global warming have become very important subjects for analysts to solve.

Over the last 50 years in particular, the health effects of environmental pollution, especially air pollution, have become the center of many epidemiological studies for risk assessments issues and of Environmental Economics for policy decision-making processes. With an increasing community awareness of human health and air quality concerns, a large body of epidemiological research has emerged, showing the adverse health effects of air pollution and focusing on the damaging effects of air pollutants on public health. Based on these research findings, which demonstrate that air pollution causes different levels of risk to human health and the environment, environmental regulatory authorities in many countries have implemented stringent air quality measures (BTRE, 2005). These researches have also revealed that human health might be affected by these exposures more than was previously believed (WHO, 1999). Evidence is still emerging that the long-term exposure to low concentrations of *particulate matter* in the air is associated with mortality and other chronic effects, such as increased rates of bronchitis and reduced lung function (WHO, 2000a, 2000b; Fisher et al., 2002; BTRE, 2005; Defra, 2006). Figure 13.1 shows the various stages of health effects. These range from mild or subtle health effects to premature mortality.

According to World Health Organization (2000a), there are different stages of health effects. The sequence of the health impacts of air pollution on the affected population ranges from mild or subtle health effects (subclinical effects) to the most severe of health effects (premature mortality). In between, impaired pulmonary functions, restricted activity/reduced performance, visits to the doctor, emergency room visits, and hospital admission can be considered as different types of severity of health effects in the affected population.

What are these air pollutants, which endanger our health? According to a report produced by BTRE (2005), air pollutants are usually classified into suspended particulate matter (such as dusts, fumes, mists, and smokes), gaseous pollutants (such as

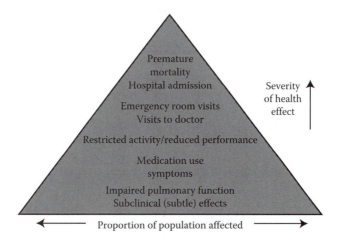

FIGURE 13.1   Air pollution health effects pyramid. (From WHO, 2000a. Quantification of the health effects of exposure to air pollution. Report of a World Health Organization working group, Bilthoven, Netherlands, November. With permission.)

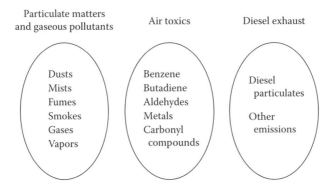

FIGURE 13.2 Polluted air mixture: Particulate matters and gaseous pollutants, air toxics and diesel exhaust. (Adapted from Health Effects Institutes, 2000. *HEI Strategic Plan for the Health Effects of Air Pollution 2000–2005*, pp. 1–26. Cambridge, MA: Health Effects Institute, 2000.)

gases and vapors), and odors. Many epidemiological studies have demonstrated that these pollutants are a risk to human health. They reported that, especially in the long term, air pollution in urban areas can cause bronchitis, respiratory diseases, lung cancer, and early deaths. Their results also suggested that carcinogenic chemicals in the smallest air particles and carcinogenic gases—such as benzene and benzopyrene—could be possible causal agents (Kjellstrom et al., 2002). Figure 13.2 shows major pollutants in the air pollution mix.

Trends in the concentrations of these pollutants in urban air, such as nitrogen dioxide or particulate matter, therefore became the focus of concern for international air pollution guidelines and many agencies such as the European Environment Agency (EEA, 2004). In addition, policy makers in developed countries in the West employed more systematic approaches to measure the economic costs of air pollution on human health and to develop effective environmental strategies for both social and economic efficiency objectives (Ad-hoc Group, 1999). The qualification of environmental-related health effects and their valuation in monetary units play a key role in policy measures (Seethaler et al., 2003).

If it is an undeniable fact that environmental pollution exists and the economic costs of pollution on human health are very high, we have to find urgent answers to a number of questions, such as: What are the fundamental responsibilities of economics and economists? What should economists do to reduce or to eliminate the level of pollution? What is the source of these valuation techniques that we use? What is the most common analyzing method for environmental management? And finally, why do we have to select the best environmental policy among alternatives?

The main objective of this study is to evaluate some of these questions, starting from the perspective of Neoclassical Environmental Economics and reaching the interdisciplinary approach. The need for this study is very clear as there are not enough economic studies to evaluate the extent of pollution on Earth. Therefore, this study is an attempt to look at the whole picture, from the angle of air pollution, economics, and policy matters.

Section 13.2 examines the main economic approaches that shaped current Neo-classical Environmental Economics in order to understand why Neoclassical Economics analyzes air pollution in a particular way and comes up with particular prescriptions. Then, three other approaches—Utilitarianism, Welfare Economics, and Efficiency Theory—are also examined as they are alternative approaches to Neoclassical environmental policies in both developed and developing countries. It is true that economics is an important subject that can shape the activities of the modern world with its criteria. Without understanding the main assumptions of Neoclassical Environmental Economics however, it is not possible to understand the rationality of international air pollution guidelines or decision-making processes regarding pollution control.

Section 13.3 assesses economic valuation approaches within Neoclassical Environmental Economics in order to measure the intensity of individual wants and analyze the impact of air pollution on health. The *Willingness-to-Pay (WTP) approach* and the *Human Capital approach* are the most used techniques for decision makers when evaluating the impact of air pollution on health, particularly in urban areas. The function of this section is also to prepare us for the next one, which is cost–benefit analysis (CBA).

Section 13.4 assesses CBA within Neoclassical Environmental Economics to demonstrate how one can measure and compare individual preferences for environmental decision making and the impact of air pollution on health. The reason why CBA was selected is because it is the most well-known technique used by policy makers and is also one of the most controversial. In this section, the rationality of CBA, the structure of CBA, and its conceptual framework will be looked at closely.

Section 13.5 examines alternative approaches to Neoclassical Environmental Economics, such as *Austrian Economics, Green Economics*, and *Ecological Economics*. After examining Neoclassical Environmental Economics and its analyzing techniques, this section is essential as we need to substitute existing assumptions with more realistic ones. We must do that not only to eliminate the problems of Neoclassical Environmental Economics, but also to have a better economic understanding so as to deal with environmental issues.

The final section, Section 13.6, concludes our findings and makes a number of proposals for decision makers.

13.2 THEORETICAL BASIS OF NEOCLASSICAL ENVIRONMENTAL ECONOMICS

Economics plays a central role in shaping the activities of the modern world by imposing criteria of what is economic and what is uneconomic. So far, there are no other criteria that exercise a greater influence over the actions of individuals, groups, and governments than economic criteria. Neoclassical Environmental Economics was, and still is, a major influence on the theories of natural capitalism and environmental finances, which are the two sub-branches of Environmental Economics concerned with resource conservation in production and the value of biodiversity of humans. With increasing influence, Neoclassical Environmental Economics moved into the very center of public concern on environmental issues. There are three very important economic approaches within Neoclassical Environmental Economy, which are still very dominant views on environmental policy matters in both developing and developed countries. These are Utilitarianism, Neoclassical Efficiency Theory, and Welfare Economics.

13.2.1 UTILITARIANISM

Utilitarianism is an early nineteenth-century ideology and differs from ethical theories that make the rightness or wrongness of an act dependent on the motive of the agent. According to the utilitarian, it is possible for the right thing to be done with a bad motive. Among the well-known utilitarians, Jeremy Bentham and John Stuart Mill should be mentioned as leading figures.

Utilitarians commonly agree that *happiness* should be the objective of individuals, and they take happiness as the basis of judgments about actions (Mill, 1965). As the ultimate objective is to maximize aggregate utility or welfare, for the same reason they believe that utility has to be something that can be easily quantified or measured. On one level the decision on whether one state of affairs is better than another depends on the sum of personal utilities and nothing else. For example, inequalities in distribution are bad because they lower the sum of utilities. More production is good if it increases total enjoyment. Utilitarians believe that low levels of pleasure count as much as higher levels of pleasures and the economic as much as the noneconomic. In their idealistic world, if utility is to be measured, it should be possible to tell how many units an individual enjoys and how many more the same individual enjoys in one situation compared to another (Bonner, 1995).

It is also believed that individuals pursue their own self-interests because they are the best judges of their own welfare or happiness. Therefore, there should be no interference with their choices on the individual level. In contrast, utilitarians also argue that social and private utility can be merged by governments for public policy programs in order to maximize the aggregate utility. It is for this reason that utilitarianism was, and still is, very influential in economics, public policy, and government regulations, and plays a significant role in environmental policy (Desjardins, 2001). According to Bonner (1995), "because many believe that utilitarianism gave reasons why one is better than another it made discussion of policy possible" (p. 4).

As the total of individual happiness itself should be the objective of public policy, redistribution of income will be desirable if total utility, which is the sum of all individual utilities, increases (Bonner, 1995). Consequently, there is no basis for the condemnation of the existence of extreme inequality as the main goal is to satisfy as many individual preferences (Desjardins, 2001).

A number of challenges have been raised against utilitarian thinking. The most important ones are *measurement* of utilities, *comparison* of individual welfares, and *aggregation* of interpersonal utility information. In fact, measuring, comparing, and aggregating individual utilities is not easily done and in some cases is impossible. How is it possible to measure or to scale choices, or to find their origins? How can we compare the satisfaction of individuals? How should we aggregate individual welfare to obtain social utility? How can we quantify pleasure, happiness, desire, and so forth? These are key questions to which even utilitarians cannot find easy answers (Bonner, 1995). Firstly, we cannot simply assume that all desires or pleasures are qualitatively the same. In the words of Desjardins (2001), "Is the pleasure that I received from breathing clean air equal to the pleasure that you receive from smoking" (p. 27). If this is the case, how can we measure them? According to what scale? What do utilitarians do when they cannot quantify pleasure, happiness, or satisfaction?

For critics, one challenge is based on the *measurement* issue. Utilitarians substitute for the good something that can be quantified in *money* terms, and this is seen as a major mistake. Utilitarians measure and compare the health consequences of pollution control decisions by using some quantifiable variables such as life expectancy, infant mortality, and per capita expenditure on health care as proxies for health; however, these variables cannot cover all aspects of the *value of life*. To be able to overcome this problem, utilitarians propose that, in the marketplace, everything should have a price in order to make them exchangeable, and there is nothing that does not have a price. According to critics, this is exactly what goes wrong when environmental regulation is subjected to CBA. For environmental policy, regulators measure different economic factors associated with health as proxies as they are unable to measure the value of health itself. Then, they assume that comparing the cost of health to the cost of eliminating the source of pollution would be enough to make policy decisions. Even simple noneconomic values like beauty, cleanliness, and health can survive only if they are proved to be "economic" and subject to CBA. How easy is it to measure something that is not measurable?

Another challenge is based on the nature of utilitarian *judgments*. Because for utilitarians there is no act, in and of itself, that is ever right or wrong, they do not take into account the consequences of a particular act. Critics claim that this approach is incomplete and is unable to account for certain ethical issues. Each society has its own rights and wrongs, and actions can be altered according to the value each society places on certain things.

In summary, utilitarianism is one of the most influential approaches that shape our public policy decisions on environmental issues, but it is also one of the most controversial ones. This is the reason why environmental debates today are so widely cast in utilitarian terms.

13.2.2 NEOCLASSICAL EFFICIENCY

The second influential approach to the environmental policy decision-making process is Neoclassical Efficiency Theory, which can also be seen as an extension of utilitarianism. What is Neoclassical Efficiency Theory? How can we relate it to the environmental decision-making process? This subsection will deal with these questions.

The three main assumptions of Neoclassical Economy are (1) existence of perfect information, (2) existence of transitivity of alternate choices, and (3) positive marginal utility/product if and only if nonzero inputs exist. The main reason for the perfect information assumption is to enable rational choices (Choudhury, 1995). Neoclassical theorists assume that consumers are rational agents who act rationally and who make rational consumption decisions in order to maximize their own self-interests and maximize on their individual rational choices. Hence, utility maximization, which is constrained by income and prices, is the ultimate goal of consumption for a rational consumer (Hanley and Spash, 1993). Therefore, the same consumer will always choose the most preferred bundle of goods from a set of feasible alternatives, and the same consumer will be able to compare various bundles of goods.

Consequently, "economic efficiency" appears to be the utilitarian goal of providing the greatest good for the greatest number for the whole economy. If the goal of economic efficiency is to achieve the optimal satisfaction of consumer preferences, "an efficient market is one in which more people get more of that for which they are most willing to pay" (Desjardins, 2001, p. 59). This leads to independence among rational agents, thus establishing a causal relation between economic competition of any kind and the degree of agent-specific independence attained. That is, competition is realized in the presence of key assumptions. Besides, if competition exists, then optimal information must be available to agents (Choudhury, 1995).

For Neoclassicists, the basic reason for social inefficiency lies behind the fact that social costs associated with external effects, such as health impacts of air pollution, are not incorporated into the cost of producing the pollution-generating product or its market price. From this perspective, the key solution is to increase the overall value of production to a level that would be generated if the pollution costs were being reflected in its price. Under such circumstances, there would be an efficient reallocation of resources. When production and consumption are arranged in such a way that all air pollution costs are accurately reflected in product prices within competitive markets, the market is said to be Pareto efficient, that is, society, on net, cannot be made better off (Cordato, 2004).

It is on this point that Neoclassical Theory has been subjected to the most criticism. One such criticism comes from environmental ethics, which claims that if an efficient market is one in which more people get more of that for which they are most willing to pay, why should we take the satisfaction of individual preferences as an overriding goal? How do we know that individual preferences are right or wrong for us in the long term? Especially when we are concerned with environmental issues, why should the satisfaction of individual preferences be the goal of public policy? Critics claim that these are the fundamental questions that Neoclassical economists should answer urgently in order to deal with today's environmental problems.

These critics also claim that it is true that Neoclassical economic analysis plays a key role in many contemporary environmental policy issues. In particular, CBA is the major public policy methodology used in reaching environmental decisions and shaping environmental regulations at national or international levels. However, when the economic efficiency idea becomes so dominant for policy makers, we cannot simply accept the criteria of the satisfaction of individual preferences.

Sagoff (1990) argues that much economic analysis rests on a serious confusion between, on the one hand, wants or preferences and, on the other, beliefs and values. Indeed, Neoclassical economics deals only with wants and preferences because these are what get expressed in an economic market in monetary terms, but excludes beliefs and values because they are not accountable. According to Sagoff (1981),

> Economic methods cannot supply the information necessary to justify public policy. Economics can measure the intensity with which we hold our beliefs; it cannot evaluate those beliefs on their merits. Yet such evaluation is essential to political decision making. This is my single greatest criticism of cost-benefit analysis. (p. 18)

It is true that the market can measure the intensity of our wants by our *WTP* (by price), measure and compare individual preferences through CBA, and determine efficient means for *optimally fulfilling wants*. However, markets cannot measure or quantify our beliefs or values. Because many environmental issues also involve our beliefs and our values, economic analyses become incomplete. In particular, when Neoclassical Economics is involved in environmental policy, it treats our beliefs as if they were mere wants, and thereby seriously distorts the issue.

13.2.3 WELFARE ECONOMICS

Pigou (1956) developed a formal welfare theory that could be applied to economic policy. His study was to highlight the question of whether perfectly competitive markets lead to an optimum allocation of resources. Therefore, Pigou showed that firms' marginal cost functions may not accurately reflect the social costs of production and the demand curves of individuals may not accurately reflect the social benefits from consumption. Thus, in this study, Pigou (1956) examined the divergences between private benefits and social benefits, and between private costs and social costs. Then, he called these divergences externalities, spill-over effects, and third-party effects, which are often used to justify government actions. The costs that a firm considers in its profit maximization decisions are private costs borne by the firm. But social costs, such as pollution, are not borne by the firm; thus there is a divergence between private cost and social cost at the margin. A free market will therefore result in the production of an excessive quantity of goods whose marginal social cost exceeds their marginal private cost. When this is the case, governments intervene and correct the externalities.

Before Neoclassical Theory, Classical Economic Theory also used the concept of welfare. While classical economists considered welfare as an increasing output, Neoclassical economists perceived welfare as more than an increase of output with the help of the marginal utility concept (Colander, 1989; Roll, 1992). Then, by distinguishing economic theory from policy, Welfare Economics became an integral part of the Neoclassical Economics policy decision-making process. When economic theory became more formal, welfare economy also specialized as a separate field.

As with previous approaches, Welfare Economics is also subject to criticism. As its conceptual framework is based on both Classical and Neoclassical Economics, it uses the same conceptual framework; however, the literature on Welfare Economics, according to Choudhury (1995), is seen as more of the reformulation of underlying Neoclassical methodology. The Neoclassical welfare approach might be seen as offering a new concept, such as externalities; however, it does not noticeably challenge the idea of the ultimate objective, which is to maximize aggregate utility or welfare. In other words, it is still not very clear how one can determine aggregate happiness, satisfactions, and beliefs in order to maximize aggregate welfare.

13.2.4 CONCLUSION-I

There is no question that these three approaches have had a considerable impact on public policy and environmental regulations. Because they were massively dominant in decision-making processes on environmental issues, many valuation and analyzing techniques, which assess the health impacts of environmental pollution, such as WTP, human capital, and CBA, have been developed and applied. Over the last 30 years, there has been reasonable epidemiological progress in the field of air pollution and significant changes in international air pollution guidelines, mostly based on these analyzing techniques and valuation methods.

The main intention of the next section is to link the economic basis for environmental policy with the health impacts of air pollution and their economic valuation techniques. Even though these valuation approaches and analyzing techniques stem from the same Neoclassical theoretical approach, and hardly reflect the real values, according to critics, we still use them to *measure the intensity of our wants* by our WTP (by price), *measure* and *compare individual preferences* (through CBA), and *determine efficient means* for optimally fulfilling wants. Section 13.3 will concentrate on the existing economic valuation approaches and how they assess the health impacts of air pollution.

13.3 ECONOMIC VALUATION APPROACHES WITHIN NEOCLASSICAL ENVIRONMENTAL ECONOMICS: MEASURING THE INTENSITY OF INDIVIDUAL WANTS AND THE ASSESSMENT OF HEALTH IMPACTS OF AIR POLLUTION

In accordance with Neoclassical Environmental Economics, environmental resources provide goods and services that have either no apparent markets or very imperfect markets. The absence of a market for the environment means that unlike man-made products, these goods and products are not priced; therefore, their monetary value to people cannot be readily observed. To deal with this problem, the main challenges for Neoclassical economists are to first *identify* the environmental pollution and then *estimate* the value of these environmental changes. To be able to estimate these changes, several direct and indirect valuation techniques have been developed (Viscusi, 1993; Lvousky, 1998; Pearce and Howarth, 2000).

Because the economic approach to valuating environmental challenges is based on people's preferences in their environment, the underlying principle for the economic valuation of environmental degradation, just as for man-made products, is to measure the intensity of people's WTP for an environmental benefit or, conversely, their willingness to accept compensation for environmental degradation (Pearce and Howarth, 2000). It is assumed that if these quantities can be measured, then economic valuation allows environmental impacts to be compared on the same basis as financial costs and benefits of the different scenarios for environmental pollution control. Consequently, it will be feasible to evaluate the social costs and benefits for the different environmental issues in relation to public policy.

The quantification of environmental-related health risks, such as mortality and morbidity and the valuation of their costs in monetary units, plays a key role in policy measures. The calculation of air pollution-related health costs is based on epidemiological information to *identify* the relevant health effects related to air pollution, and to *estimate* the different cost components related to the health impacts in monetary terms. Three different methods are available for the assessment of the health costs of air pollution: (a) the WTP approach, (b) the Human Capital approach, and (c) the Economics of Abatement approach with Pigovian taxes and tradable permits.

13.3.1 WTP Approach

The WTP approach assesses how much we are willing to pay to reduce risk, which is the prevention of a statistical fatality or illness. In their study, Seethaler et al. (2003) overviewed the health cost components and stated that WTP includes individually borne costs (such as individually borne treatment costs, individually borne loss of production due to illness, individually borne avertive expenditures, and individually borne intangible costs due to averting behavior) but does not consider collectively borne costs (such as collectively borne treatment costs, collectively borne loss of production due to illness, and collectively borne avertive expenditures due to averting behavior). Figure 13.3 shows that the WTP approach includes individual material costs and intangible costs but does not consider material costs that are collectively borne, for example, due to insurance contributions.

Apart from attempting to capture trade-offs between wealth and risk, this approach also reflects the value of intangible elements such as quality of life and joy of living. The WTP approach uses people's preferences to ascertain the value they place on reducing the risk to life. Then, other costs such as net loss output and medical and administrative costs are added to this base cost (BTRE, 2005). The literature on the

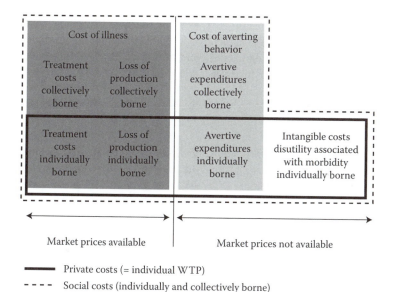

FIGURE 13.3 Economic health cost components. (From Seethaler, R. K., et al., 2003. *Clean Air and Environment Quality*, 37(1): 35–43. With permission.)

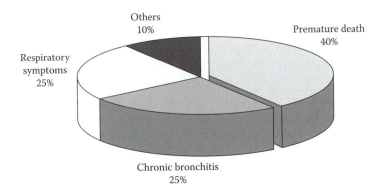

FIGURE 13.4 The composition of the health costs in the six cities due to air pollution by cause. (From Lvovsky et al., 2000. World Bank Environment Department Paper 78, Washington, DC. With permission.)

WTP approach, in order to avoid a statistical premature death, is relatively well developed (Viscusi and Evans, 1990; Viscusi, 1992, 1993; Lvovsky et al., 2000; Fisher et al., 2002). For example, by applying the WTP approach, Seethaler et al. (2003) found that across Austria, France, and Switzerland (with a total of 74 million inhabitants), health costs due to traffic-related air pollution for the year 1996 amounted to some 27 billion euros. This amount translates to approximately 1.7% GDP and an average of 360 euros per capita per year. In all three countries, premature mortality is predominant, accounting for about 70% of the costs.

Lvovsky et al. (2000) estimated the composition of health costs due to air pollution by cause, and found that premature deaths accounted for about 40% of the health costs and various illnesses made up 10%. Chronic bronchitis (25%) and acute respiratory symptoms (25%) are the largest contributors to economic costs associated with morbidity. Figure 13.4 shows the composition of air pollution-related health costs in six cities (Bangkok, Krakow, Manila, Mumbai, Santiago, and Shanghai) in more detail.

Factors that determine the cost of sickness within the WTP approach are as follows:

i. *Valuing sickness, premature death, and disability-adjusted life years (DALYs):*
   The DALYs technique is a standard measure of the burden of disease (Murray and Lopez, 1996; WDR, 1999) that combines *life years lost due to premature death* with *fractions of years of health life lost as a result of illness and disability* (Lvovsky et al., 2000). A weighting function that incorporates discounting is used for years of life lost at each age to reflect the different social weights that are usually given to illness and premature mortality at different ages. Therefore, it is possible to link the Value of Statistical Life with DALYs. This means that the value of an average death caused by air pollution (or willingness-to-pay (VTP) of the chronically sick) and the mean value of statistical life can be interconnected through DALYs.

ii. *Age effects underlying health conditions and social costs:* In general, age effect is an important variable in determining the value of a statistical life for labor market studies. It has been found that death from air pollution reduces life years by less than 35 years on average (for more detail, Moore and Viscusi, 1988; Farley, 1990; Schwetz and Dockey, 1992; Viscusi, 1993; Saldiva, 1995; Ostro, 1996; see Sunyer, 1996; Cropper, 1997; and Lvousky, 1998).

iii. *Valuation of acute morbidity effect:* Air pollution causes human morbidity; therefore, the valuation of illness and disability is recognized as a very important topic. To assess the social costs of air pollution and conduct a CBA for control measures, the valuation of illness and disability method is commonly accepted. However, the literature on the WTP approach, in order to avoid morbidity effects, is very limited in scope. An alternative method is the Cost of Illness approach, which is estimated as the economic costs of health and lost output up to recovery or death This compares some of the direct costs (hospital, medical care, etc.) and indirect costs (value of output lost—wage rate times lost hours) (Krupnick and Cropper, 1992).

iv. *Valuation of changes in life expectancy:* The value that people place on the avoidance of risk depends on the nature of the risk. In general, the life expectancy approach covers "i) estimating a change in life expectancy by age group implied by the change in ambient particulates, ii) establishing a WTP for the change in life expectancy by age group, and iii) multiplying these two values by each other and by population in each age group, and adding up" (Thurston et al., 1997, p. 65). According to Lvovsky et al. (2000), Johannessan and Johansason's (1996) study is the only one that deals extensively with changes in life expectancy and air pollution.

v. *Income effect:* One of the most fundamental issues of valuing reductions in risk is that WTP rises with income. The literature on the income elasticity of WTP to reduce the risk of insults to health is extremely limited. A simple average of the three available studies yields an income elasticity of 0.7. In other words, a one unit increase in income causes a 0.7 unit increase in WTP (see Jones-Lee, 1985; Biddle and Zarkin, 1988; Viscusi and Evans 1990; Lvousky et al., 1998).

vi. *Valuation of chronic bronchitis:* This is the most severe case of morbidity, in which once the dose–response relationship is established, it may last from the beginning of the illness for the rest of the individual's life. Estimates of WTP to avoid chronic bronchitis as a result of air pollution use the contingent valuations analysis (Krupnick and Cropper, 1992; Viscusi, 1992; Lvousky, 1998). Recently, the economic value of preventing adverse health effects related to air pollution has been estimated using contingent valuation in many developing countries. Generally, values are chosen to estimate for three health problems: cold, chronic bronchitis, and fatality. Alternative statistical models are tested to study their impact on estimated WTP and the relationship between WTP and respondent characteristics (Hammitt and Zhou, 2006).

13.3.2 Human Capital Approach

The Human Capital approach is a partial method that looks at the present value of future labor income. It is estimated by valuing medical costs and the loss of income, production, or consumption arising due to premature mortality and morbidity, but it only covers material cost elements. This is generally done by calculating the present value of a person's potential future output as measured by the discounted anticipated stream of earnings. Essentially, this method calculates and values the years lost due to mortality. To this base other costs are added, such as monetary estimates of lost quality of life, value of nonmarket output, and resource costs including medical and hospital costs (BTRE, 2005). The literature on the valuation of life according to the Human Capital approach is well developed (see Cropper and Sussman, 1990; Lvousky, 1998; Kjellstrom et al., 2002).

13.3.3 Economics of Abatement

Neoclassical economists introduced a set of assumptions based on the "optimality" condition of the Paretian premise. The social optimum, which is also known as "economic efficiency," requires both distribution and exchange efficiency and can only be reached in perfect competition when price equals marginal cost. Nevertheless, the market fails if there are externalities, such as pollution, because then price does not equal marginal cost. For example, in the presence of negative externalities, the social costs of health exceed the private costs of health, and this means that the market outcome is not efficient anymore in the case of pollution. In order to maintain "economic efficiency," pollution costs have to be internalized; in other words, they have to be taken into account by the polluter. However, for goods such as air or water, the property right structure is not clear for setting the compensation mechanism and an externality issue arises.

In the presence of negative externalities, the social cost of a market activity would exceed the private cost of the activity; therefore, the market outcome will not be efficient. When the market outcome is not efficient and the market tends to over-supply the product, a Pigovian tax, equal to the negative externality, is levied on a market activity in order to correct the market failure. In the presence of positive externality, as the market would tend to under-supply the product, the creation of Pigovian subsidies would increase the market activity. Neoclassical economists also developed the tradable permits idea because markets do not internalize pollution and failure, which also makes government intervention inevitable. Therefore, it is the duty of governments to internalize the cost and reestablish efficiency. In order to internalize costs, governments have two political instruments: (i) centralized taxes based on the Pigovian tax and (ii) decentralized tradable permits based on the Coase Theorem.

i. *Pigovian tax:* A Pigovian tax is considered as one of the oldest forms of pollution control and was developed by Arthur Cecil Pigou in 1912 and 1920. However, William Baumol was responsible for framing Pigou's work beautifully in modern economics in 1972. The Optimal Pigovian Tax idea is very simple and claims that the costs of pollution should be internalized.

In order to do so, governments should impose a tax on each unit produced and should bring the economy back to its efficient level. Although it is based on a simple idea and is one of the "traditional" means of dealing with market failure, the "knowledge problem" can be vital for the application of Pigovian tax in real life. When there is asymmetric information or not enough information, the Pigovian tax does not work. Also, in the case of international activities, taxes cannot be applied internationally. Pigou himself confessed that it can be very difficult for the State to interfere with individual preferences (Pigou, 1954). Nevertheless, economists insist that the levying of a Pigovian tax would be better than no tax at all.

ii. *Tradable permits:* Since the late 1970s in the United States and since the 1980s in other developed nations, an alternative to Pigovian taxation has been developed, which is the tradable permits idea (or the creation of a market for "pollution rights." The government issues the exact number of permits needed to produce the desired emission level to give or sell its holder the right to pollute for a certain amount. As they are freely transferable, they can be bought and sold on the market and they act as property rights, even internationally. Under a tradable permit system, an allowable overall level of pollution is established and allocated among firms in the form of permits. Firms that keep their emission levels below their allotted level may sell their surplus permits to other firms or use them to offset excess emissions in other parts of their operations (Stavins, 2003).

Even though this method is not more efficient than the Pigovian tax method, it is more appealing to policy makers, because giving out rights free of charge allows the polluter to lose less profits or even gain profits relative to the unaltered market case. In theory, tradable permits might achieve the same cost-minimizing allocation of the control burden as a Pigovian tax (see Coase, 1960; Crocker, 1966; Dales, 1968; Montgomery, 1972; Tietenberg, 1980, 1984, 2003; Hahn and Noll, 1982).

13.3.4 CONCLUSION-II

All these approaches have some common deficiencies because they involve a partial equilibrium approach, in the sense that they ignore the wider consequences of extending life. For example, both approaches ignore intergenerational costs, such as a heavier social security burden on younger members of society. A basic criticism of the WTP approach is that it fails to provide market value because it reflects what people are prepared to pay in order to reduce the risk of being unhealthy (Cropper, 1997, Lvovsky et al., 2000, BTRE, 2005). In addition to that, reliability can be a real problem when estimates are based on stated preferences (survey responses) rather than on market transactions. Fundamental criticisms of the Human Capital approach are that it assumes full employment in the economy and does not provide an accurate measurement of the intrinsic value in cases where there is loss of life or suffering. In fact, there is no full employment in the economy. Finally, even though providing tradable permits is more appealing to policy makers than levying a Pigovian tax, it is not more efficient.

13.4 CBA WITHIN NEOCLASSICAL ENVIRONMENTAL ECONOMICS: MEASURING AND COMPARING THE INDIVIDUAL PREFERENCES FOR ENVIRONMENTAL DECISION MAKING AND THE HEALTH IMPACTS OF AIR POLLUTION

CBA, which originates from Neoclassical Economic Theory, is used as an analytical tool to provide a means for systematically comparing *the value of outcomes* with *the value of resources achieving the outcomes* required. After Alfred Marshall conceived some of the formal concepts that are considered at the foundation of CBA, it was first created in 1936 to aid the environmental decision-making process in the USA (Hanley and Spash, 1993). Since then, policy makers in many countries have continued to apply CBA to environmental problems, such as air pollution and its impacts on health, in order to determine the costs and benefits of an air pollution control strategy. To do so, they have placed monetary value on health benefits and marginal damage figures and have obtained a comprehensive, but generally qualitative, review of the health effects of particles, nitrogen dioxide, and ozone. In addition, these policy makers have conducted quantitative meta-analysis, regarding mortality from time series studies, hospital admissions, and coughs among people with chronic respiratory symptoms.

In general, the key assumptions to conduct a CBA are to identify sources and quantify emissions of NO_x and volatile organic compounds (VOCs), calculate the dispersion of processors and ozone concentration, apply exposure–response functions to estimate yield loss, and value yield loss.

Nevertheless, policy makers face many challenges in relation to the treatment of long-term effects, such as irreversibility, incomplete information, risk, and uncertainty, when they use this technique. The fundamental reason for these challenges is that society has evolved and become more complex, and the character of environmental issues has become even more complex than we can imagine.

13.4.1 RATIONALITY OF CBA

Because CBA is a technical tool that is created to aid the public policy decision-making process, its rationality stems from the legitimacy of public policy. It is an established fact that welfare measures of public policy have two fundamental purposes: to improve *efficiency* and to improve *equity*. A policy is said to be efficient if it maximizes the total net benefits available to society. However, a policy, in general, is independent of who receives the net benefits. In other words, equity is not concerned with the "size of the pie," but on how the pie is distributed among members of society (Hanley and Spash, 1993).

For example, in the United States, CBAs of air pollution regulations by USEPA have demonstrated large net benefits from reducing fine particles. Title IV of the 1990 Clean Air Act, which has reduced SO_2 from power plants, costs only $2 billion per year but yields over $60 billion per year in health benefits.

i. *Equity and CBA:* Indeed, CBA is not specifically designed as a tool for evaluating equity within the Neoclassical Economic view. However, there is a general view that CBA should also pursue the distribution of costs and benefits among the various sections of society. To conduct a CBA analysis, if a researcher needs to collect data (of age, sex, income, race, geographic location, and time) then it should be possible for him/her to attempt to determine how benefits are distributed. Hence, it would not be impossible for an analyst to determine the imbalances between benefits and costs for those segments of the population that are the most vulnerable. Nevertheless, in real life, equity is not seen as CBA's main concern because efficiency improvement is the key target.

With regard to the air pollution control program, critics claim that while efficiency seeks aggregate gains, equity seeks to determine whether costs and benefits are systematically reallocated in ways that discriminate against citizens least able to protect themselves, or in favor of citizens who already enjoy advantage. Thus some potential Pareto improvements may be considered undesirable, no matter how large the difference between gains and losses (Hanley and Spash, 1993).

Once both equity and efficiency have been considered for the same project, the final decision will be able to include whether readily identifiable subgroups of the population, like the poor or the elderly, systematically occupy the most polluted areas. If they do, the social desirability of protecting vulnerable citizens might give more weight than the delivering of the greatest net benefits. These concerns have become much more visible as it is recognized that market forces, combined with environmental policies, can have unforeseen and undesirable consequences. Such considerations are often referred to as environmental justice.

On the other hand, Neoclassical economists defend their technique and claim that CBA can not only, to some degree, identify gross imbalances in the distribution of benefits and costs but may also provide special weighting for certain issues of equity, such as imbalanced impacts on readily identifiable and disadvantaged social groups. Nevertheless for critics, as it is not designed for equality, it cannot measure the multidimensional aspects of overall project desirability that may include factors such as sustainability, altruism, ethics, public participation in the decision process, and other social values (Desjardins, 2001). Thus, CBA informs the decision-making process, but it does not by itself make decisions on how to measure the desirability of the project on air pollution and air pollution control policies. For this reason, policy makers should not automatically pursue the most economically efficient project without weighing efficiency against the other important criteria that affect overall social desirability (Hanley and Spash, 1993).

ii. *Efficiency and CBA:* Although there could be many definitions of what constitutes the best outcome for efficiency, economists have focused on two particular criteria. The first of these criteria is called *Pareto improvement,* while the second is called *potential Pareto improvement*. CBA has traditionally focused on efficiency in order to help policy makers identify *Pareto improvements* and measure the magnitude of the difference between gains and losses. For example, CBA measures the economic efficiency of the proposed policy or project on air pollution control. Then a decision is reached (Roll, 1992). A project to control air pollution is said to constitute a Pareto improvement if it improves the quality of life for some people but does not make anybody worse off.

In reality, most policy changes make some people better off and some people worse off at the same time. Air pollution control policies, for example, simultaneously impose costs and benefits due to a change in the allocation of resources (Hanley and Spash, 1993). It is inevitable that every project or policy will disadvantage some segment of society and advantage others. If policy makers rigidly applied the concept of a Pareto improvement for screening potential projects or policies, it is unlikely that any policies or projects would meet these criteria and be implemented.

Consequently, Neoclassical economists have suggested a less rigid criterion for determining whether a project or policy improves the welfare of society as a whole, which is the *potential Pareto improvement*. A policy or project on air pollution control is said to constitute a *potential Pareto improvement* if those who benefit as a result of the project or policy gain by more than the losses of those who are made worse off as a result of the project or policy. This type of arrangement of costs and benefits is called a *potential Pareto improvement* because those who gain could compensate the losers for their losses and still be better off. In fact, if the winners did compensate the losers, the potential Pareto improvement would become an *actual* Pareto *improvement*.

In terms of air pollution control policies, most Neoclassical economists strongly support the use of the *potential Pareto improvement* criterion because it helps determine whether a project or policy on air pollution control improves the welfare of society as a whole. They argue that because there are so many people and so many projects and policies, it is certain that many people benefit from some policies and face costs from other policies. Consequently, policy decisions should just search for the highest net gains for the majority (Colander, 1989). Eventually, over a large number of decisions and a large number of citizens, everybody will begin to benefit because resource allocation decisions systematically seek to obtain the greatest benefit at the least cost.

13.4.2 Structure of CBA

For a CBA to determine the health impacts of air pollution, several stages must be conducted, such as defining the project, identifying the impacts of the project, finding which impacts are economically relevant, physically quantifying the impacts, and calculating a monetary valuation sensitivity analysis (Hanley and Spash, 1993). In many cases, in order to be able to measure the impact of air pollution on health, the data on population can be obtained from national censuses; death rates from vital statistics data and morbidity incidence from national health surveys can be obtained in many countries.

i. *Definition of the project:* The reasons for defining the project vary, but in general we need clear definitions for two reasons: (i) a project cannot be appraised unless what is to be appraised is known and the boundaries of the analysis are drawn and (ii) the population over which costs and benefits are to be aggregated should be determined. Once the boundaries are defined and the population is determined, the project is ready to be analyzed.

For example, if the project is about air pollution and its impacts on health for a specific city, then an air quality model for the city should be identified. For that, ambient concentrations of key pollutants (e.g., PM_{10}, $PM_{2.5}$) with and without the control strategy should be determined. The model should include inputs such as an emissions inventory, meteorological data, and other information (e.g., ammonia concentrations); also output equal to the ambient pollution concentration for each location should be obtained.

ii. *Identification of the project:* Once the project is defined, the next step is to identify all those impacts resulting from its implementation. For example, when governments appraise the introduction of lower air pollution limits, this benefit should consider the unemployment level, productivity gains, reduction in health costs, and lower mortality rates. Implementation of a project on air pollution control might have an impact on reducing health costs more than unemployment, or on mortality rates more than productivity gains, but ultimately the project should identify all these important impacts in order to allow policy makers to reach the best decision.

For example, an emissions inventory must describe emissions by location (grid square) with and without the control strategy.

iii. *Deciding on which impacts are economically relevant:* The aim of CBA is to select projects that add to the total of social utility. Once analysts select the best efficient project through CBA, the project will reflect the highest social benefit from a list of alternatives. The most economically sound project will also reflect the highest social benefit. For example, if our intention is to find the health impact of air pollution, then we need to obtain information in order to calculate the following: health effects, such as size of the population in each location (grid square); baseline incidence of each health endpoint to be evaluated (death rate by cause, incidence of chronic bronchitis); and slope of the concentration–response function for each endpoint.

iv. *Physical quantification of relevant impacts:* This stage involves determining the physical amounts of cost flows and benefit flows for a project. Identifying when they will occur in time will be the main task for a decision maker. By estimating the time and the physical amounts, expected health impacts can be targeted. For example, Environmental Impact Analysis is one of the most commonly used techniques for measuring relevant impacts.

v. *Monetary valuation of relevant impacts:* To be able to measure the impacts, they must be valued in common units. According to Neoclassical Environmental Economics, the common unit for CBA is *money*. Because prices are used as a proxy for their current relative scarcity, it is commonly believed that markets generate the relative values of all traded goods and services, such as *relative prices*. If the same tool is used for CBA, selecting the best project will be an easier task in order to obtain the value of each health endpoint, which can be obtained by estimating the value of avoided illness and death from data on earnings and medical costs. WTP to avoid illness and reduce the risk of dying can be valued by making benefits transferable.

13.4.3 Technical Conceptual Framework for CBA

Once a decision is made to conduct a CBA in order to measure the health impacts of air pollution, the conceptual framework should be properly examined so as to reach the most significant decisions. Key concepts related with the health impact of air pollution are the following:

i. *Net present value of the project:* In the air pollution application of CBA, the analyst must measure the net present value of the project over a period of time and compare the latest costs and benefits with the costs and benefits that often occur in different time periods. Even though this increases the complexity of the analysis and it is not very easy to directly compare the costs or benefits today with the costs or benefits 10 years from today, it is essential for an analyst to have a better view of the content of the project. Because comparisons require a common metric, CBA uses a process called *discounting* to express all future costs and benefits in their present value equivalent. This takes place by discounting costs and benefits in each future time period and summing them to arrive at a present value. In general, the longer the time frame and the higher the discount rate, the smaller the impact of any given year on the total net benefits.

However, the net present value method is also considered as one of the most fundamental weaknesses of CBA. This is because this calculation method is based on the decisions of the present generation and excludes the decisions of future generations. It is for this reason that costs that occur far into the future are given little weight in traditional CBA. In reality, while we should consider discounting to express all future costs and benefits in their present value equivalent for present generations, we should also be concerned with intertemporal equity issues, and factor the costs to future generations into the equation.

ii. *Choice of input values:* According to Neoclassical Environmental economists, the choices of input values method is a comparatively better technique than the net present value method, as the former ultimately determines the results of the analysis. Within this method, parameter values, which are associated with costs and benefits, represent choices. Parameter choices include future rates of economic growth, the discount rate, future rates of population growth, future rates of inflation, and future rates of technological change. Therefore, the choices of input values method is considered to be more detailed and comprehensive.

iii. *Dealing with uncertainty:* Uncertainty is considered to be the most important obstacle facing CBA and has a direct negative impact on the estimation of costs and benefits. For this reason, it is important that a CBA should include *sensitivity and scenario analysis* to deal with the problem of uncertainty. These two analyses should be conducted in order to illustrate how the results change with different analytical choices and with variation in the uncertain levels of key costs and benefits.

iv. *Choosing between alternatives:* Choosing between a discrete set of alternatives is the key for CBA. For example, there may be several alternatives for controlling air pollution. For example, imposing a tax on air pollution, modifying the transportation system, upgrading the sewer systems, or switching to new production technologies might compete with one another in terms of reducing air pollution. Both the prevention alternative and the remediation alternative may be further divided into alternatives based on options available in the technologies to accomplish each goal. Therefore, deciding between alternatives is a difficult task when a CBA is being conducted, and a decision maker should be very careful so as to select the most sensible one.

v. *Choosing from a continuum:* Some environmental decisions might involve a choice from a continuous distribution of an environmental variable. For example, a continuous distribution of potential levels of health impacts of air pollution might be the case and decision makers must choose a single level, which ideally would maximize quality of life. *Marginal Damage Function Analysis* is a tool that is related to CBA, which is designed to help decision makers with this type of choice. For example, the marginal damage estimates of Holland et al. (2005) were applied directly to emissions at the national level and were simply multiplied by the national average damage per ton estimates to find out the total damage for the EU25.

13.4.4 PROBLEMS WITH CBA

CBA is a very useful technique and an advanced analytical tool, which involves numerous choices and calculations. However, there are still some problems and risks with the technique, as follows:

i. Risk must often be considered as a factor in making the decision.
ii. In more strategic investments, the intangible benefits clearly outweigh the financial benefits.
iii. The costs are tangible, hard, and financial, while the benefits are not only hard and tangible but also soft and intangible.
iv. In order to calculate costs and benefits, if nondiscounted amounts are used in the CBA method, the results do not reflect the real values. Therefore, the net present value should be used to consider the time value of money.

13.4.5 CONCLUSION-III

The main reason for the existence of CBA is to inform the decision makers and make the decision-making process much easier. Even though CBA provides useful information to the political process, it should not, however, be used as a one-dimensional test of desirability. Other decision-making concerns, including sustainability, equity, ethics, and social values, should also be included in the decision-making process.

There are many criticisms of CBA and its origin in Neoclassical Economic Theory, as it does not reflect values and sustainability. As mentioned above, one of the most important criticisms stems from the fact that it is not specifically designed as a tool for evaluating equity within the Neoclassical Economic view. However, there is a general belief that CBA should deal with equity.

13.5 ALTERNATIVE APPROACHES TO THE NEOCLASSICAL ENVIRONMENTAL ECONOMIC VIEW

13.5.1 AUSTRIAN ECONOMIC THEORY

Austrian economists such as Rothbard (1977, 1982), Menger (1981), Lewin (1982), Kirzner (1988), Cordato (1992a, 1992b, 1995, 1997, 1998, 2001, 2004), Krecke (1996), and Mises (1998), argue intensively as to why they are against the Neoclassical efficiency approach and list their problems with this standard approach. Austrian School economists reject standard Neoclassical theories as they have conceptually specious assumptions that do not reflect the real world. The Austrians claim that even though these unrealistic assumptions have led to policy prescriptions for the last 200 years, in reality they are completely nonoperational (Cordato, 2001). In particular, the theory of externalities, which is one of the most important concepts of Neoclassical Environmental Economics, has been subjected to criticisms by many Austrian economists. For the Austrians, the key problems with the conceptual framework of Neoclassical Environmental Economics are the following: (i) efficiency is an individual goal-seeking problem, not a value maximization problem; (ii) costs are subjective and therefore social costs and social values do not exist as either measurable or even theoretical concepts; and (iii) Pareto optimality is irrelevant as a real-world efficiency benchmark (Kneese et al., 1973; Cordato, 2004).

Fundamentally, the Austrians claim that the Neoclassical approach and its analysis do not explain the methodological foundation for identifying what is and is not a pollutant (Kirzner, 1988; Cordato, 1998) unless interpersonal conflicts are removed. To suggest a better definition, Austrian economists formulized their views on positive and normative analyses of environmental problems and defined pollution, environmental costs, and policy matters from their conceptual framework. Therefore, unlike Neoclassical Economic solutions such as the use of taxes and tradable permits to remove inefficiencies, the Austrians' emphasis is on eliminating interpersonal conflicts, such as privatization, the "polluter pays" principle, and the "first come first served" principle.

 i. *Their definition of environmental pollution:* According to the Austrians, pollution is the kind of problem that creates an interpersonal conflict over the use of means, rather than only being a "social cost" issue as Neoclassical Environmental economists claimed. For this reason, the Austrians focus on how to compensate victims whose health is affected by air pollution, rather than on how to deal with the costs to restore the environment. The Austrians claim that once the concept of costs is separated from individual human beings, it loses its ground to conduct any economic analysis. In short, they believe that pollution, such as air pollution and its impacts on human health, cannot be separated and costs should be compensated (Lewin, 1982; Rothbard, 1982; Cordato, 2004).

ii. *Property rights and minimizing the interpersonal conflict:* The Austrians recognize the importance of property rights in resolving environmental problems and they also believe that, unless the concept of property rights is clearly defined, environmental problems will continue to persist and people will suffer as a result of pollution. Menger (1981) argued that all goods are to be classified as economic and noneconomic. While *economic goods* must come under the rule of private property in order to avoid conflicts of interest regarding their usage, *noneconomic goods*, such as air and water, are the cause for conflict of interest. Menger (1981) also proposed that the only practical solution to this conflict is to look at the economic aspects of these otherwise noneconomic resources from the private property point of view and solve the problem.

In terms of health impacts of air pollution, the Austrians first define the characteristic of air pollution as a consequence of human conflict over the use of this resource. Then they claim that the reasons for the impact of air pollution on human health must be found in property rights, which are neither clearly defined nor enforced in the existing economic view (Rothbard, 1977; Mises, 1998). To rectify the problem, the Austrians suggest *conflict resolution*. This means that once the source of conflict is described, possible ways of resolving the issue can be identified by focusing on issues related to property rights (Posner, 1973; Cordato, 1995; North, 2002). For example, if the tax is collected only to bring about the correct price/output combination and an "optimal level of pollution," leaving the initial conflict unresolved, there would be no reason to consider the solution as efficient from an Austrian perspective. Some of the solutions suggested by the Austrians are discussed below.

iii. *Property rights, public policy, and solutions:* Unlike Neoclassical Environmental economists, the Austrians suggest that public policy decisions on environmental issues must focus on resolving conflicts over the use of resources that cause pollution, not on obtaining an ultimately unobtainable "efficient" allocation of resources. For Austrians, the conflicts will be resolved in a much better way once we focus on clarifying titles to property and the enforcement of rights.

If a pollution problem and its health consequences exist, then its solution must be found either in a clearer definition of property rights to the relevant resources or in the stricter enforcement of rights that already exist. This has been the common approach, which has been taken regarding environmental problems by nearly all key Austrians (see Lewin, 1982; Rothbard, 1982; Cordato, 1997; Mises, 1998). In the Neoclassical Environmental policy, there are two solutions to compensate for environmental pollution: the polluter pays principle and the first come first served principle. The Austrians believe that both solutions have their own strengths and weaknesses. Therefore, they look at them closely and examine their solutions critically.

iv. *The polluter pays principle:* The polluter pays principle is a distinctive Neoclassical Economic welfare solution. As can be seen from the title, the main principle is to ask a polluter to compensate for the pollution. According to the Austrians, there are two fundamental problems with this approach: (a) it is fundamentally a form of market socialism and promotes central planning (Cordato, 1997) as it gives too much power to the central authority, who is expected to set in advance what the efficient outcome will be, and (b) the conflict or harm generated through pollution may still not be eliminated even if an efficient price/output combination is achieved (see Pearce and Turner, 1992; McGee and Block, 1994).

Even if authorities agree on the efficient level of pollution and issue permits to potential polluters, in the aggregate, for the Austrians, after implementing such a policy you are still likely to be left with a pollution problem. The level of pollution will be less than before, but will still be there. Nevertheless, they suggest that once the property rights issue is redefined exactly as the Austrian economists pointed out, then the polluter pays principle could make sense. The polluter will be forced to compensate for the costs of his/her pollutive activities to those whose property usage is being limited or whose health is being endangered. This might be done by eliminating the emissions, confining them to the polluter's own property, or compensating the victims of the polluting activity by an amount that fully addresses the grievance.

v. *First come first served principle:* The second solution is the first come first served principle. This solution is used when the pollution problem arises and when property titles, and therefore property rights, are unclear (Rothbard, 1982). However, in the case of air pollution, it might not be as efficient a technique as the polluter pays principle. In order to reduce the uncertainty in the formulation process, the amount and quality of information is captured in relative prices, but this might not be an efficient way of dealing with air pollution (Cordato, 1998).

13.5.2 Green Economics

Green Economics is an influential approach in which an economic system is considered to be a component of the ecosystem. The main contributors to Green Economic Theory are E. F. Schumacher, Murray Bookchin, Lewis Mumford, Miriam Kennet, Rachel Carson, Brian Tokar, Robert Costanza, David Korten, Buckminster Fuller, Herman Daly, Paul Hawken, Amory Lovins, Jane Jacobs, and Robin Hanson.

Like the Austrian economists, Green economists also claim that their view is fundamentally different from the Neoclassical Economic view. They argue that even though Neoclassical Economics represents the main body of modern economics today, Green Economics shares broader ecological and social concerns, including rejecting capitalism itself. For this reason, Green Economics goes beyond the narrower concerns of Neoclassical Environmental Economics, Resource Economics, and Sustainable Development, which are considered as subsets of Green Economics. Many Green economists have been heavily influenced by Marxian views to develop an understanding of ecological issues and ecological economic alternatives. Now their main differences and key principles will be examined.

i. *Main difference between Green Economics and Neoclassical Environmental Economics:* One of the most important differences between Neoclassical Economics and Green Economics is based on localization. While Green Economy generally favors *local measures* and localized *grassroots* institutions (Colin, 2000) over paternalistic and elite-driven global institutions such as the International Monetary Fund (IMF), World Bank, and World Trade Organization (WTO), Neoclassical Economics goes for global measures and global institutions.

Another very important difference between these two views is that Green economists put the ecosystem at the center rather than classifying the ecosystem as an externality as defined by Neoclassical economists. For this reason, Green economists believe that finite space cannot be expandable forever and finite resources cannot be used forever, and everything in this ecosystem is interconnected (McRobie, 1981; Michael, 1993; Cato and Kenneth, 1999; Brian, 2000; Woodin and Lucas, 2004; Derek, 2005). Hence almost all Green economists regard *economic growth* as a delusion. The idea of economic growth not only contradicts the idea that it is impossible to expand forever into a finite space, but also causes destruction and degradation in the life support capacity of the natural ecosystem: air and water filtering, food production, and fiber growth. For them, nature is a service producing natural capital and life on the Earth depends on this basic capital. Therefore, we have to maintain, preserve, and protect nature. According to them, if we look after nature, there will be no pollution and there will be no negative impact of pollution on human health. To be able to control our damaging activities, they recommend that we should focus on local measurements rather than on global ones because they are much more achievable.

For example, Schumacher (1973) was one of those scholars who emphasized the value of localization. In his argument, he says that activities such as gardening would require *use-value* in the economic process and would de-emphasize the value of resource, commodity, or product measures. In addition, Schumacher also looked critically at the concept of economic growth and agreed to the impossibility of expanding forever. Many other Green economists contributed significantly to a green microeconomics, and proposed establishing an educational network that both formalizes its educational tasks and systemizes connections with the rest of the community.

ii. *Their key principles:* To be able to eliminate environmental pollution and its negative impacts, such as health-related problems, Green economists suggest a number of key principles:

a. We need to focus simultaneously on both human and environmental needs, not *materialistic well-being* itself. We should understand that matter exists to satisfy our needs, and it is not the main purpose for our existence. However, human beings need more than material well-being. They also need values, peace, harmony, and so on.

b. For millions and millions of years, nature itself has not created as much waste as we have over the last 100 or 200 years. In nature there is no waste, as every process output is an input for some other process. Therefore, we can copy nature in terms of our economic activities, such as production, which can be nontoxic food for some other process.

c. Each regenerative activity should have its own matching appropriate scale of operation. Even the smallest activities have larger impacts on nature and human health. Therefore, we have to be very careful when we decide what to produce and how much to produce so as not to cause any irreversible action.

d. We should protect the diversity of life on Earth, as our existence depends on it. Each day, more than 100 species are becoming extinct, and our natural resources that sustain life on the planet—air, water, and soil—are becoming polluted or depleted at an alarming scale together with exponentially increasing human population growth. Even though we cannot reverse this extinction, we can stop it. We should realize that our existence depends on clean water, air, and soil.

e. Self-reliance is a very important ingredient to being independent. In an economy that moves with ecosystem processes, there is tremendous scope for local response. Design and adaptation must be provided for these local and regional responses for more flexible and holistic interdependence. This will bring about greater success.

f. Participation in environmental-related decisions is fundamental for direct and widening democracy. Pluralistic societies with established democracies will take better environmental-related decisions than nondemocratic societies.

In short, Green Economy has been affected by the Marxian Economic perspective and favors "local measures" and localized "grassroots" institutions over paternalistic and elite-driven global institutions, such as the IMF, World Bank, WTO, and multinational organizations. This idea directly opposes that of the Neoclassical Economic view.

13.5.3 Ecological Economics

Ecological Economic Theory is a newly adopted branch of economics that addresses the interdependence and coevolution between human economies and natural ecosystems. The main scholars in this field are Robert Costanza, Herman Daly, Nicholas

Georgescu-Roegen, David Harvey, and John Bellamy Foster. It has similarities to Green Economics, but it also differs from this theory in its distinctive objective, which combines economic thinking, knowledge of biology, and the laws of physics. In other words, it is a mixture of social science and scientific realities. Therefore, its goal is to improve human welfare through economic development, which is based on a balance between ecology and human needs. Similarly, the main differences, principles, and solutions of Ecological Economics will be examined closely in order to understand the basic conceptual framework.

  i. *Main difference between Ecological Economics and Neoclassical Environmental Economics:* Like other schools and approaches, Ecological Economics criticizes Neoclassical Environmental Economics for being myopic and close-minded to environmental facts and for believing that the environment is a subset of human economy. Ecological economists claim that it is unfair for Neoclassicists to suggest that economic pollution and its harmful impacts on human health can be eliminated easily by paying compensations. Nevertheless, the Ecological Economic Theory combines ecology with human economy, and Ecological economists suggest that it offers better solutions to the problems. For them, while the ecology side deals with the energy and matter transactions of life and the Earth, human economy is by definition contained within Ecological Economics.

     Similar to the Green Economic Theory, Ecological economists also believe that infinite economic growth is not possible and not desirable because our resources are limited and cannot be expanded forever. Even though nature is limited, it is claimed that it already provides us with what we need and there is no need to destroy nature to get more. According to some estimations, the price of services provided by the environment, in looking at the price to filter water and other such services, is something around 33 trillion dollars.

     Rather than focusing on economic growth, Ecological economists suggest sustainable development. Furthermore, they believe that sustainable development is the only means of improving the standard of living for citizens worldwide. According to them, while quantitative economic growth emphasizes per capita consumption, which can have harmful effects on the environment and even on the broader societal well-being, sustainable development concentrates on improving the quality of life.

 ii. *As an interdisciplinary approach:* In addition, Ecological economists are inclined to acknowledge that much of what is important for human well-being is not analyzable from a strictly economic standpoint and they suggest an interdisciplinary approach to address complex issues such as pollution. They claim that both social and natural sciences, as a means of addressing issues, should form a platform to solve problems.

iii. *Solution of Ecological Economics to protect the environment:* Costanza and Perrings (1990) provide an example of how to combine what we now know about the uncertainties of environmental protection with what we also know about the difficulties of more direct forms of social control such as regulation or outright prohibition. In order to develop more cost-effective, less intrusive, and generally more positive stimuli to protect and/or manage environmental use, they evaluated a flexible assurance bonding system. This bond would be required by developers and would be set equal to the largest estimated potential environmental damage that might occur from the proposed action. The bond would be kept in an interest-bearing account and would be returned to the developer with some of the interest as soon as the firm proved that the damage would or could not occur. If the catastrophe did occur, the bond would be used to compensate those harmed or to help repair the damage. But no further payment would be required from the developer.

They also suggest that a better way of achieving a sustainable ecological and societal system is by educating consumers about the need for living in harmony with nature. This will prevent government intervention and allow consumers and producers to act in the interest of the ecological economy.

13.5.4 Conclusion-IV

Three very distinct views on environmental problems, such as Austrian Economics, Green Economics, and Ecological Economics, are discussed in this study so as to highlight their key differences with Neoclassical Environmental Economics as well as their differences with one another. However, they strongly agree on the fact that each of them is against Neoclassical Environmental Economics assumptions, and they all criticize this standard view from different angles. The efficiency problem, as typically seen by the Austrians, is the generation of human conflict and disruption to inter- and intrapersonal plan formulation and execution. This is in contrast with Pigouvian Environmental Economics, which defines pollution problems primarily in terms of resource allocation. On the other hand, for Green economists, the highest efficiency is achieved through self-sufficiency and optimal scale of operation. Finally, for Ecological Economics, efficiency means protecting the social and ecological system.

It is also demonstrated that the property rights approach to policy analysis taken by the Austrians is different from the Neoclasssical view in the sense that the social purpose of private property is to resolve interpersonal conflicts and allow for the peaceful pursuit and fulfilment of plans. However, as Green economists object to capitalism and defend socialism, the property rights issue has a different dimension in their arguments. Some Green economists have begun to look at more holistic and internally consistent aspects, even though they are led by materialistic philosophies.

Nevertheless, with the constructive criticism of alternative approaches, the formulation of environmental policy within the standard approach has recently experienced some improvements as there is a general understanding among Neoclassical economists that their indefinable concepts of social cost and general equilibrium might not be implemented in the real world. In addition, they accept the idea that their role is to devise efficient methods for achieving politically determined pollution or emission targets (Orr, 1981).

13.6   CONCLUSION: WHAT MUST THE ROLE OF ECONOMICS BE TO ELIMINATE AIR POLLUTION IN THE FUTURE?

Colander (1989) once stated that "Economics is a relatively young discipline" (1989, p. 1), but a very powerful tool. Even though the history of economics as a distinctive subject does not go back beyond 1500 AD (and the quantity of economic literature only increased significantly in Western Europe between 1500 and 1750 and a body of economic knowledge only began to evolve during the period from 1776 to 1876 with an increasing interest in the discipline of political economy), Economics plays a central role in shaping the activities of the modern world by imposing criteria of what is economic and what is uneconomic (Roll, 1992). If there is no other set of criteria that exercises a greater influence over the actions of individuals, groups, and governments more than economic criteria, there is no reason for us not to understand the role of economics in the context of environmental pollution. The influencing power of Economics over policy decision mechanisms and over economic education is so powerful that we need to use its pragmatic solutions for environmental problems before it is too late (IEEP, 2005).

The birth of Neoclassical Economics was very impressive. In particular, the final three decades of the nineteenth century witnessed the beginning of the modern microeconomic theory. Notably, the introduction of a new set of analytical tools, such as marginal analysis, helped transform Classical Economics into Neoclassical Economics. The development of marginal analysis was significant because it initiated an appreciable increase in the use of mathematics in economic analysis. Even today, Neoclassical Economic analysis plays a fundamental role in any economic decision, such as consumption, production, and policy making. For example, as mentioned in Section 13.3, CBA is the major public policy methodology used in reaching environmental decisions and shaping environmental regulations at any national, international, or supranational level.

Despite the fact that Neoclassical Economics is very influential over the decisions of individuals, firms, and governments, it has received a great deal of criticism from different economic schools, environmentalists, philosophical approaches, and religious ethics. As discussed in previous sections, Austrian Economics, Green Economics, and Ecological Economics are among the most influential critics of Neoclassical Economics. Although all these critics have their differences, they agree, however, on the ethical side of the misuse of the environment. For example, from an ethical point of view, some scholars claim that the methodological individualism assumption is the key to a better understanding of the mental structure of the Neoclassical school. For Choudhury (1995),

> Neoclassics present the individual as self-seeking individual, who chooses himself/
> herself against others for optimal share of resources under the motive of enhanc-
> ing the goals of economic efficiency in the midst of market consequentialism. The

resulting competition explained by the Neoclassical principle of substitution, in fact, can be interpreted as a picture in duality of being. In respect to the trade-off between economic efficiency and distributive equity, there is duality between these moral and material aspects of human welfare (p. 17).

Unfortunately, the moral aspect of human welfare is the one that, in most cases, is sacrificed in exchange for material aspects of human welfare.

It is widely accepted that distributive equity is not the priority for Neoclassical Economic order and, as long as there are rational choices, the economic motive would be enough to maximize profits, utility, output, and productivity. So the ethical goal of distributive equity will inevitably be less attractive, and more costly to attain, than economic efficiency. As a matter of fact, it is inevitable that individuals will face a very important dilemma here. If less equity is chosen over more equity, it implies an unethical choice. Because the demand and need for equitable distribution remain unsatisfied, there will be imbalances in the economy. On the other hand, if more equity is chosen over less, then a sacrifice of economic efficiency must be made somewhere in the economic system. Nevertheless, in much more complicated economic systems, such a trade-off between greater distributive equity and lesser economic efficiency becomes unrealistic in a market of output-optimizing firms and utility-maximizing consumers.

Therefore, the principle of substitution, as can be observed in the context of environmental pollution, permits the choice between ethical and unethical bundles as a permanent possibility that cannot change in the long run with the advance of knowledge respecting these choices. The choice between ethical and unethical bundles, in fact, should not even be a subject for environmental issues as we depend on nature. Economics should recognize the fact that humans and their economies are parts of larger natural ecosystems. There is a material and energy basis for the relations between human economies and their ecosystems, defining not only economic structures and processes but also social structures and processes (Paul, 2005). Economies possess general ecosystem properties such as dynamism, evolution, integrity, stability, and resilience. The magnitude of potential impact on their own welfare through effects on natural systems requires that human decisions be guided by some notion of the value of their actions and the value of their impacts on ecosystems, in terms of either benefits of use or costs of abuse. Some concept of value is required for rational activities of human economies within their natural systems.

We have moral and cultural values for the natural system. These values are immeasurable and incomparable using traditional human preferences on the assumption that agents are rational, and may not be reflected in the simple summation across social members of individual values, since they are social and not wholly private. Establishing conditions on human economies would allow for the sustainability and growth of human welfare, conditioned on the sustainability of the economy's supporting ecosystem (Lord, 2003). In that sense, sustainability has developed as an additional consideration for public policy decision making precisely because of the concern that the process of discounting may steer us toward policies that overly emphasize short-term gain. However, we have to keep in mind that, like the consideration of efficiency, the consideration of sustainability provides the decision maker with additional information, but does not itself make the decision. Our main task should involve establishing a way of using the ecosystem more effectively so as to enhance human wealth and welfare.

To reduce environmental pollution, we should

1. Examine the implications of various moral systems for the sustainability of human welfare and place in bold relief those instances where there are apparent incompatibilities between moral systems and sustainability norms.
2. Understand the interdependence between economies, human beings, and natural systems. This includes understanding the tolerances of ecosystems to human-induced changes as well as the tolerances of economies to ecosystem changes.
3. Create opportunities for human economies that would allow for the sustainability and growth of human welfare, conditioned on the sustainability of the economy's supporting ecosystem and effectiveness of solutions.
4. Develop key regulatory instruments, laws, and associated institutions that assist human economies in attaining sustainable welfare development goals.

To control air pollution and its health impacts, we should specifically

1. Create an emerging technology for a greater convergence between the economics and public health approaches to assessing the burden of diseases from air pollution causes as a positive sign that should be further encouraged.
2. Establish an interdisciplinary approach from different fields and combine the effects of various experts and institutions that work in the overlapping areas of public health, environment, and economics, which can significantly contribute to influencing policy makers and the public and can leverage decisions that bring large environmental health benefits.
3. Valuate illness and premature death as consequences of air pollution, in particular in developing countries, as this issue needs more attention from international communities of economics and supporting institutions.

In summary, air pollution causes damaging impacts on people's health. From asthma to heart disease to certain kinds of cancer, the air we breathe can have a fundamental bearing on our health. Achieving tighter air quality standards through regulations could be an appropriate policy aim as a range of technologies, tools, and alternatives are available to help clean up our air, and companies and governments should regard this as a priority—especially for those areas and communities that remain the most polluted and at the most risk. However, the most important thing we need to change should be the way in which we think about what to produce, how to produce, and for whom to produce. These are the fundamental questions for Economics, and we need to change our understanding of Economics in the first place in order to reverse the damage that we have caused to nature. We should also understand the fact that the Earth is for all of us, not only for a minority. We cannot economically grow as it is impossible to expand forever into a finite space, and we cannot ignore future generations.

REFERENCES

Ad-hoc Group, 1999. *Ad-hoc Group on the Economic Appraisal of the Health Effects of Air Pollution. Economic Appraisal of the Health Effects of Air Pollution*. London: Department of Health. The Stationery Office.

Baumol, W. J., 1972. On taxation and the control of externalities. *American Economic Review*, 62(3): 307–322.

Biddle, J. and Zarkin, G. 1988. Worker preferences and market compensation for job risk. *Journal of Economic Statistics*, 70(4): 660–666.

Bonner, J., 1995. *Economic Efficiency and Social Justice: The Development of Utilitarian Ideas in Economics from Bentham to Edgeworth*. England: Edward Elgar.

Brian, M., 2000. *Designing The Green Economy: The Post-Industrial Alternative to Corporate Globalization*. Lanham, MD: Rowman & Littlefield.

BTRE, 2005. Bureau of Transport and Regional Economics (BTRE). Working Papers 2005. Health Impacts of Transport Emissions in Australia: Environmental Cost, Australia: The Department of Transport and Regional Services.

Cato, M. S. and Kenneth, M., 1999. *Green Economics: Beyond Supply and Demand to Meeting People's Needs*. Aberystwyth: Green Audit Books.

Choudhury, M. L., 1995. Ethics and economics: A view from ecological economics. *International Journal of Social Economics*, 22(3): 18–25.

Coase, R., 1960. The problem of social cost. *Journal of Law and Economics*, 3(October): 1–44.

Colander, L., 1989. *History of Economic Theory*. USA: Houghton Muffin.

Colin, H., 2000. *Localization: A Global Manifesto*. London: Earthscan.

Cordato, R., 2004. An Austrian theory of environmental economics. *Quarterly Journal of Austrian Economics*, 28: 3–36.

Cordato, R. E., 1992a. *Welfare Economics and Externalities in an Open Ended Universe*. Boston: Kluwer Academic Publishers.

Cordato, R. E., 1992b. Knowledge problems and the problem of social cost. *Journal of the History of Economic Thought*, 14(Fall): 209–224.

Cordato, R. E., 1995. Pollution taxes and the pretense of efficiency. *Journal of Private Enterprise*, 10: 105–118.

Cordato, R. E., 1997. Market-based environmentalism and the free market: They're not the same. *Independent Review*, 1(3): 371–386.

Cordato, R. E., 1998. Time passage and the economics of coming to the nuisance: Reassessing the Coasean perspective. *Campbell Law Review*, 20(2): 273–292.

Cordato, R. E., 2001. *The Polluter Pays Principle: A Proper Guide for Environmental Policy Studies in Social Cost, Regulation, and the Environment*. Washington, DC.: Institute for Research on the Economics of Taxation.

Costanza, R. and Perrings, C. H., 1990. A Flexible assurance bonding system for improved environmental management. *Ecological Economics*, 2: 57–76.

Crocker, T. D., 1966. The structuring of atmospheric pollution control systems. In: H. Wolozin (ed.), *The Economics of Air Pollution*. New York: W. W. Norton & Company, Inc.

Cropper, M., 1997. The Health Effects of Air Pollution in Delhi, India. Policy Research Working paper, no. 1860, Washington, DC: World Bank.

Cropper, M. and Sussman, F., 1990. Valuing future risks to life. *Journal of Environmental Economics and Management*, 20(2): 258–268.

Dales, J., 1968. *Pollution, Property and Prices*. Toronto: University Press.

Department for Environment, Food and Rural Affairs (Defra), 2006. *An Economic Analysis to Inform the Air Quality Strategy Review Consultation*. London: Crown.

Derek W., 2005. *Babylon and Beyond: The Economics of Anti-Capitalist, Anti-Globalist and Radical Green Movements*. London: Pluto Press.

Desjardins, J., 2001. *Environmental Ethics: An Introduction to Environmental Philosophy*. Canada: Wadsworth.

EEA, 2004. European Environment Agency (EEA). Air Pollution in Europe 1999–2000. Copenhegan: Topic Report 4/2003.

Farley, D., 1990. The relationship of daily mortality to suspended particulates in Santa Clara county 1980–1986. *Environmental Health Perspectives*, 89: 159–168.

Fisher, G. W., Rolfe, K. A., Kjellstroom, T., et al., 2002. Health effects due to motor vehicle air pollution in New Zealand. A Report Submitted to Ministry of Transport. Wellington, New Zealand.

Hahn, R. and Noll, R., 1982. Designing a market for tradeable permits. In: W. Magat (ed.), *Reform of Environmental Regulation*. Cambridge, MA: Ballinger Publishing Co.

Hammitt, J. K. and Zhou, Y., 2006. The economic value of air pollution related health risks China: A contingent valuation study. *Environmental and Resource Economics*, 33(3): 399–423

Hanley. N. and Spash, C. L. 1993. *Cost–Benefit Analysis and the Environment*. England: Edward Elgar.

Health Effects Institute, 2000. *HEI Strategic Plan for the Health Effects of Air Pollution 2000–2005*, pp. 1–26. Cambridge, MA: Health Effects Institute, 2000.

Holland, M., Pye, S., Watkiss, P., Droste-Franke, B., and Bickel, P., 2005. Damages per tonne emission of PM2.5, NH3, SO$_2$, NO$_x$ and VOCs from each EU25 Member State (excluding Cyprus) and surrounding seas. March 2005. Available at http://www.cafe-cba.org/reports/

IEEP, 2005. *Institute for European Environment Policy*. Workshop on best practices in analysing and developing environmental policy. Workshop Report, November 15, 2005. Brussels.

Johannessan, M. and Johansason, P., 1996. To be, or not to be, that is the question: An empirical study of the WTP for an increased life expectancy at an advanced age. *Journal of Risk Uncertainty*, 13: 136–174.

Jones-Lee, M., 1985. The value of safety: The results of a National Sample Survey. *Economic Journal*, 95(377): 49–72.

Kirzner, I. 1988. Welfare economics: A modern Austrian perspective. In: Walter Block and Llewellyn H. Rockwell, Jr. (eds), *Man, Economy, and Liberty: Essays in Honor of Murray N. Rothbard*, pp. 77–88. Auburn, Alabama: Ludwig von Mises Institute.

Kjellstrom, T., Neller, A., and Simpson, R. W., 2002. Air pollution and its health impacts: The changing panorama. *Medical Journal of Austria*, 177(2): 604–608.

Kneese, A., Ayres, R.U., and. Arge, R. C., 1973. Economics and the environment: A materials balance approach. In: C. Alain (ed.), *Pollution, Resources, and the Environment*, Enthoven, New York: W.W. Norton.

Krecke, E., 1996. Law and the market order: An Austrian critique of the economic analysis of law. *Journal des Economistes et des Etudes Humaines*, 7(1): 19–37.

Krupnick, A. and Cropper, M., 1992. The effects of information on health risk valuations. *Journal of Risk and Uncertainty*, 5: 29–48.

Lewin, P., 1982. Pollution externalities, social costs and strict liability. *Cato Journal*, 2(1): 205–230.

Lord, C., 2003. *A Citizens' Income: A Foundation for a Sustainable World*. Charlbury: Jon Carpenter.

Lvousky, K., 1998. Economic costs of air pollution with special references to India. Conference Paper for the National Conference on Health and Environment, Delhi, India, July 7–9, 1998.

Lvovsky, K., Hughes, G., Maddison, D., Ostro, B., and Pearce, D., 2000. Environmental costs for fossil fuels: A rapid assessment method with application to six cities. World Bank Environment Department Paper 78. Washington, DC.

McGee, R. and Block, W., 1994. Pollution trading permits as a form of market socialism and the search for a real market solution to environmental pollution. *Fordham Environmental Law Journal*, 6: 51–77.

McRobie, G., 1981. *Small is Possible*. England: Abakus.

Menger, C., 1981. *Principles of Economics*. New York: New York University Press.

Michael, J., 1993. *The Green Economy*. Vancouver: University of British Columbia Press.

Mill, J. S., 1965. In: J. M. Robson (Ed.), *Principal of Political Economy*, Vol. 2(3). UK: Oxford University Press.

Mises, L. von., 1998. *Human Action*. Scholar's Edition. Auburn, Alabama: Ludwig von Mises Institute.

Montgomery, D., 1972. Markets in licenses and efficient pollution control programs. *Journal of Economic Theory*, 5: 395–418.

Moore, M. and Viscusi, K., 1988. The quantity-adjusted value of life. *Economic Inquiry*, 26(3): 369–388.

Murray, C. and Lopez, A., 1996. *The Global Burden of Disease*. Cambridge, MA: The American Association for the Advancement of Science.

North, G., 2002. Undermining property rights: Coase and Becker. *Journal of Libertarian Studies*, 16(4): 75–100.

Orr, L. D., 1981. Social costs, incentive structures, and environmental policies. In: J. Baden and R. Stroup (Ed.), *Bureaucracy vs. Environment: The Environmental Costs of Bureaucratic Governance*. Ann Arbor: University of Michigan Press.

Ostro, B., 1996. *A Methodology for Estimating Air Pollution Health Effects*, WHO/EHG/96.5, Geneva.

Paul, M., 2005. *Energy Beyond Oil*. Leicester: Matador.

Pearce, D. and Turner, K., 1992. Packaging waste and the polluter pays principle: A taxation solution. *Journal of Environmental Management and Planning*, 35(1): 5–15.

Pearce, D. W. and Howarth, A., 2000. Technical report on methodology: Cost–benefit analysis and policy responses. RIVM report 481505020.

Pigou, A. C., 1912. *Wealth and Welfare*. London: Macmillan and Co.

Pigou, A. C., 1920. *The Economics of Welfare*. London: Macmillan and Co.

Pigou, A. C., 1954. Some aspects of the welfare state. *Diogenes*, 7: 1–11.

Pigou, A. C., 1956. *Memorials of Alfred Marshal*. New York: Kelley and Milman.

Posner, R., 1973. *Economic Analysis of the Law*. Boston: Little Brown.

Roll, E., 1992. *A History of Economic Thought* (5th edition). England: Faber and Faber.

Rothbard, M., 1977. *Toward a Reconstruction of Utility and Welfare Economics*. New York: Center for Libertarian Studies.

Rothbard, M., 1982. Law, property rights, and air pollution. *Cato Journal*, 2(1): 55–100.

Sagoff, M., 1981. Economic theory and environmental law. *Michigan Law Review*, 79: 1393–1419.

Sagoff, M., 1990. *The Economy of the Earth*. New York: Cambridge University Press.

Saldiva, P., 1995. Air pollution and mortality in elderly people in Sao Paulo, Brazil. *Archives of Environmental Health*, 50(2): 159–163.

Schumacher, E. F., 1973. *Small is Beautiful*. England: Abakus.

Schwetz, J. and Dockey, D. W., 1992. Increased mortality in Phladelphia associated with daily air pollution concentrations. *American Review of Respiratory Disease*, 145: 600–604.

Seethaler, R. K., Kunzli, N., Sommer, H., et al., 2003. Economic costs of air pollution related health impacts: An impact assessment project of Austria, France and Switzerland. *Clean Air and Environment Quality*, 37(1): 35–43.

Stavins, R. N., 2003. Market-based environmental policies: What can we learn from U.S. experience (and related research)? Discussion Paper 03–43. Resources for the Future.

Sunyer, J., 1996. Air pollution and mortality in Barcelona. *Journal of Epideomiology and community Health*, 50: 877–880.

Thurston, G., Bates, D., Burnett, R., et al., 1997. Health and Environmental Impact Assessment Report, Joint Industry/Government Study of Sulphur in Gasoline and Diesel Fuels, June 25, 1997. Ottawa: Health Canada.

Tietenberg, T., 1980. Transferable discharge permits and the control of stationary source air pollution: A survey and synthesis. *Land Economics*, 56: 391–416.

Tietenberg, T. H., 1984. *Environmental and Natural Resource Economics*. USA: Foresman and Co., Glenview, Il.

Tietenberg, T. H., 2003. The tradable-permits approach to protecting the commons: Lessons for climate change. *Oxford Review of Economic Policy*, 19(3): 400–419.

Viscusi, K., 1992. Pricing health risks: Survey assessments of risks–risk and risk-dollar tradeoffs. *Journal of Environmental Economics and Management*, 21(1): 32–51.

Viscusi, K., 1993. The value of risks to life and health. *Journal of Economic Literature*, 31: 1912–1946.

Viscusi, K. and Evans, W., 1990. Utility functions that depend on health status estimates and economic implications. *American Economic Review*, 80(3): 353–374.

WDR, 1999. World Development Report. *Investing in Health*. Washington, DC: The World Bank.

WHO, 1999. Overview of the environment and health in Europe in the 1990s. Third Ministerial Conference on Environment and Health. London, June 16–18, 1999, Copenhagen: World Health Organization. WHO Regional Office for Europe.

WHO, 2000a. Quantification of the health effects of exposure to air pollution. Report of a World Health Organization working group, Bilthoven, Netherlands, November.

WHO, 2000b. *Air Quality Guidelines for Europe* (2nd edition). Geneva: WHO Regional Publications, European Series, 91.

Woodin, M. and Lucas, C., 2004. *Green Alternatives to Globalization: A Manifesto*. London: Pluto Press.

14 Elements of Air Quality Management

Atmospheric Science Tools for Developing Effective Policy

Jeffrey R. Brook, Michael D. Moran,
William Pennell, and Lorraine Craig

CONTENTS

14.1 INTRODUCTION

Throughout the world, reducing the impacts attributed to environmental exposures is an important public health objective and one with significant societal benefits. Poor air quality (AQ) is recognized as one of the leading contributors to the global environmental burden of disease. There is extensive scientific evidence of population health effects, even in countries with relatively low concentrations. Air pollution also damages terrestrial and aquatic resources, including those of direct economic importance, and it is interwoven with the causes and consequences of climate change and many other pressures confronting society, such as urbanization, energy production, and waste management. Consequently, the development of AQ policy is an active and relatively mature part of the environmental management process. Although this process is complex and requires that the problems be addressed through several different, but complementary, mechanisms, many countries have realized significant improvements in AQ over the past several decades.

The objective of this chapter is to discuss some of the scientific elements associated with AQ management. Current health effects evidence is reviewed first, which provides a basis for an ongoing need to improve AQ. This section is followed by a discussion of the scientific foundation of AQ management, which consists of emission inventories, ambient measurements, and atmospheric models. These tools provide a quantitative understanding of how atmospheric chemistry, meteorology, and natural emissions influence the fate of human-generated emissions and ultimately population exposures. They are essential for identifying AQ problems, tracking progress, and developing effective policies and programs to improve AQ. Therefore, a key goal of this chapter is to discuss how models, emissions inventories, and measurements support AQ management and then to address technical issues and uncertainties in the application of these tools. Awareness of these issues helps to ensure the successful application of these tools and to guide effective communication of their results to AQ managers in all levels of government and in the private sector. For a more comprehensive discussion of the topics covered in this chapter as well as for guidance in AQ management policy strategies, readers are referred to Craig et al. (2008).

14.2   AIR POLLUTION'S EFFECT ON HUMAN HEALTH

The World Health Organization (Cohen et al., 2005) has identified ambient air pollution as a high priority in its Global Burden of Disease initiative, estimating that outdoor particulate matter (PM) air pollution is responsible for 1.2% of all premature deaths and 6.4 million (0.5%) years of life lost (YLL). Although the magnitude of the estimated increased risk might appear to be small, the numbers of the people affected are large when extrapolated to the entire population.

The primary objective of AQ management is to protect human health and the environment. From a policymaker's perspective, one key question on the issue of health effects is: What is currently known about the impacts of air pollution on public health? In recent years, a number of reviews addressing this question, largely regarding PM effects, have been published by the World Health Organization (WHO, 2002, 2004a,b, 2005a,b,c, 2006), the U.S. Environmental Protection Agency (U.S. EPA, 2004, 2005a, 2006a), The American Heart Association (Brook et al., 2004), and The Air and Waste Management Association (Pope and Dockery, 2006).

Some of the key points from these reviews and from other recent publications are as follows:

There is little evidence to indicate a threshold concentration below which air pollution has no effect on population health or to indicate that the relationship between concentration or exposure and health effects is nonlinear.

There is consistency among the many time-series studies conducted throughout the world relating day-to-day variation in air pollution to health, and a smaller number of longer-term cohort studies also find that air pollution increases the risk of mortality.

It is estimated that the shortening of life expectancy of the average population that is associated with long-term exposure to PM is 1–2 years.

Improved exposure classifications have tended to lead to increases in the magnitude of the risk attributed to air pollution.

The number of adverse health endpoints found to be associated with air pollutant exposure has been increasing and it is widely recognized that the cardiovascular impacts are as significant as, if not greater than, those due to respiratory conditions.

Human clinical and animal experimental studies have identified a number of plausible mechanistic pathways of injury, including systemic inflammation that could lead to the development of atherosclerosis and alter cardiac autonomic function so as to increase susceptibility to heart attack and stroke.

Intervention studies examining sharp reductions in air pollution levels have provided strong evidence of the benefits to health of lower air pollutant concentrations.

Despite continuing uncertainties, the evidence overall tends to substantiate the fact that PM effects are at least partly due to ambient PM acting alone or in the presence of other covarying gaseous pollutants.

The amount of evidence implicating traffic-related air pollution in a range of
adverse effects is large and is increasing rapidly, suggesting that this source
requires significant attention from AQ managers.

14.2.1 Brief Review of the Evidence

Sources of evidence from which the health effects associated with air pollution exposures can be assessed include observational epidemiological, toxicological, and clinical
studies. The findings of these different lines of investigation are complementary and
each has well-defined strengths and weaknesses. The findings of epidemiological
studies have been assigned the greatest weight in standard-setting for airborne particles because they characterize the consequences of the exposures that are actually
experienced in the community setting.

There is a substantial body of epidemiological evidence linking exposure to current levels of air pollution, especially airborne PM, with increased mortality and
morbidity, including a wide range of adverse cardiorespiratory health outcomes. This
evidence includes countries or cities with relatively low levels of air pollution.
Associations between air pollution exposure and mortality have been assessed
mainly through two types of epidemiological studies. Cohort studies follow large
populations for years and typically relate mortality to an indicator of average exposure to PM over the follow-up interval. Time-series studies investigate the association between daily mortality and the variation in recent PM concentrations. To
establish standards for short-term exposures, regulatory agencies rely on the findings
of time-series studies, whereas findings of cohort studies are used to establish annual
standards. Long-term cohort studies of PM and mortality are fewer in number than
those of day-to-day variations. They are typically expensive to carry out and require
a substantial number of participants, lengthy follow-ups, and information on PM
exposure as well as on potential confounding and modifying factors. Most of the
studies have been carried out in the United States, but findings have also been
reported for two European studies.

Two cohort studies of the health effects of long-term exposure to air pollution in
large populations have had the greatest influence in the development of ambient AQ
standards for PM_{10} and $PM_{2.5}$: (1) The Harvard Six Cities Study (Dockery et al.,
1993) and (2) The American Cancer Society (ACS) Cancer Prevention Study (Pope
et al., 1995). As indicated above, these types of studies are rare, but a number of
other past and recent studies support the Six Cities and ACS basic conclusions that
PM has a significant effect on mortality (Abbey et al., 1999; Chen et al., 2005;
Gehring et al., 2006; Lipfert et al., 2006; Rosenlund et al., 2006).

The Six Cities Study was the first large, prospective cohort study to demonstrate
the adverse health impacts associated with long-term air pollution exposures. This
study demonstrated that chronic exposure to air pollutants is independently related
to cardiovascular mortality. In the group of 8111 adults with 14–16 years of follow-up,
the increase in overall mortality for the most polluted city versus the least polluted
city was by 26%. The range of exposure to PM across the six cities was 11–29.6 µg/m³
for fine particles. Laden et al.'s (2006) extended follow-up of the Six Cities Study
found effects of long-term exposure to particulate air pollution that were consistent

with previous studies. Total cardiovascular and lung cancer mortalities were positively associated with ambient $PM_{2.5}$ concentrations. They examined the effect of a decrease in the annual average $PM_{2.5}$ concentrations, from 18 to 14.8 $\mu g/m^3$, and found a statistically significant reduction in mortality risk for cardiovascular and respiratory causes, but not for lung cancer, suggesting that reduction in fine PM pollution yields positive health benefits.

The ACS study linked chronic exposure to multiple air pollutants to mortality over a 16-year period and found a robust association between long-term exposure to $PM_{2.5}$ and mortality. An extended follow-up indicated that the long-term exposures were most strongly associated with mortality from ischemic heart disease, dysrhythmias, heart failure, and cardiac arrest (Pope et al., 2004). For these cardiovascular causes of death, a 10 $\mu g/m^3$ elevation of $PM_{2.5}$ was associated with an 8–18% increase in risk of death. Mortality attributable to respiratory diseases had relatively weak associations. More recent analysis of the U.S. ACS data focusing on neighborhood-to-neighborhood differences in urban air pollution in Los Angeles with more precise exposure assessment methods have found death rates from all causes and cardiopulmonary diseases that are at least two times higher than previously reported in analyses of the ACS cohort (Jerrett et al., 2006). The highest rate estimated from the original ACS study (Pope et al., 2002) for all causes of mortality was 6%, but taking into account neighborhood confounders, the estimated risk was about 11% for a 10 $\mu g/m^3$ increase in the annual average $PM_{2.5}$.

Daily time-series studies examine variations in day-to-day mortality counts in relation to ambient PM concentration measured by AQ monitoring networks. In general, the evidence from daily time-series studies shows that elevated PM exposure of a few days is associated with a small increased risk of mortality. Large multicity studies in Europe (APHEA2: Air Pollution and Health: A European Approach 2) and the United States [National Morbidity, Mortality Air Pollution Study (NMMAPS) based on the largest 90 U.S. cities] indicate that the increase in daily all-cause mortality risk is small (0.21% per 10 $\mu g/m^3$ PM_{10}; 95% CI, 0.09–0.33), but consistent (Dominici et al., 2003; Katsouyanni et al., 2003). A review of time-series studies conducted in Asia also indicates that short-term exposure to air pollution is associated with increases in daily mortality and morbidity (HEI, 2004).

A major methodological issue affecting epidemiological studies of both short-term and long-term exposure effects relates to the use of appropriate methods for evaluating the extent to which gaseous copollutants (e.g., ozone [O_3], NO_2, SO_2, and CO), air toxics, and/or bioaerosols may confound or modify PM-related effects estimates (U.S. EPA, 2004). Gaseous copollutants are candidates for confounders because all are known to cause at least some adverse health effects that are also associated with particles. In addition, gaseous pollutants and primary PM may be emitted from common sources and dispersed by common meteorological factors. For example, both CO and particles are emitted from motor vehicles, and SO_2 and $PM_{2.5}$ are both emitted from coal-fired power plants. Krewski et al. (2000) found significant associations for both PM and SO_2 in their reanalysis for the Health Effects Institute of the Pope et al. (1995) study. Numerous new short-term PM exposure studies continue to report significant associations not only between various PM indices and mortality but also between gaseous pollutants and mortality. In some cities, the estimated PM effect is relatively stable when the copollutant is included in the model, whereas the estimated PM effect in other cities changes substantially when certain copollutants are included, particularly NO_2 (Burnett et al., 2004; Qian et al., 2007; Brook et al., 2007a). Despite continuing uncertainties, the evidence overall tends to substantiate that PM effects are at least partly due to ambient PM acting alone or in the presence of one or more covarying gaseous pollutants (Burnett et al., 1997b; U.S. EPA, 2004).

Independent from the challenge of isolating PM and gaseous copollutant effects, there is a large body of evidence demonstrating that ground-level O_3 has a significant impact on human health (Burnett et al., 1997a). Consequently, there is a relatively long history of AQ management activities focusing on reducing O_3 (Schere and Hidy, 2000) and continuing efforts are required in North America and worldwide. There are serious concerns about future O_3 levels in developing countries because with increase in wealth, the number of automobiles has also increased (e.g., China), which emit the O_3 precursors, NO_x, and volatile organic compounds (VOCs).

While O_3 and related oxidants have traditionally been viewed as powerful respiratory irritants leading to a wide range of morbidities, new studies provide evidence of an elevated risk of mortality associated with acute exposure to O_3, especially in the summer or warm season when O_3 levels are typically high. For example, studies carried out in 95 U.S. communities [U.S. National Morbidity, Mortality Air Pollution Study (NMMAPS)] (Bell et al., 2004) and in 23 European cities (APHEA) (Gryparis et al., 2004) reported positive and significant O_3 effect estimates for all-cause (nonaccidental) mortality. Three recent meta-analyses evaluated potential sources of heterogeneity in O_3-mortality associations, and these studies provide evidence of a robust association between ambient O_3 and nonaccidental and cardiopulmonary-related mortality, especially for the warm O_3 season. The risk estimates are consistent across studies and robust to control for potential confounders.

The U.S. EPA's recent O_3 criteria document (U.S. EPA, 2006b) indicates that effects due to O_3 exposure are independent of PM. This separate effect has been relatively easy to detect due to the fact that O_3 forms in the atmosphere (i.e., is a secondary pollutant), and so its temporal variability is generally not correlated with other pollutants (e.g., PM_{10}, CO, SO_2, and NO_2). However, O_3 can be more highly correlated with secondary fine particles, especially during the summer months. Nonetheless, multicity and single-city studies indicate that O_3-mortality associations do not appear to be substantially altered in multipollutant time-series models including PM_{10} or $PM_{2.5}$ (Burnett et al., 2004). These results suggest that the effects of O_3 on respiratory health outcomes appear to be robust and independent of the effects of other copollutants.

The evidence on cardiovascular outcomes has grown, leading to the American Heart Association acknowledging that the association of airborne particles with adverse cardiovascular outcomes is causal (Brook et al., 2004). Recent epidemiological, clinical, and toxicological studies report new evidence linking long-term exposure to fine particles with the development of atherosclerosis (Künzli et al., 2005; Sun et al., 2005). Numerous new studies have reported associations between ambient $PM_{2.5}$ and subtle cardiovascular effects such as changes in the cardiac rhythm or heart rate variability, blood pressure, and vascular function (Urch et al., 2005; U.S. EPA, 2006a; Auchincloss et al., 2008).

Much of the past epidemiological research on the impacts of short-term and long-term air pollution exposures has focused on premature death, hospital admissions, and emergency room visits. However, more recent studies have examined a broader scope of health concerns such as exacerbations of chronic respiratory diseases, including chronic obstructive pulmonary disease (COPD) and asthma, the respiratory health of children, and several adverse cardiac and reproductive outcomes. Impacts on susceptible subpopulations, such as on children and older adults, have also been examined more closely, including those with preexisting cardiopulmonary and metabolic illnesses. Furthermore, the number of health endpoints reported to be associated with PM exposures has expanded to include (1) indicators of the development of atherosclerosis with long-term PM exposure; (2) indicators of changes in the cardiac rhythm and blood pressure; (3) increased risk of making atherosclerotic plaques more vulnerable to rupture, clotting, and eventually causing heart attack or stroke; (4) effects on developing children and infants; (5) markers of inflammation such as exhaled NO; and (6) effects on organ systems outside the cardiopulmonary systems (U.S. EPA, 2006a). The long-range implications for individuals of some of the intermediate markers of outcome remain to be established, but at the very least, they offer valuable insight into potential biological mechanisms.

People with diabetes have been found to be more sensitive to cardiovascular effects from air pollution (Goldberg et al., 2001, 2006; Zanobetti et al., 2001; O'Neill et al., 2005), and there are mechanistic explanations for these results (Proctor et al., 2006). Goldberg et al. (2006) reported significant associations between $PM_{2.5}$ and diabetes deaths, as well as total mortality in people with previous diagnoses of diabetes. The acute risk of cardiovascular events in patients with diabetes mellitus may be twofold higher than that in nondiabetics. These findings are of particular concern, given the increasing incidence of diabetes in North America, and there is some prevalence-based evidence that chronic air pollutant exposure may contribute to the development of type II diabetes (Brook et al., 2008).

There is substantial evidence to indicate that PM exposure in children is associated with adverse effects on lung function, aggravation of asthma, and increased incidence of cough and bronchitis. In addition, there is evidence to suggest an increased risk of postneonatal respiratory mortality as concentrations of PM increase (adjusted odds ratio of 1.16 for a 10 μg/m³ increase of PM_{10}). Studies on birth weight, preterm births, and intrauterine growth retardation also suggest a link with air pollution (WHO, 2005b; Brauer et al., 2008; Siddiqui et al., 2008; Tang et al., 2008).

The World Health Organization report on the "Health Effects of Transport-Related Air Pollutants" (WHO, 2005c) concluded that the evidence indicates that pollutants related to vehicular traffic contribute to an increased risk of death, particularly from cardiopulmonary causes, and increase the risk of nonallergic respiratory symptoms and disease. New research, supporting this conclusion and extending the health endpoints to other morbidity outcomes, continues to be published (Finkelstein et al., 2004; Lwebuga-Mukasa et al., 2004; Grahame and Schlesinger, 2005; Mills et al., 2005, 2007; Brauer et al., 2006, 2007, 2008; Hoffmann et al., 2006; Smargiassi et al., 2006; Gauderman et al., 2007; Kan et al., 2007; Maynard et al., 2007; Slama et al., 2007). Consequently, of the myriad of sources contributing to the mix of urban air pollutants the population is exposed to, the case against traffic-related air pollution is one of the strongest. In addition to this large and diverse amount of evidence, there are two key reasons for the strength of this case: (1) specific chemical compounds and/or physical–chemical characteristics of traffic-related pollutants are known, in their own right, to be acutely toxic, mutagenic, and/or carcinogenic; (2) the potential for individuals or the population to be highly exposed to traffic pollutants is high due to the proximity of roads to places where people spend time and because people spend a considerable amount of time traveling in traffic.

Beyond these reviews, it is important to recognize that air pollution is a broad public health problem with implications for children and adults worldwide, and while much of the epidemiological evidence linking air pollution exposures to health impacts focuses on measures of AQ and health in North America and Europe, for millions of people living in developing countries, indoor pollution from the use of biomass fuel occurs at concentrations that are orders of magnitude higher than what is currently seen in the developed world. Deaths due to acute respiratory infection in children resulting from these exposures are estimated to be over 2 million per year (Brunekreef and Holgate, 2002). Regardless of the magnitude of the exposure and, in general, the source, there remains considerable evidence that further reductions in air pollutant levels, including those in relatively clean countries/cities, will lead to significant public health benefits with subsequent benefits to the economy.

14.3   AQ MANAGEMENT FRAMEWORK

The lack of health effect thresholds complicates the process of AQ management. It implies that the appropriate amount of reduction (i.e., the ambient target or emission controls needed to reduce impacts to an acceptable level) can only be determined through a quantitative comparison of risks and benefits, potentially expressed in economic terms (i.e., cost–benefit analysis). Assuming that the incremental costs of attaining a certain lower ambient concentration increase as this concentration decreases, there will be a point where costs outweigh benefits. In theory, this approach sounds practical, but the number of factors to be considered is large and includes information related to AQ science, health science, engineering, and socioeconomic factors, where the latter can vary considerably from country to country and across time. For this reason, the cost–benefit analysis approach used to identify an acceptable level of risk and the subsequent emission reduction targets can only strive to be as comprehensive as possible, given the current information, and must involve stakeholders* from all sides of the issue. This is equally true with respect to applying integrated risk models for quantifying the benefits of different policy options (U.S. National Research Council, 1983).

---

* To an increasing degree, the air-quality management process is recognized as a "science–policy–stakeholder interplay" (U.S. Presidential/Congressional Commission on Risk Assessment and Risk Management, 1997; van Bree et al., 2007). It is important, within this interplay, to be aware that not only the content of air pollution and health issues is considered important by the players, but also the process and mechanisms by which the interface operates. All stakeholders play a role in the generation, dissemination, and evaluation of policy options. Without substantive involvement by all stakeholders, it is more difficult to develop and successfully implement AQ management programs.

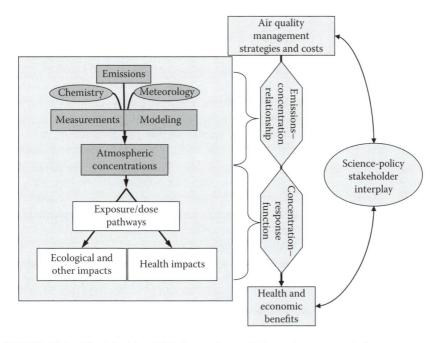

FIGURE 14.1   The left side of this figure depicts linkages between emission inventories, ambient measurements, AQ models, and epidemiological studies. These are the science tools (inside the box on the left half of the figure) that allow the risks or costs and benefits of various AQ management options to be evaluated and communicated to stakeholders for acceptance or refinement of strategies/policies. The right half of the figure depicts, in a simplified fashion, the approach to developing policy.

The development of policies, with or without cost–benefit analysis, for AQ risk management also needs to consider multiple pollutants, including different size and chemical fractions of PM, as well as current and future linkages to climate change and other atmospheric issues, such as acidic and toxic depositions (NRC, 2006). Examining multiple issues in an integrated fashion is often referred to as the "one atmosphere" approach. The Clean Air for Europe report (CEC, 2005) represents one of the first attempts to examine multiple air issues quantitatively. Ideally, this approach could be extended to include long-term-sustainability and whole-life-cycle-management concepts in the assessment of AQ management options.

Figure 14.1 illustrates a generalized AQ management framework from both a science perspective and a policy perspective. The left side of the schematic in Figure 14.1 depicts the connection between emissions and the observed levels of outdoor air pollution. The role of inventories, measurements, and modeling tools, along with analysis and interpretation (e.g., source apportionment), is to provide a quantitative understanding of how atmospheric chemistry, meteorology, and natural emissions influence the fate of human-generated emissions and ultimately population exposures. This understanding, via conceptual and numerical models, then leads to the capability of predicting future concentrations due to changes in the atmospheric environment and/or changes in emissions. This capability is essential to AQ management because it allows the comparison of a variety of potential emission reduction strategies. These strategies can be compared with respect to how much and where they lead to reductions in air concentrations, which can be extended to estimates of how exposure or dose will change and/or what will the health benefits be due to these changes.

In a benefit or a cost–benefit analysis, the first step is to propose specific emissions reductions and then use the predicted emission-to-concentration relationships (i.e., the appropriate numerical AQ model) to determine how air concentrations representative of the population exposure will change in response to these emission reductions. The second step is to use concentration-to-response relationships (or concentration–response functions, CRFs) to determine health benefits expected from the emission reductions. These typically link ambient concentrations at population-representative locations to health endpoints and, as indicated above, are derived from epidemiological studies that rely on ambient measurements.

As depicted in Figure 14.1, a strong scientific knowledge base from health, atmospheric, and emissions research is important because it provides the means to link emissions to human health or to environmental responses. This supports the development of policy for AQ management via the science–policy–stakeholder interplay. Such policies may be informed solely in terms of AQ standards and their benefits (health and/or environmental) or in terms of the costs of specific emission reductions and the economic benefits of improved human and ecosystem health or public welfare.

Given their importance in AQ management, much of the rest of this chapter will focus on the effective application of AQ models and the determination of emission inventories. The latter are a critical model input and must be known before appropriate emission reduction scenarios can even be considered. AQ measurements are essential to the entire process and their role is also discussed in more detail near the end of the chapter.

14.4   AQ MODELS AND EMISSION INVENTORIES

AQ models provide policymakers with the means to obtain predictions of *future* air concentration and deposition patterns based on possible future emission rates. In practice, quantifying the link between emissions of primary pollutants or precursors of secondary pollutants and ambient pollutant concentrations involves a modeling system as depicted in Figure 14.2. At its heart is the representation of emissions, transport, diffusion, and removal processes. The emissions and meteorological data that are needed to "drive" an AQ model are provided by an emissions processing system (Dickson and Oliver, 1991; Houyoux et al., 2000) and a numerical weather prediction model (Seaman, 2000), respectively. For most applications, atmospheric *chemical* transformations also need to be considered. This requires that the models consider time scales from fractions of seconds to days, leading to model domains that extend at least several hundreds of kilometers in the horizontal and from the Earth's surface up to at least the middle of the troposphere in the vertical (~5–6 km).

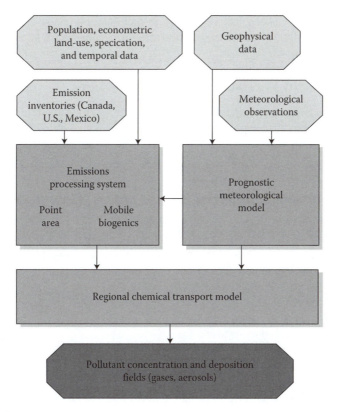

FIGURE 14.2 Schematic description of the components of an AQ modeling system. [Reprinted from Craig L., et al. 2008. *Journal of Toxicology and Environmental Health Part A* 71: 588–698. With permission of the publisher (Taylor & Francis Ltd, http://www.tandf. co.uk/journals).]

The main outputs of the AQ modeling system are pollutant concentration fields as functions of space and time for a given set of pollutant emissions and meteorological conditions (Peters et al., 1995; Seinfeld and Pandis, 1998; Jacobson, 1999; Russell and Dennis, 2000; Reid et al., 2003). This information can then be extended to other physiologically, environmentally, and optically important characteristics of the ambient pollutant mix.

For AQ models to be useful there must already be information available about emissions and atmospheric measurements of the region of interest. If such data are not available, then AQ model applications for that region seldom provide much useful guidance for policymakers because of the greater uncertainties associated with model predictions that arise from the uncertainties in model inputs.

AQ models have been used for decades to support AQ management. During this time, they have undergone continued and rapid development. Given this history, there have been a number of overviews of AQ models and AQ modeling over the years, including textbooks such as by Jacobson (1999) and review articles such as by Peters et al. (1995), Russell and Dennis (2000), Seigneur (2001), and Seigneur and Moran (2004).

14.4.1 APPLICATIONS OF MODELS FOR AQ MANAGEMENT

The main ways in which AQ models (hereafter referred to as models) are applied, directly or indirectly, to support AQ management and policy formulation are listed in the text box below.

AREAS OF APPLICATION OF MODELS IN AQ MANAGEMENT

Evaluation of the impact of emissions changes

Simulation of emission scenarios and quantification of the resulting benefits and disbenefits by prediction of ambient concentrations at multiple time and space scales

Base case (e.g., current emissions)

Emission levels when policies currently "on-the-books" are fully implemented (business-as-usual)

New emission-reduction scenarios

Estimation of emission changes required to attain AQ objectives or standards

Input to conceptual model development

Characterization of governing chemical regimes and limiting reactants for current and future conditions

Source apportionment and source attribution

Emission inventory evaluation

AQ forecasting

Measurement network and field experiment design

Data "fusion" (i.e., techniques to merge measurements with model predictions)

Testing current understanding of science

The use of models to assess the impact on AQ of emission changes due to pollutant abatement strategies, new pollution sources, population and economic growth, and so on, has been the most common application. The first step involved running the model for a "base-case" simulation, for which the emissions used are either historical or current and for which AQ measurements are available with which model performance can be evaluated. In the second step, the model is run again for the same time period and hence the same meteorological conditions, but with different emissions that represent "scenarios" of interest to policymakers. Conversely, the results from a suite of emissions scenarios can also be considered together in order to estimate the emission changes that will be needed in order to obtain a specific AQ threshold or an objective.

An AQ *conceptual* model is a qualitative mental model for a geographic region that is based on a synthesis and simplification of available AQ information obtained from analysis of emissions, measurements, and model results to distill the primary contributing factors, including key emission sources, terrain characteristics, and local weather and climate. Model results from a suite of emissions scenarios, from source apportionment studies sensitivity studies, can contribute significantly to the development or refinement of a conceptual model. One example of the development of a conceptual model is the analysis described by Pun and Seigneur (1999) for PM pollution in California's San Joaquin Valley. In areas with limited availability of emissions data and measurements, model results can contribute significantly to the development of conceptual models, as demonstrated by Zunckel et al. (2006) for southern Africa.

Source–receptor relationships are often an important element of an AQ conceptual model for an airshed of interest. The simplest way of using AQ models to investigate and quantify source–receptor relationships is the "zero-out" approach. That is, emissions from a particular jurisdiction or from a particular source sector (e.g., from electric power generation and from heavy-duty diesel trucks) are set to zero while all other emissions are left unchanged. Predictions from this case can then be subtracted from predictions from a base run in which all emission sources are included in order to estimate the impact of the targeted source sector or jurisdiction. Two more sophisticated approaches are "source tagging" and inverse methods. In the former, pollutant emissions from particular source sectors or geographic locations are tracked simultaneously in the model as separate ("tagged") species (Kleinman, 1987; McHenry et al., 1992; Kleeman and Cass, 1999a,b; Zhang et al., 2005). In the latter, the adjoint of the AQ model can be constructed, to quantify the sensitivity of the model to emission inputs, and then combined with ambient measurements, or else initial attribution results can be refined based on the synthesis inversion technique (Uliasz, 1993; Pudykiewicz, 1998; Mendoza-Dominguez and Russell, 2001; Mallet and Sportisse, 2005; Knipping et al., 2006).

The comparison of model predictions with measurements can also give some indication of the accuracy of the input emissions. For example, if model predictions are systematically biased high or low in a certain region compared with measurements, one possible cause may be a corresponding high or low bias in the input emissions for that region. This approach was followed in comparing predictions and emissions between rural and urban areas by Yu et al. (2004). Inverse modeling analyses, in which enhanced models are combined with ambient measurements, provide another, more quantitative, approach for estimating emissions strengths on either a regional or a global basis (Pudykiewicz, 1998; Mendoza-Dominguez and Russell, 2001; Martin et al., 2002; Palmer et al., 2003; Gilliland et al., 2006).

A model can be run in "real time" to provide an AQ forecast for the next one or two days. This information can warn sensitive individuals to take measures to reduce their exposure, and short-term interventions on selected sources can be implemented. Routine dissemination of forecasts helps maintain or raise public awareness about AQ and evaluations of forecasts can provide insights into modeled weaknesses (Eder et al., 2006) over a broad range of conditions.

To assist in optimizing network configurations, modeled pollutant fields can be assumed to represent the "true" spatial pattern. Differing numbers and locations of grid cells can then be sampled in order to investigate the impact of adding or removing hypothetical stations to a network on the estimation of the actual pollutant spatial pattern. Similarly, scientists planning and designing a field experiment can use model predictions to help in the identification of measurement locations for stationary instrumentation and sampling tracks for the deployment of aircraft and other mobile sampling platforms.

Models can also be employed as sophisticated interpolation schemes since their results are based on consistent and comprehensive representations of physical and chemical laws. An example of a "fused" model and observed O_3 concentrations is shown in Figure 14.4. Ultimately, a goal of models is to be running continuously in forecast mode in real time with four-dimensional data assimilation. The source of the measurement data used will likely be diverse, including a range of space-based (satellites) and surface-based remote sensing information and *in situ* measurements. The predicted fields will be archived as the optimal picture of the state of the atmosphere, both chemically and physically.

Lastly, AQ models provide a means to represent and link in a single package our best understanding of all of the chemical and physical processes relevant to AQ. When this knowledge synthesis is evaluated by comparing model predictions with enhanced measurement datasets from dedicated field campaigns (Dennis et al., 1993; Berkowitz et al., 1998; Heald et al., 2003, 2005; Frost et al., 2006; Hodzic et al., 2006; Pun et al., 2006; Smyth et al., 2006a), new insights are often obtained, thereby advancing knowledge. Ultimately, this leads to improved models, both numerical and conceptual, which can then provide more reliable information on AQ risk management.

14.4.2 TECHNICAL ISSUES AND UNCERTAINTIES TO CONSIDER IN AQ MODELING

There are many choices to be made and issues to be considered when using an AQ modeling system to study a problem. As discussed in the following, it is very important for the users of model results to be aware of these choices when judging the robustness and reliability of the guidance provided.

The first step in applying a model is to define the questions that need to be answered. Then, if possible, a conceptual model should be identified or developed to narrow down the set of numerical AQ models that might be used to answer the question. For example, a detailed but computationally expensive model might not be the best choice for performing a multiyear simulation, or a model with complex chemical reactions may not be needed for very local impacts by a point or line source. Another consideration is that a model designed to address one issue (e.g., photochemical smog) may not include representations of all of the processes necessary to address another issue (e.g., deposition of acidic species). And a model designed for highly polluted atmospheres may not be appropriate to model a clean atmosphere and vice versa (e.g., regional atmospheric chemistry in emission source regions versus background global chemistry).

There are also many choices to be made in setting up a model run, including (a) the location and (b) the size, in both the horizontal and the vertical dimensions, of the model domain, (c) the map projection to be used, (d) the grid spacing in both the horizontal and the vertical, (e) the integration time step, (f) the simulation period, including any required "spin-up" time (the time for atmospheric concentration fields to reach an equilibrium between emissions and removal processes), (g) the "refresh" rate (the length of time in which the *meteorological* model will be run before being reinitialized using a new set of meteorological analyses), (h) the set of chemical species to be considered, (i) the choice (in some cases) of parameterizations to be used for different chemical and physical processes, (j) the specification of initial chemical conditions, (k) the treatment of chemical lateral and upper boundary conditions, and so on.

Each choice has its implications. For example, the use of large horizontal grid spacing may "average out" a suspected hot spot or not represent small-scale meteorological circulations forced by local terrain features. See U.S. EPA (2005b) for a discussion about the choice of horizontal grid spacing, Berge et al. (2001) for a discussion about the specification of chemical initial conditions, and Brost (1988) for a discussion about the specification of chemical lateral boundary conditions.

One key limitation for models can be gaps in the scientific understanding of the pollutants of interest. For example, it is well known that the sources of much of the carbonaceous component of atmospheric $PM_{2.5}$ are not presently known in spite of the fact that this component typically contributes to ~40–50% of the total $PM_{2.5}$ mass. Another example is our limited understanding of nighttime NO_x chemistry (Brown et al., 2006). A second limitation is the use in current models of process parameterizations of limited fidelity to the real atmosphere. For example, Dabberdt et al. (2004) recently identified the need for improved treatments of the influence on AQ of the planetary boundary layer and clouds and cloud processes. Nonlinear responses can further complicate the interpretation of model predictions and the formulation of possible emission control measures. The possibility of a nonlinear response in sulfate deposition to SO_2 emission reductions due to oxidant limitations was identified in the 1980s as a potential concern for acid deposition (Misra et al., 1989). Nonlinearities in O_3 photochemistry are well known (Seinfeld and Pandis, 1998), but PM chemistry possesses even more nonlinearities. Meng et al. (1997) presented model predictions for O_3 control scenarios for a Los Angeles smog episode. VOC emission reductions reduced O_3 levels but caused increases in $PM_{2.5}$ mass. Reductions in SO_2 emissions can also increase $PM_{2.5}$ concentrations in some cases due to "nitrate substitution" (West et al., 1999).

Even when AQ processes are well understood scientifically, they must still be represented mathematically in AQ models by parameterizations, and then the complex, coupled system of governing equations that comprise the model must be solved numerically. Both steps have limitations and can introduce errors. In many cases, a number of different parameterizations have been developed to describe the same chemical or physical process, and these different parameterizations will produce different results (Kuhn et al., 1998; Zhang et al., 2000, 2001; Mallet and Sportisse, 2006). Typically, more sophisticated (and complex) parameterizations have a greater number of parameters and coefficients that must be specified, but measurements to do so may be scarce or lacking completely. This implies that while a more sophisticated scheme may have the potential to do a better job in describing a process, there is no guarantee that it will actually do a better job in practice because of input data limitations.

The numerical integration of the model also introduces errors since it usually requires the solution of large coupled systems of both ordinary and partial differential equations. As discussed by Pielke (1984) and Jacobson (1999), among others, finite-difference methods are usually employed in both time and space. Any time-stepping scheme used to integrate the AQ model in time will have truncation errors that depend on both the order of the scheme and the magnitude of the chosen time step. Operator splitting is usually employed on the right-hand side of the governing equations in order to allow each process parameterization to be calculated separately, but operator splitting also introduces errors that depend on the order of the splitting and the overall time step. Advection is well known to be a difficult process to solve, and literally hundreds of numerical schemes have been developed for advection. All suffer to varying degrees from some or all of truncation errors, numerical diffusion, phase errors, lack of positive definiteness, and violation of mass conservation.

Model "resolution" is another important consideration. The choice of a discrete model time step (Δt) and grid-cell size (Δx) implicitly imposes numerical filtering on the model solution. In essence, no temporal feature shorter than $2\Delta t$ in duration and no spatial feature smaller than $2\Delta x$ in size can be predicted by the model; however, $4\Delta t$ and $4\Delta x$ are probably a more realistic threshold (Pielke, 1984; Grasso, 2000). This has important implications for processes operating at smaller temporal and spatial scales (Pielke and Uliasz, 1998). As a consequence, many parameterization schemes have been developed to represent the influence of subgrid-scale processes at grid scales. One obvious example is the representation of point source emissions. In any Eulerian (i.e., grid) model, all or most point sources will be represented as volume sources since the emissions are assumed to be well-mixed throughout at least one grid cell, thus introducing large numerical (i.e., artificial) diffusion in the vicinity of major point sources.

Numerical AQ models require a number of input datasets to be supplied in order to run properly, and the accuracy and representativeness of these input datasets are a key concern. Even for a perfect model, the well-known aphorism "garbage in, garbage out" will apply. Model input datasets include the following:

1. Emission rates of a number of gaseous and particulate species must be specified for each model time step at each model grid cell at all levels. There are significant uncertainties due to errors in the emission inventories themselves and due to additional uncertainties introduced by the emissions processing systems that perform the chemical speciation for PM and VOCs and by spatial and temporal disaggregation steps needed to create model-ready emission files (Hogrefe et al., 2003). For large point sources, ancillary information about smokestack characteristics such as stack height, stack diameter, stack-gas exit velocity, and stack-gas exit temperature is also needed. If a future-year scenario is being considered, current emission inventories must be manipulated and modified to account for all assumptions built into the scenario. Since emissions data are an extremely critical

model input, the development of inventories and their main sources of uncertainty are discussed further below.

2. Meteorology is very important as it influences every aspect of the Air Quality System (AQS):
   a. Natural sources, such as windblown dust, sea salt, and biogenic VOCs
   b. Plume rise
   c. Transport and diffusion
   d. Gas-phase and heterogeneous-phase chemistries (via temperature and humidity effects)
   e. Cloud shading
   f. Aqueous-phase chemistry
   g. Dry removal
   h. Wet removal
3. Chemical initial conditions must be supplied for all model species for every grid cell, chemical upper boundary conditions must be specified for all model species at the top model level, and, for a limited-area model, chemical lateral boundary conditions must also be specified for all model species.
4. Geophysical fields, including terrain height, land-use type, vegetation type, aerodynamic surface roughness, albedo, sea surface temperature, and soil texture, must also be provided.

The resources required is another issue. AQ modeling is typically resource-intensive in terms of model input data, people, calendar time, and computer power (Reid et al., 2003). In order to apply a model for a particular case, the input datasets described above must be prepared for the model configuration selected, including emissions, meteorological, and geophysical files, the model must be run, and then the model results must be processed, analyzed, and interpreted. Typically, a minimum of three highly trained modelers is required to contribute, namely an emissions-processing specialist, a meteorological-modeling specialist, and an AQ-modeling specialist. The required calendar time from start to finish, including configuring and testing the model for the application, would likely be a minimum of weeks but more likely months. The minimum computer resources needed would be a high-end PC with multiple processors, large internal memory and disk space, and off-line archiving hardware to save numerous large model output files. Access to emissions data, meteorological data, geophysical data, and AQ measurement data is another prerequisite.

14.4.3 JUDGING A MODEL'S ACCURACY, SENSITIVITY, AND UNCERTAINTY

For an AQ model's predictions to be used by policymakers, it must be shown, to the extent possible, to give the right answers for the right reasons. This requirement can be very challenging, for it is important to remember the dictum (from statistical modeling) that "All models are wrong, but some are useful" (Box, 1979). *How does one judge if an AQ model will be useful for considering alternatives to reduce concentrations of one or a combination of atmospheric pollutants?* This question is addressed by model performance evaluations, in which model predictions are compared with measurements. There are some fundamental issues related to model accuracy, sensitivity, and uncertainty that need to be kept in mind when evaluating model performance. For one thing, how is the "right answer" determined? For another, how can model uncertainty be determined?

There are also a number of issues that arise when comparing model predictions with ambient measurements. The biggest one is *incommensurability*, which arises due to the fact that model predictions correspond to grid-volume averages, whereas measurements are typically made at points in space or along lines (e.g., aircraft flight tracks, DIAL). For example, for a regional-scale model whose smallest grid volume is 20 km by 20 km by 50 m, how representative would a single point measurement be of the 20 km^3 of air contained in that grid volume?

In eastern North America, the subgrid-scale (<80 km × 80 km) daily variability was found to be approximately linearly related to the mean concentration. Among the pollutants measured at 3–5 stations within a grid cell, the largest variability was associated with SO_2 and minimum O_3 (~ ±70%), the intermediate variability was associated with p-SO_4 (~ ±30%) and t-NO_3 (~ ±40%), and the smallest variability was associated with maximum O_3 (~ ±20%) (Seilkop, 1995a,b). At these scales, the uncertainties that arise in comparing a single point measurement to a model value thus clearly overwhelm uncertainties associated with the instrument error. Within smaller circular areas (25 km radius), the 1987 SCAQS study (Los Angeles basin) found that local inhomogeneities for O_3, NO_2, and CO had normalized gross errors in the 25–45% range (McNair et al., 1996).

A second issue related to the comparison of model predictions and ambient measurements is the need to compare "apples with apples." For example, for gas-phase species, AQ model predictions correspond to ambient conditions whereas some networks report measurements at standard temperature and pressure (STP). For PM comparisons, model PM predictions are calculated based on the Stokes diameter whereas PM measurements are reported using the aerodynamic diameter. PM measurements, unlike model predictions of PM, can also suffer from artifacts related to volatile species such as nitrate, some organic compounds (Fan et al., 2003), and aerosol-bound water, and the distinction between elemental carbon (EC) and organic carbon (OC) is measurement-technique-based and can vary from network to network (Seigneur and Moran, 2004).

Model uncertainty is dependent not only on many factors, some of them dependent on the particular application, but also on the *interactions* of these factors. Thus, it is not possible to quantify overall uncertainty (Reid et al., 2003). There are three main sources of model uncertainty (Venkatram, 1988):

Errors in model inputs
Errors in model formulation
Inherent uncertainty associated with the stochastic nature of turbulence, which constitutes a lower limit on the model uncertainty (Moran, 2000)

Contributing factors include errors and uncertainties in input data such as emissions, meteorology, and boundary conditions, uncertainties in our scientific understanding and in process parameterizations, errors associated with numerical methods, and uncertainties associated with required parameters such as reaction rates. Some individual sources of uncertainty can be quantified, particularly for numerical methods, to identify model sensitivity to various inputs and parameters, and finally to compare results from parameterizations and even entire models in order to try to characterize the range of uncertainty. Error characterization is generally reported as part of the description of new numerical methods and parameterization techniques. A wide range of sensitivity analysis techniques exist, including DDM, ADIFOR, FAST, variational techniques, perturbation theory techniques, Green's function techniques, and stochastic techniques, that can be used to understand which model parameters and input variables most influence the selected model outputs [see Zhang et al. (2005) for a useful literature review].

Besides being compared side by side outside of models, the impact of different parameterization schemes can also be compared when embedded in a host model (Padro et al., 1993; Mallet and Sportisse, 2006). At a higher level, the impact of different modeling system components can also be evaluated. For example, Hogrefe et al. (2003) compared the impact of using emissions files constructed by two different emissions processing systems from an identical emission inventory on the predictions of one model. They found differences on the order of ±20 ppb in predicted daily maximum 1-hour O_3 concentration. Smyth et al. (2006b) compared the outputs from one emissions processing system and one regional PM model for two sets of meteorological input files for the same period that were provided by two different meteorological models. On an average, their performance (operational evaluation) was essentially equivalent, as was the performance of the AQ model for the two sets of meteorological files, but when grid cells were matched for the same time, large variability was observed, particularly in aerosol quantities influenced by relative humidity. The performance of seven models in predicting O_3 was compared for the same period (summer 2004) and region (eastern United States). The range of predictions generally bracketed the measurements, and, interestingly, none of the models individually could match the skill of a weighted average of the seven forecasts (McKeen et al., 2005).

Another approach for assessing uncertainty is by synthesizing expert opinion. For example, Seigneur and Moran (2004) prepared a table that presented qualitative ratings of PM modelers' level of confidence in major aspects of the predictions of current PM AQ models. Only a few model aspects (SO_2, NO_x, and p-SO_4 air concentrations) were judged to have a "high" level of confidence. Most aspects were assigned "medium" or "low" ratings, and a few aspects, such as secondary OC and PM ultrafine mass and number concentrations, were assigned "very low" ratings. These ratings were based on an assessment of all contributing uncertainties, including the uncertainties associated with the emissions of different pollutants and with scientific understanding.

14.4.4 GENERATING EMISSION INVENTORIES

Accurate data on the emissions of many gaseous and particulate chemical species are a critical model input. Without accurate information on the sources of air pollutants—what they are, where they are located, what they are emitting, and how much—it is impossible to identify which sources are most important to control, to predict the AQ consequences of these emissions, or to monitor the effectiveness of emission reduction programs. Timely and historically consistent emission inventories are also essential for assessing the current emission environment (and for keeping abreast of economic conditions and changes in technology) and for tracking progress in improving the AQ. Inventories need to be widely accessible, transparent, and easy to be updated with standard procedures for doing so.

Obtaining emissions information requires cooperation at many different levels, and both government and industry must bear some responsibility. Ideally, specific industries should be continuously measuring and reporting their emissions, national and local governments should be compiling information and running models for extrapolation of the available information across sources, time, and space, and AQ modelers and field measurement specialists should be evaluating the data.

To be of the most value, emission inventories and models need to provide quantitative estimates at national, state or provincial, and county (or their equivalent) levels for many source categories. At a minimum, this information is needed for NO_x, SO_2, CO, fine particle mass, coarse particle mass, VOCs, NH_3, and CO_2. Additionally, the particles and the VOCs need to be broken down into several classes that describe chemical composition in a manner that interfaces with chemical mechanisms in models and that may allow for some assessment of health or environmental risks.

Emission inventories are usually developed using the following equation:

$$E = EF \cdot A \cdot (1 - ER), \tag{14.1}$$

where E is the emission rate (e.g., kg/h or tones/year) of a given pollutant, EF and A are the emission factor and the activity factor, respectively, and ER $(0 < ER < 1)$ is an emission reduction factor that accounts for any emission control devices that may be applied to the source. The emission factor, EF, is the mass of a given pollutant or chemical species emitted per unit process variable. The activity factor, A, is the related process variable, such as the mass of fuel consumed, vehicle kilometers traveled, and so on, in a given amount of time. In reality, emission factors and emission reduction factors can vary from source to source as well as with the value of the activity factor, type of fuel, operating conditions, age of the source, geographical location, time of the year, and so forth. Not all of these complexities can be accurately represented in such a simple relationship, and more sophisticated emission models have been developed for very complex categories such as mobile source emissions (Miller et al., 2006).

Except in rare instances, emission factors or their equivalents are based on measurements. The easiest source class to characterize is large point sources, such as electric-generating units or stack emissions from large industrial operations. Emissions from these facilities can be measured by direct sampling of flue gases—as long as reliable sensors and methods are available (they can be *in situ* or remote) and appropriate sampling techniques are used.* Using these techniques, gaseous emissions from large point sources, such as CO_2, SO_2, and NO_x, can be estimated to better than ±20% over time periods as short as 1 h (NARSTO, 2005).

---

* For example, when measuring PM emissions it is necessary to mimic the cooling and flue (or exhaust) gas dilution processes that occur immediately after the emissions enter the atmosphere, since many "primary" particles are formed in this near-source region.

Emissions from more dispersed and numerous individual sources are more difficult to evaluate. Examples include fugitive (i.e., inadvertent) emissions from industrial sources, natural emissions from vegetation, agricultural emissions, emissions from small industrial or commercial sources, emissions from residential sources (e.g., PM emissions from cooking or space heating), and emissions from large-scale biomass burning. For these types of emission sources, direct measurements may be difficult or they may be feasible for only a small sample of the sources in question. All of these factors lead to emission estimates that are more uncertain than for large point sources. These uncertainties can range from a factor of two to a complete neglect of an unknown source or chemical precursor which turns out to be significant upon later analysis.

On-road and non-road mobile sources (such as automobiles, trucks, aircraft, locomotives, construction equipment, ships, etc.) are a good example of an important, but widely dispersed and variable, component of pollutant emissions. The traditional approach for estimating automobile and truck emissions is to measure emissions from dynamometer tests of representative vehicles in the laboratory following standard driving cycles. These measurements are used as input to complex mobile source emission models that attempt to simulate vehicle fleet operating conditions for a wide range of urban, suburban, and rural settings. The problems with dynamometer tests are that the number of sampled vehicles may be too small to represent a statistically valid sample, and they may not represent the range of fuels used, driving cycles or conditions, environmental factors, and states of repair of the actual vehicles in use.

Because most sources are not equipped with continuous emission monitors to measure actual emissions, Equation 14.1 forms the basis of most data reported in inventories. Although the focus is often placed on the value of the emission or the emission reduction factor, the activity factor is equally important. Activity factors can be developed from continuously monitored process data, but as with continuous emission monitors, these data are generally scarce. More frequently, activity factors are developed from economic activity data or activity surveys. Fuel consumption data are a good example of the use of economic activity data that are collected for reasons other than emissions, but can be used in the development of emission inventories. Information on construction activities can be used to develop emissions from off-road construction sources. Population densities coupled with activity surveys can provide inventory developers with information on emissions from residential woodstove, fireplace, and open burning. Land-use maps and satellite data are useful for estimating the types and densities of vegetative cover, which in turn are used to estimate biogenic emissions. In each case, emission estimates can be developed using data that have been collected for other purposes, such as for tax estimation, economic development, or land-use planning.

Projections of future emissions also depend on these data and estimates of their growth rates in future years. In the absence of significant technological change, past relationships between population growth and the types of activity factors noted above provide a good starting point for estimating future activity levels, and subsequently, future emission levels.

14.4.4.1   Costs of Emission Inventory Development and Improvement

The cost of developing an emissions inventory depends on the level of detail that the final product considered in its determination and in the output files. In addition, quantifying emissions of specific hazardous air pollutants, such as mercury and other metals or speciation of organic compounds that are present in flue gases at very low concentrations (NARSTO, 2005), adds considerable cost, but will improve the capabilities of AQ models. If an inventory is derived using previously existing information on the emission factors, as opposed to a systematic assessment of the representativeness of the available factors, then it will be less expensive. In North America, the U.S. federal government currently invests approximately $25 million per year to develop and update emission inventories (NARSTO, 2005). This does not include the amounts spent by state and local agencies (estimated at about $10 million per year) or the additional costs that would be required to address the shortcomings identified in the current inventories, which are estimated to be on the order of an additional $35 million per year (NARSTO, 2005). In Canada, about $6 million (U.S.) per year is invested in compiling its national inventory, not counting local and provincial efforts, and Mexico has spent about $600,000 (U.S.) per year in developing its National Emission Inventory.

14.4.4.2   Weaknesses of Current Emission Inventories

In many countries, AQ managers now have a good understanding of the emissions from major point sources, and they have used this knowledge in developing effective actions for reducing them. Models for estimating emissions from mobile sources have continuously improved and the importance of natural and biogenic emissions has been recognized. However, several key weaknesses remain:

Development of mobile source inventories, particularly regarding the speciation of VOCs.

Emissions for important categories such as biogenic emissions, ammonia, fugitive emissions, open biomass burning, and many other area sources.

Emission estimates for air toxics (e.g., the 187 hazardous air pollutants designated by the U.S. EPA) are particularly uncertain since there are so many of these compounds, so many potential sources (many of them area sources), and so little data for establishing emission factors or speciation profiles.

Emissions of PM, and more importantly its chemical constituents, size distribution, and key volatile and semivolatile precursors, are in need of improvement across many source categories. Carbonaceous particles (OC and EC) are a large contributor from many sources and there is limited information on a number of them.

Emission estimates are only based on a limited number of emission measurements, and their representativeness to real-world activity needs to be examined closely.

The process of developing information on emissions with the spatial and temporal resolutions needed for location-specific AQ modeling and intraurban-scale exposure estimation is problematic and a source of unquantified uncertainty in model results.

Differences in how emission inventories are developed in adjacent countries create difficulties for jointly managing the AQ.

Detailed recommendations on how to address these and other shortcomings are provided in: Improving Emission Inventories for Effective Air Quality Management Across North America: A NARSTO Assessment (NARSTO, 2005).

Reducing known uncertainties in an inventory will provide a more accurate starting point for AQ management strategy development, which should result in more cost-effective approaches. Typically, management actions are initially focused on large point and mobile source emissions. Large point sources are the easiest to characterize and frequently the easiest to control. Mobile sources may be more difficult to completely characterize, but there are few points of manufacture. Thus, control devices can be readily installed during the manufacturing process. As emissions from these sources decline, however, remaining emissions are more evenly distributed across source categories that are even more difficult to characterize, model, and control. These remaining sources will also grow as both population and economic activity increase, and errors in emission estimates from smaller individual sources will have greater consequences. These consequences could range from wrongly identifying a pollutant that should be controlled to overlooking source categories whose control could result in more cost-effective emission reductions.

14.4.4.3   Evaluating Uncertainty in Emission Estimates

Quantifying uncertainty is an essential "best practice" in inventory development and it is more efficient in obtaining the information needed to assess uncertainty at the time the emissions data are developed. Some key points to keep in mind:

Emission factors do not usually account for variability in emissions due to changes in source operating conditions, or across the individual sources that make up a source category.

New technologies can change processes and emissions, and such changes may not be reflected in emission factors that were based on the original process.

Emission factors that are based on idealized operations, such as the use of vehicle operating cycles, do not accurately capture the actual operation and therefore the actual emissions.

Measurement biases or errors introduce additional uncertainties into the reported inventory data. These differences can be associated with the location, time, or composition of emissions, leading to uncertainties in the spatial, temporal, or chemical data used in AQ models.

In general, emission uncertainties tend to have a smaller impact on understanding and AQ management decisions when models are used to estimate changes in AQ over longer periods of time and across geographical areas that are of the same order of magnitude as the spatial scale covered by the inventories. In other words, national annual average pollutant concentration estimates are likely to have a lower uncertainty than concentration estimates for a specific urban area over the course of a single day, because the differences between actual and estimated emissions based on emission factors are more likely to average out over a longer period of time and over a broader area.

However, as AQ models become more sophisticated in order to meet the demand of more specific AQ management questions, more detailed emissions information is required. Modeling atmospheric processes over the course of a single hour with more detailed chemical reaction mechanisms and in smaller areas cannot be done with confidence with emission estimates based on an annual average emission factor. This would have a high potential of leading to misidentification of the most important sources within a given area or erroneous estimates of the specific emissions that need to be controlled for a given source type, as well as incorrect AQ forecasts and population exposure estimates.

Uncertainties associated with non-point-source emissions are most problematic, and two general approaches are usually taken—bottom-up and top-down—in estimating them. In the bottom-up approach, uncertainties (bias and random error) in the individual measurements or parameters that make up the emission model are estimated (e.g., from field measurements similar to those described in the previous section) and represented as a probability distribution function (pdf). These uncertainties are then propagated, often using a Monte-Carlo approach, through the model in order to provide an estimate of the uncertainty in the emission estimate. An alternative bottom-up approach is the sensitivity analysis. A simple form of sensitivity analysis is to evaluate the sensitivity of emission-model output to its various input values in terms of the partial derivative of the model output to the input parameter in question. This approach indicates the relative sensitivity of the emission model to its various inputs and enables a crude estimate of uncertainty by providing a measure of how much the emission estimate would change for a given change in an input parameter.

In top-down evaluations of emission inventories or emission models, ambient measurements or other independent data are used to evaluate the accuracy of the emission estimates. The most effective applications of top-down evaluations are those that are combined with concurrent examination of the original bottom-up inventory data, so that the source of the errors can be identified rather than simply stating that the inventory is in error (Miller et al., 2006). One top-down method is to compare temporal trends in emission estimates with past trends or to compare trends in the ambient concentrations of a pollutant (or in ratios of pollutants) with the trend in estimated emissions under conditions in which the effects of transport, chemical transformation, and removal can be neglected or accounted for. Results from of this kind of analysis are reported in Parrish et al. (2002) and CRC (2004). A description of this approach is also provided in NARSTO (2005).

Other top-down methods for evaluating emission inventory uncertainties include using alternative methods for estimating emissions (such as comparing vehicular emission estimates based on vehicular distance traveled with those based on total fuel consumption), source apportionment techniques, and inverse modeling. Source apportionment (or receptor modeling) techniques use various multivariate statistical methods to infer source types, source location, and relative contribution from ambient measurements (Watson and Chow, 2005). These methods have been used to evaluate inventories of $PM_{2.5}$ and VOCs (Watson et al., 2001; McMurry et al., 2004).

Inverse modeling involves reformulating source-based AQ models so that emission source strengths are expressed in terms of the observed concentrations. In other words, the model is used to deduce the temporal and spatial emissions that explain the observed concentration field. Because of the limited spatial resolution of AQ models, this technique is most commonly used to deduce area sources (Petron et al., 2002; Gilliand et al., 2003; Park et al., 2003).

14.5 REVIEW OF BEST PRACTICE FOR USING MODELS FOR AQ MANAGEMENT

In applying models for AQ management, we must accept from the start that no model is perfect. How then can we account for the resulting uncertainty and apply models in a reasonable and defensible way in order to inform AQ management?

For a model to be useful, presumably it must be *credible*. That is, it must have demonstrated sufficient skill and reliability so that its predictions can be used with some confidence by analysts and policymakers in the formulation of AQ management strategies. Confidence can, in turn, be built in two ways: first, through *model verification* to assess the consistency, completeness, and correctness of the model and through *model performance evaluations* to characterize its performance and quantify its errors; and second, by applying the model in as appropriate, transparent, and defensible a manner as possible for the AQ issues being considered.

14.5.1 MODEL VERIFICATION AND MODEL EVALUATION

Model verification and model performance evaluations should always be the required steps before a model is applied in the policy arena. Model verification is an assessment of the accuracy, reality, or truth of a model (Fox, 1981; Russell and Dennis, 2000) and involves assessing the consistency, completeness, and correctness of a model's design, science, process representations, algorithms and numerical methods, inputs, and source code. This is best done through peer review and, as well, any interested party should have unrestricted access to the model source code.

Model performance evaluation is the process of examining and appraising model performance through comparison of model predictions with measured AQ data and/ or predictions from other models (Fox, 1981; Dennis et al., 1990; Russell and Dennis, 2000). At least six main types of model performance evaluation have been identified: (i) operational; (ii) diagnostic; (iii) mechanistic; (iv) comparative; (v) dynamic; and (vi) probabilistic (Seigneur and Moran, 2004).

An *operational* evaluation requires the statistical evaluation of model predictions of a few key pollutants of interest with atmospheric measurements over time and space scales consistent with the intended applications of the model. An operational evaluation is intended to answer the basic question: "Are we getting the right answers?"

A *diagnostic* evaluation is more of a research-level evaluation and involves an examination of model performance at the process level for all relevant species. A diagnostic evaluation addresses the basic question: "Are we getting the right answers for the right reasons?"

A *mechanistic* evaluation involves testing individual model components (i.e., process representations) in isolation against field or laboratory measurement. Such evaluations address the question: "Are we using good parameterizations?"

A *comparative* evaluation involves a side-by-side comparison with another model or model component for identical or similar inputs. A comparative evaluation addresses the basic question: "Are we getting comparable answers from comparable models?"

A *dynamic* evaluation examines model responses to changes in one of two key model inputs: meteorology and emissions. A dynamic evaluation addresses the basic question: "Are we getting the correct model response to a change in model input?"

A *probabilistic* evaluation involves an examination of model sensitivity to changes in model inputs and to model parameters, algorithms, and numericals. A probabilistic evaluation addresses the basic question: "What is our confidence in model-predicted values?"

It is also important to consider which aspects of model performance need to be evaluated. Most AQ model evaluations involve case studies in which a model is run for a particular period using input emissions and meteorology suitable for that period and then the model performance is examined using measurements from that same period. The most common AQ model application, however, is to evaluate the impact of emissions *changes* on AQ. The key aspect of model performance in this instance is how well the model predicts the atmospheric *response* to the change in input emissions, and the approach to the corresponding performance evaluation is necessarily somewhat different. For a direct evaluation of model response, AQ measurements are required for *two* different periods so that an atmospheric response can be calculated, which means that the AQ model must be run for the same two periods using different input emissions corresponding to each of the two periods. In this process, it is important to be aware of issues such as (i) the need to use emissions for two different periods estimated using a consistent methodology and (ii) the additional variability introduced by interannual meteorological variability. Published model-response evaluations are uncommon, but a few are available (Moran and Zheng, 2006). Note also that in terms of U.S. regulatory modeling terminology (U.S. EPA 2001, 2005b), a model-response evaluation is equivalent to the evaluation of model-predicted *relative reduction factors*.

Reid et al. (2003) suggested that "... the model predictions should be good enough that model uncertainty does not affect the decisions that are based on the predictions." In the real world, of course, this may not always be the case. How then should models be used given such uncertainties?

14.5.2 MODEL APPLICATIONS

In their review of photochemical models and modeling, Russell and Dennis (2000) discussed the *modeling process* as a separate topic. By this, they meant the set of steps required to apply a model, including the selection of the model domain, grid resolution, and model configuration, preparation of model input files, model execution, and post processing and analysis of model predictions. All of these steps may influence the results provided by the modeling system, and thus it is important to work through the modeling process in a reasonable and defensible way.

Figure 14.3 links the eight steps of best practice for AQ modeling based on guidance from two EPA reports (U.S. EPA, 2001, 2005b). Most of these steps, which are described in more detail below, are relevant to any AQ model application. Given the open-endedness of some of these steps and the reality of limited resources, it may not be possible to do as thorough a job as policymakers and modelers desire. The penalty for "cutting corners," however, could (but may not) be incorrect predictions. At a minimum, it will be a greater degree of uncertainty and lower confidence in the predictions.

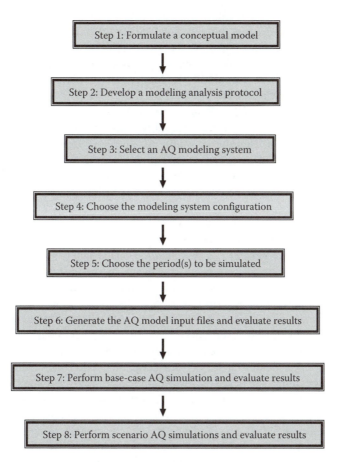

FIGURE 14.3 Eight-step set of best practices for AQ modeling applications for emission control strategies. [Reprinted from Craig L., et al. 2008. *Journal of Toxicology and Environmental Health Part A* 71: 588–698. With permission of the publisher (Taylor & Francis Ltd, http://www.tandf.co.uk/journals).]

Step 1. Formulate a conceptual model   Both modeling specialists and modeling "clients" should have a conceptual understanding of the AQ issue to be considered with an AQ model. A conceptual model will provide useful guidance for all of the remaining modeling-process steps, including the identification of stakeholders, the selection and configuration of the AQ model, the development of candidate emission control scenarios, and the assessment of model results.

Step 2. Develop a modeling/analysis protocol   It is desirable from the onset to identify all interested parties and to obtain agreement on (a) which questions should be addressed, (b) what assumptions are reasonable to make (e.g., What processes can be neglected? How large does the model domain need to be?), (c) how should the modeling work be performed, (d) what sorts of results should be generated, (e) who should review them and how, (f) what should the timetable be, and so on.

Step 3. Select an AQ modeling system   The selected AQ modeling system must be capable of answering the agreed-upon questions and be viewed as credible and skillful through peer review and performance evaluations. The time and resources available for the model application are practical considerations that may also affect the choice of the modeling system. McKeen et al. (2005) describe seven current PM AQ modeling systems. Other AQ modeling systems are described by EMEP (2005), Heald et al. (2005), Hodzic et al. (2006), Mallet and Sportisse (2005), Meng et al. (1997), and Zunckel et al. (2006).

Step 4. Choose the modeling-system configuration   The modeling domain and map projection, horizontal and vertical grid resolutions, level of nesting, if any, an integration time step, a "spin-up" or "ramp-up" time, a "refresh" rate for the meteorological model, methods to specify chemical initial and boundary conditions, and, where choices are available, the particular physics and chemistry process parameterizations to be used in the meteorological and AQ models (e.g., convective parameterization, gas-phase chemistry mechanism, secondary organic aerosol scheme) are all important decisions. Sections 12 and 13.2 of U.S. EPA (2005b) provide some useful discussions about some of these choices.

Step 5. Choose the period(s) to be simulated   This step is strongly constrained by the question(s) to be answered and, if relevant, by the form of the AQ standard (e.g., daily or annual, average or maximum) or the exact wording of the legislation of interest. In the case of short-term effects or AQ standards, the conceptual model should provide useful guidance, particularly related to the meteorological conditions that are associated with AQ exceedances. For the latter, they should ideally include a set of periods (a) for which extensive emissions, meteorological, and AQ datasets exist, (b) that correspond to a variety of relevant synoptic conditions, and (c) that provide enough samples to have statistical significance, and (d) in which each period is long enough to span a full synoptic cycle (~5–15 days) and includes a relevant exceedance. By considering full synoptic cycles, the model is forced to simulate the conditions before and after an exceedance as well, allowing the confirmation that the model can forecast non-exceedances as well as exceedances (Biswas et al., 2001).

In the case of long-term effects or AQ standards, correspondingly longer simulation periods will be required. Continuing advances in computer technology have meant that running AQ models for periods as long as a year or more has become feasible (Eder and Yu, 2006), but even then there are still likely to be representativeness issues due to interannual meteorological variability (Brook and Johnson, 2000). Choosing periods that satisfy the short-term selection criteria (a), (e), and (f) is desirable.

Step 6. Generate the AQ model input files and evaluate results   This step involves (a) preparing geophysical fields for the selected domain and grid, (b) running a prognostic meteorological model with some type of data assimilation, and (c) running an emissions processing system for the same simulation periods to prepare emissions input files for a base case and often a number of emission scenarios. For any regional (i.e., limited-area) AQ modeling system, it may also be necessary (d) to run both global meteorological models and AQ models or to analyze available chemical climatologies (Logan, 1999) in order to provide chemical boundary conditions. In preparing the input emission files, the size of the model domain will dictate how many emission inventories will need to be processed. Current emissions processing systems produce log files and summary tables that can be checked for warning and error messages and for continuity, consistency, and plausibility. Visualization tools can also be applied to check the spatial and temporal patterns contained in the processed emission files. The emission files for various emissions scenarios should probably receive even greater scrutiny since extensive manipulations were likely required to transform current inventories to account for various socioeconomic projections and control measures. The inclusion or exclusion and the treatment of natural emissions such as wildfires and windblown dust should also be checked.

It is also important to ensure that the inputs provided to the AQ model are as accurate and credible as possible. At a minimum, an operational evaluation should be performed against meteorological measurements: the suite of meteorological parameters considered should include temperature, humidity, wind speed, wind direction, cloud-related fields, precipitation, and, if possible, planetary-boundary-layer depth (Hogrefe et al., 2001, Smyth et al., 2006b).

Step 7. Perform base-case AQ simulation and evaluate results   The AQ model's performance for the base-case time periods needs to be evaluated so as to characterize and quantify the overall modeling system's performance (i.e., including the treatment of emissions and meteorology) and to determine whether that performance is acceptable. Given the known model limitations, errors, and uncertainties, Russell and Dennis (2000), Reid et al. (2003), Seigneur and Moran (2004), and the U.S. EPA (2001, 2005b) have all argued that this performance evaluation for the base case should not be restricted to just a basic operational evaluation against surface measurements of one or two pollutants, but instead should include a broad set of analyses that all feed into a "weight-of-evidence" judgment. This evidence, not all of which may be possible, can be provided by

A more comprehensive operational evaluation of more chemical species in
  both the horizontal and vertical dimensions
Sensitivity tests based on alternate configurations of the modeling system
Bounding tests in which emissions inputs are either increased or decreased to
  reflect the magnitude of uncertainties related to those inputs
Comparisons with results from other models that have been run for the same
  region and time period(s), with receptor-based model results (Marmur et al.,

2006), and with observation-based models or analyses for chemical regimes, including indicator species ratios and gas ratios (Sillman et al., 1998; Stein and Lamb, 2002; Martin et al., 2004)

Use of model probing techniques, including process analysis and the direct decoupled method (Zhang et al., 2005).

Step 8. Perform scenario AQ simulations and evaluate results   Diagnostic or comparative evaluations for examining the reasonableness of the AQ model's response to specified emission changes are generally similar to the range of tests suggested above in step 7. Running more than one model is very valuable in gaining confidence with the results from different emission scenarios. In particular, when there is model-to-model directional consistency among multiple, interrelated pollutants, then confidence is increased.

Application of an AQ modeling system always entails compromise. However, the eight-step set of modeling best practices described above should be viewed as a goal to be approached as closely as possible if AQ modelers are to provide their clients with the best possible guidance. Overall, the best AQ modeling practice is easiest to achieve and maintain when a qualified core of experts, with past experience for the jurisdictions of concern, is retained and well supported in terms of resources and national and international exchange. If such a group is active, then a conceptual model for many circumstances likely exists; some AQ modeling may have been performed, input datasets will routinely be available, and there is often past experience in studying emission scenarios. Thus, the necessary level of credibility is demonstrated on an ongoing basis and the information infrastructure will be constantly evolving and improving so that most new AQ management questions can be examined starting at Step 5. If the base scenario has been considered before, then Step 7 may not be required; hence the completion of Step 8 is effectively the minimum requirement for a new modeling study.

14.6   AQ MEASUREMENTS

This section discusses several issues related to how measurements best support AQ management. Protecting human health is an important motivation and thus particular attention is paid to that application. This section concludes with some general guidance on technical issues to be considered in establishing a robust measurement program.

Measurements play several roles in AQ management:

Effects research: Providing information needed to derive quantitative relationships between ambient concentrations and human health (or other adverse effects on the environment, climate, or visibility) and between ambient concentrations and source contributions.

Tracking progress: Demonstration of the efficacy of past and present policies (e.g., emission reductions) or detection of emerging AQ problems.

Model development: Prediction of ambient pollutant concentrations from knowledge of emissions and emission changes.

Public information: Provision of current concentrations or air quality index (AQI) values to allow citizens to reduce their exposure or to become aware of the problems in their area.

Risk assessment: Describing current risks and detecting potential future risks to human and environmental health.

14.6.1 ASSESSING HEALTH EFFECTS

Measurement data and, most often, routine monitoring data are central to AQ health studies and the CRFs derived from this research. Ultimately, measurements can lead to new insights into the specific pollutants or sources posing the greatest risk to health and the environment. A range of health research approaches and detailed AQ and exposure measurements are needed to continue to advance knowledge. Strategies and methods for providing information on ambient AQ that is adequate for health impact assessment were reviewed by the WHO (1999).

For AQ risk management, CRFs based on direct links between ambient observations of a range of air pollutants, as measured by standard monitoring networks, and acute and chronic impacts occurring within the general population are crucial. Demonstrating "real-world" associations also establishes the fact that air pollutant effects are relevant to actual conditions. This necessitates the fact that ambient measurements continue to be obtained in order to support both acute and chronic exposure health studies. One of the key challenges in working with ambient pollutant data is to derive CRFs for individual pollutants and for pollutant mixtures that are appropriate for risk management or cost–benefit analysis. They must be scientifically defensible, and their uncertainties and/or their strengths and weaknesses need to be understood in detail. As with AQ models, expert solicitation represents a valuable approach in dealing with these uncertainties when CRFs need to be applied for AQ management (Cooke et al., 2007). Significant challenges arise from confounding and differential exposure error among pollutants. Among other things, there may be an association between a health endpoint and a given pollutant because that pollutant acts as an indicator of an unmeasured pollutant (gas or particle) or a mix of pollutants (Brook et al., 2007a; Goldberg, 2007). In this situation, applications of the CRF (e.g., cost–benefit analysis) must be done with clearly stated caveats.

The level of detail that the air pollutant measurements should provide, including direct personal exposure studies or other extrapolations related to exposure (e.g., intake fraction and population-weighted concentrations), will depend on the type of health study they are intended to support. There is also considerable demand for better information on the specific pollutants, mixtures, and/or sources that have the greatest impact on health. This demand can only be satisfied by enlisting a variety of health study designs, including toxicology, in different locations, enhancing ambient measurement and exposure research activities, capitalizing on recent advances in measurement capabilities (Wexler and Johnston, 2008) and detailed analysis and interpretation of the measurements, such as the use of receptor models (Thurston et al., 2005; Sarnat et al., 2008).

14.6.2 Obtaining Representative Ambient Measurements

No matter how detailed they may be, ambient measurements cannot reflect what a person or most members of a population are truly exposed to. These measurements are an indicator for some aspects of the air pollutant stresses that the population is confronted with (Brook et al., 2007a). In reality, the relationship between these indicators and the actual exposures of the population are likely to differ from pollutant to pollutant and from one monitoring site to the next. Site locations relative to the population and selection of the pollutants to measure at each site are therefore critical issues.

Acute exposure–effect studies and chronic exposure–effect studies have different requirements regarding site location. The former requires that the measurements accurately reflect temporal variations in the population's exposure whereas the latter is interested in how exposure levels vary across space. This could range from differences between one location in a large city and another (intraurban) or from differences between cities (interurban). For both types of studies, the ideal measurement data are rarely, if ever, available for both financial and technical reasons. Therefore, compromises are necessary and it is important to understand the limitations of and implication in using the available measurements for health studies or risk assessment.

The best practice for both acute and chronic studies is to measure multiple pollutants at any site that is established and to operate more than one site in the region containing the population of interest. European criteria for site coverage and representativeness are discussed in Kuhlbusch et al. (2004). In terms of site locations, some of the key criteria to consider in network evaluation are as follows:

Sites are established which provide data on the highest concentrations

The network comprises both hot spot and urban background sites

Hot spot sites are representative for at least 200 m^2

Urban background sites are representative for several square kilometers

Urban background sites are representative for similar locations not in their vicinity

Sites are established which are representative for the exposure of the general population

The spatial distribution of the population, local physical features (e.g., topography and shorelines), and prevailing meteorological conditions are all important to study when selecting site locations or choosing sites to be used for health studies. With this information, a better understanding of the link between each set of measurements and the population will be realized and population-weighted concentrations can be derived with more confidence. This perspective also helps identify and address weaknesses in the monitoring network via new sites or special studies.

For the most part, comparisons of monitoring site data within cities show that day-to-day variations in outdoor pollutant concentrations are correlated across a city. This behavior is advantageous for acute effect studies. For example, average intersite correlations among all available pairs of sites in Toronto, Canada, for SO_2,

NO_2, O_3, $PM_{2.5}$, and PM_{10} are 0.78, 0.77, 0.90, 0.96, and 0.84, respectively. For typical monitoring networks, the pair of sites with the minimum correlation provides a good indication of the lower limit of the representativeness of using a single site to estimate day-to-day changes in population exposure. The representativeness of a given site's measurements to conditions in another area decreases with distance. Consequently, combining the time series from a group of sites spanning a city improves the representativeness of a time series for population health studies. For NO_2, the average and minimum correlation between the downtown Toronto site and the sites within the city are 0.83 and 0.75. These values increase to 0.89 and 0.86 when each of the sites is compared with the average time series generated by a combination of the sites. Over a broader area that considers suburban sites, the minimum correlation increases from 0.58 to 0.66 when the day-to-day time series is represented by the combined sites as opposed to just the central downtown site. Often, $PM_{2.5}$ varies most consistently across a region, as indicated in the correlations for Toronto. However, the degree of correlation for $PM_{2.5}$ (i.e., total mass) is not likely to be the case for the different PM chemical constituents. OC, for example, has many urban sources, ranging from traffic to cooking, and can be highly variable across space, exhibiting different temporal patterns.

Assessing intersite correlations is an important step before measurements are used in health effect studies or before sites are added or removed from a network. Examination of the correlations provides insight regarding representativeness of a time series and can help explain differences among pollutants in observed health effects from time-series studies or in measured personal exposures. Logically, this source of exposure error should attenuate the significance of true health associations, but in multipollutant analyses there is the potential to "transfer" the association to the pollutant that can be monitored at fixed sites with greater spatial representativeness and/or with a stronger link to actual personal exposures (Goldberg, 2007).

Within limits, monitoring site locations can be optimized to best represent the potential exposure of the population. When combining information from sites, the data from each site can also be weighted according to the size of the population within a certain radius. Measurements at sites with a greater surrounding population are thus counted more heavily. A potentially stronger link to exposure can be derived by considering the intake fraction (Marshall et al., 2003). This approach is more appropriately used for weighting the pollutant exposure risks posed by emissions from a variety of sources. Thus, assigning an intake fraction weighting to an ambient measurement also requires knowledge of the sources contributing to the observations and how the measurement, the source location, and the population of interest are related, including estimates of pollutant concentrations in a range of microenvironments. Probabilistic exposure models, such as the REHEX model (Winer et al., 1989; Fruin et al., 2001) and SHEDS (Burke et al., 2001), also provide an approach for estimating, from ambient concentrations, the range of exposures expected among the population. Among other things, these models require more detailed spatial information on concentrations, either by interpolating the observations or by using the AQ models.

Short-term field studies can help identify the optimum site placement and the nature of the relationship with personal exposures. There have been several exposure studies examining how well outdoor measurements correspond to actual personal exposures. Brauer and Brook (1997) showed that day-to-day variations in outdoor central site O_3 measurements agreed reasonably well with the temporal variability in average personal exposures, but not necessarily for all people. In addition, the amount of agreement has been found to vary from city to city and among pollutants (Sarnat et al., 2001; Kim et al., 2006).

$PM_{2.5}$ has been the primary focus of most recent exposure studies and the results generally show that each individual's day-to-day personal exposure time series is correlated with the variations in the outdoor levels. However, there is wide variability from person to person depending on their time activity, the predominant indoor microenvironments in which they spend their time, and the indoor sources of $PM_{2.5}$ in these locations. Attempts have also been made to separate personal $PM_{2.5}$ exposures into particles of indoor and outdoor origins (Ebelt et al., 2005).

Intraurban variations in exposure are now recognized as an important signal to exploit in chronic-exposure studies. However, over the long term it is not possible to operate enough monitors. Short-term saturation monitoring studies, where many sites are operated simultaneously to characterize spatial patterns, can help fill the gap (Henderson et al., 2007; Wheeler et al., 2008). Mobile measurement platforms (Bukowiecki et al., 2002; Kolb et al., 2004; Polina et al., 2004; Westerdahl et al., 2005; Yli-Tuomi et al., 2005; Xu et al., 2006) also assist in studying spatial patterns and in optimizing site placement. Mobile platforms are becoming more common as a facility for monitoring agencies, and current advances in technology are allowing even more sophisticated measurements to be obtained. However, measurements alone cannot provide all the spatial details desired. Thus, alternate sources of information and/or a range of spatial models, statistical or physical, are becoming increasingly important to develop. These will be discussed further below.

14.6.3 TRACKING PROGRESS

One of the primary objectives of monitoring networks is to track progress toward achieving standards and to ensure that good AQ is maintained. Most mature networks have multiple sites that are in standardized locations; several, if not all, of the criteria pollutants are being monitored and long data records have been and are continuing to be collected. Thus, they are well suited to satisfy this primary objective. Beyond selecting the right locations and pollutants to measure, best practice dictates that for studying trends, a long, unbroken record is important to maintain. This is because AQ changes can be small and gradual and are obscured by meteorological variability. Lack of continuity in measurements, lack of sensitive enough measurements, and under-representation in some geographic areas hinders trend analysis.

Monitoring networks tend to focus on urban areas, leading to a deficiency of rural measurements. For example, worldwide, there are few long-term trends on rural NO_x levels, despite its importance to understanding regional O_3 and $PM_{2.5}$ and the impact nitrogen deposition can have on ecosystems. Rural sites will continue to grow in importance for assessing how the growth in the size and density urban areas (e.g., sprawl) is impacting the proximate regional AQ. Ideally, urban–rural pairs of measurement sites should be operated since they are essential for untangling the contributions and trends attributed to local/urban sources versus upwind sources from regional-scale transport (Brook et al., 2007b).

Demonstrating progress in direct response to implemented policies is referred to as *accountability*. Ultimately, accountability should extend beyond just documenting AQ improvements to demonstrating that the desired benefits have been realized. This would include, for example, improved public health and recovering ecosystems. Tracing back along the full accountability chain (HEI, 2003) is challenging and the implications it has for how AQ measurements are undertaken need to be considered and addressed.

If the accountability goal is to detect the expected AQ improvements resulting from a new policy or a specific intervention with relatively large and abrupt changes in emissions, then time-series length may not be as important as for detecting gradual trends. In this case, however, high-quality baseline measurements in advance of the emission reduction(s) are critical as is the continuation of these measurements after the reductions have been implemented.

AQ models can play a role in informing monitoring activities for accountability purposes. They can be used to simulate the magnitude of the change at the sites being used for tracking progress. This can help determine the measurement sensitivity required, given the anticipated changes, and/or provide insight regarding the need for additional measurements at existing sites or at new locations. Such extra measures, guided by models would help to more effectively and more rapidly demonstrate accountability.

14.6.4   Use of Measurements in Model Development

Monitoring network data play a valuable role in model evaluation and this application is becoming more important as many models are being run continuously for AQ forecasting. These new long-term modeled datasets are offering new opportunities to learn a great deal about how the models perform and the quality of the emissions information. Although the network data tend to lend themselves more to operational evaluations of the model, there are opportunities for diagnostic evaluations as well (see below). If feasible, deployment of continuous or semicontinuous particle composition instruments for monitoring should be considered as these data to provide greater opportunities for model evaluations. Network data are also critical to define the model's initial conditions. Advances in rapid data assimilation (Ménard and Robichaud, 2005) have been occurring for this purpose. There are many potential applications of these assimilated datasets, as will be discussed below for the case of surface O_3 concentrations.

Very detailed measurements, which can only be sustained for relatively short field studies, are ultimately needed to study atmospheric processes (dynamical, chemical, and physical) and for more-detailed diagnostic model evaluations. The scope of these studies can vary greatly from a small team collecting measurements to study one process (Padro et al., 1993; Makar et al., 1998) related to one model module to large collaborations across institutions (EMEFS ICARTT; Dennis et al., 1993; Frost et al., 2006). Due to continued technological improvements, the ease with which highly comprehensive measurements can be undertaken during field studies is much greater than what it was 5–10 years ago. This is beneficial for obtaining, quickly, much more data at more locations from which one can study, in detail, source apportionment and atmospheric processes. However, the risk has never been greater for valuable data to be underinterpreted. Best practice obviously dictates that this be avoided. Thus, with these new capabilities comes an increased need for highly qualified and creative experts in data analysis along with careful advance thinking regarding the underlying hypotheses motivating any venture into the field.

14.6.5 PUBLIC INFORMATION

Informing the public is critical to the process of AQ management because when a large enough majority decides that an issue is important, elected officials are more likely to respond. This can create opportunity for progress on the state of air quality. Providing routine, reliable, and understandable AQ information to susceptible members of the population also allows them to reduce their own exposure, thereby managing their own risk. In terms of public information, a common communication tool is an AQI, which is derived from measurements in real time or from AQ predictions to provide a forecast. In cities or regions where there are frequent "bad air days," the public is generally more aware of AQ issues, at least partly due to the publicity of the AQI. The form of the AQI is similar in many countries, reporting AQ using descriptive terms such as good, moderate, poor, very poor, unhealthy, and so on (http://www.msc-smc.ec.gc.ca/aq_smog/aqcurrent_e.cfm; http://airnow.gov/index.cfm?action=static.aqi; http://www.lml.rivm.nl/data/smog/index.html; and Mexico: http://www.ine.gob.mx/). This approach is easier for the public to understand and act upon as opposed to reporting actual pollutant concentrations. Often the AQI is designed to identify the worst effects that may result from the mixture of pollutants currently being measured and to describe the prevailing AQ. The increase in knowledge regarding air pollutant health effects has led to the Canadian Air Quality Health Index (AQHI) (Stieb et al., 2005). The unique feature of the AQHI is that it is based on recent epidemiological results from across Canada. In addition, it considers multiple air pollutants simultaneously and they all contribute to the index value in every case. Indices are also being used to rank cities or countries based on their AQ in an attempt to provide more-consistent comparisons to inform local, national, and global environmental managers regarding the key areas of concern. Gurjar et al. (2008) presented an approach to combine AQ and emissions information and compared several megacities using their proposed measure.

14.7 TECHNICAL ISSUES IN ESTABLISHING A MEASUREMENT PROGRAM

Air pollutant measurement requires a strong commitment to a recognized standard of quality and a plan for data archival and for interpretation. Furthermore, knowledge and technology are continually improving. Thus, to the extent possible, new measurements should seek to use the most current, accepted methods, increasing the probability that the data are acceptable far into the future. In the long run, a small amount of high-quality measurements will be of more value than many measurements collected with insufficient documentation, quality assurance (QA), and interpretation.

AQ measurement activities generally fit into one of two categories:

1. *Monitoring:* A core set of systematic measurements at well-selected locations that are maintained indefinitely for trend analyses (i.e., to evaluate the effectiveness of current policies), to determine if an area is complying with or achieving an official AQ standard or guideline, and to identify emerging problems as soon as possible, which may involve ongoing environmental health studies (e.g., epidemiological studies).
2. *Field studies:* A relatively short period (<2 years) of more-detailed or more-specific measurements collected within a well-defined geographical area or at a given location or for a given population. These data are essential for the development of conceptual models, source-oriented models, more-refined source apportionment studies, and for understanding the relationship among emissions, ambient concentrations, and personal and/or population exposure. A variety of prospective health studies may also derive their exposure information from AQ field studies.

From monitoring data, a standard set of reports documenting current conditions, trends, and comparison with other geographical areas need to be produced and promoted on an ongoing basis. This helps to keep AQ management issues in the forefront. Software tools that can process the AQ data along with meteorological data are an important component of the monitoring program, enabling AQ scientists and managers to examine some of the causes of pollution events in near real time.

Field study data are usually more complex and less standardized. Sufficient time and resources need to be dedicated to working with the data after the study. At a minimum, 1–2 years is likely necessary. To guide the planning of the study and subsequent data analysis, there needs to be a set of testable hypotheses in place before the study begins.

Addressing these technical issues, which will be expanded upon below, before measurements start helps ensure that the data are of the greatest value. In establishing a measurement program, it may also be relevant to consider the potential applicability of the data for evaluating models and also for integrating the data with the model output and other information in order to improve the detail and coverage of ambient concentration information (see below). Linkages to personal exposures and the ability to quantify the degree of exposure error associated with using the measurements for health research may also be necessary to consider.

14.7.1   What to Measure and How Often

For a wide range of both gaseous and particulate pollutants, Chapter 5 of the PM assessment published by NARSTO (McMurry et al., 2004) provides considerable detail on what can be measured, the methods available, how reliable they are, and the reasons as to why such measurements might be needed (e.g., for health effects studies, compliance monitoring, visibility, scientific understanding, etc.).

When feasible, a greater frequency of measurement (i.e., a finer time resolution) is preferred because this permits a much better understanding of source contributions and atmospheric processes (Wexler and Johnston, 2008). If a standard exists, then its "form" or "metric" (e.g., hourly maximum, 8 h maximum, 24 h, annual) will dictate that a certain resolution be achieved. Inclusion in a real-time data reporting program, such as what may be needed for AQI and AQ advisory purposes or for AQ forecasting, will also likely demand that data be available on a frequent basis (e.g., hourly).

Although AQ standards or other types of regulations/guidelines require that several common pollutants are monitored indefinitely, several other pollutants or trace gases are important to measure in support of AQ management. This wide range of trace atmospheric chemicals can be classified in a variety of ways. In Table 14.1, four classifications are considered, however, with any such attempt, the distinctions are blurred.

14.7.2 SITING CRITERIA AND WHERE TO MEASURE

On a broad geographic scale, sites can be classified as either being remote, regional, urban background, urban exposure hot spot, or industrial. The spatial scale or "footprint" that each type of site can represent clearly varies from being nearly continental for remote sites to less than neighborhood scale for industrial sites. Kuhlbusch et al. (2004) show a breakdown, by country, of what type of station—traffic, industrial, or background—monitoring sites contributing to the European AirBase dataset fall into. The U.S. EPA describes four categories of networks and sites and lays out the general purposes of the overall program and of each network on their website: http://www.epa.gov/air/oaqps/qa/monprog.html. Similarly, a range of site categories, their purposes and criteria for inclusion in EuroAirNet are described at http://air-climate.eionet.europa.eu/databases/EuroAirnet/euroairnet_criteria.html. Standardization is important to consider for monitoring networks and this becomes more challenging when they span multiple countries. The height of measurement is also an important siting criterion. Measurements made near the surface can be affected by local dry deposition, local surface emissions, or resuspension of dust. At regional and remote locations, 10 m is often the standard height as long as there are no nearby (~50 m) obstructions. In urban areas, the concept of breathing zone enters into consideration, but there is a much greater risk that a very local emission will influence measurements made close to the surface. Alternatively, in cities, higher measurement heights, such as rooftop locations, tend to provide a better indication of the urban background and its temporal variation (Brook et al., 1999).

14.7.3 QUALITY ASSURANCE

Documentation of QA measures and expectations that are or will be followed throughout a measurement program helps to ensure the data's value and to ensure that they are not misinterpreted by other users. Providers of the funds for measurements should expect that evidence of a QA plan is available before a program proceeds. Measurement methods and types and model numbers of the instrumentation

TABLE 14.1
Air Pollutant Classes, Measurement Capabilities, and Other Issues to Consider

Pollutant Class	Examples	Comments
Combustion emissions	Nitrogen oxides (NO_x or NO_2) Carbon monoxide (CO) Sulfur dioxide (SO_2) Carbon dioxide (CO_2) Fine particles ($PM_{2.5}$) Black carbon (BC) Ultrafine particles (UFP)	NO_x, CO, SO_2, and some form of particles (e.g., total suspended particulate-TSP, PM less than 10 μm in diameter-PM_{10}) are generally referred to as criteria pollutants. Particles are monitored according to the total mass below a specific size, but in the case of ultrafines, measurements are based on the total number. Combustion particles are also composed of many different chemical compounds. All of the pollutants mentioned above can be measured with instruments that provide hourly or better time resolution and that are capable of reliable, real-time data storage and transmission. With the exception of the particle-related measurements, regular, automated QA or calibration can be included in monitoring routines. Calibration of particle instruments is generally not possible because absolute standards are not available. Other approaches to ensure data quality are thus necessary. Information on $PM_{2.5}$ mass measurement and $PM_{2.5}$ sampling and chemical analysis is provided by Chow (1995)
Noncombustion surface or fugitive releases	Ammonia (NH_3) Methane (CH_4) Pesticides (persistent organic pollutants, POPs) Resuspended dust ($PM_{10-2.5}$ or $>PM_{10}$) Total reduced sulfur (TRS) or H_2S	Related to human activities and natural emissions. Agricultural emissions are important and releases from a range of industrial processes and waste management are also important sources. Road dust, windblown dust from agricultural practices and from certain types of industrial facilities (e.g., tailings associated with mining and smelting) can contribute to particle concentrations. Particles from these sources are typically in the coarse ($PM_{10-2.5}$ or PM_{coarse}) particle and giant particle (i.e., $>PM_{10}$) size ranges. Very few of these pollutants are measured routinely due to the lack of reliable, cost-effective techniques, because there are no AQ standards that necessitate monitoring and/or because they are only a problem in localized areas. For example, TRS or H_2S can be serious issues in the vicinity of pulp mills, sour gas flaring, and certain agricultural practices. In addition to country-specific AQ standards, international agreements often necessitate some level of monitoring. For example, POPs are routinely measured at several locations in support of the Stockholm Convention (http://www.pops.int/)

continued

TABLE 14.1 (continued)

Air Pollutant Classes, Measurement Capabilities, and Other Issues to Consider

Pollutant Class	Examples	Comments
Secondary pollutants	O_3 and other oxidants $PM_{2.5}$ Some VOCs and semivolatile organic compounds (SVOCs)	Form in the atmosphere. Sulfate, nitrate, ammonium, and some organic species found on $PM_{2.5}$ (fine particles) are secondary (i.e., formed via reactions that occur in the atmosphere). Besides O_3, other secondary oxidants, acids, and VOCs are important due to their potential to cause health or environmental effects and/or their role in atmospheric chemistry. Examples are hydrogen peroxide, the hydroxyl radical, peroxyacetyl nitrate (PAN), nitric acid, nitrous acid, hydrochloric acid, formic acid, acetic acid, formaldehyde, acetaldehyde, 1–3 butadiene, and acrolein. Although O_3, $PM_{2.5}$, and some specific secondary constituents of $PM_{2.5}$ (e.g., BC, total OC, sulfate, and nitrate) can be measured with relative ease, it is difficult for most of the other secondary compounds because the methods are expensive, experimental, and challenging. Their measurement can be very important for understanding atmospheric chemistry. More routine approaches for the measurement of PAN and formaldehyde requires highly trained operators and/or very capable analytical laboratories
VOCs and SVOCs	100s of individual compounds Examples: Benzene Toluene Ethylbenzene Xylene 1,3,Butadiene Isoprene Polycyclic aromatic hydrocarbons (PAHs)	Associated with both combustion and fugitive emissions as well as secondary formation. SVOCs are also found in both gas and particle phases depending on their properties and ambient conditions. Total gas-phase nonmethane hydrocarbons are often measured with relatively simple instrumentation, but provide limited information. Measurement of individual compounds is necessary to provide insight into their contributions to O_3 and $PM_{2.5}$ formation. In addition, they are useful for source apportionment and for characterizing "hot spots" of high exposure to toxics. Air samples can be collected in the field and analyzed with a good degree of accuracy and precision for the lower-molecular-weight compounds (i.e., fewer than about 10 carbon atoms). Larger molecules and the more-oxygenated species are more difficult to measure with confidence. VOC and SVOC measurement is labor-intensive and expensive. Although there are commercially available systems capable of time-resolved, real-time measurement of some compounds, these require highly trained operators and/or very capable analytical laboratories, and careful consideration of the uses of the data and their subsequent storage is necessary

used should be recorded, as well as any changes during the program. Details of the measurement site, such as latitude, longitude, elevation (above MSL), inlet height and design, and proximate emissions and obstructions (photos), are also necessary to document. This is referred to as site "meta data." Precision and accuracy targets for the measurements need to be quantified and the actual values being achieved should be determined routinely to ensure that a failing piece of equipment is identified and replaced. Duplicate or repeated measurements of the same samples and analysis of standards with known, traceable concentrations are, therefore, critical to undertake routinely to track precision and accuracy, respectively. QA measurements should be based on concentration levels that are typical for the site and that are within the measurement range of interest to the program. QA samples should also be introduced into the instrument or the analytical procedure (i.e., for laboratory analysis) in a manner that mimics the real measurement process and the conditions during measurement as much as possible.

Measurements that involve sample collection in the field followed by chemical or gravimetric analysis in the laboratory require QA in both the field and the lab. With respect to the field, one of the most important QA measures is the collection of field blanks. Whatever be the approach to capturing the sample (e.g., filters, denuders, canisters, passive sorbants, traps, or cartridges), at least 10% of the samples analyzed in the lab should be field blanks. These procedures should be clearly stated in the QA plan and the resulting data need to be rigorously analyzed to ensure that data quality objectives are upheld and/or adjusted.

14.7.4 Data Archiving and Reporting

Long-term storage of the final measurement data, including the QA data and the site metadata, is of utmost importance. Whenever possible, data should be archived at their native temporal resolution since averaging to longer time periods will likely lead to unrecoverable loss of information. Faster turnaround time can be expected to increase the value of the data assuming that the more current the information, the greater the number of interested users and the greater the impact of publicizing what is being observed. There are a number of recognized national archives or portals for data access, each with their own criteria. Examples of national archives of standard monitoring data are the National Air Pollutant Surveillance Network (NAPS) maintained by Environment Canada's Environmental Technology Centre (http://www.etcentre.org/NAPS/index_e.html), the AQS, which is the U.S. EPA's repository of ambient AQ data (http://www.epa.gov/ttn/airs/airsaqs/), Instituto Nacional de Ecología's archive of Mexico's AQ data (http://www.ine.gob.mx/), and the Air Quality Archive (AQA) for data for the United Kingdom (http://www.airquality.co.uk/archive/data_and_statistics_home.php). Data from many European countries are also available from AirBase, which is under the European Topic Centre on Air and Climate Change (http://etc-acc.eionet.europa.eu/databases/airbase/airbasexml/index_html). Joint North American data from specialized field studies are archived by NARSTO (http://eosweb.larc.nasa.gov/PRODOCS/narsto/table_narsto.html).

14.8   NEW DIRECTIONS IN THE APPLICATION OF MEASUREMENTS AND MODELS

More resolved spatial and temporal information on pollutant concentrations is needed to continue to advance knowledge on human health and environmental effects and to assess the benefits of more-specific AQ management options. This detail cannot be provided by direct measurement alone, and thus alternate sources of information utilizing a range of spatial/temporal models, statistical or physical, are becoming necessary. New methods are rapidly being developed that combine (or "fuse" or "assimilate") model outputs with measured concentrations, including satellite-based measurements (Liu et al., 2004; Martin, 2008), to help provide more detail (Sariglannis et al., 2004). These take advantage of improvements in emission inventories, AQ modeling systems, and measurement technologies that have resulted from the past decade or more of scientific research.

Data assimilation routines using real-time data and model output are now being applied on a continuous basis to characterize large-scale patterns across North America. The amount and quality of information available varies from pollutant to pollutant and geographically. At present, in North America, O_3 is the most advanced, while routines for $PM_{2.5}$ are being developed. Figure 14.4 presents O_3 concentrations in the Great Lakes region of eastern North America. Concentrations are presented in the form of the metric that the current Canadian and U.S. standards are based upon. The numerical value of the standard in Canada is 65 ppb and in the United States it is 75 ppb. The left panel in the figure is a contoured map from the concentrations observed at all the available measurement sites, whereas the right panel is based on a combination of these data and output from a comprehensive AQ model.

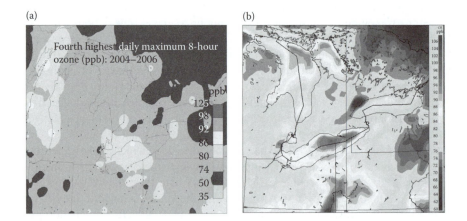

FIGURE 14.4   Comparison of surface O_3 concentration patterns based on the interpolation of available measurement data (a) and on an optimal interpolation scheme that uses AQ model predictions to assist in describing the most likely O_3 pattern (b). Concentrations are expressed as the three-year average of each year's fourth-highest daily maximum 8-hour average concentration (2004–2006). The area shown represents approximately 800 × 800 km.

Overall, it is hypothesized that the concentrations in many of the unmeasured areas, by nature of how they were determined, are more realistic in the right panel than the estimates produced by data interpolation alone. The O_3 pattern produced by combining observations and model fields contains considerably more structure. This reflects the locations of the precursor emissions, terrain features (i.e., that impact emissions and deposition), atmospheric chemistry, and the predominant meteorological conditions. At most measurement points (black dots in the left panel), the values in the two panels agree, as is imposed by the method. There are discrepancies, though, particularly in the southern parts of the region (east Ohio and west Pennsylvania), which point to aspects of the methodology needing improvement. In particular, the model's ability to resolve near-source (subgrid) atmospheric processes is limited and there are large point sources in the southern part of the region. However, the main cause of the discrepancy is that the version of the model-assisted interpolation demonstrated here is predominantly for data assimilation to support AQ forecasting and thus it only utilizes O_3 data available in real time. This leads to the inclusion of fewer sites compared to the number available to produce the interpolated map shown in the left panel. One of the most dramatic features produced by the model is the higher concentrations over Lakes Erie and Huron. This is hypothesized to be due to low O_3 deposition velocities over water and enhanced formation within more-concentrated pockets or layers of O_3 precursors that persist as a result of greater atmospheric stability over water.

Much of the population exposure to ambient air pollution occurs at neighborhood scales. This is beyond the resolution of both the current AQ models and the measurement networks. Furthermore, it is unlikely that any of these will be able to resolve such scales in the future. At best, new parameterization schemes or independent emissions models will be developed to treat subgrid-scale features to enable AQ models to reliably predict at the 1–5 km resolution. However, determination of how best to use such models to predict population exposure changes and their uncertainties within these grid sizes so that the costs and benefits of local scale AQ risk management strategies can be evaluated is an ongoing topic of research.

To resolve smaller-scale concentration patterns (urban to neighborhood scale) for health studies, a variety of approaches are currently being applied. These range from interpolation of monitoring site data (Künzli et al., 2005; Jerrett et al., 2006) to exposure surrogates such as distance-to-roadway and traffic counts (Hoek et al., 2002) to small-scale dispersion models and/or combinations of both (Cyrys et al., 2005; Wu et al., 2005). The local-scale exposures of interest have generally been associated with traffic since data on road networks are readily available. A wider range of emission sources have been included in some ambient air pollutant exposure modeling efforts (Gram et al., 2003). However, while all of these efforts are offering better spatial resolution, most are not strongly linked to direct concentration measurements. Geographic Information Systems (GIS) are useful for mapping exposure patterns, integrating different sources of information, and developing land-use regression (LUR) models (Briggs, 2005). Intraurban chronic exposure estimates have been derived using LUR for several cities (Brauer et al., 2003; Gilbert et al., 2005; Kanaroglou et al., 2005; Luginaah et al., 2006; Sahsuvaroglu et al., 2006; Setton et al., 2006; Silva et al., 2006; Henderson et al., 2007; Rosenlund et al., 2007). Cyrys et al. (2005) compared both LUR and dispersion model estimates for NO_2 and $PM_{2.5}$

and reported that for their cohort of interest in Munich, Germany, the two approaches led to similar exposure classifications. These results and most other LUR efforts have also focused on traffic-related pollutants (e.g., NO_2 and sometimes $PM_{2.5}$). However, recent studies in Windsor, Ontario, have expanded the dependent variables in LUR to include SO_2, benzene, and toluene (Wheeler et al., 2008).

The small-scale spatial variability (i.e., neighborhood scale or better), produced by applying the LUR for all points in a GIS database, appears more realistic compared to the pattern obtained using interpolation, with respect to the known distribution of traffic (Jerrett et al., 2007). Future research in this area is aimed at techniques to add temporal information to the LUR maps and to consider more pollutants. Ultimately, such approaches at resolving very fine urban-scale patterns should be able to combine or "fuse" with the more regional-scale concentration patterns, as displayed in Figure 14.4. Continued improvements in emission inventories, AQ modeling systems, and *in situ* and remote measurement methodologies and applications can be expected to advance this area of scientific research and to assist in more effective AQ management at all scales ranging from local to continental and even global.

14.9   SUMMARY

In this chapter, some of the key scientific elements associated with AQ management have been discussed. The current evidence indicating that air pollutants, especially PM, have an effect on humans was briefly reviewed and the risk management framework utilizing this information and AQ models was described. This framework requires quantitative links between emissions and air concentrations and between these concentrations and specific health endpoints. AQ measurements play a central role in the process. They allow us to identify health effects and then quantify the relationship between changes in concentrations and resulting changes in health responses, and they are critical in the development of AQ models. Measurements are also essential for identifying problems and tracking progress, as well as for educating and warning the public.

AQ models quantify the links between emissions of primary pollutants or precursors of secondary pollutants and ambient pollutant concentrations and other physiologically, environmentally, and optically important properties. They are the only tool available for detailed predictions of *future* air concentration and deposition patterns based on possible future emission levels and climatic conditions.

Accurate emission inventories provide essential information to understand the effects of air pollutants on human health and ecosystem health, to identify which sources need to be controlled in order to protect health and the environment, and to determine whether or not actions taken to reduce emissions have been effective.

AQ measurements are essential for public health protection and are the basis for determining the current level of population health risk and for prioritizing the need for reductions. They are also critical for evaluating the effectiveness of AQ management strategies and for altering such strategies if the desired outcomes are not being achieved (i.e., by adaptive management).

AQ problems typically become more difficult to address as the more obvious and less costly emission control strategies are implemented. This increases the demand on scientific and technological tools with a clearer picture of uncertainty and its implications. Multipollutant emission inventories, ambient measurements, and AQ models represent three of the most essential tools for AQ management. Over the past 20 years, tremendous advances have been and continue to be made in each of these areas, as well as in the analysis, interpretation, and integration of the information they provide. Despite these advances, including improved understanding of the impacts of poor AQ, the pressure to identify the most cost-effective policies that provide the maximum benefit to public health pushes our current tools and knowledge to their limits. For example, the highly specific control options that target specific chemical compounds found on fine particles, specific sources, or source sectors, or that lead to subtle changes in the overall mix of chemicals in the air (gases and particles), remain extremely difficult to evaluate in terms of which benefits public health the most. Lack of a complete understanding of exposure and health impacts of the individual components in the mix and their additive or synergistic effects further increases this challenge. As scientists gain an improved understanding of the environmental effects and the implications of climate change on AQ, it will be increasingly important to embrace a broader perspective in the development of future strategies to improve AQ, including comanagement of air pollutants and greenhouse gases.

REFERENCES

Abbey D.E., Nishino N., Mcdonnell W.F., et al. 1999. Long-term inhalable particles and other air pollutants related to mortality in nonsmokers. *American Journal of Respiratory and Critical Care Medicine* 159: 373–382.

Auchincloss A.H., Diez Roux A.V., Dvonch J.T., et al. 2008. Associations between recent exposure to ambient fine particulate matter and blood pressure in the multi-ethnic study of atherosclerosis. *Environmental Health Perspectives* 116: 486–491.

Bell M.L., McDermott A., Zeger S.L., Samet J.M., and Dominici F. 2004. Ozone and short-term mortality in 95 US urban communities, 1987–2000. *Journal of the American Medical Association* 292: 2372–2378.

Berge E., Huang H.-C., Chang J., and Liu T.-H. 2001. A study of the importance of initial conditions for photochemical oxidant modeling. *Journal of Geophysical Research* 106: 1347–1363.

Berkowitz C.M., Fast J.D., Springston S.R., et al. 1998. Formation mechanisms and chemical characteristics of elevated photochemical layers over the northeast United States. *Journal of Geophysical Research* 103: 10631–10647.

Biswas J., Hogrefe C., Rao S.T., Hao W., and Sistla G. 2001. Evaluating the performance of regional-scale photochemical modeling systems: Part 3. Precursor predictions. *Atmospheric Environment* 35: 4175–4188.

Box G.E.P. 1979. Robustness in the strategy of scientific model building. In: *Robustness in Statistics*, R.L. Launer and G.N. Wilkinson (Eds). Academic Press, New York, p. 202.

Brauer M. and Brook J.R. 1997. Ozone personal exposures and health effects for selected groups residing in the Fraser Valley. *Atmospheric Environment* 31: 2113–2121.

Brauer M., Gehring U., Brunekreef B., et al. 2006. Traffic-related air pollution and otitis media. *Environmental Health Perspectives* 114: 1414–1418.

Brauer M., Hoek G., Smit J., et al. 2007. Air pollution and the development of asthma, allergy and respiratory infections in a birth cohort. *European Respiratory Journal* 29: 879–888.

Brauer M., Hoek G., van Vliet P., et al. 2003. Estimating long-term average particulate air pollution concentrations: Application of traffic indicators and geographic information systems. *Epidemiology* 14: 228–239.

Brauer M., Lencar C., Tamburic L., Koehoorn M., Demers P., and Karr C. 2008. A cohort study of traffic-related air pollution impacts on birth outcomes. *Environmental Health Perspectives* 116: 680–686.

Briggs D. 2005. The role of GIS: coping with space (and time) in air pollution exposure assessment. *Journal of Toxicology and Environmental Health* 68: 1243–1261.

Brook J.R. and Johnson D. 2000. Identification of representative warm season periods for regional air quality (ozone) model simulations. *Atmospheric Environment* 34: 1591–1599.

Brook J.R., Burnett R.T., Dann T.F., et al. 2007a. Further interpretation of the acute effect of nitrogen dioxide observed in Canadian time series studies. *Journal of Exposure Science and Environmental Epidemiology* 17: S36–S44.

Brook J.R., Dann T.F., and Bonvalot Y. 1999. Observations and interpretations from the Canadian fine particle monitoring program. *Journal of the Air and Waste Management Association* 49: 35–44.

Brook R.D., Frankin B., Cascio W., et al. 2004. Air pollution and cardiovascular disease. A Statement for Healthcare Professionals from the Expert Panel on Population and Prevention Science of the American Heart Association. *Circulation* 109: 2655–2671.

Brook R.D., Jerrett M., Brook J.R., and Finkelstein M.M. 2008. The relationship between diabetes mellitus and traffic-related air pollution. *Journal of Occupational and Environmental Medicine* 50: 32–38.

Brook J.R., Poirot R.L., Dann T.F., Lee P.K.H., Lillyman C.D., and Ip T. 2007b. Assessing sources of $PM_{2.5}$ in cities influenced by regional transport, *Journal of Toxicology and Environmental Health Part A* 70: 191–199.

Brost R.A. 1988. The sensitivity of input parameters of atmospheric concentrations simulated by a regional chemical model. *Journal of Geophysical Research* 93: 2371–2387.

Brown S.S., Ryerson T.B., Wollny A.G., et al. 2006. Variability in nocturnal nitrogen oxide processing and its role in regional air quality. *Science* 311: 67–70.

Brunekreef B. and Holgate S.T. 2002. Air pollution and health. *Lancet* 360: 1233–1242.

Bukowiecki N., Dommen J., Prévôt A.S.H., Richter R., Weingartner E., and Baltensperger U. 2002. A mobile pollutant measurement laboratory—measuring gas phase and aerosol ambient concentrations with high spatial and temporal resolution. *Atmospheric Environment* 36: 5569–5579.

Burke J.M., Zufall M.J., and Özkaynak H. 2001. A population exposure model for particulate matter: Case study results for $PM_{2.5}$ in Philadelphia, PA. *Journal of Exposure Analysis and Environmental Epidemiology* 11: 470–489.

Burnett R.T., Brook J.R., Yung W.T., Dales R.E., and Krewski D. 1997a. Association between ozone and hospitalization for respiratory diseases in 16 Canadian cities. *Environmental Research* 72: 24–31.

Burnett R.T., Cakmak S., Brook J.R., and Krewski D. 1997b. The role of particulate size and chemistry in the association between summertime ambient air pollution and hospitalization for cardio-respiratory diseases. *Environmental Health Perspectives* 105: 614–620.

Burnett R.T., Stieb D., Brook J.R., et al. 2004. The short-term effects of nitrogen dioxide on mortality in Canadian cities. *Archives of Environmental Health* 59: 228–237.

CEC (Commission of the European Communities). 2005. Commission Staff Working Paper. Annex to the Communication on Thematic Strategy on Air Pollution and the Directive on "Ambient Air Quality and Cleaner Air for Europe." Impact Assessment, http://ec.europa.eu/environment/air/cafe/pdf/ia_report_en050921_final.pdf

Chen I.H., Knutsen S.F., Shavlik D., et al. 2005. The association between fatal coronary heart disease and ambient particulate air pollution: Are females at greater risk? *Environmental Health Perspectives* 113: 1723–1729.

Cohen A.J., Anderson H.R., Ostro B., et al. 2005. The global burden of disease due to outdoor air pollution. *Journal of Toxicology and Environmental Health Part A* 68: 1301–1307.

Cooke R.M., Wilson A.M., Tuomisto J.R., Morales O., Taintio M., and Evans J.S. 2007. A probabilistic characterization of the relationship between fine particulate matter and mortality: Elicitation of European experts. *Environmental Science and Technology* 41: 6598–6605.

Craig L., Brook J.R., Chiotti Q., et al. 2008. Air pollution and public health: A guidance document for risk managers. *Journal of Toxicology and Environmental Health Part A* 71: 588–698.

CRC. 2004. Evaluation of the U.S. EPA MOBILE6 Highway Vehicle Emission Factor Model, Final Report CRC Project E-64, ENVIRON International Corp., Novato, CA.

Cyrys J., Hochadel M., Gehring U., et al. 2005. GIS-based estimation of exposure to particulate matter and NO_2 in an urban area: Stochastic versus dispersion modeling. *Environmental Health Perspectives* 113: 987–992.

Dabberdt W.F., Carroll M.A., Baumgardner D., et al. 2004. Meteorological research needs for improved air quality forecasting. *Bulletin of the American Meteorological Society* 85: 563–586.

Dennis R.L., Binkowski F.S., Clark T.L., McHenry J.N., Reynolds S., and Seilkop S.K. 1990. Selected applications of RADM (Part II). In *National Acid Precipitation Assessment Program: State of Science and Technology*, P.M. Irving (Ed.), NAPAP SOS/T Volume 1, Report 5 (App. 5F). National Acid Precipitation Assessment Program, Washington, DC, 37pp.

Dennis R.L., McHenry J.N., Barchet W.R., Binkowski F.S., and Byun D.W. 1993. Correcting RADM's sulfate underprediction: Discovery and correction of model errors and testing the corrections through comparisons against field data. *Atmospheric Environment* 27A: 975–997.

Dickson R.J. and Oliver W.R. 1991. Emissions models for regional air quality studies. *Environmental Science and Technology* 25: 1533–1535.

Dockery D.W., Pope C.A., Xu X., et al. 1993. An association between air pollution and mortality in Six U.S. Cities. *New England Journal of Medicine* 329: 279–285.

Dominici F., McDermott A., Daniels M., Zeger S.L., and Samet J.M. 2003. Mortality among residents of 90 cities, in revised analyses of time-series studies of air pollution and health. Special Report. Health Effects Institute, Boston, MA, pp. 9–24, http://www.healtheffects.org/Pubs/TimeSeries.pdf (May12, 2004).

Ebelt S., Wilson W.E., and Brauer M. 2005. Exposure to ambient and nonambient components of particulate matter: A comparison of health effects. *Epidemiology* 16: 396–405.

Eder B. and Yu S. 2006. A performance evaluation of the 2004 release of Models-3 CMAQ. *Atmospheric Environment* 40: 4811–4824.

Eder B., Kang D., Mathur R., Yu S., and Schere, K. 2006. An operational evaluation of the Eta-CMAQ air quality forecast model. *Atmospheric Environment* 40: 4894–4905.

EMEP. 2005. Transboundary acidification, eutrophication and ground level ozone in Europe in 2003. Status Report 1/2005, ISSN 0806-4520, August, Norwegian Meteorological Institute, Oslo, Norway, 229pp, http://www.emep.int//publ/common_publications.html.

Fan X., Brook J.R., and Mabury S. 2003. Sampling semivolatile organic species associated with $PM_{2.5}$ using an integrated organic gas and particle sampler. *Environmental Science and Technology* 37: 3145–3151.

Finkelstein M.M., Jerrett M., and Sears M.R. 2004. Traffic air pollution and mortality rate advancement periods. *American Journal of Epidemiology* 160: 173–177.

Fox D.G. 1981. Judging air quality model performance. *Bulletin of the American Meteorological Society* 62: 599–609.

Frost G.J., McKeen S.A., Trainer M., et al. 2006. Effects of changing power plant NO_x emissions on ozone in the eastern United States: Proof of concept. *Journal of Geophysical Research* 111: D12306, doi:10.1029/2005JD006354.

Fruin S.A., St Denis M.J., Winer A.M., Colome S.D., and Lurmann F.W. 2001. Reductions in human benzene exposure in the California South Coast Basin. *Atmospheric Environment* 35: 1069–1077.

Gauderman W.J., Vora H., McConnell R., et al. 2007. Effect of exposure to traffic on lung development from 10 to 18 years of age: A cohort study. *Lancet* 369: 571–577.

Gehring U., Heinrich J., Kramer U., et al. 2006. Long-term exposure to ambient air pollution and cardiopulmonary mortality in women. *Epidemiology* 17: 545–555.

Gilbert N., Goldberg M.S., Beckerman B., Brook J.R., and Jerrett M. 2005. Assessing spatial variability of ambient nitrogen dioxide in Montreal, Canada, with a land-use regression model. *Journal of the Air and Waste Management Association* 55: 1059–1063.

Gilliland A.B., Appel K.W., Pinder R.W., and Dennis R.L. 2006. Seasonal NH$_3$ emissions for the continental United States: inverse model estimation and evaluation. *Atmospheric Environment* 40: 4986–4998.

Gilliand A.B., Dennis R.L., Roselle S.J., and Pierce T.E. 2003. Seasonal NH$_3$ emission estimates for the Eastern United States using ammonium wet concentrations and an inverse modeling method. *Journal of Geophysical Research* 108: 4477.

Goldberg M.S. 2007. On the interpretation of epidemiological studies of ambient air pollution. *Journal of Exposure Science and Environmental Epidemiology* 17: S66–S70.

Goldberg M.S., Burnett R., Bailar J.C., et al. 2001. The association between daily mortality and ambient air particle pollution in Montreal, Quebec: 2: Cause-specific mortality. *Environmental Research* 86: 26–36.

Goldberg M.S., Burnett R.T., Yale J.F., Valois M.F., and Brook J.R. 2006. Associations between ambient air pollution and daily mortality among persons with diabetes and cardiovascular disease. *Environmental Research* 100: 255–267.

Grahame T.J. and Schlesinger R.B. 2005. Evaluating the health risk from secondary sulfates in Eastern North American regional ambient air particulate matter. *Inhalation Toxicology* 17: 5–27.

Gram F., Nafstadb P., and Haeheim L.L. 2003. Estimating residential air pollution exposure among citizens in Oslo 1974–1998 using a geographical information system. *Journal of Environmental Monitoring* 5: 541–546.

Grasso L.D. 2000. The differentiation between grid spacing and resolution and their application to numerical modeling. *Bulletin of the American Meteorological Society* 81: 579–580.

Gryparis A., Forsberg B., Katsouyanni K., et al. 2004. Acute effects of ozone on mortality from the "air pollution and health: A European approach" project. *American Journal of Respiratory and Critical Care Medicine* 170: 1080–1087.

Gurjar B.R., Bultler T.M., Lawrence M.G., and Lelieveld J. 2008. Evaluation of emissions and air quality in megacities. *Atmospheric Environment* 42: 1593–1606.

Heald C.L., Jacob D.J., Fiore A.M., et al. 2003. Asian outflow and trans-Pacific transport of carbon monoxide and ozone pollution: An integrated satellite, aircraft, and model perspective. *Journal of Geophysical Research* 108-D24: 4804, doi:10.1029/2003JD003507.

Heald C.L., Jacob D.J., Park R.J., et al. 2005. A large organic aerosol source in the free troposphere missing from current models. *Geophysical Research Letters* 32: L18809, doi:10.1029/2005GL023831.

HEI. 2003. Assessing the health impact of air quality regulations: Concepts and methods for accountability research. Report#: Communication 11. Health Effects Institute, September, http://pubs.healtheffects.org

HEI. 2004. Health effects of outdoor air pollution in developing countries of Asia: A literature review. Special Report 15. Health Effects Institute, Boston, MA.

Henderson S.B., Beckerman B., Jerrett M., and Brauer M. 2007. Application of land use regression to estimate long term concentrations of traffic-related nitrogen oxides and fine particulate matter. *Environmental Science and Technology* 41: 2422–2428.

Hodzic A., Vautard R., Chazette P., Menut L., and Bessagnet B. 2006. Aerosol chemical and optical properties over the Paris area within ESQUIF project. *Atmospheric Chemistry and Physics—Discussion* 6: 401–454.

Hoek G., Brunekreef B., Goldbohm S., Fischer P., and van den Brandt P.A. 2002. Association between mortality and indicators of traffic-related air pollution in the Netherlands: A cohort study. *Lancet* 360: 1203–1209.

Hoffmann B., Moebus S., Stang A., et al. 2006. Residence close to high traffic and prevalence of coronary heart disease. *European Heart Journal* 27: 2696–2702.

Hogrefe C., Rao S.T., Kasibhatla P., et al. 2001. Evaluating the performance of regional-scale photochemical modeling systems: Part 1. Meteorological predictions. *Atmospheric Environment* 35: 4159–4174.

Hogrefe C., Sistla G., Zalewsky E., Hao W., and Ku J.-Y. 2003. An assessment of the emissions inventory processing systems EMS-2001 and SMOKE in grid-based air quality models. *Journal of the Air and Waste Management Association* 53: 1121–1129.

Houyoux M.R., Vukovich J.M., Coats Jr. C.J., and Wheeler N.J.M. 2000. Emission inventory development and processing for the Seasonal Model for Regional Air Quality (SMRAQ) project. *Journal of Geophysical Research* 105: 9079–9090.

Jacobson M.Z. 1999. *Fundamentals of Atmospheric Modeling.* Cambridge University Press, Cambridge, 656pp.

Jerrett M., Arain M.A., Kanaroglou P., et al. 2007. Modelling the intra-urban variability of ambient traffic pollution in Toronto, Canada. *Journal of Toxicology and Environmental Health Part A* 70: 200–212.

Jerrett M., Burnett R.T., Ma R., et al. 2006. Spatial analysis of air pollution and mortality in Los Angeles. *Epidemiology* 17: 727–736.

Kan H., Heiss G., Rose K.M., Whitsel E., Lurmann F., and London S.J. 2007. Traffic exposure and lung function in adults: The atherosclerosis risk in communities study. *Thorax* 62: 873–879.

Kanaroglou P.S., Jerrett M., Morrison J., et al. 2005. Establishing an air pollution monitoring network for intra-urban population exposure assessment: A location-allocation approach. *Atmospheric Environment* 39: 2399–2409.

Katsouyanni K., Touloumi G., Samoli E., et al. 2003. Sensitivity analysis of various models of short-term effects of ambient particles on total mortality in 29 cities in APHEA2. In *Revised Analyses of Time-Series Studies of Air Pollution and Health.* Special Report. Health Effects Institute, Boston, MA, pp. 157–164, http://www.healtheffects.org/news.htm (May 16, 2003).

Kim D., Sass-Kortsak A., Purdham J.T., Dales R.E., and Brook J.R. 2006. Associations between personal exposures and fixed-site ambient measurements of fine particulate matter, nitrogen dioxide and carbon monoxide in Toronto, Canada. *Journal of Exposure Analysis and Environmental Epidemiology* 15: 172–183.

Kleeman M.J. and Cass G.R. 1999a. Identifying the effect of individual emissions sources on particulate air quality within a photochemical aerosol processes trajectory model. *Atmospheric Environment* 33: 4597–4613.

Kleeman M.J. and Cass G.R. 1999b. Effects of emissions control strategies on the size and composition distribution of urban particulate air pollution. *Environmental Science and Technology* 33: 177–189.

Kleinman L.I. 1987. Source–receptor relations from nonlinear atmospheric models. *Atmospheric Environment* 21: 1219–1225.

Knipping E.M., Kumar N., Pun B.K., Seigneur C., Wu S.-Y., and Schichtel B.A. 2006. Modeling regional haze during the BRAVO study using CMAQ-MADRID: 2. Source region attribution of particulate sulfate compounds. *Journal of Geophysical Research* 111: D06303, doi:10.1029/2004JD005609.

Kolb C.E., Herndon S.C., McManus J.B., et al. 2004. Mobile laboratory with rapid response instruments for real-time measurements of urban and regional trace gas and particulate distributions and emission source characteristics. *Environmental Science and Technology* 38: 5694–5703.

Krewski D., Burnett R.T., Goldberg M.S., et al. 2000. Reanalysis of the Harvard Six Cities Study and the American Cancer Society Study of particulate air pollution and mortality. Special Report. Health Effects Institute, Cambridge, MA.

Kuhlbusch T., Astrid J., Hugo A., et al. 2004. Analysis and design of local air quality measurements: Towards European Air Quality Health Effect Monitoring. Service Contract 070501/2004/389487/MAR/C1, EURAQHEM Final Report, http://ec.europa.eu/environment/air/cafe/activities/pdf/euraqhem_final_report.pdf

Kuhn M., Builtjes P.J.H., Poppe D., et al. 1998. Intercomparison of the gas-phase chemistry in several chemistry and transport models. *Atmospheric Environment* 32: 693–709.

Künzli N., Jerrett M., Mack W.J., et al. 2005. Ambient air pollution and atherosclerosis in Los Angeles. *Environmental Health Perspectives* 133: 201–206.

Laden F., Schwartz J., Speizer F.E., and Dockery D.W. 2006. Reduction in fine particulate air pollution and mortality: Extended follow-up of the Harvard Six Cities Study. *American Journal of Respiratory and Critical Care Medicine* 73: 667–672.

Lipfert F.W., Wyzga R.E., Baty J.D., and Miller J.P. 2006. Traffic density as a surrogate measure of environmental exposures in studies of air pollution health effects: Long-term mortality in a cohort of U.S. veterans. *Atmospheric Environment* 40: 154–169.

Liu Y., Park R.J., Jacob D.J., Li Q., Kilaru V., and Sarnat J.A. 2004. Mapping annual mean ground-level $PM_{2.5}$ concentrations using multiangle imaging spectroradiometer aerosol optical thickness over the contiguous United States. *Journal of Geophysical Research* 109: D22206, doi:10.1029/2004JD005025.

Logan J.A. 1999. An analysis of ozonesonde data for the troposphere: Recommendations for testing 3-D models and development of a gridded climatology for tropospheric ozone. *Journal of Geophysical Research* 104: 16115–16149.

Luginaah I., Xu X., Fung K.Y., et al. 2006. Establishing the spatial variability of ambient nitrogen dioxide in Windsor, Ontario. *International Journal of Environmental Studies* 63: 487–500.

Lwebuga-Mukasa J.S., Oyana T., Thenappan A., and Ayirookuzhi S.J. 2004. Association between traffic volume and health care use for asthma among residents at a U.S.-Canadian border crossing point. *Journal of Asthma* 41: 289–304.

Makar P.M., Wiebe H.A., Staebler R.M., Li S.-M., and Anlauf K.G. 1998. Measurement and modeling of nighttime particle nitrate formation. *Journal of Geophysical Research* 103: 13095–13110.

Mallet V. and Sportisse B. 2005. A comprehensive study of ozone sensitivity with respect to emissions over Europe with a chemistry-transport model. *Journal of Geophysical Research* 110: D022302, doi:10.1029/2005JD006234.

Mallet V. and Sportisse, B. 2006. Uncertainty in a chemistry-transport model due to physical parameterizations and numerical approximations: An ensemble approach applied to ozone modeling. *Journal of Geophysical Research* 111: D01302, doi:10.1029/2005JD006149.

Marmur A., Park S.-K., Mulholland J.A., Tolbert P.E., and Russell A.G. 2006. Source apportionment of $PM_{2.5}$ in the southeastern United States using receptor and emissions-based models: Conceptual differences and implications for time-series health studies. *Atmospheric Environment* 40: 2533–2551.

Marshall J.D., Riley W.J., McKone T.E., and Nazaroff W.W. 2003. Intake fraction of primary pollutants: Motor vehicle emissions in the South Coast Air Basin. *Atmospheric Environment* 37: 3455–3468.

Martin R.V. 2008. Satellite remote sensing of surface air quality. *Atmospheric Environment* 42: 7823–7843.

Martin R.V., Fiore A.M., and Van Donkelaar A. 2004. Space-based diagnosis of surface ozone sensitivity to anthropogenic emissions. *Geophysical Research Letters* 31: L06120, doi:10.1029/2004GL019416.

Martin R.V., Jacob D.J., Chance K., Kurosu T.P., Palmer P.I., and Evans M.J. 2002. Global inventory of nitrogen oxide emissions constrained by space-based observations of NO_2 columns. *Journal of Geophysical Research* 108.17: 4537, doi:10.1029/2003JD003453.

Maynard D., Coull B.A., Gryparis A., and Schwartz J. 2007. Mortality risk associated with short-term exposure to traffic particles and sulfates. *Environmental Health Perspectives* 115: 751–755.

McHenry J.N., Binkowski F.S., Dennis R.L., Chang J.S., and Hopkins D. 1992. The Tagged Species Engineering Model (TSEM). *Atmospheric Environment* 26A: 1041–1052.

McKeen S., Wilczak J., Grell G., et al. 2005. Assessment of an ensemble of seven real-time ozone forecasts over Eastern North America during the summer of 2004. *Journal of Geophysical Research* 110: D21307, doi:10.1029/2005JD005858.

McMurry P.H., Shephard M., Vickery J. (Eds). 2004. *Particulate Matter Science for Policy Makers: A NARSTO Assessment.* Cambridge University Press, New York, NY.

McNair L.A., Harley R.A., and Russell A.G. 1996. Spatial inhomogeneity in pollutant concentrations, and their implications for air quality model evaluations. *Atmospheric Environment* 30: 4291–4301.

Ménard R. and Robichaud A. 2005. The Chemistry-Forecast System at the Meteorological Service of Canada. *Proceedings of ECMWF Seminar on Global Earth-System Monitoring*, September, 5–9. ECMWF 2005—Annual Seminar Proceedings, http://www.ecmwf.int/publications/library/do/references/show?id=86891, http://www.ecmwf.int/publications/library/ecpublications/_pdf/seminar/2005/sem2005_menard.pdf

Mendoza-Dominguez A. and Russell A.G. 2001. Estimation of emission adjustments from the application of four-dimensional data assimilation to photochemical air quality modeling. *Atmospheric Environment* 35: 2879–2894.

Meng Z., Dabdub D., and Seinfeld J.H. 1997. Chemical coupling between atmospheric ozone and particulate matter. *Science* 277: 116–119.

Miller C.A., Hidy G., Hales J., et al. 2006. Air emission inventories in North America: A critical assessment. *Journal of the Air and Waste Management Association* 56: 1115–1129.

Mills N.L., Törnqvist H., Gonzalez M.C., et al. 2007. Ischemic and thrombotic effects of dilute diesel-exhaust inhalation in men with coronary heart disease. *New England Journal of Medicine* 357: 1075–1082.

Mills N.L., Tornquist H., Robinson S.D., et al. 2005. Diesel exhaust inhalation causes vascular dysfunction and impaired endogenous fibrinolysis. *Circulation* 112: 3930–3936.

Misra P.K., Bloxam R., Fung C., and Wong S. 1989. Non-linear response of wet deposition to emissions reduction: A model study. *Atmospheric Environment* 23: 671–687.

Moran M.D. 2000. Basic aspects of mesoscale atmospheric dispersion. In: *Mesoscale Atmospheric Dispersion*, Z. Boybeyi (Ed.). WIT Press, Southampton, pp. 27–119.

Moran M.D. and Zheng Q. 2006. Modelling long-term sulphur and nitrogen deposition using Lagrangian chemical transport models. *Proceedings of the 14th Joint AMS/A&WMA Conference on the Applications of Air Pollution Meteorology*, Atlanta, January 30 to February 2, American Meteorological Society, Boston, 23pp, http://ams.confex.com/ams/pdfpapers/101260.pdf

NARSTO. 2005. Improving emission inventories for effective air quality management across North America: A NARSTO Assessment. NARSTO-05-001, Oak Ridge, TN.

NRC. 2004. *Air Quality Management in the United States.* The National Academies Press, Washington, DC, ISBN 0-309-08932-8.

O'Neill M.S., Veves A., Zanobetti A., et al. 2005. Diabetes enhances vulnerability to particulate air pollution-associated impairment in vascular reactivity and endothelial function. *Circulation* 111: 2913–2920.

Padro J., Puckett K.J., and Woolridge D.N. 1993. The sensitivity of regionally averaged O_3 and SO_2 concentrations to ADOM dry deposition velocity parameterizations. *Atmospheric Environment* 27A: 2239–2242.

Palmer P.I., Jacob D.J., Jones D.B.A., et al. 2003. Inverting for emissions of carbon monoxide from Asia using aircraft observations over the western Pacific. *Journal of Geophysical Research* 108: D21, doi:10.1029/2003JD003397.

Park R.J., Jacob D.J., Chin M., and Martin R.V. 2003. Sources of carbonaceous aerosols over the United States and implications for natural visibility. *Journal of Geophysical Research* 108: 4355.

Parrish D.D., Trainer M., Hereid D., et al. 2002. Decadal change in carbon monoxide to nitrogen oxide ratio in U.S. vehicular emissions. *Journal of Geophysical Research* 107: 4140.

Peters L.K., Berkowitz C.M., Carmichael G.R., et al. 1995. The current state and future direction of Eulerian models in simulating the tropospheric chemistry and transport of trace species: A review. *Atmospheric Environment* 29 : 189–222.

Petron G., Granier C., Khattatov B., et al. 2002. Inverse modeling of carbon monoxide surface emissions using CMDI network observations. *Journal of Geophysical Research* 107: 4761.

Pielke R.A. 1984. *Mesoscale Meteorological Modeling*. Academic Press, Orlando, 612pp.

Pielke R.A. and Uliasz M. 1998. Use of meteorological models as input to regional and mesoscale air quality models—limitations and strengths. *Atmospheric Environment* 32: 1455–1466.

Polina B.M., Offenberg J.H., Clemente J., Blaustein M., Thurston G.D., and Chen L.C. 2004. Ambient pollutant concentrations measured by a mobile laboratory in South Bronx, NY. *Atmospheric Environment* 38: 5283–5294.

Pope C.A. and Dockery D.W. 2006. Health effects of fine particulate air pollution: Lines that connect. 2006 Critical Review. *Journal of the Air and Waste Management Association* 56: 709–742.

Pope C.A., Burnett, R.T., Thurston, G.D., et al. 2004. Cardiovascular mortality and long term exposure to particulate air pollution. *Circulation* 109: 71–77.

Pope C.A., Burnett R.T., Thun M.J., et al. 2002. Lung cancer, cardiopulmonary mortality, and long-term exposure to fine particulate air pollution. *Journal of the American Medical Association* 287: 1132–1141.

Pope C.A., Thun M.J., Namboodiri M.M., et al. 1995. Particulate air pollution as a predictor of mortality in a prospective study of U.S. adults. *American Journal of Respiratory and Critical Care Medicine* 151: 669–674.

Proctor S.D., Dreher K.L., Kelly S.E., and Russell J.C. 2006. Hypersensitivity of prediabetic JCR: LA-cp rats to fine airborne combustion particle-induced direct and noradrenergic-mediated vascular contraction. *Toxicological Science* 90: 385–391.

Pudykiewicz J.A. 1998. Application of adjoint tracer transport equations for evaluating source parameters. *Atmospheric Environment* 32: 3039–3050.

Pun B.K. and Seigneur C. 1999. Understanding particulate matter formation in the California San Joaquin Valley: Conceptual model and data needs. *Atmospheric Environment* 33: 4865–4875.

Pun B.K., Seigneur C., Vijayaraghavan K., et al. 2006. Modeling regional haze in the BRAVO study using CMAQ-MADRID: 1. Model evaluation. *Journal of Geophysical Research* 111: D06302, doi:10.1029/2004JD005608.

Qian Z., He Q., Lin H.-M., et al. 2007. Short-term effects of gaseous pollutants on cause-specific mortality in Wuhan, China. *Journal of the Air and Waste Management Association* 57: 785–793.

Reid N., Misra P.K., Amman M., and Hales J. 2003. Air quality modelling. Paper presented at Third NERAM International Colloquium on Health and Air Quality, Santo Spirito Hospital, Rome, Italy, November 5–7, 2003, http://www.irr-neram.ca/rome/rome.html#proceed

Rosenlund M., Berglind N., Pershagen G., Hallqvist J., Jonson T., and Bellander T. 2006. Long-term exposure to urban air pollution and myocardial infarction. *Epidemiology* 17: 383–390.

Rosenlund M., Forastiere F., Stafoggia M., et al. 2007. Comparison of regression models with land-use and emissions data to predict the spatial distribution of traffic-related air pollution in Rome. *Journal of Exposure Science and Environmental Epidemiology* 17: 1–8.

Russell A. and Dennis R. 2000. NARSTO critical review of photochemical models and modeling. *Atmospheric Environment* 34: 2283–2324.

Sahsuvaroglu T., Arain A., Kanaroglou P., et al. 2006. A land-use regression model for predicting ambient concentrations of nitrogen dioxide in Hamilton, Canada. *Journal of the Air and Waste Management Association* 56: 1059–1069.

Sariglannis D.A., Soulakellis N.A., and Sifakis N.I. 2004. Information fusion for computational assessment of air quality and health effects. *Photogrammetric Engineering and Remote Sensing* 70: 235–245.

Sarnat J.A., Marmur A., Klein M., Kim E., Russel A.G., Sarnat S.E., Mulholland J.A., Hopke P.K., and Tolbert P.E., 2008. Fine particle sources and cardiorespiratory morbidity: An application of chemical mass balance and factor analytical source apportionment methods. *Environmental Health Perspective* 116: 459–466.

Sarnat J.A., Schwartz J., Catalano P.J., and Suh H.H. 2001. Gaseous pollutants in particulate matter epidemiology: Confounders or surrogates? *Environmental Health Perspectives* 10: 1053–1061.

Schere K.L. and Hidy G.M. 2000. Foreward: NARSTO Critical Reviews. *Atmospheric Environment* 34: 1853–1860.

Seaman N.L. 2000. Meteorological modeling for air-quality assessments. *Atmospheric Environment* 34: 2231–2259.

Seigneur C. 2001. Current status of air quality models for particulate matter. *Journal of the Air and Waste Management Association* 51: 1508–1521.

Seigneur C. and Moran M.D. 2004. *Using Models to Estimate Particle Concentration in Chapter 8 in Particulate Matter Science for Policy Makers: A NARSTO Assessment*, P. McMurry, M. Shepherd, and J. Vickery (Eds). Cambridge University Press, Cambridge, 42pp.

Seilkop S.K. 1995a. Representativeness of surface site air concentrations relative to an 80 km grid. In: *Regional Photochemical Measurement and Modeling Studies*, Vol. 1, *Results and Interpretation of Field Measurements*, A.J. Ranzieri and P.A. Solomon (Eds). VIP-48, Air & Waste Management Association, Pittsburgh, PA, pp. 197–210.

Seilkop S.K. 1995b. Characterization and usage of subgrid scale variability in model evaluation. Report prepared by Analytical Services, Inc., Durham, NC, for U.S. Environmental Protection Agency, Research Triangle Park, NC, November, 57pp.

Seinfeld J.H. and Pandis S.N. 1998. *Atmospheric Chemistry and Physics—From Air Pollution to Climate Change*. Wiley, New York, 1326pp.

Setton E., Hystad P., Keller C.P., et al. 2006. Simulating risk of exposure to traffic-related air pollution in urban and suburban working and non-working populations. Presented at the Air Quality 2006, Banff, AB, February, http://www.ec.gc.ca/aqw-aqa/program_e.html

Siddiqui A.R., Gold E.B., Yang X., Lee K., Brown K.H., and Bhutta Z.A. 2008. Prenatal exposure to wood fuel smoke and low birth weight. *Environmental Health Perspectives* 116: 543–549.

Sillman S., He D., Pippin M., et al. 1998. Model correlations for ozone, reactive nitrogen and peroxides for Nashville in comparison with measurements: Implications for VOC–NO$_x$ sensitivity. *Journal of Geophysical Research* 103: 22629–22644.

Silva C., Gould T., Larson T., et al. 2006. A tale of two cities: Comparing land use and traffic related pollution exposure models in Seattle and Vancouver. Presented at the Air Quality 2006, Banff, AB, February, http://www.ec.gc.ca/aqw-aqa/program_e.html

Slama R., Morgenstern V., Cyrys J., Zutavern A., Herbart O., Wichmann H.-E., Heinrich J., and the LISA Study Group, 2007. Traffic-related atmospheric pollutants levels during pregnancy and offspring's term birth weight: A study relying on a land-use regression exposure model. *Environmental Health Perspective* 115: 1283–1292.

Smargiassi A., Berrada K., Fortier I., and Kosatsky T. 2006. Traffic intensity, dwelling value and hospital admissions for respiratory disease among the elderly in Montreal (Canada): A case-control study. *Journal of Epidemiology and Community Health* 60: 507–512.

Smyth S., Yin D., Roth H., Jiang W., Moran M.D., and Crevier L.-P. 2006b. The impact of GEM and MM5 meteorology on CMAQ air quality modeling results in eastern Canada and the northeastern United States. *Journal of Applied Meteorology* 45: 1525–1541.

Smyth S.C., Jiang W., Yin D., Roth H., and Giroux E. 2006a. Evaluation of CMAQ O_3 and $PM_{2.5}$ performance using Pacific 2001 measurement data. *Atmospheric Environment* 40: 2735–2749.

Stein A.F. and Lamb D. 2002. Chemical indicators of sulfate sensitivity to nitrogen oxides and volatile organic compounds. *Journal of Geophysical Research* 107(D20): 4449, doi:10.1029/2001JD001088.

Stieb D.S., Smith-Doiron M., Blagden P., and Burnett R.T. 2005. Estimating the public health burden attributable to air pollution: An illustration using the development of an alternative air quality index. *Journal of Toxicology and Environmental Health Part A* 68: 1275–1288.

Sun Q., Wang A., Jin X., et al. 2005. Long-term air pollution exposure and acceleration of atherosclerosis and vascular inflammation in an animal model. *Journal of the American Medical Association* 294: 3003–3010.

Tang D., Li T., Liu J.J., Zhou A., Tao Y., Chen Y., Rauh V.A., Xie J., and Perera F., 2008. Effects of prenatal exposure to coal burning pollutants on children's development in China. *Environmental Health Perspective* 116: 674–679.

Thurston G.D., Ito K., Mar T., et al. 2005. Workgroup report: Workshop on source apportionment of particulate matter health effects—intercomparison of results and implications. *Environmental Health Perspectives* 113: 1768–1774.

Uliasz M. 1993. The atmospheric mesoscale dispersion modeling system. *Journal of Applied Meteorology* 32: 139–149.

Urch B., Silverman F., Corey P., et al. 2005. Acute blood pressure responses in healthy adults during controlled air pollution exposures. *Environmental Health Perspectives* 113: 1052–1055.

U.S. EPA. 2001. Guidance for demonstrating attainment of air quality goals for $PM_{2.5}$ and regional haze. U.S. Environmental Protection Agency Report Draft 2.1, January, 273pp., http://www.epa.gov/scram001/guidance/guide/draft_pm.pdf (July28, 2006).

U.S. EPA. 2004. Air Quality Criteria Document for particulate matter, Vol. II. EPA/600/P-99/002bF, Office of Research and Development, USEPA, Research Triangle Park, NC, http://cfpub2.epa.gov/ncea/cfm/recordisplay.cfm?deid=87903

U.S. EPA. 2005a. Review of the National Ambient Air Quality Standards for particulate matter: Policy Assessment of Scientific and Technical Information. OAQPS Staff Paper, http://www.epa.gov/ttn/naaqs/standards/pm/data/pmstaffpaper_20051221.pdf

U.S. EPA. 2005b. Guidance on the use of models and other analyses in attainment demonstrations for the 8-hour Ozone NAAQS. Report No. EPA-454/R-05–002, U.S. Environmental Protection Agency, October, 128pp., http://www.epa.gov/scram001/guidance/guide/8-hour-o3-guidance-final-version.pdf (July 28, 2006).

U.S. EPA. 2006a. Provisional assessment of recent studies on health effects of particulate matter exposure. EPA/600/R-06/063, National Center for Environmental Assessment, Office of Research and Development, Research Triangle Park, NC, http://epa.gov/pm/pdfs/ord_report_20060720.pdf

U.S. EPA. 2006b. Air quality criteria for ozone and related photochemical oxidants (final). EPA/600/R-05/004aB-cB, Washington, DC, http://cfpub.epa.gov/ncea/cfm/recordisplay.cfm?deid=137307

U.S. National Research Council. 1983. *Risk Assessment in the Federal Government: Managing the Process*. National Research Council, National Academy of Sciences, National Academy Press, Washington, DC.

U.S. Presidential/Congressional Commission on Risk Assessment and Risk Management. 1997. Framework for environmental health risk management. Final report, Vols 1 and 2, Washington, DC.

van Bree L., Fudge N., Tuomisto J.T., and Brunekreef B. 2007. Closing the gap between science and policy on air pollution and health. *Journal of Toxicology and Environmental Health Part A* 70: 377–381.

Venkatram A. 1988. Inherent uncertainty in air quality modeling. *Atmospheric Environmental* 22: 1221–1227.

Watson J.G. and Chow J.C. 2005. Chapter 16B *Receptor Models, Air Quality Modeling: Theories, Methodologies, Computational Techniques, and Available Databases and Software*, Vol. II, P. Zannetti (Ed.). EnviroComp Institute and Air & Waste Management Association, Pittsburg, PA, pp. 455–501.

Watson J.G., Chow J.C., and Fujita E.M. 2001. Review of volatile organic compound source apportionment by chemical mass balance. *Atmospheric Environment* 35: 1567–1584.

West J.J., Ansari A.S., and Pandis S.N. 1999. Marginal $PM_{2.5}$: Nonlinear aerosol mass response to sulfate reductions in the Eastern United States. *Journal of the Air and Waste Management Association* 49: 1415–1424.

Westerdahl D., Fruin S., Sax T., Fine P.M., and Sioutas C. 2005. Mobile platform measurements of ultrafine particles and associated pollutant concentrations on freeways and residential streets in Los Angeles. *Atmospheric Environment* 39: 3597–3610.

Wexler A.S. and Johnson M.V. 2008. What have we learned from highly time resolved measurements during the EPA supersite program and related studies? *Journal of the Air and Waste Management Association* 58: 303–319.

Wheeler A.J., Smith-Doiron M., Xu X., Gilbert M.L., and Brook J.R. 2008. Intra-urban variability of air pollution in Windsor, Ontario—measurement and modeling for human exposure assessment. *Environmental Research* 106: 7–16.

WHO. 1999. *Monitoring Ambient Air Quality for Health Impact Assessment*. WHO Regional Publications, Geneva, Switzerland, European Series, No. 85, ISBN 92 890 1351 6/ISSN 0378-2255.

WHO. 2002. *The Health Effects of Indoor Air Pollution Exposure in Developing Countries*. WHO/SDE/OEH/02.05, Geneva, Switzerland, http://whqlibdoc.who.int/hq/2002/WHO_SDE_OEH_02.05.pdf

WHO. 2004a. *Meta-analysis of Time-Series Studies and Panel Studies of Particulate Matter (PM) and Ozone (O_3)*. WHO Regional Office for Europe, Copenhagen, http://www.euro.who.int/document/E82792.pdf

WHO. 2004b. *Health Aspects of Air Pollution: Results from the WHO Project "Systematic Review of Health Aspects of Air Pollution in Europe."* Publications WHO Regional Office for Europe, Scherfigsvej 8 DK-2100, Copenhagen, Denmark.

WHO. 2005a. WHO Air Quality Guidelines global update 2005. Report on a Working Group meeting, Bonn, Germany, October 18–20, http://www.euro.who.int/Document/E87950.pdf

WHO. 2005b. *Effects of Air Pollution on Children's Health and Development*. A Review of the Evidence. European Centre for Environment and health, Bonn Office, http://www.euro.who.int/document/E86575.pdf

WHO. 2005c. Health Effects of Transport Related Air Pollution, M. Krzyzanowski, B. Kuna-Dibbert, and J. Schneider (Eds), Publications WHO Regional Office for Europe, Copenhagen, Denmark. http://www.euro.who.int/document/e86650.pdf

WHO. 2006. *Health Risks of Particulate Matter from Long Range Transboundary Air Pollution*. European Centre for Environment and Health, Bonn Office, http://www.euro.who.int/document/E88189.pdf

Winer A.M., Lurmann F.W., Coyner L.A., and Colome Poe M.P. 1989. Characterization of air pollutant in the California South Coast Air Basin: Application new regional human exposure (REHEX) model. Report to the South Coast Air Quality Management District and the California State University Foundation, Contract Number TSA 106-01-88.

Wu J., Lurmann F., Winer A., Lu R., Turco R., and Funk T. 2005. Development of an individual exposure model for application to the Southern California children's health study. *Atmospheric Environment* 39: 259–273.

Xu X., Brook J.R., and Guo P. 2006. A statistical assessment of mobile sampling strategy. *Journal of the Air and Waste Management Association* 57: 1396–1406.

Yli-Tuomi T., Aarnio P., Pirjola L., Mäkelä T., Hillamo R., and Jantunen M. 2005. Emissions of fine particles, NO_x, and CO from on-road vehicles in Finland. *Atmospheric Environment* 39: 6696–6667.

Yu S., Dennis R., Bhave P.V., and Eder B.K. 2004. Primary and secondary organic aerosols over the United States: Estimates on the basis of observed organic carbon (OC) and elemental carbon (EC), and air quality modeled primary OC/EC ratios. *Atmospheric Environment* 38: 5257–5268.

Zanobetti A. and Schwartz, J. 2001. Are diabetics more susceptible to the health effects of airborne particles? *American Journal of Respiratory and Critical Care Medicine* 164: 831–833.

Zhang L., Moran M.D., and Brook J.R. 2001. A comparison of models to estimate in-canopy photosynthetically active radiation and their influence on canopy stomatal resistance. *Atmospheric Environment* 35: 4463–4470.

Zhang Y., Seigneur C., Seinfeld J.H., Jacobson M.Z., Clegg S.L., and Binkowski F.S. 2000. A comparative review of inorganic aerosol thermodynamic equilibrium modules: Similarities, differences, and their likely causes. *Atmospheric Environment* 34: 117–137.

Zhang Y., Vijayaraghavan K., and Seigneur C. 2005. Evaluation of three probing techniques in a three-dimensional air quality model. *Journal of Geophysical Research* 110: D02305, doi:10.1029/2004JD005248.

Zunckel M., Koosailee A., Yarwood G., et al. 2006. Modelled surface ozone over southern Africa during the Cross Border Air Pollution Impact Assessment Project. *Environmental Modelling and Software* 21: 911–924.

Section V

Environmental Impacts
of Air Pollution

15 Assessing Ground-Level Ozone (O$_3$) Impacts to Crops in Parts of Asia and Southern Africa

The Regional Air Pollution in Developing Countries (RAPIDC) Crops Project

Lisa D. Emberson, Patrick Büker, Magnuz Engardt, Anna M. van Tienhoven, Madhoolika Agrawal, Mark Zunckel, Kevin Hicks, Håkan Pleijel, Nguyen T. K. Oanh, Lal P. Amgain, Towhid Islam, Syed R.A. Shamsi, G. Anoma D. Perera, Gert H.J. Krüger, and Pieter R. Smit

CONTENTS

15.1   INTRODUCTION

Ground-level ozone (O_3) is the most widespread and phytotoxic pollutant that frequently exceeds the World Health Organization (WHO) air quality guidelines (AQGs) for agricultural crops across many parts of the globe. Elevated O_3 levels have been found to cause declines in the yield of many crop species such as wheat, rice, soybean, and cotton (e.g., Fuhrer et al., 1997; Fuhrer and Booker, 2003). Such yield losses have been attributed to the reduced photosynthetic rate (e.g., Lehnherr et al., 1997; McKee et al., 1997) and altered carbon allocation (e.g., Grantz and Yang 2000) as well as an indirect effect of accelerated leaf senescence (e.g., Grandjean and Fuhrer, 1989). O_3 has also been shown to induce visible injury, which may, for example, reduce the economic value of leafy crops, such as spinach and lettuce (e.g., Emberson et al., 2003); such damage generally occurs after acute O_3 exposures. A rather limited number of studies have also shown that O_3 affects crop quality (e.g., nitrogen content of grains, tubers, etc. and nutritive quality of forage crops) although the direction of the effect (i.e., improvement or degradation of quality) is not always consistent between different quality aspects (e.g., Pleijel et al., 1999a; Vorne et al., 2002; Sanz et al., 2005).

Many of the studies that have identified O_3 effects similar to those described above have been made using controlled fumigation techniques in order to define exposure–response relationships. A common experimental tool for such studies has been open top chambers (OTCs) (e.g., Heagle et al., 1979), which allow controlled O_3 fumigations under near-field conditions.

Two major programs employed these techniques during the 1980s. The first was the National Crop Loss Assessment Network (NCLAN) that applied standardized protocols at locations across the United States (Heck et al., 1988) and the second was the European Open Top Chamber (EOTC) Program (Jäger et al., 1992). Both studies led to the development of a variety of O_3 characterization indices capable of summarizing the seasonal O_3 exposure (summarized in Mauzerall and Wang, 2001). These indices have been applied in modeling studies to assess the risk of crop yield losses from ground-level O_3, as well as associated economic losses.

In Europe, economic losses for more than 20 arable crops were estimated to be in the region of US$8 billion per year by Holland et al. (2006). In the United States, Adams et al. (1988) predicted losses of US$ 3 billion for nine different arable crop species. Wang and Mauzerall (2004) estimated economic losses for wheat, rice, maize, and soybean for China, South Korea, and Japan, totaling US$ 5 billion. These losses were represented by percentage yield losses of up to 9% for the cereal crops and 23–27% for soybean (a species recognized as very sensitive to O_3). The magnitude of these losses was in agreement with a similar study (although this only estimated productivity rather than economic losses) conducted by Aunan et al. (2000), where current yield losses of between 1% and 3% were estimated for selected cereals in China.

However, all these studies have used dose–response relationships established for either European or North American species and cultivars, since these have been all that are available. Fundamental uncertainties lie in the necessary assumption that the species and cultivars grown in Asia will respond to O_3 similarly to those grown in

Europe and North America. Factors that may alter plant response to O_3 include those associated with climate, agricultural management practices, crop phenology, genetically based tolerance, and pollutant exposure patterns. As such, to enable more realistic modeling studies to be performed, there is an urgent need to establish dose–response relationships for locally grown species and cultivars under local environmental conditions and management practice regimes.

In a first attempt to assess the significance of this shortcoming, Emberson et al. (2009) collated and analyzed dose–response data for Asian crops (wheat, rice, and legumes) and compared these against dose–response relationships derived in North America. The results suggested that Asian-grown cultivars of wheat and rice are considerably more sensitive to O_3 than their North American counterparts; hence, the studies by Wang and Mauzerall (2004) and Aunan et al. (2000) may have substantially underestimated the effect O_3 has on crop productivity in Asia. However, the modeling studies conducted using European and North American dose–response relationships should still be viewed as providing valuable information on the relative spatial magnitude of risk across regions and give an indication of the potential damage that is likely to occur according to current knowledge.

The need to assess the potential risk posed to agricultural productivity by O_3 is all the more urgent given the projected trends for future O_3 concentrations. Figure 15.1 provides an indication from Prather et al. (2003) of the projected global increase in O_3 concentration over the next 100 years assuming the IPCC 2001 A2x SRES global emission scenarios (Nakićenović and Swart, 2000). Although this figure shows projections for 2100 and uses one of the more extreme Intergovernmental Panel on Climate Change (IPCC) global emission scenarios, it serves to provide a clear indication of the geographical regions that may be expected to see increases in ground-level O_3 in the future; these include South and South East Asia and southern Africa. Modeling studies conducted over a shorter time frame by Dentener et al. (2006) using current legislation emission scenarios support the identification of South Asia as the region that is most likely to suffer the largest increases in O_3 concentration with projections of increases in mean surface O_3 concentrations of 7.2 ppb between 2000 and 2030; these studies strongly suggest that conditions may significantly worsen in the near future.

For both South and South East Asia and southern Africa, there is an obvious need to extend, improve, and evaluate modeling studies to assess present-day and future impacts of air pollutants on agriculture. This chapter introduces a framework methodology developed as one component of the Swedish International Development Co-operation Agency (Sida)-funded Regional Air Pollution In Developing Countries (RAPIDC) Program for assessing the risks to crops caused by ground-level O_3 in parts of Asia and southern Africa.

15.2 THE RAPIDC CROPS PROGRAM

The RAPIDC Crops Program was established in 1997 to better understand the potential impacts of air pollutants on agriculture in both South Asia and southern Africa. The program is currently coordinated by the Stockholm Environment Institute (SEI) and is carried out in collaboration with research organizations across Europe, Asia, and southern Africa together with intergovernmental agencies. The aim of RAPIDC

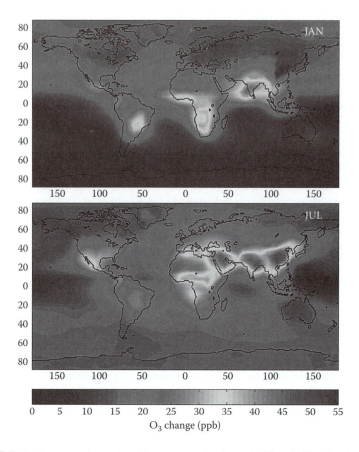

FIGURE 15.1 Increases in surface O_3 concentration from 2000 to 2100. (Reprinted from Prather, M. 2003. *Geophysical Research Letters* 30, 1100. With permission.)

is to facilitate the development of agreements and protocols and methods to implement measures that will prevent and control air pollution.

Initially, the RAPIDC Crops Project focused on identifying the current "state of knowledge" on air pollution's impacts on both agricultural productivity and forest health by collating evidence from 10 countries around the globe. A number of air pollutants were investigated including sulfur dioxide (SO_2), nitrogen oxides (NO_x), ground-level O_3, suspended particulate matter (SPM), and hydrogen fluorides (HF). Evidence of air pollution effects was demonstrated by various observational and experimental techniques including the recording of visible injury in the field, transect studies along pollution gradients, chemical protectant studies, and filtration and controlled fumigation studies. The information collected from the developing countries demonstrated that some countries (e.g., China and India) were significantly more advanced in terms of collating large databases from which air pollution impacts could be assessed. However, most of the air pollution impact assessment work performed to date had investigated effects caused by SO_2 (making up more than half of all studies), with O_3 and SPM assessments together making up the majority of the remaining studies. The results of these investigations have been reported in Emberson et al. (2003) and Emberson et al. (2001). Table 15.1 summarizes some of the key results describing ambient O$_3$ concentration impacts on yield losses (i.e., data are only collated from filtration, ethylenediurea (EDU), and transect studies) for the more important crops in South and South East Asia. Southern Africa is omitted, since, to our knowledge, there have been no studies assessing crop yield loss at ambient O$_3$ concentrations in this region. Table 15.1 also provides details of the crop-growing seasons since this determines the O$_3$ concentration (which varies seasonally) to which the crop will be exposed (e.g., in South Asia, depressions in O$_3$ concentrations occur during the monsoon period because of reduced photochemical activity). As such, this information is crucial for identifying which crops are likely to be at risk from O$_3$ pollution.

Additional crops may be appropriate for future targeted O$_3$ impact studies as inferred from existing evidence of O$_3$-induced visible injury in the regions of

TABLE 15.1
Details of Experimental Evidence of Crop Yield Losses under Ambient O$_3$ Concentrations and Associated Growing Seasons Collated from Studies Conducted in South Asia and South East Asia

Crop	Ambient O$_3$ Concentration Range, 4–8 h Growing Season Means (ppb)	Yield Losses (%)	Growing Season	Reference
Rice (*Oryza sativa* L.)	33–60	6–47	May to November	Maggs et al. (1995), Wahid et al. (1995b, 1997)
Wheat (*Triticum aestivum* L.)	36–72	13–48	November to May	Maggs et al. (1995), Nasim et al. (1995), Rai et al. (2007), Tiwari et al. (2005), Wahid (2006), Wahid and Maggs (1999), Wahid et al. (1995a), Ishii et al. (2004)
Soybean (*Glycine max* L.)	40–75	32–65	August to May	Wahid et al. (2001)
Mung bean (*Vigna radiata* L.)	33–67	30–50	July to September	Agrawal et al. (2005), Bajwa et al. (1997), Singh et al. (in press)
Pea (*Pisum sativum* L.)	42	38	November to May	Rajput and Agrawal (2004)

concern, yield losses in studies conducted in other regions of the world (e.g., Europe and North America), and their importance as staple crops in the targeted region. These crops could include maize (*Zea mays* L.), potato (*Solanum tuberosum* L.), cabbage (*Brassica oleracea* L.), cotton (*Gossypium* spp.), sugar cane (*Saccharum* spp.), and peanut (*Arachis hypogaea* L.).

It was also noted that certain perennial crop species, such as peach, apple, mango, guava, and tea, were also of importance and should be considered for future studies. Effect variables of significance for these types of crops would likely include yield, yield components, harvest index, and nutritional quality (including aspects such as protein/nitrogen as well as the content of sugars and certain acids).

A key conclusion from this review was that although a number of different observational and experimental methods existed and had been used to assess air pollution impacts, these methods had never been applied according to any kind of standard protocol. As such, the data could not be easily combined to produce a large dataset from which it might be possible to derive the dose–response relationships necessary to perform regional-scale risk assessments. The work conducted under RAPIDC aimed to address this problem and, having established the current level of understanding of air pollution impacts, attention turned to the establishment of methods and infrastructures to try to fill knowledge gaps.

This work concentrated on South Asia and southern Africa since these were the regions where air pollution was most likely to increase (in terms of both spatial extent and magnitude) in future years given projected trends in emissions resulting from economic growth (e.g., Prather et al., 2003; Emberson et al., 2003; Dentener et al., 2006).

15.2.1 THE APCEN NETWORK

To help achieve these goals, an Air Pollution Crop Effect Network (APCEN) (http://www.sei.se/apcen/index.html) was established in 2002. This network of air pollution effects practitioners aimed to facilitate effective communication between air pollution scientists across different regions of the world and to help coordinate observational and investigative studies. APCEN has developed experimental protocols designed specifically for application in developing country regions to allow the application of standardized methods for air pollution impact assessments. Methods for socioeconomic analysis of air pollutant impacts were also considered.

The APCEN network meets regularly at international workshop meetings to develop and consolidate methods to assess the air pollution impacts on crop yields and consider how best to communicate research findings to policy makers. To date, the APCEN network consists of more than 80 air pollution effects practitioners from ~30 countries across the world but predominantly from Asia, reflecting the evolving regional foci of the RAPIDC Crops Project (Table 15.2).

These practitioners are experts in assessing air pollution impacts in their respective countries, with technical expertise and access to experimental and air pollutant-monitoring facilities or proficient in modeling and mapping air pollution and air pollution impacts for provision of information to policy makers. In this respect, the potential for the network to conduct coordinated air pollution assessment studies is

Table 15.2
The APCEN Network Members by Country and Global Region

Region	Number of Network Members	Countries Represented
Africa	13	Botswana, Egypt, Kenya, Mozambique, South Africa, Tanzania, Zimbabwe, Zambia
Asia	58	Bangladesh, India, Iran, Japan, Nepal, Pakistan, Peoples Republic of China, Philippines, South Korea, Sri Lanka, Taiwan, Tajikistan, Thailand, Vietnam
The Americas, Europe, and Australia	19	Argentina, Australia, Brazil, Chile, France, Germany, Norway, Sweden, UK, USA

impressive, as is evident from the variety and geographical coverage of experimental infrastructure associated with the APCEN network (Figure 15.2).

Continued expansion of the APCEN network will ensure good spatial coverage of network facilities (expertise, monitoring, and experimental equipment) and maintenance and expansion of scientific linkages, especially in regions that are currently resource poor. Future expansion will also focus on the South Asian, South East Asian and southern African regions where very little is known of the likely regional scale of air pollution impacts.

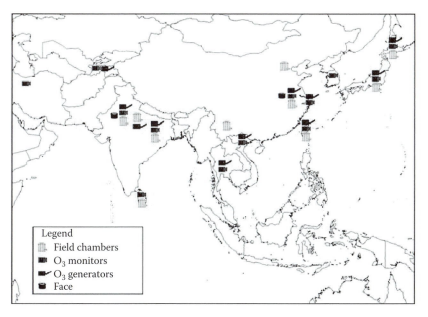

FIGURE 15.2 Map showing the geographical distribution of key experimental facilities: Free-air concentration enrichment (FACE), field chambers, O₃ generators, and O₃ monitors associated with the APCEN network.

The primary role of the APCEN network within the RAPIDC Crops Project has been to provide scientific and technical support for risk assessments conducted across South Asia and southern Africa. In this respect, APCEN has built APCEN has built and strengthened capacity to perform such risk assessments by providing advice on the selection and application of appropriate AQGs for use in regional risk assessments and has developed and tested experimental and observational protocols.

To ensure optimum use of APCEN resources to provide information that can be used to drive policy in the most effective manner, it is imperative that development of the network role will be considered fully in terms of the goals and research interests of related international fora. In particular, this will involve strengthening links with ongoing policy processes within both the Malé Declaration countries of South Asia and with the emerging policy process in the southern African countries, the Air Pollution Information Network for Africa (APINA). These two science-to-policy networks have been active since 1998 with the aim of having the best available scientific knowledge on air pollution impacts effectively integrated into policy making (see Hicks et al., 2001). The Malé Declaration was adopted by the Council of Ministers of the South Asia Cooperative Environment Programme (SACEP) in 1998 and is now nearing the stage where the countries are prepared to discuss the form of a regional Convention or other instruments to limit their air pollution problems. In southern Africa, the Lusaka Agreement (2008)—Southern African Development Community (SADC) Regional Policy Framework on Air Pollution—has been submitted by APINA to SADC for possible adoption by the SADC Ministers of Environment and Sustainable Development. Importantly, the work of APCEN is recognized by these policy processes.

Finally, to ensure the longevity of the network, it is considered that the coordination of the network should eventually be managed within the regions under an appropriate international body. This aspect is considered below in connection with the emergence of the Global Atmospheric Pollution Forum (GAP Forum).

15.2.2 PROVISIONAL RISK ASSESSMENT MODELING AND MAPPING

Provisional risk assessment modeling and mapping has been performed according to the processes defined in Figure 15.3 (cf. Emberson et al., 2003). Here, regionally mapped concentration data may either be derived from interpolated monitoring data or from atmospheric transfer models. The modeling of concentrations requires a detailed emission inventory, meteorological data, and an appropriate atmospheric transfer model. Once concentrations have been characterized in a manner appropriate to assess impacts on crops, receptor information is required to show the species location—preferably as a map with an associated relevant database. Overlaying these data will produce maps that show locations where concentrations exceed critical levels and hence where there may be a risk of yield reductions for sensitive crops.

Provisional risk assessments have been performed across South and South East Asia (the latter region is included to broaden the regional scope of the modeling-based risk assessment studies) and southern Africa with the intention of identifying areas where elevated O_3 concentrations may be high enough to cause crop damage. These assessments have been performed using modeled O_3 concentration data

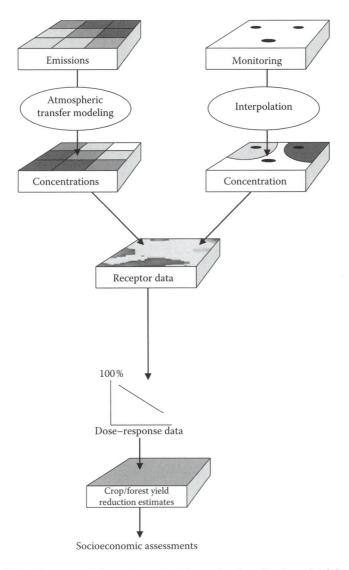

FIGURE 15.3 Process and datasets required for regional application of AQGs or dose–response approaches. (From Emberson, L.D. et al. (Eds). 2003. *Air Pollution Impacts on Crops and Forests.* Imperial College Press, London, pp. 309–355. With permission.)

combined with appropriate AQGs derived from pollutant dose–crop response relationships as recommended by APCEN. The examples of the modeling work described here make use of existing AQGs that have been developed in Europe in the absence of data representing the response of staple Asian and southern African crops. These AQGs have been developed from European dose–response relationships that characterize O₃ according to hourly O₃ concentrations Accumulated Over a Threshold of 40 ppb (AOT40), this being an O₃ metric developed in Europe from which those "critical levels" below which damage would not be expected to occur can be identified (Fuhrer et al., 1997). The European critical level for wheat (a crop recognized as being particularly O_3 sensitive) is set at 3 ppm.h (LRTAP Convention, 2004). The areas across the region exceeding this critical level would indicate locations where agricultural crops may be at risk of damage from O_3; the magnitude of the exceedance would provide an indication of the relative degree of risk.

15.2.2.1 Provisional Risk Assessment for South Asia

The provisional risk assessment for South Asia was performed using output from the MATCH model (Engardt, 2008). This is an off-line Eulerian dispersion model driven by meteorological data from the European Centre for Medium-Range Weather Forecasts (ECMWF) > reanalysis. Anthropogenic emissions of SO_x, NO_x, NH_x, nonmethane volatile organic compounds (NMVOC), and CO are taken from Streets et al. (2003), biogenic volatile organic compounds (VOC) emission estimates come from Guenther et al. (1995). MATCH (Langner et al., 1998; Robertson et al., 1999) uses a photochemical scheme including approximately 60 chemical species, based on Simpson et al. (1993). The horizontal resolution in MATCH is 50 km; the vertical resolution increases from 20 m near the surface to 400 m at 5 km above the surface.

The MATCH model was used to simulate three-month AOT40 at 3 m above the surface during three different periods for the year 2000 (Engardt, 2008). The results of these simulations are provided in Figure 15.4.

Table 15.1 indicates that many crops have a winter growing season; as such, it is useful to analyze the maps in Figure 15.4 from September to July since this provides an indication of the surface O_3 to which the plants may be exposed. During the September-to-November period, exceedance of the critical level occurs over large parts of India (especially the western Maharashtra province and the north and north eastern provinces including Haryana, Uttar Pradesh, and Bihar) as well as north eastern Pakistan, much of Bangladesh, and a large swathe of the Tibetan plateau; the latter exceedances are due to intrusion of free tropospheric air and low deposition to snow-covered surfaces. Perhaps most importantly, these higher O_3 exposures are experienced across the fertile and agriculturally productive Indo-Gangetic plain. Any adverse impacts of surface O_3 on agriculture would be significant across such an agriculturally important region. AOT40 values in these exceedance regions commonly reach 10–20 ppm.h. This would suggest potential yield losses for sensitive crops such as wheat of up to 30% assuming that the European dose–response relationship is applicable under South Asian conditions.

During the period from January to March, exceedance of the critical level increases spatially with most of India and Bangladesh now also showing some exceedance. However, the magnitude of the exceedances declines somewhat with most areas registering AOT40 values of less than 10 ppm.h. In contrast, O_3 exposures on the Tibetan plateau increase further with exceedances above 20 ppm.h; again this results from intruding free tropospheric air and very slow deposition to snow-covered high-altitude regions. Finally, the period May–July shows the lowest exceedances of the AOT40 of all periods, although exceedance still occurs across large parts of eastern India, most of Bangladesh, and north eastern Pakistan. The high surface O_3 of the Tibetan plateau has almost entirely disappeared.

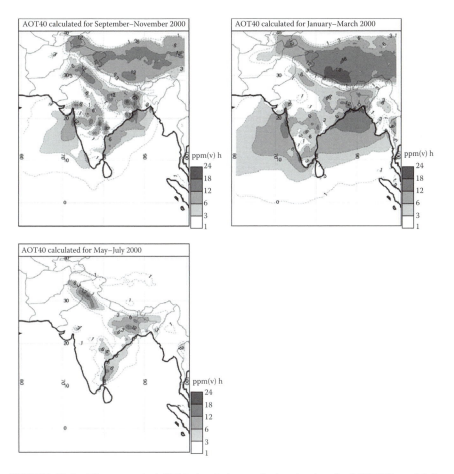

FIGURE 15.4   Three-month AOT40 simulations calculated using the MATCH model for three different periods for the year 2000.

15.2.2.2   Provisional Risk Assessment for South East Asia

The provisional risk assessment for South East Asia was performed using output from the Model-3 Community Multi-scale Air Quality (CMAQ) model. CMAQ, "one atmosphere," is an Eulerian-type model developed by the United States Environmental Protection Agency (US EPA) to simulate tropospheric O_3 and a range of other atmospheric phenomena taking into consideration the complex interactions between atmospheric pollutants at both regional and urban scales (Byun and Ching, 1999). The meteorological fields necessary as input to CMAQ were generated from the Fifth-Generation Pennsylvania State University/National Center for Atmospheric Research (PSU/NCAR) Mesoscale Meteorological Model (MM5) (Grell et al., 1994).

Anthropogenic emissions of NO_x, CO, VOCs, and SO_2 for the year 2000 were obtained from Streets et al. (2003), projected for the modeling year of 2004 using average regional Gross Domestic Product (GDP) growth rates (Nghiem and Oanh, 2008). The biogenic emissions of NO_x and VOCs were obtained from Guenther et al. (1995). The Carbon Bond Mechanism IV (Gery et al., 1989) was used with updated reaction rates (Atkinson et al., 1997). The horizontal grid size is the same for both CMAQ and MM5, which is 56 km (~0.5°). Vertically, for both the Continental Southeast Asia (CSEA) and the central region of Thailand (CENTHAI) simulations, the domains extend from mean sea level (0 m) to about 16,000 m. The MM5 has 30 vertical layers, while the CMAQ has 15 layers, with the vertical resolution of these two models increasing from 38 m near the surface to nearly 2 and 3 km at the top of the domain, respectively. The performance of the model system was satisfactorily evaluated on selected historical episodes (Nghiem and Oanh, 2008) before being applied to estimate hourly O_3 concentration fields.

The MM5–CMAQ model system was used to produce hourly O_3 concentrations for the period from November 2003 to April 2004 over the CSEA region. This period was selected because high O_3 concentrations are observed during these months and these months also correlate with important crop-growing seasons (e.g., the second rice cropping period lasts from November to April). The three-month AOT40 values (ppb.h) based on the model output for the modeling layer are presented in Figure 15.5.

High values of AOT40 are observed over the main urban areas within the modeling domain with the maximum values predicted over the CENTHAI (around

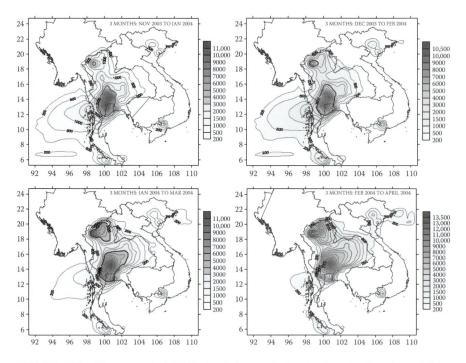

FIGURE 15.5 Three-month AOT40 simulations calculated using the CMAQ model for three different periods for the year 2003–2004 (Nov–Jan, Dec–Feb, Jan–March, and Feb–April) for Southeast Asia.

Bangkok). However, plumes of elevated O_3 concentrations are modeled to extend rather substantially away from the urban centers into rural areas and thus may affect agricultural crops. The areas for which AOT40 exceeds the critical level of 3000 ppb.h are greatest in Thailand, suggesting a high risk for crops, although it is important to be aware that studies suggest a reduced O_3 sensitivity of rice as compared to wheat (Mills et al., 2007). There are also possible transboundary effects with plumes of elevated O_3 concentrations, whose origin can be traced to Thailand, extending into other parts of South East Asia, in particular Burma. Finally, the modeling also indicates a seasonality to the elevated O_3 concentrations with the magnitude and spatial extent of critical level exceedance changing with time of year.

15.2.2.3 Provisional Risk Assessment for Southern Africa

The provisional risk assessment performed for southern Africa is described in more detail in van Tienhoven et al. (2006) and tried to establish whether surface O_3 concentrations may pose a threat to crop productivity and livelihoods across the region. This modeling study used near-surface O_3 concentrations (Zunckel et al., 2006) generated on the application of the Comprehensive Air Quality Model with extensions (CAMx) (ENVIRON, 2006). The modeling domain incorporated the southern African countries of Botswana, Mozambique, South Africa, Zambia, and Zimbabwe. The modeling was based on spatially resolved emissions of CO, SO_2, NO_x, and total hydrocarbons resulting from anthropogenic activities (industry, transport, and domestic burning) (Fleming and van der Merwe, 2002) and biogenic emissions of VOCs (Otter et al., 2003) that reflected hourly and seasonal variations (Zunckel et al., 2006). Biomass-burning emissions after the onset of summer rains were considered negligible and, hence, are not considered in the October-to-April modeling period.

The staple arable crop maize (*Zea mays* L.) was selected for investigation since this crop is important for both commercial and subsistence farming across the southern African region and has been found to be moderately sensitive to O_3 (e.g., Mulchi et al., 1995; Rudorff et al., 1996; Mills et al., 2007). Important southern African maize production areas in relation to the CAMx modeling domain are shown in Figure 15.6. The growing season for maize is assumed to last from October to April, although the actual planting and harvest dates will vary with climate and, hence, with location across the region. The CAMx model was run for seven months and for five days during each month. The output O_3 concentrations were used to estimate AOT40 O_3 concentrations (assuming that the five-day period was representative of the entire month) to provide an indication of the potential for damage. It should be noted that as for rice, maize is considered less sensitive to O_3 than wheat (Mills et al., 2007); this should be considered when interpreting the results of the risk assessment.

Figure 15.7 shows that much of southern and central Zimbabwe experiences an AOT40 of 1000 ppb.h or more each month from October to April, but there are no exceedances in Zambia and infrequent exceedances in the northern parts of South Africa. Over a three-month growing season, these values accumulate to exceed the 3000 ppb.h critical level during every month of the growing season in the central parts of Zimbabwe with the exception of October.

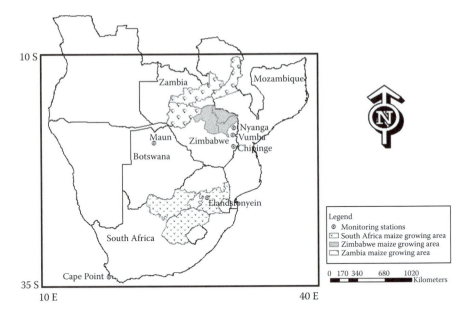

FIGURE 15.6   Map of southern Africa showing the CAMx modelling domain, maize growing regions, and locations of O_3-monitoring sites. (Reprinted from van Tienhoven, A.M. et al.
2006. *Environmental Pollution* 140, 220–230. With permission.)

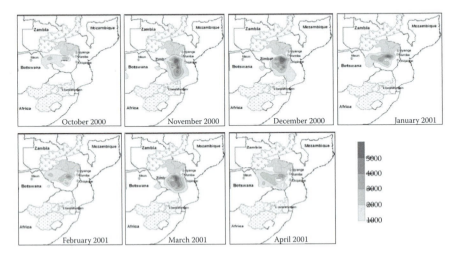

FIGURE 15.7   Monthly AOT40 values for southern Africa for October 2000 to April
2001, based on projecting the five-day modeled concentrations to represent a full month.
(Reprinted from van Tienhoven, A.M. et al. 2006. *Environmental Pollution* 140, 220–230.
With permission.)

There are common limitations in all three provisional risk assessments presented here, of which perhaps the most obvious is the use of the European AOT40 concentration-based risk assessment method. The critical level of 3000 ppb.h is based on European-grown wheat varieties, and it is questionable how representative of risk this level is for Asian- and African-grown crops. These studies also necessitate an assumed accuracy in the photo-oxidant modeling since rural monitoring of O_3 concentrations in parts of Asia and southern Africa is either extremely limited or unavailable, making the evaluation of modeled concentrations impossible. For the southern African risk assessment, the extrapolation of the five-day O_3 concentrations is another large source of uncertainty given the fact that O_3 episodes are associated with specific meteorological conditions (e.g., sunshine hours, temperature, the prevailing wind direction from precursor pollutant sources, etc.) and, hence, will ordinarily have a high variability from week to week. These studies should therefore be viewed as providing an indication of potential risk, but further work is urgently needed to improve our understanding of the potential for damage to crops across these regions.

Modeling-based risk assessments are useful, but capacity needs to be built to undertake studies to verify whether the impacts predicted by the modeling studies are actually occurring on crops on the ground. To this end, simple experimental studies are a useful technique to help demonstrate the magnitude and geographical distribution of crop damage from O_3.

15.2.3 BIOMONITORING AND CHEMICAL PROTECTANT STUDIES IN SOUTH ASIA AND SOUTHERN AFRICA

Although there is substantial experimental evidence of O_3 affecting crop yield losses across Asia (Emberson et al., 2003), these studies have tended to be performed in isolation from each other. As such, the experimental work has not followed standardized protocols, making it difficult to pool together data from different studies for the derivation of robust dose–response relationships. As such, there is a need to develop a coordinated program of work whereby experimental studies that have followed standard protocols can produce comparable results.

A number of different protocols for assessing the risks posed by a range of air pollutants (e.g., O_3, SO_2, HF, and heavy metals) to vegetation are being developed and piloted by APCEN. Two methods were identified for priority development to "ground truth" the provisional modeling-based risk assessments and provide evidence of the real impacts of O_3 on crop biomass and yield. These methods were (i) a clover clone biomonitoring method that recorded visible injury and biomass loss and (ii) an EDU chemical protectant method that allowed quantification of yield losses of selected crops. Both methods assess damage under ambient O_3 concentrations.

APCEN trained users from both the Malé Declaration and APINA crop experimental networks in the application of these standardized protocols across South Asia and southern Africa, respectively. A combination of one or both of these methods was then performed within these regional monitoring networks at five sites in south Asia (Bangladesh, India, Nepal, Pakistan, and Sri Lanka) and six sites in southern Africa (Botswana, Mozambique, South Africa, Tanzania, Zambia, and Zimbabwe), respectively. To aid the interpretation of the results, passive samplers were used to provide four-weekly mean ambient O_3 concentrations and microloggers to monitor temperature and relative humidity at all study sites.

15.2.3.1   The Clover Clone Biomonitoring Method

The clover clone method was initially developed in the humid subtropical climate of North Carolina, USA (cf. Heagle et al., 1994) and has been used extensively since 1996 in Europe and North America by the International Cooperative Programme on Effects of Air Pollution on Natural Vegetation and Crops (ICP Vegetation) of the United Nations Economic Commission for Europe (UNECE) (e.g., Harmens et al., 2005). The method uses O_3-sensitive and O_3-resistant white clover genotypes (*Trifolium repens* cv. Regal) and works on the principle that the difference in plant foliar injury as well as the biomass ratio between the O_3-sensitive (NC-S) and O_3-resistant (NC-R) clover genotypes can be directly related to the prevalent O_3 concentrations during the exposure period. The existing protocol [as developed and applied by the UNECE ICP Vegetation described in Mills et al. (2000)] was piloted during the 2005/2006 growing season at a study site at the North-West University, Potchefstroom, South Africa (Krüger, G.H.J., Personal Communication) and during the 2006/2007 growing season at a periurban site close to Lahore in Pakistan (Shamsi, S.R.A., Personal Communication). The successful testing of the clover biomonitoring protocol suggested that the method was "fit for purpose" for southern African and South Asian conditions. Hence, during the 2007/2008 growing season, the method was applied at selected sites across southern Africa and South Asia. The first results from South Asia suggested that biomass losses in the region of 10–20% are experienced by the sensitive clover genotype as compared to the resistant clover genotype, providing evidence that ambient O_3 concentrations in these regions are capable of causing damage to plants in the region. However, issues related to the establishment of the clover plants at the different sites across both regions made reliable comparisons of the results difficult; it was concluded that, in the future, the plants would require establishment over a full growing season to ensure that they were suitable for use as biomonitors.

15.2.3.2   The EDU Chemical Protectant Study

A chemical protectant protocol using EDU has been developed to quantitatively assess the actual yield losses for specific crops resulting from elevated O_3 concentrations.

The antioxidant EDU (*N*-[2-(2-oxo-1-imidazolidinyl) ethyl]-*N*-phenylurea) is an O_3 chemical protectant that has been used successfully in a number of experimental campaigns for agricultural crops in Europe and North America (e.g., Eckardt and Pell, 1996; Godzik and Manning, 1998; Pleijel et al., 1999b; Elagöz and Manning, 2005), in Asia (e.g., Bambawale, 1989; Varshney and Rout, 1998; Wahid et al., 2001; Agrawal et al., 2005; Tiwari et al., 2005), and to a lesser extent in North Africa (e.g., Hassan et al., 1995; Hassan, 2006) to assess the damage caused by ambient O_3 concentrations in a range of physiological and crop growth parameters [including quantity (e.g., yield) and quality (e.g., nutritional content)]. Details of the studies conducted in Asia and Africa are presented in Table 15.3 to give an indication of the range of species for which this technique has been successfully applied to date. In

TABLE 15.3

Details of Agricultural Species and Cultivars for which EDU has been Successfully Applied in Selected Asian and African Studies

Country	Species/ Cultivar	EDU Application	EDU Concentration (mg); EDU Applied per Plant	No. of Applications/ Application Time	Effect of EDU Application (Reference: Nno-EDU-Treated Plants)	Reference
Egypt	Radish, turnip; Egyptian cultivars	Soil drench	500 ppm; 300 mg	3; 10, 20, and 30 days after seeding	Increase in growth of radish and turnip plants exposed to ambient levels of O₃	Hassan et al. (1995)
Egypt	Potato; Kara	Foliar spray	300 ppm; 810 mg	9; 10-day intervals with first application 48 days after sowing	Reduction of foliar injury and increase in tuber weights of plants exposed to ambient air	Hassan (2006)
India	Tomato; Pusa Ruby	Soil drench	400 ppm; not specified	6–7; 12-day intervals	Increase in shoot and root length, as well as shoot and root biomass of plants exposed to ambient air	Varshney and Rout (1998)
India	Mung bean; Malviya Jyoti	Soil drench	500 ppm; 2750 mg	11; weekly intervals with first application 1 week after seedling emergence	Maintenance of high levels of photosynthetic pigments, soluble protein, and ascorbic acid in foliage, as well as growth increment of plants exposed to ambient air	Agrawal et al. (2005)
India	Wheat; Malviya 533 and 234	Soil drench	150, 300, and 450 ppm; 270, 540, or 810 mg	11; 10-day intervals with first application 10 days after germination	Increase in shoot and root length, number of tiller per plant and total biomass of plants exposed to ambient air	Tiwari et al. (2005)
Pakistan	Soybean; NARC-1	Soil drench	400 ppm; 1200 mg	5; 10-day intervals with first application 2 weeks after seedling emergence	Increase of seed weight per plant exposed to ambient air	Wahid et al. (2001)

addition, this table provides certain experimental details (i.e., frequency, duration, and magnitude of the EDU applications used in the treatments) since the protective effect of EDU is known to vary from species to species (and even cultivar to cultivar). Collating this type of information is important in designing chemical protectant studies for standardized application across regions with specific local crops and cultivars to assess the real O_3 impacts (both yield loss and nutritional implications) on agriculture.

EDU protocols were developed by APCEN for a variety of species [mung bean (*Vigna radiata* L.), wheat (*Triticum aestivum* L.), spinach (*Spinacea oleracea* L.), potato (*Solanum tuberosum* L.), and pea (*Pisum sativa*)], using pilot studies carried out at locations in India and South Africa. These species were selected as they are known to be O_3 sensitive (Agrawal, S.B., Personal Communication; Agrawal et al., 2005; cf. Tiwari et al., 2005) and are of economic importance in either South Asia or southern Africa. Application of this protocol for mung bean was performed at selected sites across South Asia during the 2006/2007 and 2007/2008 growing seasons.

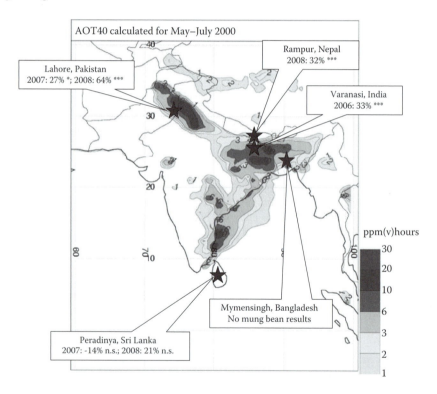

FIGURE 15.8 The provisional risk assessment for South Asia prepared using MATCH modelled O_3 concentrations for the year 2000 presented as three-month AOT40 concentrations for May to July. Also shown are the results of the EDU chemical protectant study for mung bean conducted at the South Asian experimental sites (indicated by blue stars) during equivalent months. The significance of the experimental results is presented (n.s. denotes not significant; * denotes moderately significant; and *** denotes highly significant).

Preliminary results of these data are shown in relation to O_3 concentrations presented as AOT40 for the period May, June, and July, which coincides with the important "pod-filling" mung bean growth period (Figure 15.8). The results found that those areas identified from the modeling study as being at greater risk from prevailing O_3 concentrations (see Section 2.2.1) correlated well with those sites where statistically significant damage was recorded during the experiments. The sites with the greatest O_3 damage are those in Pakistan, India, and Nepal where between 2006 and 2008 robust statistically significant yield losses for mung bean ranged from 32% to 64%. In contrast, statistically significant yield losses were not recorded in Sri Lanka.

15.2.3.3 Other Risk Assessment Techniques

Other experimental, and also observational, methodologies are being investigated for development by the ACPEN network to identify additional tools that could be used for standardized assessment of the risks posed by air pollutants to agriculture. Additional methods that may be used include (i) passive biomonitoring methods; requiring the development of a common photoguide to identify the extent of visible injury occurring in the field for key species of the different regions; (ii) transect studies that would assess damage to selected crops (or active biomonitors) along a pollution gradient, such as from a city or from a major highway or industrial complex with large emissions of pollution to the countryside; (iii) fumigation/filtration studies that would involve the establishment of filtration/fumigation facilities, most likely based on OTCs, to perform experimental investigations for derivation of dose–response relationships and to screen for sensitivities of local species and cultivars.

15.3 CONCLUSIONS

In order to estimate the extent and magnitude of local and regional impact of O_3 on agriculture in parts of Asia and southern Africa, a framework methodology for assessing the risks caused by tropospheric O_3 has been developed and applied within the RAPIDC Crops Program. This methodology includes the expansion of the global APCEN network, the performance of provisional risk assessments, and the coordination of field experimental campaigns. These field campaigns are carried out according to standardized experimental protocols developed within APCEN by researchers in South Asian and southern African countries who have direct affiliation with the policy processes of the Malé Declaration and APINA.

Although this work has established methods that enable increased understanding of present-day air pollution impacts in these regions, there still remain a large number of future challenges, for example, continuing these studies to obtain multiple years worth of data and applying the standardized methods to other regions (e.g., South East Asia) where O_3 is a potential current or future threat.

The future sustainability of cereal production per capita in South Asia is rather uncertain. South Asia's Indo-Gangetic Plains benefited from the 1960s Green Revolution. Using improved wheat and rice varieties, irrigation, and higher doses of fertilizer, South Asian farmers were able to double rice production and boost wheat output by almost five times in just three decades. However, the area under rice and wheat cultivation has stabilized, and further expansion seems unlikely. In addition, evidence suggests that growth in cereal yields have begun to slow down in many high-potential agricultural areas, with variability in trends occurring between countries of South Asia (Timsina and Connor, 2001). Factors such as soil nutrient mining, declining levels of organic matter, increasing salinity, falling water tables, and the buildup of weed, pathogen, and pest populations will all have contributed to this decline. Given the magnitude and extent of yield losses found for key crops across the South Asian region in this and other studies, it would seem that O_3 pollution might well be an additional and significant stress on agroecosystems. A comprehensive understanding of the relative importance of all stresses facing current and future agricultural production in the South Asian regions is vitally important given the challenge of the region to provide sustainable increases in productivity to balance reduced per capita area harvested.

In southern Africa, access to fertile lands for agriculture is already severely limited due to a variety of factors including land degradation, soil nutrient limitations, climatic suitability, and urbanization pressures (e.g., McCarthy et al., 2001). Climate change has been identified as a key factor that will alter the suitability of land for different uses. Of the few observations of O_3 concentrations made across the southern African region, most record regular exceedances of 40 ppb (the cut-off concentration above which O_3 accumulation causes damage to vegetation).

Global modeling of ground-level O_3 concentrations for 2030 suggest that O_3 annual mean surface averages for South Asia and southern Africa could vary from -5.9 to $+11.8$ ppb for South Asia and from -2.5 to $+7$ ppb for southern Africa, with the range being dependent on the emission scenario applied (e.g., Dentener et al., 2006). The possible increases in mean annual averages would translate into disproportionately higher growing season average concentrations when the climatic conditions (namely temperature and solar radiation) are enhancing the chemical formation of O_3 in the atmosphere. As such, it seems more than likely that O_3 concentrations, which are already at concentrations capable of causing yield and productivity losses across many parts of South Asia and southern Africa, will continue to worsen over the next 20–50 years. It would therefore seem prudent to consider this pollutant in future research to assess the effect of multiple stresses on sustainable crop production across Asia and Southern Africa.

To promote convergence of approaches for assessing impacts of these multiple stresses on agriculture across the world, APCEN is now affiliated with the GAP Forum, an initiative of the International Union of Air Pollution Prevention Associations (IUAPPA) and SEI, which since 2004 has been supporting cooperation and development of common practice among scientific and policy networks around the world, primarily concerned with the abatement of air pollution at the regional scale.

It is proposed under the GAP Forum that the APCEN network be used to further the global understanding of the threat that tropospheric O_3 (and other relevant air pollutants) pose to crop yield and quality. It will also be important to expand this work to incorporate work to identify the social groups (e.g., urban, periurban) and agricultural practices (e.g., subsistence, small-scale, and large-scale agroforestry) that may be particularly vulnerable to the impacts of air pollution damage on crops.

APCEN will continue to work with the research institutes in its network and through the different regional initiatives (e.g., the Malé Declaration and APINA). Specifically, it will aim to set up and coordinate collaborative projects that cover Asia, Africa, and Latin America making appropriate linkages with the expertise and methods developed within the UNECE Convention on Long-Range Transboundary Air Pollution.

Future assessments related to crop impacts from O_3 would ideally incorporate the effects of climate change and seek to involve specialists on adaptation options. Ideally, key decision makers from governments would come together to discuss likely combined impacts, measures to reduce vulnerability of end users, national risk assessments, and policy options to reduce the threat from this environmental problem. The opportunity for cobenefits for air pollution and climate change in emission reduction policy (e.g., Ramanathan and Feng, 2008) is of particular importance in many developing regions where there is suspicion of climate change policies, but where the importance of abating air pollution is recognized. Most work on cobenefits has, however, been undertaken with developed world perspectives and priorities. This now needs to be widened to reflect the perspectives and priorities of developing countries.

ACKNOWLEDGMENTS

The work described in this chapter was part of the RAPIDC Crops Program funded by Sida; SEI institutional programme support from Sida funded the preparation of this chapter.

REFERENCES

Adams, R.M., Glyer, J.D., and McCarl, B.A. 1988. The NCLAN economic assessment: Approach, findings and implications. In: W.W. Heck, D.T. Tingey, and O.C. Taylor (Eds). *Assessment of Crop Loss from Air Pollutants*. Elsevier Science Publishers, London.

Agrawal, S.B., Singh, A., and Rathore, D. 2005. Role of ethylene diurea (EDU) in assessing impact of ozone on *Vigna radiata* L. plants in a suburban area of Allahabad (India). *Chemosphere* 61, 218–228.

Atkinson, R., Baulsch, D.L., Cox, R.A., et al. 1997. Evaluated kinetics, photochemical and heterogeneous data. *Journal of Physical and Chemical Reference Data* 26, 521–1012.

Aunan, K., Berntsen, T.K., and Seip, H.M. 2000. Surface ozone in China and its possible impact on agricultural crop yields. *Ambio* 29, 294–301.

Bajwa, R., Ahmad, S., Uzma, M., Nasim, G., and Wahid, A. 1997. Impact of air pollution on mung bean, *Vigna radiata* (L.) Wilczek grown in open top chamber system in Pakistan. I. Effect on vegetative growth and yield. *Scientific Khyber* 10(2), 37–50.

Bambawale, O.M. 1989. Control of ozone injury on potato. *Indian Phytopathology* 42, 509–513.

Byun, D.W. and Ching, J.K.S. (Eds). 1999. *Science Algorithms of the EPA Models-3 Community Multi-Scale Air Quality (CMAQ) Modeling System*. NERL, Research Triangle Park, NC.

Dentener, F., Stevenson, D., Ellingsen, K., et al. 2006. The global atmospheric environment for the next generation. *Environmental Science and Technology* 40, 3586–3594.

Eckardt, N.A. and Pell, E.J. 1996. Effects of ethylenediurea (EDU) on ozone-induced acceleration of foliar senescence in potato (*Solanum tuberosum* L.). *Environmental Pollution* 92, 299–306.

Elagöz, V. and Manning, W.J. 2005. Factors affecting the effects of EDU on growth and yield of field-grown bush beans (*Phaseolus vulgaris* L.), with varying degrees of sensitivity to ozone. *Environmental Pollution* 136, 385–395.

Emberson, L.D., Ashmore, M.R., Murray, F., et al. 2001. Impacts of air pollutants on vegetation in developing countries. *Water, Air and Soil Pollution* 130, 107–118.

Emberson, L.D., Ashmore, M.R., and Murray, F. (Eds). 2003. A*ir Pollution Impacts on Crops and Forests: A Global Assessment*. Imperial College Press, London, pp. 309–355.

Emberson, L.D., Büker, P., Ashmore, M.R., et al. 2009. A comparison of North American and Asian exposure-response data for ozone effects on crop yields. *Atmospheric Environment* 43, 1945–1953.

Engardt, M. 2008. Modeling near-surface ozone over South Asia. *Journal of Atmospheric Chemistry* 59, 61–80.

ENVIRON. 2006. *CAMx, Comprehensive Air Quality Model with Extensions. User's Guide*, Version 4.30, ENVIRON International Corporation, Novato, California, U.S.A, http://www.camx.com/

Fleming, G. and van der Merwe, M. 2002. Spatial disaggregation of greenhouse gas emissions inventory data for Africa south of the equator, http://gis.esri.com/library/userconf/proc00/professional/papers/PAP896/p896.htm.

Fuhrer, J., Skärby, L., and Ashmore, M.R. 1997. Critical levels for ozone effects on vegetation in Europe. *Environmental Pollution* 97, 91–106.

Fuhrer, J. and Booker, F. 2003. Ecological issues related to ozone: Agricultural issues. *Environment International* 29, 141–154.

Gery, M.W., Whitten, G.Z., Killus, J.P., and Dodge, M.C. 1989. A photochemical kinetics mechanism for urban and regional scale computer modeling. *Journal of Geophysical Research Atmospheres* 94, 12925–12956.

Godzik, B. and Manning, W.J. 1998. Relative effectiveness of ethylenediurea, and constituent amounts of urea and phenylurea in ethylenediurea, in prevention of ozone injury to tobacco. *Environmental Pollution* 103, 1–6.

Grandjean, A. and Fuhrer, J. 1989. Growth and leaf senescence in spring wheat (*Triticum aestivum* L. cv. Albis) grown at different ozone concentrations in open-top field chambers. *Physiologia Plantarum* 77, 389–394.

Grantz, D.A. and Yang, S.Y. 2000. Ozone impacts on allometry and root hydraulic conductance are not mediated by source limitation or developmental age. *Journal of Experimental Botany* 51, 919–927.

Grell, G., Dudhia, J., and Stauffer, D. 1994. *A Description of the Fifth-Generation Penn State/NCAR Mesoscale Model (MM5) NCAR Tech*. Note NCAR/TN-398 + STR, National Center For Atmospheric Research, Boulder, CO.

Guenther, A., Hewitt, C.N., Erickson, D., et al. 1995. A global model of natural volatile organic compound emissions. *Journal of Geophysical Research* 100, 8873–8892.

Harmens, H., Mills, G., Hayes, F., et al. 2005. Air pollution and vegetation. Annual Report 2003/2004. ICP Vegetation Coordination Centre, Centre for Ecology and Hydrology, Bangor, UK, http://icpvegetation.ceh.ac.uk

Hassan, I.A. 2006. Physiological and biochemical response of potato (*Solanum tuberosum* L. cv. Kara) to O_3 and antioxidant chemicals: Possible roles of antioxidant enzymes. *Annals of Applied Biology* 148, 197–206.

Hassan, I.A., Ashmore, M.R., and Bell, J.N.B. 1995. Effect of ozone on radish and turnip under Egyptian field conditions. *Environmental Pollution* 89, 107–114.

Heagle, A.S., Miller, J.E., and Sherrill, D.E. 1994. A white clover system to estimate effects of tropospheric ozone on plants. *Journal of Environmental Quality* 23, 613–621.

Heagle, A.S., Philbeck, R.B., Rogers, H.H., and Letchworth, M.B. 1979. Dispensing and monitoring ozone in open-top field chambers for plant-effects studies. *Phytopathology* 69, 15–20.

Heck, W.W., Taylor, O.C., and Tingey, D.T. 1988. Assessment of crop loss from air pollutants. *Proceedings of the International Conference*, Raleigh, NC, Elsevier Applied Science, London.

Hicks, W.K., Kuylenstierna, J.C.I., Mathur, V., et al. 2001. Development of the regional policy process for air pollution in South Asia, Southern Africa and Latin America. *Water, Air and Soil Pollution* 130, 211–216.

Holland, M., Kinghorn, S., Emberson, L., et al. 2006. Development of a framework for probabilistic assessment of the economic losses caused by ozone damage to crops in Europe. Part of the UNECE International Cooperative Programme on Vegetation. Contract Report EPG 1/3/205, CEH Project No: C02309NEW, 49pp.

Ishii, S., Marshall, F.M., Bell, J.N.B., and Abdullah, A.M. 2004. Impact of ambient air pollution on locally grown rice cultivars (*Oryza sativa* L.) in Malaysia. *Water, Air and Soil Pollution* 154, 187–201.

Jäger, H.J., Unsworth, M., De Temmermann, L., and Mathy, P. (Eds). 1992. Effects of air pollution on agricultural crops in Europe—results of the European open-top chamber project. Air Pollution Research Report 46, Commission of the European Communities, Brussels.

Langner, J., Bergström, R., and Pleijel, K. 1998. European scale modeling of sulfur, oxidised nitrogen and photochemical oxidants. Model development and evaluation for the 1994 growing season. Swedish Meteorological and Hydrological Institute, RMK No. 82, Norköpping, Sweden.

Lehnherr, B., Grandjean, A., Mächler, F., and Fuhrer, J. 1997. The effect of ozone in ambient air on ribulose bisphophate carboxylase/oxygenase activity decreases photosynthesis and grain yield in wheat. *Journal of Plant Physiology* 130, 189–200.

LRTAP Convention. 2004. Manual on methodologies and criteria for modelling and mapping critical loads and levels and air pollution effects, risks and trends. (2007 revision). Available from www.icpmapping.org

Maggs, R., Wahid, A., Shamsi, S.R.A., and Ashmore, M.R. 1995. Effects of ambient air pollution on wheat and rice yield in Pakistan. *Water, Air and Soil Pollution* 85, 1311–1316.

Mauzerall, D.M. and Wang, X. 2001. Protecting agricultural crops from the effects of tropospheric ozone exposure: Reconciling science and standard setting in the United States, Europe and Asia. *Annual Review of Energy and the Environment* 26, 237–268.

McCarthy, J.J., Canziani, O.F., Leary, N.A., Dokken, D.J., and White, K.S. (Eds). 2001. *Climate Change 2001: Impacts, Adaptation, and Vulnerability*. Intergovernmental Panel on Climate Change, Working Group II. Cambridge University Press, Cambridge.

McKee, I.F., Bullimore, J.F., and Long, S.P. 1997. Will elevated CO_2 concentrations protect the yield of wheat from O_3 damage? *Plant Cell and Environment* 20, 77–84.

Mills, G., Buse, A., Gimeno, B., et al. 2007. A synthesis of AOT40-based response functions and critical levels for ozone for agricultural and horticultural crops. *Atmospheric Environment* 41, 2630–2643.

Mills, G., Ball, G., Hayes, F., et al. 2000. Development of a multi-factor model for predicting the effects of ambient ozone on the biomass of white clover. *Environmental Pollution* 109, 533–542.

Mulchi, C., Rudorff, B., Lee, E., Rowland, R., and Pausch, R. 1995. Morphological responses among crop species to full-season exposures to enhanced concentrations of atmospheric CO_2 and O_3. *Water, Air and Soil Pollution* 85, 1379–1386.

Nakićenović, N. and Swart, R., (Eds). 2000. *Special Report on Emission Scenarios*. Intergovernmental Panel on Climate Change. Cambridge University Press, Cambridge.

Nasim, G., Saeed, S., Wahid, A., and Bajwa, R. 1995. Impact of air pollution on growth, yield and vesicular arbuscular mycorrhizal status of wheat, *Triticum aestivum* var. Pak-81. *Biota* 1(1&2), 91–111.

Nghiem, L.H. and Oanh, N.T.K. 2008. Evaluation of mesoscale meteorological model (MM5)-community multi-scale air quality model (CMAQ) system performance in hindcast and forecast of ground level ozone. *Journal of the Air and Waste Management Association* 58 (10), 1341–1350.

Otter, L., Fleming, G., Wiedinmyer, C., Guenther, A., Greenberg, J., and Harley, P. 2003. Spatial and temporal variations in biogenic VOC emissions for Africa south of the equator. *Journal of Geophysical Research Atmospheres* 108(D13), 8505.

Pleijel, H., Almbring Norberg, P., Selldén, G., and Skärby, L. 1999b. Tropospheric ozone decreases biomass production in radish plants (*Raphanus sativus*) grown in rural southwest Sweden. *Environmental Pollution* 106, 143–147.

Pleijel, H., Mortensen, L., Fuhrer, J., Ojanperä, K., and Danielsson, H. 1999a. Grain protein accumulation in relation to grain yield of spring wheat (*Triticum aestivum* L.) grown in open-top chambers with different concentrations of ozone, carbon dioxide and water availability. *Agriculture, Ecosystem and Environment* 72, 265–270.

Prather, M., Gauss, M., Berntsen, B., et al. 2003. Fresh air in the 21st century? *Geophysical Research Letter* 30, 1100, doi:10.1029/2002GL016285.

Rai, R., Agrawal, M., and Agrawal, S.B. 2007. Assessment of yield losses in tropical wheat using open top chambers. *Atmospheric Environment* 41, 9543–9554.

Rajput, M. and Agrawal, M. 2004. Physiological and yield responses of pea plants to ambient air pollution. *Indian Journal of Plant Physiology* 9, 9–14.

Ramanathan, V. and Feng, Y. 2008. On avoiding dangerous anthropogenic interference with the climate system: Formidable challenges ahead. *PNAS* 105, 14245–14250.

Robertson, L., Langner, J., and Engardt, M. 1999. An Eulerian limited-area atmospheric transport model. *Journal of Applied Meteorology* 38, 190–210.

Rudorff, B.F.T., Mulchi, C.L., Lee, E.H., Rowland, R., and Pausch, R. 1996. Effects of enhanced O_3 and CO_2 enrichment on plant characteristics in wheat and corn. *Environmental Pollution* 94, 53–60.

Sanz, J., Muntifering, R.B., Bermejo, V., Gimeno, B.S., and Elvira, S. 2005. Ozone and increased nitrogen supply effects on the yield and nutritive quality of *Trifolium subterraneum*. *Atmospheric Environment* 39, 5899–5907.

Simpson, D., Anderss-Skjöld, Y., and Jenkin, M.E. 1993. Updating the chemical scheme for the EMEP MSC-W oxidant model: Current status. EMEP MSC-W Note 2/93, Oslo, Norway.

Singh, S., Agrawal, M., Agrawal, S.B., Emberson, L., and Büker, P. Use of ethylene diurea for assessing the impact of ozone on mung bean plants at a rural site in a dry tropical region of India. *International Journal of Environment and Waste Management* (in press).

Streets, D.G., Bond, T.C., Carmichael, G.R., et al. 2003. An inventory of gaseous and primary aerosol emissions in Asia in the year 2000. *Journal of Geophysical Research* 108(D21), 8809, doi:10.1029/2002JD003093.

Timsina, J. and Connor, D.J. 2001. Productivity and management of rice-wheat cropping systems: Issues and challenges. *Field Crops Research* 69, 93–132.

Tiwari, S., Agrawal, M., and Manning, W.J. 2005. Assessing the impact of ambient ozone and growth and productivity of two cultivars of wheat in India using three rates of application of ethylenediurea (EDU). *Environmental Pollution* 138, 153–160.

van Tienhoven, A.M., Zunckel, M., Emberson, L.D., Koosailee, A., and Otter, L. 2006. Preliminary assessment of risk of ozone impacts to maize (*Zea mays*) in southern Africa. *Environmental Pollution* 140, 220–230.

Varshney, C.K. and Rout, C. 1998. Ethylene diurea (EDU) protection against ozone injury in tomato plants at Delhi. *Bulletin of Environmental Contamination and Toxicology* 61, 188–193.

Vorne, V., Ojanperä, K., De Temmerman, L., et al. 2002. Effects of elevated carbon dioxide and ozone on potato tuber quality in the European multiple-site experiment CHIP-project. *European Journal of Agronomy* 17, 369–381.

Wahid, A. 2006. Influence of atmospheric pollutants on agriculture in developing countries: A case study with three new wheat cultivars in Pakistan. *Science of the Total Environment* 371, 304–313.

Wahid, A. and Maggs, R. 1999. The effects of air pollution on crops in developing countries—a case study in Pakistan. *Acta Scientia* 9(2), 51–63.

Wahid, A., Maggs, R., Shamsi, S.R.A., Bell, J.N.B., and Ashmore, M.R. 1995a. Air pollution and its impact on wheat yield in the Pakistan Punjab. *Environmental Pollution* 88, 147–154.

Wahid, A., Maggs, R., Shamsi, S.R.A., Bell, J.N.B., and Ashmore, M.R. 1995b. Effects of air pollution on rice yield in the Pakistan Punjab. *Environmental Pollution* 90, 323–329.

Wahid, A., Milne, E., Shamsi, S.R.A., Ashmore, M.R., and Marshall, F.M. 2001. Effects of oxidants on soybean growth and yield in the Pakistan Punjab. *Environmental Pollution* 113, 271–280.

Wahid, A., Shamsi, S.R.A., Bell, J.N.B., and Ashmore, M.R. 1997. Effects of ambient air pollution on the yield of some rice varieties grown in open-top chambers in Lahore, Pakistan. *Acta Scientia* 7(2), 141–152.

Wang, X. and Mauzerall, D.L. 2004. Characterizing distributions of surface ozone and it's impact on grain production in China, Japan and South Korea: 1990 and 2020. *Atmospheric Environment* 38, 4383–4402.

Zunckel, M., Koosailee, A., Yarwood, G., et al. 2006. Modeled surface ozone over southern Africa during the cross border air pollution impact assessment project. *Environmental Modeling & Software* 21, 911–924.

16 Impacts of Air Pollution on the Ecosystem and Human Health
A Sustainability Perspective

Ioan Manuel Ciumasu and Naela Costica

CONTENTS

16.1 ECOSYSTEMS, SOCIETY, AND ECONOMY

Environmental sustainability of human activities is nowadays of commonplace interest that comes along the nowadays economic and social issues. An economy is not a totally self-regulated system (Van Griethuysen, 2002). On the contrary, economic systems are subsystems of social systems, which are themselves subsystems of natural systems (Giddings et al., 2002). At the planetary scale, taking into account the

447

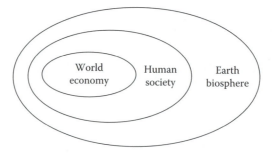

FIGURE 16.1 Relationship between world economy, human society, and the Earth's biosphere. Any economic system is part of a social system, which is part of a natural system.

phenomenon of globalization, the above relationship can be described as follows: world economy is a subsystem of humanity, which is a subsystem of the Earth's biosphere (Figure 16.1).

Any economic system is part of a social system, which is part of a natural system. Consequently, human health and welfare cannot be perpetuated in ignorance of the health state of ecosystems. Rather, human health begins and ends with the health of the surrounding ecosystems (Waltner-Towes, 2004), passing through social equilibrium and socially responsible economy (Scholtens, 2005). Sustainable health care is dependent on environmental health and sustainability. Caring for ecosystem health is taking care of our own health, thus preventing much higher health risks and expenses on medical treatments (Jameton and Pierce, 2001).

The Earth's ecosystems are integrated by global biogeochemical cycles through natural circuits involving the atmosphere, the hydrosphere, and the lithosphere. Human economic activities influence natural energy fluxes and biogeochemical cycles, and hence induce various stresses on ecosystem dynamics and health.

In ecosystems, stress is a syndrome manifested through changes in ecosystem energetics, nutrient cycling, and the structure and function of natural species and communities. In this context, human activity may be understood as a "stressor," that is, an activity with "detrimental or disorganizing influence" (Odum, 1985; Rapport et al., 1985). Terrestrial and aquatic ecosystems are linked mainly through the gravitational movement of materials from land to water and from the atmosphere to land and water. This general dynamics accelerates anthropogenic pollution.

Intelligent management of natural resources must hold a holistic view on such air–land–water interactions (Likens and Bormann, 1974). This approach is based on the broad concept of *sustainable development* of society, meaning precisely the kind of development that satisfies the needs of the present generation without compromising the possibility of future generations to satisfy theirs. The concept was officially launched by the World Commission on Environment and Development, which was established in 1983 by the United Nations General Assembly, in its report to the United Nations called *Our Common Future*. The report is also known as the Brundtland Report, after the name of its chair, the Norwegian Prime Minister Mrs Gro Harlem Brundtland (WCED, 1987). In other words, in order to mitigate global risks, the long-term socioeconomic value of nature must be preserved.

16.2   ENVIRONMENTALLY CONSCIOUS DEVELOPMENT: HOW TO FATHOM THE CAUSES AND EFFECTS OF AIR POLLUTION

Various pollutants have attracted the interest of society. Basically, pollution appears to follow the different stages and patterns of the historical development of human society and economy. In this respect, air pollution was at the heart of the scientific and public concerns about the relationship between economic activity and environmental and human health.

Perhaps the most comprehensive scientific debate, and the most relevant for the public and the decision makers, is the one around the *Environmental Kuznets Curve* (EKC) hypothesis. According to the EKC theory, pollution intensity (and the depletion of natural resources) increases with socioeconomic development (often expressed in terms of per capita income) up to a turning point, and decreases thereafter. This inverted-U relationship is visible in Figure 16.2, which extends to entire societies an earlier relationship described by Panayotou (1993) in relation to economic stages, but restricts it to environmental pollution.

Drawing on the experience of developed countries, typical EKC estimates correlate income with aerial emissions of certain pollutants, most famously with sulfur dioxide (SO_2) (Stern et al., 1996; Richard et al., 2004). Yet, that this pattern can be generalized to all pollutants and to a cross section of countries has not been resolved (Jha and Bhanu Murthy, 2003).

Some examples of EKC-type relationships in a series of European countries, calculated on the basis of recently available data on emissions for SO_2, NO_2, and CO_2, are presented below (Figures 16.3 and 16.4). By selecting a few countries representing different stages of development (industrialization), an EKC-type tendency is detectable (Figure 16.3). In this case, Hungary roughly represents the preindustrial stage of development, Greece and Spain the industrial stage, and the UK and Germany the postindustrial stage. Considering these data, an inverted-U relationship is more easily visible with SO_2 and NO_2 data than with CO_2 data. However, an N-shaped relationship can also be detected, depending on the fitted equation.

Several explanations are possible for the inverted-U EKC pattern. The most cited one is that increasing income with time is usually associated with structural economic changes. Thus, initial development is realized through massive industrialization, typically with highly polluting technologies. Afterwards, new economic

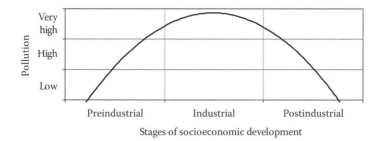

FIGURE 16.2   Hypothetical inverted-U relationship between environmental pollution and stages of socioeconomic development.

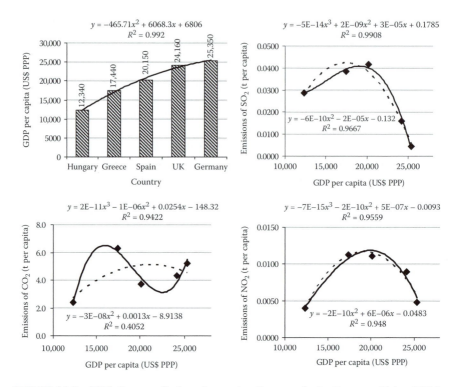

FIGURE 16.3 EKCs for a set of selected countries: Germany [in the European Union (EU) by 1957], UK (in the EU by 1973), Spain (in the EU by 1986), Greece (in the EU by 1981), and Hungary (in the EU by 2004): the relationship between emission of SO_2, NO_2, and CO_2 and income per inhabitant (per capita gross domestic product—GDP) calculated at purchasing power parity (PPP). Emission data are from the recently established European Pollutant Register—EPER (data for 2001), http://eper.cec.eu.int/eper/introduction.asp?i=, whereas GDP data (data for 2001) are from the UNDP, Human Development Report 2003 (HDR, 2003).

sectors develop, notably services, reducing the relative share of industries. In addition, new fast developing countries tend to be less resource efficient and therefore more polluting. For instance, emerging economies use more than double the amount of oil used by developed economies to produce one dollar of GDP (*The Economist*, 2005b).

At the same time, once people begin to escape from the basic economic constraints, they demand life in a healthier environment and pressurize for the adoption of less polluting activities and technologies (Barbier, 2005). In short, environmental degradation is dynamically linked to general human development, generally described by the human development index (HDI), an aggregate measure of economic well-being, health, and education (HDR, 2003).

However, when most countries for which data are available are added (mainly industrial and postindustrial countries), the inverted-U tendency is not so easily detectable (as reflected by a weak R^2 of the fitted binomial equation), indicating a much more complex situation than that suggested by the inverted-U EKC theory (Figure 16.4).

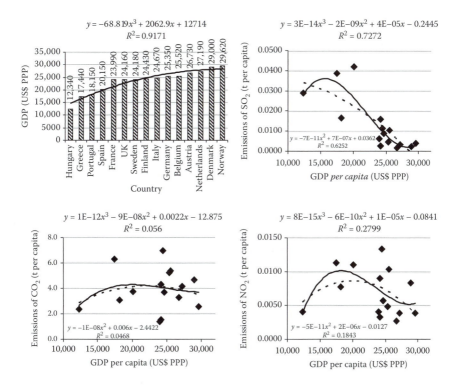

FIGURE 16.4   EKCs: the relationship between emissions of SO_2, NO_2, and CO_2 and income per inhabitant (per capita GDP) calculated at PPP. Emission data are from the recently established European Pollutant Register—EPER (data for 2001), http://eper.cec.eu.int/eper/introduction.asp?i=, whereas GDP data (data for 2001) are from the UNDP, Human Development Report 2003 (HDR, 2003).

As already observed in Figures 16.3 and 16.4 and more clearly represented in Figure 16.5, the relationship between income and emissions is sometimes better described by an N-shaped graph. An N-relationship signifies that after an initial turning point and a decrease of emissions with increasing income, emissions may again start to increase due to increasing energy consumption in highly affluent societies. This is visible when the two most affluent countries in Europe are added to the pictures (Figure 16.5). This tendency is probably due to fresh industrial momentum brought in by new technologies and means of industrial production, and is most likely supported by intense domestic consumption.

Taken to the extreme, the EKC hypothesis suggests that, afterwards, we should not care too much about the environment, because environmental problems will be overcome naturally with further economic development. It has even been proposed that the only way of resolving environmental problems is by becoming rich, and any limitation imposed on economic growth compromises efforts to safeguard the environment (Beckerman, 1992).

However, it is very unlikely that things happen that way. First, despite being a valuable tool in scientific and public debates, EKC estimates can be misleading,

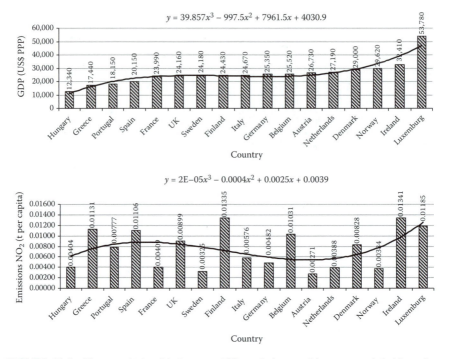

FIGURE 16.5 N-type relationship between NO_2 emissions and income per inhabitant (per capita GDP) calculated at PPP, indicating a restart of emissions with highly affluent societies. NO_2 emission data are from the recently established European Pollutant Register—EPER (2001), http://eper.cec.eu.int/eper/introduction.asp?i=, whereas GDP data (data for 2001) are from the UNDP, Human Development Report 2003 (HDR, 2003).

because they only use certain pollutants or simple pollution indicators, which cannot substitute for descriptions of real ecosystems. For example, the decreasing air pollution with SO_2 in rich countries does not mean that the problem has been solved: it may simply mean that such polluting activities are no longer the central motor of these national economies and that they have been relocated to poorer countries, for example Asia. Indeed, the trend of global SO_2 emissions is actually increasing (Stern et al., 1996; Dasgupta et al., 2002). As significant modifications in the patterns of atmospheric circulation are expected with the current climate changes, atmospheric circulation may spread SO_2 all over the Earth.

Indeed, according to the latest comprehensive report on aerial emissions by the United Nations Intergovernmental Panel on Climate Change, the inverted-U relationship is possible but so are other emission scenarios. Particularly for CO_2, economic development is not itself a trigger of less emissions (SRES, 2000; SR, 2007).

In a wider perspective, many specific environmental indices exist for all the major environmental issues. Examples include those used by the Organisation for Economic Co-operation and Development (OECD) for measuring greenhouse gas (GHG) emissions, apparent consumption of ozone-depleting substances (ODSs), SO_x, and NO_x emission intensities, municipal waste generation intensities, wastewater treatment connection rates, intensities of use of energy and water, forest, and fish resources, and threatened species (OECD, 2004). But they do not offer an integrated picture of the state of the environment.

Therefore, there is a strong need to use composite indices, which relate global environmental degradation to economic development. But no matter how efficient they are in encapsulating complex realities in simple terms, aggregated indices must be used critically, since the more complex the reality they concentrate, the higher the danger that they do not apply to the intended conditions.

Many authors have expressed the opinion that environmental degradation will inevitably increase with increasing economic activity, and so lead to a decline in general human welfare despite increasing incomes, because of ever higher production and consumption and hence larger material and energy inputs and larger quantities of waste by-products (Georgescu-Roegen, 1971; Daly, 1991). Following this logic, this tendency amounts to placing economic activity itself at risk, because of the overexploitation of irreplaceable natural capital (Jansson et al., 1994; Giddings et al., 2002).

16.3 ENVIRONMENTALLY CONSCIOUS DEVELOPMENT: NATURAL CAPITAL IS IRREPLACEABLE

Valuation of natural resources can only be based on the properties of natural systems, including the often nonlinear behavior as well as unpredictable and irreversible processes (Straton, 2006). Natural capital, together with physical capital and human capital, forms the total capital stock (Barbier, 2005).

Natural capital relates to a plethora of ecosystem goods and services that lay the foundation for the further development of physical and human capital. These benefits arise from both the general and the peculiar properties of each ecosystem. With hardly any exception, these benefits are essential to human health and general welfare (De Groot et al., 2002):

- Atmospheric gas regulation (biogeochemical cycles)
- Climate regulation (biologically mediated processes; local and global climate)
- Disturbance prevention (dampening)
- Water regulation (land cover and river collection basins)
- Water supply (filtering, retention, and storage)
- Oil formation and retention
- Nutrient regulation (recycling)
- Waste treatment (biodegradation and bioassimilation of pollutants)
- Pollination
- Biological control of pests (via natural food chains and food pyramids)
- Food and medicinal resources
- Raw materials
- Genetic resources
- Ornamental resources
- Recreation and inspiration, and education

Ecosystem functions and services are dependent on key processes of biochemical cycles, such as biomass production and biomass decay by specific organisms called primary producers and decomposers, respectively. Decomposers are invertebrate detritivores, which break down litter into small pieces that can be further reduced chemically, and bacterial and fungal decomposers, which decompose the small pieces into inorganic molecules such as ammonium, phosphate, water, and carbon dioxide (CO_2) (Aerts, 2006). The biomass and community composition of detritivores are functionally correlated with plant chemistry and productivity and, through them, with herbivores (consumers) (Wurst et al., 2003) and bacterial chemical decomposers (Scheu et al., 2002).

In brief, species and the functional groups of organisms are interdependent, so that the integrity of the natural composition of vegetal, animal, and bacterial communities is an essential condition for the preservation of all ecosystem structure and function (for in-depth ecological descriptions, see Chapman and Reiss, 1999).

There are voices claiming that gains in physical and human capital should be able to compensate for a great deal of, or even the entire, natural capital loss. But compensations, if they exist, can only be taken into account in relation to some critical natural capital (thresholds), after which compensations cannot function anymore. But such critical capital cannot be identified because of the complexity of the natural systems, particularly because of the inherently nonlinear dynamics of ecosystems, and because of their historical character (Ekins, 2003). Given this peculiar complexity, it is hard to conceive (to say the least) that knowledge and technology can make up for ultimate losses of the natural capital (Ehrlich et al., 1999; Barbier, 2005). In this situation, blind action annihilating natural capital is the worst possible course of action, and should be avoided.

16.4 ENVIRONMENTALLY CONSCIOUS DEVELOPMENT: THE CARRYING CAPACITY AND THE ECOLOGICAL FOOTPRINT

Environmental degradation by humans is basically caused by three factors: the size of the human population supported by local ecosystems, the average individual consumption, and the inappropriate use of technology (Ehrlich and Ehrlich, 1981). The overall relationship between humans and ecosystems is captured by two interrelated concepts from ecological economics: the ecosystem *carrying capacity* and the *ecological footprint* of humans.

Depending on bioproductivity and other biological parameters, support ecosystems afford a certain amount of exploitation above their own regenerative capacity: renewable resources. Each ecosystem has its particular, lower or higher "carrying capacity." The impact of human activity on ecosystems can be estimated in its basic parameters and is referred to as the "ecological footprint." This usually comprises nourishment, shelter, mobility, and goods and services, and is strongly correlated with income (Wackernagel and Rees, 1996; Barrett and Simmons, 2003).

According to preliminary calculations, the global ecological footprint of man already equaled the regenerative capacity of the biosphere in the late 1980s, and exceeded it by 20% in around the year 2000 (Wackernagel et al., 2002). Some authors already locate this event in the late 1970s (Wilson, 2002).

Typical high-income cities use the productive and assimilative capacity of a wide hinterland, and their ecological footprint is several hundred times larger than the areas they physically occupy (Folke et al., 1997). The ecological footprint of highly populated and urbanized countries may represent an area of over several times the national territory, for example, three times in the case of UK (Barrett and Simmons, 2003).

Such numbers dismiss the idea that environmental problems are being overcome when countries get richer. On the contrary, national economies in developed countries appear to have strong ecological footprints, low flexibility in mitigating it (Richard et al., 2004), and unequally distributed effects, including a real impact on other countries' support ecosystems (Stern et al., 1996).

Ecological footprint estimations signal that fundamentally new approaches and solutions are needed for future socioeconomic development. The ultimate challenge is to transfer true sustainability knowledge and practice into the world business community (Gray and Milne, 2002; Boron and Murray, 2003). For this, one tool is the ecological footprint index (EFI), which measures the amount of natural resources needed by people (Wackernagel and Rees, 1996). This index is essential in linking economy to ecology because it proposes a financial valuation of ecosystem goods and services, which actually makes the concepts of natural capital and ecosystem goods and services economically understandable (and financially amenable).

Ecosystem resources should be viewed necessarily, but not exclusively, as a natural capital that should be managed and developed as a limited and irreplaceable resource for the human economy and that should be continuously studied and monitored (Ekins, 2003). In addition, ecological education (EE) must be fully committed toward economic agents, and included in programs and university curricula of economical education.

16.5 CO₂ EMISSIONS AND CLIMATE CHANGE: RISKS TO ECOSYSTEMS AND HUMAN HEALTH AND WELFARE

Probably the most ample effects of human activities on the environment are climate changes. These appear to be caused mainly by industrial emissions of CO_2, but also by emissions of CH_4, N_2O, and other GHGs constituting "aggregate GHG" (UNDESA-DSD, 2006; SR, 2007).

Climate changes modify many functional patterns of natural systems, often endangering the very existence of the more fragile ecosystems and the material security of human populations depending on them. Climate changes may have important feedback effects with ecosystems, basically because of repercussions on energy fluxes in the ecosystems: food chains and food webs formed by biomass producers, biomass consumers, and decomposers. For example, global warming has consequences on the decomposition of plant litter (including root litter), a key process that was estimated to contribute to about 70% of the total annual carbon flux (Reich and Schlesinger, 1992). Such consequences are mediated in the short term by the physicochemical properties of soils (including temperature and humidity), and in the long term by litter quality resulting from climate-induced changes in plant community composition (Aerts, 2006).

Local climate warming and shifts in precipitation regimes force communities to change and species to relocate so that they remain within the limits of their adaptation capacity and optimal physiological and ecological requirements. Global analyses documented shifts in species range boundaries toward the poles averaging 6.1 km per decade (Parmesan and Yohe, 2003). Such pressures are particularly critical for specialized species, for example with narrow limits of temperature tolerance—stenotherm species. But to do so, the species must overcome two other major obstacles. First, because of massive emissions of GHGs in a relatively short time, the speed of climatic changes is so high that species may not have the time to move the way they naturally do. Second, even if species could theoretically relocate fast, they would not have sufficient moving corridors between the present and the would-be geographical ranges, since most of the natural habitats have been destroyed through conversion to agriculture and other uses (Honney et al., 2002; Armstrong, 2005; Jump and Penuelas, 2005).

This situation raises serious concerns about the large-scale consequences of greenhouse emissions on ecosystem integrity and the natural resources on which societies depend. Limiting emissions of GHGs is of paramount importance for ecosystem sustainability.

Some areas on the Earth are more susceptible to suffer as a consequence of climate changes. For example, high-latitude areas are very sensitive to global warming. Arctic and subarctic tundra ecosystems, as well as high-altitude ecosystems, are most valuable indicators of the possible effects of global warming on the Earth's ecosystems (Rusek, 1993). Thus, tundra regions are known to function as CO_2 sinks, because cold winter temperatures prevent the decomposition of organic matter produced during the arctic summer. Lately, scientists have begun to study the possibility that global warming may trigger a positive feedback where huge stores of organic matter in tundra soils begin to decompose (provided there is enough moisture), and thus begin to behave like carbon sources contributing to GHG emissions and global warming (Oechel et al., 1993).

In addition, one must point out that plants and vegetation in all biogeographic provinces are not only CO_2 sinks (through photosynthesis) but also CO_2 sources (through respiration). Increasing global and local mean temperatures may shift the balance of CO_2 emission/fixation, and not necessarily toward CO_2 fixation.

Elevated CO_2 concentrations and global warming determine changes in the composition of plant communities, either directly or mediated, and hence determine fundamental changes in the ecological mechanisms regulating those ecosystems (Aerts, 2006). For example, climatic changes at high latitudes have impacts on both the reproductive ecology of flowering plants (Aerts et al., 2004) and the dynamics of nutrients (Owensby et al., 1993). Because such effects are many and influence each other, relatively sudden climatic changes will have unpredictable and disturbing effects on ecosystem functioning.

Climatic changes are also important in tropical areas. For example, a potential increase of rainfall in the Indian subcontinent by up to 50% is expected to result in important stresses on the livelihood sustainability of populations through stresses on ecosystem functions. In such cases, studying possible scenarios is imperative (Ramakrishnan, 1998).

Another major concern related to air-pollution-driven climate changes is the changing epidemiological pattern in both ecosystems and human populations. Climate warming may change the host–pathogen relationships, and favor the spread of pests that would otherwise be kept in control by low temperatures. For example, insect pests could increase the number of generations/life cycles per year, and so provoke widespread damage to forests and crops that are already under direct stress from climate warming. This would favor population declines and extinctions of many plant and animal species, with dramatic consequences on ecosystem stability and productivity (Rossignol et al., 2006). In Arctic areas, there is already evidence that the dynamics of host–parasite systems has begun to change, mainly through effects on the transmission of parasites (Kutz et al., 2005).

Synthetic indices might be used to estimate the degradation of natural resources. For example, the environmental degradation index (EDI) proposed by Jha and Bhanu Murthy (2003) provides policy makers and environmental managers with a practical end of the EKC theory. EDI is based on calculations of a global EKC (GECK) out of common, separated EKC analyses.

An environmental sustainability index (ESI) has been proposed by Esty et al. (2005) and is supported by a consortium led by the Yale Center for Environmental Law and Policy of the Yale University. ESI ranks countries according to their performances in environmental stewardship. Scores are calculated on the basis of five main components: current health state of natural systems, level of anthropogenic stresses on the environment, level of human and societal vulnerability to environmental disturbances affecting human well-being, societal and institutional capacity to respond to environmental challenges, and capacity to address environmental issues in the international arena. It is worth noting that no country scores particularly high on the five components, which means that all countries have much to learn from the experience of others. Another interesting observation made by the ESI team is that the best-scoring countries all have strong natural endowment and low population density. This means that those high calculated standards still need to be demonstrated in the future with challenges of higher orders of magnitude, for example with future climate-related environmental changes.

For human populations across the globe, climate warming will mostly have adverse effects, with higher rates of illness due to direct and indirect causes. In its 2002 World Health Report entitled *Reducing Risks, Promoting Healthy Lives*, the World Health Organization (WHO) estimated that warming and precipitation trends due to climate changes in the past three decades are responsible for about 150,000 deaths annually (WHO, 2002).

Many direct causes are linked to meteorological extremes. According to the World Health Organization Europe (WHOE, 2004), classical heat-related illnesses are skin eruptions, heat fatigue, heat cramps, heat syncope, heat exhaustion, and heat stroke. At higher body temperatures caused by deficient air regulation in warm air, the circulatory volume dilates, causing decreases in blood pressure and oxygen supply to the brain, leading to heat syncope. Blood pressure decreases are accelerated by dehydration and lack of muscle activity. Muscle activity helps increase blood pressure and prevents syncope, leading to a further increase in body temperature and heat exhaustion. If the body temperature rises above 40.5°C, like with intense activity during heat waves, the body enters a heat stroke, in which damage may occur at the level of the cellular structure and the thermoregulatory system, with a high risk of mortality.

Complications of heat stroke include severe functional impairment such as respiratory distress syndrome, kidney and liver failure, and disseminated intravascular coagulation. Lesser fit persons (such as some of the elderly) have a lower heat tolerance. In their case, heat-related illness can even occur at less or no physical effort. Many occasional situations that are usually not regarded as posing great health risk can be predisposing factors for heat-related illness: poor acclimatization, intestinal problems, use of diuretics, alcohol abuse, use of medications affecting temperature regulation systems, overweight, fatigue, sleep deprivation, prolonged and intense exercise, and clothing.

In addition, vasodilatation and dehydration may exacerbate other health problems such as cardiovascular and respiratory diseases. Heat waves increase the mortality of cardiovascular and respiratory illnesses. During the summer of 2003, with average temperatures of 3.5°C above normal, it was estimated that about 22,000–45,000 people died in a two-week-long heat wave in Europe (WDR, 2004; Kosatsky, 2005; Patz et al., 2005).

Climate warming and heat-related illnesses are particularly frequent in cities, because cities warm up the local climate, a phenomenon referred to as the "urban heat island" effect. The temperature difference between the city and its surrounding area increases with the number of inhabitants and the building density: from up to 2.5°C for towns of 1000 inhabitants to 12°C for cities of one million inhabitants in America. This effect is caused by many factors related to the modification of local natural ecosystems, such as (WHOE, 2004) increased exposure to radiation, less radiant heat loss in the urban vegetation, lower wind velocities (meaning lower rates of heat dissipation), changes in the water balance, and higher domestic and industrial heat production (including waste heat and emission of air pollutants). In cities, the health stress induced during severe heat waves is not relieved at night, which boosts casualty numbers (Epstein, 2005).

Increasing intensities and frequencies of extreme events such as hurricanes and floods increase the rates of injury and death. However, a word of caution is necessary: the precise links between climate warming and hurricanes are still under debate (Pielke Jr. et al., 2005), and climate changes should not be taken as an excuse for casualties caused by human mismanagement of emergency situations (SR, 2007).

Indirect causes are linked to ecosystem stress and the subsequent decrease or removal of ecosystem benefits. Some examples of perturbations of natural mechanisms and phenomena with repercussions on human health are the following (Kalkstein and Smoyer, 1993; Haines et al., 2000; Patz, 2002; WHO, 2002; Epstein, 2005; Patz et al., 2005):

- Prolonged drought depletes rodent predators, which may lead to increases in the populations of rodents that can spread viruses and trigger outbreaks of various diseases.
- Mosquitoes, which are vectors of many diseases, are very sensitive to temperature changes. Higher temperatures increase their rate of reproduction and the number of blood meals they take, prolong their breeding season, and enlarge

their area of living (for example, they move up the highlands). Higher temperatures also shorten the maturation period of the microbes they spread.

- Higher temperatures allow diseases such as cholera, malaria, dengue, and viral encephalitis to shorten incubation times, achieve higher infection incidence, and spread over new geographical areas.
- In the aftermath of very destructive hurricanes, the incidence of malaria, dengue fever, cholera, and leptospirosis soars.
- Floods are often followed by disease clusters, because of disturbances in the populations of pests and vectors and in the living conditions of people in the affected area, notably deficient nutrition, hygiene, and medication.
- Higher temperatures may trigger algal blooms and subsequent intoxication of fishes and humans.
- Higher temperatures also favor the higher production and dissemination of allergens such as pollen and fungal spores.
- Dust clouds emanate from expanding deserts, and increased temperature gradients into the atmosphere cause faster dust transport at the intercontinental scale, bringing respiratory irritants to the population.
- Climate extremes may reduce the success in crops and food production, and hence increase the incidence of alimentation-related diseases and malnutrition-related susceptibility to infections in certain countries and regions.
- The overexploitation of local natural resources in the wake of natural disasters such as hurricanes and floods leads to the lowering of human living conditions and vulnerability to further natural or socioeconomic shocks.
- Climate stresses lower the resistance of human beings to pollution. Such effects hinder both population health status and costs of public and private health care.

In addition, the effects of epidemiological and other types of climate-driven changes on human health do not equally affect human populations. In reality, such health risks contribute to widening the health inequalities among humans (WHO, 2002; Epstein, 2005; Sunyer and Grimalt, 2006). Actually, vulnerability is currently deepening in underdeveloped countries and within disfavored human populations and social categories. The true consequences for the good functioning of society as a whole are hard to anticipate.

An index to capture social vulnerability to climate changes was proposed by Vincent (2004). Social vulnerability is understood in terms of both exposure to risks and social capacity to cope with risks. This social vulnerability index (SVI) makes special reference to African countries and results from a weighted averaging of five composite subindices: economic well-being and stability (20%), demographic structure (20%), institutional stability and strength of public infrastructure (40%), global interconnectivity (10%), and natural resource dependence (notably water availability) (10%). SVI is meant to be used in conjunction with appropriate indicators of biophysical vulnerability. The main uses of SVI are to orientate efforts of capacity building for adaptive management to those countries that are most in need, and to proceed to this in a structured, and hence efficient, manner (Vincent, 2004). Obviously, it is advisable to use SVI in conjunction with the earlier mentioned HDI of the United Nations Development Programme.

The issue of social dilemmas and fair adaptation to climate change is a complex one, where general commitments are easy to take but responsibilities are ambiguous. Solutions have been proposed, where the accent is on the need to remove obstacles to fair participation of planning and decision making in adaptation (Paavola and Adger, 2005). This again raises the problem of poverty and human development. Any solution begins with education, particularly education on environmental changes and sustainable development.

16.6 AIR POLLUTION AND UV RADIATION: RISKS TO ECOSYSTEMS AND HUMAN HEALTH

Since already several decades ago, the release of organic compounds such as chlorinated fluorocarbons (CFCs) and other reactive pollutants into the atmosphere has led to another major issue: stratospheric ozone is being broken down, so that the ozone layer of the stratosphere is becoming thinner, allowing more ultraviolet (UV) radiation to reach the surface of the Earth (Herman et al., 1996).

In the end, anthropogenic air pollution is causing organisms to be exposed to increased natural doses of UV radiation. This affects biological processes and ecosystem functioning, including agricultural, grassland, and forest exploitation. Some direct effects on individual organisms (including humans) are production of active oxygen species and free radicals, photochemical lesions of DNA and proteins, and partial inhibition of photosynthesis. But many other negative effects of UV radiation are indirect and very complex, with wide and far-reaching consequences for ecosystem stability and human welfare.

Such effects concern various mechanisms of ecosystem structure and function (Caldwell et al., 1998):

- Plant biomass production
- Plant tissue changes, leading to different plant susceptibility to infection by pathogens and plant consumption by herbivores (including insects)
- Seed production
- Disease incidents in plants and animals (such as skin cancer and eye damage in humans and other vertebrates, from fish to mammals)
- Accumulation effects of UV radiation over time in sexually reproducing organisms
- Plant and animal population fluctuations (due, for example in insects, to the fact that different development stages such as eggs, larvae, pupae, and adults have different degrees of vulnerability to UV)
- Changes in species composition (including plants, animals, and microorganisms) due to various responses of species to UV effects
- Mineral nutrient cycling

Some effects are very subtle, but no less important for ecosystems. For example, UV radiation hinders the symbiotic association between various plant species and arbuscular mycorrhizal fungi (AMF), possibly because of UV-induced changes in plant hormone levels. Such effects have important consequences on the biodiversity, productivity, and nutrient dynamics of ecosystems (Van de Staaij et al., 2001). As another example, UV radiation can trigger entire biochemical pathways, which can act as protective responses of plants. Thus, induction of the phenylpropanoid pathway appears as a universal adaptation mechanism to UV radiation in the plant kingdom. Phenolics such as flavonoids seem to lower the effect of UV radiation on photosynthetic tissues (Meijkamp, 1999).

Currently, ODSs are being controlled by the *Montreal Protocol on Substances that Deplete the Ozone Layer*, from 1987, several times adjusted and/or amended since then under the Ozone Secretariat of the United Nations Environment Programme (UNEP) (http://www.unep.org/ozone/pdfs/Montreal-Protocol2000.pdf). The Protocol was negotiated under the frame of the *Vienna Convention for the Protection of the Ozone Layer* (http://www.unep.ch/ozone/pdfs/viennaconvention2002.pdf). Under the Montreal Protocol, ODSs started being phased out and replaced with other substances. Since 1991, developing countries are being financially assisted to comply with the Montreal Protocol through a mechanism called the *Multilateral Fund for the Implementation of the Montreal Protocol* (http://www.multilateralfund.org/). Although there are differences in the emissions and control of ODSs between developed and developing countries, significant uncertainties persist in all countries (SOLGCS, 2005). See Chapter 17 for more discussion about CFCs and stratospheric ozone depletion.

16.7   AIR POLLUTION AND ACID DEPOSITIONS: RISKS TO ECOSYSTEMS AND HUMAN HEALTH

What actually are the consequences of SO_2 pollution on ecosystems and human health that we should expect and deal with in future? Such knowledge is already available from ecological studies in the regions and countries that have already experienced this type of pollution, namely Europe and North America. The problem is known as acid deposition or more popularly as *acid rain*, and involves the dry or wet deposition of not only sulfur oxides (SO_x) but also other oxides, notably NO_x. At present, acid depositions are already declining in Europe and North America, but the phenomenon is not over yet. Moreover, in other regions of the Earth, acid depositions may soar, depending on the local emissions of SO_x and NO_x into the atmosphere and on regional and local climate.

In the first place, the acid deposition problem was acknowledged as such by decision makers when ecosystem health and productivity, and implicitly human health, were proved to be at risk. Notably, noxious effects of acid depositions were first observed in forests and fish (Overrein et al., 1980; Sprinz, 1992). Afterwards, more data were gathered on the affected ecosystems. For example, the geographical declines of birds correlate with the geographical occurrence of acid depositions and with thinner eggshells. This suggests that the acid-induced depletion of calcium in the environment hinders the nutrition, fitness, and reproduction success of birds; however, the involved mechanisms are not straightforward, nor always clear (Hames et al., 2002).

The superposition of atmospheric acid depositions on natural acidification processes in soils determines the acidification of soils and surface waters (Galloway et al., 1984; Reuss et al., 1987). This has negative consequences on ecosystems, either indirectly or directly. For example, mountain forests are directly exposed to acid clouds, with sulfate being more harmful than nitrate (Cape, 1993).

Acid depositions influence biogeochemical cycles, which drive the good functioning of ecosystems. First, they can hinder (individually or combined with other factors) photosynthesis, and hence primary production and consumers. For example, acid rain and subsequent streams acidification could locally diminish or even destroy entire populations of Atlantic salmon in Nova Scotia, in the absence of treatments with limestone (Watt, 1986). Second, acid depositions affect complex soil biotas, and hence biomass decomposition and nutrient cycling. For example, they affect populations of decomposer species such as enchytraeid worms (Oligochaeta) (Abrahamsen, 1983) and also biotic relationships between soil arthropods (Haagvar, 1984). Acid depositions also affect microbial activity in soils, mainly through changes in the nitrogen supply (Killham et al., 1983). Similarly, entire populations of lichens may be damaged to the point of extinction, probably through the effect of low pH on nitrogen fixation by their algal symbiont (Gilbert, 1986).

Today, acid depositions are tending to decrease in Europe and North America, but emissions of SO_x are increasing in Asia and worldwide (Galloway, 1995; Barbier, 2005). Therefore, it is worth knowing whether various natural processes may come to the rescue in buffering acidic deposition. Alkaline soil dust and fly ash may prevent the spread of acid rains in certain regions, such as India (Khemani et al., 1989). Biological buffering mechanisms are also known. Insufficient inputs of calcium (e.g., from atmospheric deposition and mineral weathering of silicate rocks) in base-poor forest ecosystems can be compensated for by weathering of apatite (calcium phosphate) through the activity of fungi living in symbiosis with the trees (forming ectomycorrhizae on tree roots). The fungi absorb the released ions directly and provide it to the trees. Calcium ions are further circulated into ecosystems through litter decomposition in soil. However, not all tree species have such mycorrhizae, and the process is dependent on specific local conditions (Blum et al., 2002).

There are long-term effects of acid rains in ecosystems. Because of acid rains and declines in the atmospheric deposition of base cations, large quantities of Ca and Mg can be lost from soil complexes and transported by drainage waters, which delays the recovery of soil and water chemistry following decreases of acid deposition. Such lingering effects can cause episodic acidification of streams in regions previously affected by acid deposition (Likens et al., 1996; Krajick, 2001; Lawrence, 2002).

Lake and stream acidification is chemically reversible. Biological recovery is also possible, but only to a certain extent, particularly when acidic input never stops completely. Reinstallation of typical predisturbance communities is a rather optimistic perspective (Keller et al., 1998). This has been called "chronic disturbance" of the environment. Although there is a tendency to discuss more about "acute disturbances," the two are equally destructive (Singh, 1998).

16.8 AIR POLLUTION: COMBINED EFFECTS ON ECOSYSTEMS AND HUMAN HEALTH

Seldom do natural systems undergo isolated stresses; rather they face a combination of stressor factors, with combined effects (Brydges, 2001). For example, acid deposition can interact with climatic changes and can have combined effects on terrestrial and aquatic ecosystems. Sometimes, acid deposition may not have started the problem, but it prevents its resolving. Acidification of certain lakes can be induced solely by climatic changes: for example, the acid-neutralizing capacity (ANC) of lakes can decrease after significant droughts modified local hydrology (Webster et al., 1990) or after local climate warming modified the rates of biological activity (Sommaruga-Wograth et al., 1997).

Acid depositions may have various effects on crops, including significant yield loss in many species (Lee et al., 1981). Organic acids in soils moderate pH changes following acid deposition reduction or cessation (Wright, 1989), but excessive bio-mass removal in certain terrestrial ecosystems (e.g., agricultural landscapes) may deplete nutrients and the ANC of soils, so that acidification may occur even without any acid deposition. When acid deposition appears, it only worsens the situation and hinders recovery from previously overexploited ecosystems (Glatzel, 1991).

Besides SO_x, NO_x depositions may induce major disturbances in natural ecosystems. Even small increases in nitrogen deposition can determine observable changes in ecosystem properties, because the nitrogen cycle is essential for ecosystems. According to some estimates, anthropogenic alteration of the nitrogen cycle has roughly doubled the nitrogen input rate into the terrestrial nitrogen cycle, increased the global concentration of N_2O (a GHG), increased the concentration of NO_x, which contributes to the formation of ozone in the troposphere (in smog), contributed significantly to the loss of nutrients that are essential for long-term soil fertility (e.g., calcium and potassium), and increased the transfer of nitrogen in coastal and estuarine areas.

Similarly, acidification of soils through dry and wet deposition of airborne oxides may increase the mobility of trace elements and heavy metals in soils and waters. Subsequent cation nutrient depletion may also increase the aluminum toxicity in trees, which contributes to forest declines (Johnson and Taylor, 1989).

In addition, other effects of NO_x deposition may occur as a result of nitrogen fertilization of soils and waters (eutrophication) (Vitousek et al., 1993; Bobbink et al., 1998):

- Higher rates of plant growth, and subsequent higher carbon storage in ecosystems
- Accelerated loss of biodiversity by local competitive exclusion of plants adapted to efficient uses of nitrogen, and of the animals and bacteria dependent on them
- Changes in estuarine and coastal ecosystems, followed by the long-term decline of fisheries

A combined effect of acid deposition is also the decrease of dissolved organic carbon in lakes, allowing increased penetration of incident solar radiation, notably UV, into the upper water column. Particularly in clear shallow lakes and streams, such effects can be more effective than depletion of the stratospheric ozone in increasing the exposure of aquatic organisms to biologically effective UV radiation (Schindler et al., 1996; Yan et al., 2002).

Acid deposition may have opposite effects on ecosystems, with complex outcomes. Thus, on the one hand, aerosols tend to increase haze, and hence alter the radiation balance, which has harmful consequences on photosynthesis and plant growth. This adds to the negative effects of pH changes. On the other hand, increased nitrogen deposition in nitrogen-limited ecosystems causes increased plant growth. This increases sequestration of CO_2 from the atmosphere, which slows down the increase of atmospheric levels of CO_2 and global warming. Global warming may cause higher frequencies and severities of droughts, with effects on nitrogen mineralization and oxidation of sulfur compounds. Effects on lake biotas may include reduced cold water refugia for cold stenotherm species, lower nutrient concentration, and greater penetration of harmful UV radiation into lake ecosystems (Wright and Schindler, 1995).

The effects of SO_x and NO_x pollution combine with those caused by other atmospheric pollutants, such as NH_3, metals, and so on. Some pollutants such as organochlorine compounds and mercury tend to circulate from warm regions to higher altitudes and latitudes, because of factors such as climate warming, increased UV levels, temperature-dependent volatility, and greater condensation in cold regions. There (but not only there), the bioaccumulation of pollutants in plants and animals, and biomagnifications upward the food chain, seriously threaten indigenous people (Travis and Hester, 1991).

In addition, in areas of low levels of ozone in the stratosphere and with high levels of atmospheric NO_x (like in large urban areas, with many circulating cars), more ozone can form in the troposphere (i.e., at ground level), which has noxious effects on plants and animals.

Ozone represents a major pollution problem and is a matter of public health in cities (associated with smog). It is a severe irritant to the eyes, damages the lungs, and makes people more susceptible to respiratory infections. Particularly vulnerable are children and adults with existing diseases. Ozone also inhibits the growth, respiration/photosynthesis, and water-use efficiency of plants and can cause serious damage to terrestrial ecosystems (notably crops and forests) and aquatic ecosystems (where phytoplankton productivity determines general ecosystem productivity, including fish stocks) (Galloway, 1995).

These are not only technical aspects for discussion in specialist circles, but also serious issues requiring societal responses. Each country must be aware of its vulnerability to the complex effects of environmental degradation on its citizens. The environmental vulnerability index (EVI) proposed by Kaly et al. (2004) and supported by a consortium led by the South Pacific Applied Geoscience Commission (SOPAC) and the UNEP does not explicitly take into account socioeconomic parameters. Instead, it uses 50 different environmental indicators to estimate the degree of vulnerability of each country's environment (ecosystem goods and services), in the face of natural and human hazards, on a scale of five steps: resilient, at risk, vulnerable, highly vulnerable, and extremely vulnerable. The environmental indicators pertain to general issues such as climate change, exposure to natural disasters, human health, agriculture and fisheries, water, desertification, and biodiversity. According to the results, from a total of 235 countries taken into account, 14 appear to be resilient, 43 at risk, 81 vulnerable, 62 highly vulnerable, and 35 extremely vulnerable to future shocks. Particularly among the 47 small island developing countries (SIDS, most of them in the South-Eastern Asia-Pacific area), the situation is really bad: none is resilient, while three are at risk, 17 highly vulnerable, and 17 extremely vulnerable. At a global level, besides geological and geographical aspects like rainfall regime, volcanoes, earthquakes, and tsunamis, some highest risks correspond to certain socioeconomic categories: SO_2, mining, population density, and conflict (often on exploitation of natural resources) (Kaly et al., 2004; SOPAC-UNEP-EVI, 2005). If periodically updated, the EVI index can be used for the adaptive management of a country and for monitoring sustainable development.

A complementary measure is the disaster risk index (DRI) proposed by the Bureau for Crisis Prevention and Recovery (BCPR) of the United Nations Development Programme (BCPR, 2004). This index refers exclusively to the risks of human life loss in situations of medium- and large-scale disasters falling into one of the following three categories: earthquakes, cyclones, and floods. Mortality is taken as a proxy for disaster risks, and RDI was constructed on the basis of historical data, from 1980 to 2000. It does not account for the risk of hazardous events *per se*, but is a combined expression of the physical exposure to extreme events and vulnerability. Vulnerability explains why countries with the same level of exposure have different risks of death: their coping capacity and adaptive capacity vary with extreme events. Indicators of vulnerability are economic (lack of reserves and poor assets), social (weak social organization and absence of mechanisms of social support), technical (unsafe constructions and housing), and environmental (ecosystem fragility).

16.9 ENVIRONMENTALLY CONSCIOUS DEVELOPMENT: THE QUEST FOR ENVIRONMENTAL SCIENCE AND TECHNOLOGY

Air pollution must be fought by adopting cleaner, environmentally friendly technologies. Ecological modernization is a major issue in the present debates and projects of implementing sustainable development. But ecological modernization is not by itself sustainable development. Rather, it is a necessary but not sufficient condition for sustainable development (Langelle, 2000). In addition to ecological modernization, precaution must be taken with any program and activity of capital management in order to avoid causing, in the name of nature protection, irremediable losses to the natural capital. That is why the International Union for the Conservation of Nature (IUCN) adopted a resolution in 2004, which recommends that the *Precautionary Principle* should prevail in environmental decision making and management (IUCN, 2004).

In ecology, this concept corresponds to the "insurance hypothesis," stating, for example in the case of ecosystem structure, that biodiversity is good for ecosystems because diversity confers overall stability in the face of disturbances or stressful conditions. This hypothesis has actually launched a very fertile "diversity-stability debate" in ecology, and now an impressive plethora of theoretical and empirical evidence support the "insurance" view (Tilman and Downing, 1994; Yachi and Loreau, 1999; McCann, 2000).

As far as technology is concerned, application of the precautionary principle in conserving the natural capital may satisfy both "technological skeptics" (who do not see much scope of technological development for the sustainable use of natural resources) and "technological optimists" (who see technological development as eliminating all obstacles against growth and sustainable development), because ecosystem resources will be conserved for future generations. In addition, such policies will result in rising prices for natural resources, which will accelerate the arrival of environmentally friendlier technologies (Costanza and Daly, 1992). Such an approach has been named "precaution through experience," and involves substantial public involvement (Welsh and Ervin, 2006).

In this sense, the concept of *industrial hygiene* developed as "anticipation, recognition, evaluation, and control of chemical and physical stressors whose presence or action may lead to undesirable outcomes such as injury or death" (Harper et al., 1997). Science and technologies may even be able to provide spectacular solutions for mitigating global risks. For example, various mechanisms might be designed for capturing and storing atmospheric carbon and thus harness climate changes (IPCC, 2005).

However, technology alone cannot resolve much. For environmental regulations to be applicable, there is a need for international (not only national) agreements on environmental pollution. As will be described later, the Kyoto Protocol is a good example of a necessary step forward.

There is increasing understanding of the fact that technology must give up its entirely artificial character and become more hybrid, in the sense of being both man-made and ecological, in order to become environmentally friendly. Biotechnology holds this promise, provided it is used wisely (Moser, 1994). It can lead the transition from an environmentally indifferent industrial chemistry to an environmentally conscious industrial chemistry (Cano-Ruiz and McRae, 1998; Gavrilescu and Chisti, 2004). The central document of the UNEP, Agenda 21 (UN Agenda 21, 2005), dedicated an entire chapter (16th) to biotechnology.

Although Agenda 21 mainly addresses issues related to DNA engineering, the discussion is relevant for the entire area: increasing the availability of food and renewable raw materials; improving human health; enhancing protection of the environment; enhancing safety and developing international mechanisms for cooperation; and establishing enabling mechanisms for the development and the environmentally sound application of biotechnology. In the following, we will briefly describe the principles of some frontline biotechnological approaches that are essential in the quest for sustainability: bioresponse-linked analyses.

Plant and animal species respond to intensities of environmental pollution according to their capacity to tolerate various concentrations of individual pollutants or mixtures of pollutants. In the presence of pollutants, biological performance may be affected only slightly and within their limits of physiological tolerance, or may suffer fatal injuries and perturbations (Figure 16.6).

These are often compensatory mechanisms, for example the physiological adjustments in plants: pollutants shift to root-to-shoot ratio and accelerate leaf maturation. Understanding all these mechanisms through ecotoxicological and multiple-stress studies is mandatory for any good management (Winner, 1994). Plants and generalist predators connect below-ground systems (soil communities and soil chemistry) and above-ground systems (vegetation cover and associated animal and bacterial communities), that is, food webs (Scheu, 2001). Animals act as resultants of various effects on ecosystem health, regulation and productivity, chemical cycling, and genetics (Newman and Schreiber, 1984). Similarly, because of their biological sensitivity, lichens and mosses are often used in biomonitoring air pollution with trace elements (Wolterbeek, 2002).

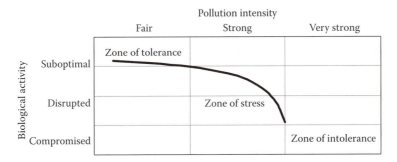

FIGURE 16.6   Effect of pollution on the biological activity of species.

Therefore, known species and populations can be used as indicators of long-term trends in ecosystem responses to air emissions and may be used in empirical calculations of critical loads of airborne pollutants (Bobbink and Roelofs, 1995). Many ecological indicators are already available, that is, measurable characteristics of the structure, composition, or function of ecological systems. Ecological indicators are primarily biological and respond to chemical, physical, and other biological phenomena. They capture intrinsic, highly complex dynamics and are useful in monitoring ecosystem health status (Niemi and McDonand, 2004).

For example, the airborne acidification of lakes, and subsequent recovery, can be traced by declines of fish populations and species (the remaining ones also having reproduction problems) and expansion of peat mosses (such as *Sphagnum*) over the lake bottom, and, respectively, by retreats and eventual disappearance of peat mosses and recovery of fish populations. In addition, the pH of the lake correlates with the population and community dynamics of *Diatom* algae in the lake sediment (Ek et al., 1995).

Ecosystems affected by excessive stress are less able to filtrate and assimilate other pollutants (Reich and Amundson, 1985; Caldwell et al., 1998), like in urban and periurban areas. Such ecosystems relate to street trees, lawns/parks, urban forests, cultivated land, wetlands, lakes/seas, and streams, and the major services they provide are air filtration (air pollutant retention), microclimate regulation, noise reduction, rainwater drainage, sewage treatment, and recreational and cultural values (Bolund and Hunhammar, 1999). Knowledge-based land use decisions by local and national authorities are essential for mitigating air pollution effects on public health.

In providing managers with knowledge about the complex effects of pollutant substances on ecosystems and humans, biological and biotechnological methods such as bioassays are frontrunners. As pollutants pose biotoxicity problems, bioassays are bioresponse-linked methods of pollution analysis, such as ecotoxicological tests and biosensor systems. Ecotoxicological tests are being used for exploring and assessing all kinds of biological–toxicological effects of known pollutants in natural and laboratory matrices (Box 16.1). Field biosensor instruments can be developed for laboratory or on-site screening of pollutants. Biosensor systems (or biosensors) employ known biological–toxicity effects for the detection and quantification of pollutants in air, water, and soil (Brenner-Weiss and Obst, 2003; Ciumasu et al., 2005; Breitholtz et al., 2006).

BOX 16.1.   BIORESPONSE-LINKED ASSAYS: MEASUREMENT OF THE MEDIAN LETHAL CONCENTRATION LC$_{50}$ OF A TOXICANT OVER GROWTH OF *DAPHNIA MAGNA* (OECD, 1997, 2004; SANTIAGO ET AL., 2002)

EC$_{50}$ is the concentration that gives rise to a 50% reduction in fecundity when compared with the control. Similarly, EC$_{20}$ and EC$_{10}$ are the concentrations associated with reductions in fecundity of 20% and 10%, respectively. The notation EC$_q$% will be used to denote a general term, namely the concentration associated with a reduction of q%.

Estimation of an ECq requires the following:

1. The assumption of a model for the relationship between fecundity and concentration—this model will have several unknown parameters;
2. A method for fitting the data to the model to derive estimates of the parameters of the model.

The model used in the analysis is given by

$$y = \frac{c}{1 + (x/x_o)^b},$$

where y is the mean fecundity at concentration x, c is the mean fecundity when x is 0, x_o is EC$_{50}$, and b is a slope parameter.

Knowledge of EC$_{50}$ and the slope parameter is sufficient to enable any EC (EC$_{10}$, EC$_{20}$, and so on) to be calculated. Because this model is nonlinear, the curve cannot be fitted by standard linear least-squares techniques. Instead, nonlinear iterative methods have to be used (Figure 16.7).

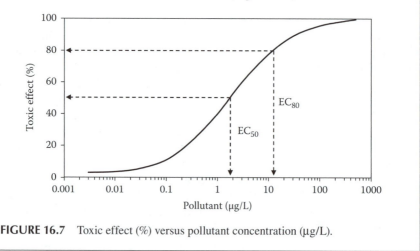

FIGURE 16.7   Toxic effect (%) versus pollutant concentration (µg/L).

Standardized ecotoxicological tests exist (Box 16.2) and they measure the biotoxicity of natural samples (e.g., waters) polluted with certain substances or mixtures of substances. Such tests employ all major classes of organisms, for example, invertebrates (e.g., *Daphnia*), algae (e.g., *Pseudokirchneriella subcapitata*), and bacteria (e.g., *Vibrio fischeri*). The usual standardized measures are "median lethal concentration" (LC_{50}) and "no observed effect concentration" (NOEC), which provide a sound basis for establishing threshold levels of pollutants in the environment. They provide a first step for various objectives of environmental risk assessments, with end-benefits for the ecosystem and human health. Further steps in ecotoxicology are controlled manipulations of natural populations (microcosm studies) and communities and ecosystems (macrocosm studies) (Kimbal and Levin, 1985; OECD, 1997, 2004; US EPA, 2002; Santiago et al., 2002).

BOX 16.2. BIORESPONSE-LINKED ASSAYS: TYPICAL STANDARD CURVES USED FOR THE DEVELOPMENT OF IMMUNOASSAYS AND IMMUNOSENSORS, WHERE A TOXICANT (ANALYTE) IS BEING DETECTED AND QUANTIFIED WITH THE USE OF MONOCLONAL ANTIBODIES (OR OTHER MOLECULAR RECOGNIZERS)

The biochemical reaction involves competition between the analyte and a labeled analogue. Inhibition of the analogue binding to (known amounts of) antibodies determines inhibition of the measured signal, as shown with a standard curve (inhibition curve, Figure 16.8). The standard curve produced by an immunoassay can be fitted with the four-parameter equation, to account for the sigmoid pattern of the signal inhibition:

$$y = \frac{A - D}{1 + (x / C)^B} + D.$$

where y is the ordinate axis (signal intensity), x is the abscissa axis (analyte concentration), A is the highest value of the signal (on the ordinate), D is the lowest value of the signal (on the ordinate), B is the slope, and C is the analyte concentration at 50% inhibition of the maximum signal (IC_{50}).

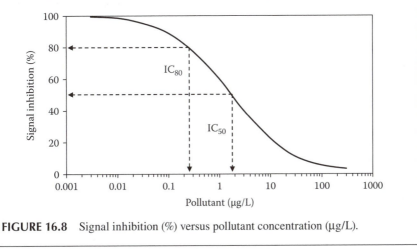

FIGURE 16.8   Signal inhibition (%) versus pollutant concentration (μg/L).

Standard curves are produced using either a specialized software program for a microtiter plate absorbance reader, for example, SoftMaxPro, or a general one, for example, MS Office Excel. On the curve, *A* corresponds to the reference, also known as the "zero dose." *D* corresponds to the highest analyte concentration. The slope of the curve is known as the dose–response region. This curve segment allows quantitative analyses in samples with unknown concentrations of analytes. The limits of the dose–response region can be used to make semiquantitative estimations (screening). Depending on the specific experimental design and setup, the dose–response curve can also be ascending, similar to that in the ecotoxicological tests described in Box 16.1.

Using standard analyte concentrations, the bioassays (immunoassays) can be optimized for a slope parameter (*B*) of 1.0, which coincides with the best dose–response resolution. At this value for the slope parameter, the analyte concentration range stretches over ca. two orders of magnitude.

The minimum concentration of analyte yielding a signal that is discernible from zero dose signals represents the detection limit (DL) of the assay. DLs may be established statistically or empirically.

Theoretically, DLs may be calculated from the standard deviation and the slope, that is, subtracting from the zero dose signal a value that equals 3 times the standard deviation. In practice, the problem with this method is that the standard deviation is heavily variable with the number of measurements. In some developments, a more empirical and conservative method was used: DL is established at 20% inhibition (IC_{20}), that is, 80% in the %-normalized control curve.

Such data, along with those from the biomedical literature, may further be taken over by ecologists and ecological economists and used in performing economical estimates of risks. Such economic analyses are needed to establish appropriate and authoritative degrees of regulations of pollutant emissions. Thus, the ecological, social, and economic efficiencies of policies are being addressed in an integrated manner. For example, based on ecotoxicological and biomedical research and on subsequent education of the population on the concrete risks of pollution, pesticide risk valuations and willingness-to-pay estimates can be used in policies of pesticide risk reduction (Travisi et al., 2006).

This is essential because there is extensive evidence that poverty strongly prevents nations from abating many types of pollution (Hilton, 2006). This further leads to entire populations being caught in a vicious circle of overexploitation of natural resources and underdevelopment, with malign effects on the health, human development, and general welfare of individuals and of the nation (Barbier, 2005). For example, biosensors are a very promising new technology and tool for addressing issues of monitoring natural resources, but technology is today restricted to uses in developed countries, with very few exceptions. Still, such scientific progress in richer countries holds great hope for the future of poorer countries too. Indeed, all biotechnologies hold great potential for improving the collaboration between countries of different levels of richness, because all are new in the field of biotechnology and all are linked by globalization (DaSilva, 1998).

Biosensors have the analytical power to detect in the environment a large spectrum of pollutants with potential or proven harmful effects on reproduction, development, and others: mutagenic, carcinogenic, hepatotoxic, nephrotoxic, immunotoxic, neurotoxic, and pharmacokinetic (Richardson, 2003). They can use molecules (e.g., antibodies, DNA, and receptors), cellular organelles (e.g., chloroplasts), entire cells (e.g., cellular algae), or even tissues (e.g., plant tissues) as biological components, with huge possibilities of combining analytical performance with specific or simple or more complex biological responses. Progress in miniaturization allows the construction of highly integrated microdevices, such as molecules- or entire cell-based *lab-on-a-chip* systems for environmental *on-site analysis*, food analysis, or medical *point-of-care diagnostics* (Ciumasu et al., 2005; El-Ali et al., 2006). For the detection of airborne substances (e.g., explosives) or pathogens (e.g., with respiratory infections), interesting biosensors (and other physicochemical sensors taking advantage of developments in micro- and nanotechnology) are being developed, such as the so-called *nose biosensors* or, more generally, *electronic noses* (Schuetz et al., 2000).

Highly miniaturized integrated instruments could replace most of the expensive scientific infrastructure and laboratory endowments and running costs that are not affordable for the developing world. With such instruments, reliable environmental and medical investigations will provide baseline information for *risk assessments* and environmental and public health management in both developing and developed countries (Yager et al., 2006).

For example, with heavy metals (e.g., Cu, Zn, Pb, Cd, Cr, and Ni), from air deposition or from other sources, ecotoxicity studies should take into account their general fluxes in ecosystems, which relate to pollution loads, soil type and acidity, vegetation, and land uses (Bergkvist et al., 1989). Another example of toxic pollutants is represented by endocrine-disrupting compounds. These are substances that are released into the environment and come into contact with plants and animals (including humans, obviously). They have the same physiological effects as hormones because of their molecular resemblance, and so perturb the normal functioning of organisms and populations. For instance, many pollutants mimic estrogenic activity, and so induce a phenomenon of "feminization" in the affected population. This has consequences on population health and dynamics. Endocrine disruptors can be analyzed with bioassays both for detection (e.g., molecular recognition) and for assessment of biological toxicity in humans and in ecosystems (Petrovic et al., 2004).

Following pollution monitoring and biological–ecological risk assessments, and together with it, ecotoxicity studies provide a basis for environmental decontamination and ecological restoration by employing natural mechanisms that biological systems use to self-recover, in the form of environmental biotechnologies. Thus, the phytoremediation of waters and soils employs the capacity of certain plant species to metabolically break down and assimilate pollutants, either alone or (usually) in combination with several plant species as part of local ecosystems formed by plant and microbial (but also animal) communities (Salt et al., 1995).

Phytoremediation has the potential to develop as a veritable green technology (Schwitzguebel, 2001), possibly helped by genetic engineering methods and by its considerable appeal to the public (Cherian and Oliveira, 2005). For example, artificial wetlands can be constructed for water decontamination (Cheng et al., 2002). Various methods can be designed for soil decontamination using plants, including on-site remediation (Cunningham et al., 1995). As we discussed the deposition of airborne SO_2, it is appropriate to mention that plant metabolization of sulfur can be employed for plant decontamination in sulfur-enriched soils (Ernst, 1998). In addition, plants can be used in phytoremediation methods for removing volatile and semivolatile pollutants directly from air (Cornejo, 1999).

As direct impacts of pollution on ecosystem and human health are more visible to, and more easily understood by, the general public, bioresponse-linked and bioremediation technologies are natural promoters of public and professional debates on pollution issues (Turner, 2005). These two scientific areas should provide platforms for, and nurture, programs of EE and education for sustainable development at all ages.

16.10 ENVIRONMENTALLY CONSCIOUS DEVELOPMENT: THE QUEST FOR ECOLOGICAL ECONOMICS

The ecological consequences of environmental changes are, in principle, investigated with methods falling into the following interrelated categories: extrapolation, experiments, phenomenolgical models, game-theory population models, expert opinion, and outcome-driven modeling and scenarios. However, several major problems arise for natural resource managers: adaptive management is desirable, but unfortunately hardly applied in practice, for reasons pertaining to economic and political logic. Much information is often not available, and much conservation practice is not based on evidence. This is often due to the high degree of complexity and uncertainty with many natural and human-driven processes (Sutherland, 2006).

But overall, managers must concentrate their action on maintaining and developing ecosystem services, including here all kinds of risks posed by greenhouse emissions and climate changes. For example, ecosystems can mitigate the effects of disasters brought about by climatic changes, notably pollution and floods. From an economic and financial perspective, it is essential that ecosystems (notably forests) be counted as water infrastructure, and valued accordingly, using cost–benefits balance (Emerton and Bos, 2004; see also IUCN, 2004).

A pioneering initiative in this direction comes from the city government of New York, who realized in 1997 that preserving the quality of the drinking water of the city meant getting involved in the agricultural practices of the city hinterland. For the same benefit, instead of newly investing 4–6 billion dollars in water-filtration plants and 250 million dollars in their annual running costs, the authorities have chosen to invest 250 million dollars in buying land to prevent development and 100 million dollars annually to minimize water pollution from all kinds of sources (*The Economist*, 2005d).

As suggested by the last example, good financial management must take into consideration environmental risks. Therefore, several initiatives in linking environmental and economic issues are being carried out by insurance companies. They are among the first main economic actors to seek a thorough understanding of the local and global risks brought by environmental changes. Just to give an example, one recent joint action of insurers and scientists was the *Workshop Climate Change and Disaster Losses: Understanding and Attributing Trends and Projections*, held on May 25–26, 2006 in Hohenkammer, Germany. At this meeting, the socioeconomic consequences of the disasters associated with climate warming were debated in terms of costs, with the aim of identifying possible solutions to minimize costs by mitigating environmental risks (WCCDL, 2006).

The involvement of insurance companies in putting environmental–financial solutions into practice may already have spectacular results. One such example comes from the state of Panama. The watershed of the Panama Canal is deforested, and hence the canal needs more expenditure for its maintenance than if the watershed had been forested. A forested watershed would release less water, but it would regulate this supply in time, which makes a lot of sense for the regular functioning of the locks of the canal. In addition, a deforested watershed allows more sediments and nutrients (including those from aerial deposition) to flow into the canal, which clog the canal either directly or through the abundant growth of waterweeds. This results in regular and expensive dredging. Forests trap sediments and nutrients. Therefore, any investment in reforesting the watershed truly meant investment in infrastructure. Even though the state authorities may not afford it, funding can be attracted from the market. Insurance companies would ask their clients (who currently insure for huge losses in case the canal closes) to sign bonds that would pay for reforestation. Such a business, in addition to economic gains, provides ecosystem services and social benefits for the local people (*The Economist*, 2005d).

In the examples above, the hub of the solution was valuation of ecosystem goods and services on the market. An example of wider concern is the so-called carbon market. This follows the entering into force, on February 16, 2005, of the Kyoto Protocol, which imposes individual target reductions of GHG emissions in developed countries. The reductions were mainly negotiated on the basis of each country's history of emissions (UNFCCC, 2005). To avoid target cuts on carbon emissions putting insupportable costs on business, European countries launched a pioneering pan-European carbon-trading system in 2005, which allows mobility and efficiency in achieving the ratified targets (Lecocq, 2005; *The Economist*, 2005a).

If the assumed target is reached, market mechanisms may prove to be the main direct contributor to getting out from what may have proved to be an impasse. For instance, many critics of the Kyoto Protocol felt that imposing greater emission-reduction responsibilities on developed countries on the grounds of their historical emissions record would trigger industry delocalization from their countries to the developing ones (Grubb et al., 2002). The general argument is that stringent regulations on pollution might harm the competitiveness of business in certain regions or countries, which would be outrun by worldwide competitors who do not need to comply with the same national or regional regulations. Economic development was often invoked by several developed countries as the main concern and reason for not ratifying the Kyoto Protocol on reducing GHG emissions, even though the target emission reductions are very small (UNFCCC, 2005). Actually, this small target was widely criticized by the scientific community as being too weak in its responsibility ascribed to developed countries. Alternative scenarios even predict that the agreed target reduction of only 5.3% of GHG emissions of the level of 1990 by 2008–2012 would have negligible consequences on the climate in the long term (Wigley, 1998). The scientific community made pressures for a much more ambitious reduction.

Such problems of economic competition, being an obstacle in taking internationally decisive action, can and should be circumvented by market mechanisms. For instance, market-driven evaluations already indicate that the concerns about unfair competition might be unfounded. Signals exist that the cost of reducing emissions might actually be lower than the overall costs of not doing so (De Leo et al., 2001; IPCC-WGIII, 2007). Moreover, pollution abatement costs need not have a significant influence on production expenditures (Shadbegian and Gray, 2005).

While about one decade ago companies simply opposed any initiative to cut carbon emissions, now they are more split; some companies have begun to size the market opportunities of investing in cleaner technologies, including in renewable energy production. Investments in environmentally friendly technology multiply because the price gap between the so-called green technology and the traditional ones is narrowing, and there are more opportunities for making money with new products. This market tendency is helped both by advances in green technology, in particular solar and wind power, and by the increasing oil prices. Since 2001, the oil price more than tripled because of increasing oil demands, in particular by emerging economies like China and India, and low spare capacity in the oil industry. New and "cleaner" products may be chosen by the consumer less because they are "green" and more because they offer practical advantages of being good and affordable (*The Economist*, 2005a,b). One excellent example of such products is the case of recycled printer ink cartridges. Although these pioneering products may not yet equal the very best, they are good enough and also cheaper. That is quite a reasonable start.

Surely enough, markets are volatile and poorly designed carbon market mechanisms may be ineffective in helping the environment. This is what recently happened with the carbon permits issued by the European Union Emissions Trading Scheme (ETS), which rather helped power companies without helping to curb emissions because ETS has only been designed for three years while companies need at least five years to plan the required investments (*The Economist*, 2006a). Another deficiency of the ETS system was the recent leaking of emission data through the European Commission website, which provoked excessive volatility on the carbon market. However, the system is technically working, being somehow similar to that which controls sulfur dioxide emissions in the United States (Schiermeier, 2006).

In countries where a mandatory carbon-trading system does not exist yet, a similar market flaw is with the idea of carbon "offsetting," meaning that a company may, on a voluntary basis, pay other companies to emit less CO_2 on its behalf. As this system is immature, there are no standards, and it is being diverted to other purposes, such as image polishing for heavy-polluting companies and for overzealous politicians (*The Economist*, 2006b).

All in all, although ecosystem goods and services are more than mere economic assets, people need financial incentives to act (*The Economist*, 2005d). As market forces provide these incentives, it is virtually impossible to exclude markets from finding solutions to any global environmental problem. Rather, alternative market mechanisms should be sedulously considered every now and then (Manne and Richels, 2001).

The following example draws directly from such market, ecological economics mindsets. The almost pure CO_2 usually released directly into the atmosphere by oil refineries as waste can actually be a valuable commodity for agriculture. With an apparently risky but promising investment, the so-called organic carbon dioxide for assimilation by plants (OCAP) project ensures that this GHG is actually being transported through pipelines and released into the greenhouses of neighboring agricultural producers (Stafford, 2006). The company has turned down its environmental concerns while upturning its profit. The owners of the greenhouses ensured high biomass production rates (which sequesters CO_2) and reduced heating expenditures, both because of higher concentrations of CO_2 in the greenhouses. In the end, the waste CO_2 contributes less to planetary climate warming, because much of it is being "captured" in revenues. This is a memorable example in which a problem is turned into an advantage by intelligent use of free-market mechanisms. Although the director of the gas supplier company would not disclose the revenue extracted from this project, his words are quite relevant: "If there had been no business case for OCAP, we would not have done it."

In fact, the most serious risks to sustainability come from the possibility that ecosystems and markets are fundamentally incompatible. They are probably not, because they obey the same general laws of energy distribution and consumption. Or, if they are, we have no choice but to create the right mechanism that makes them compatible. This is to say that there is more than fluidity in market-driven reductions of GHGs. While the crux of the CO_2 emissions issue is its direct connection with the present-day excessive dependence of the world economy on oil, the solution is creativity. In the mid-1950s, the prime minister of the United Kingdom, Churchill, observed that "safety and certainty in oil lie in variety, and variety alone." Nowadays again, energy security (and the prevention of dreadful events such as oil shocks) depends on variety. As the well-known weekly *The Economist* put it, "Now variety needs to be sought in sources of energy, rather than sources of oil alone" (*The Economist*, 2005c).

All in all, the value of the Kyoto Protocol might not be the amount of the achieved reduction itself (if achieved), but the international mobilization, the innovative and promising use of market-based instruments such as carbon-emission trading (*The Economist*, 2005d), and the "international spillover" effect of such actions, meaning worldwide diffusion of novelties and developments in technology, systems, and policies (Grubb et al., 2002).

16.11 ENVIRONMENTALLY CONSCIOUS DEVELOPMENT: THE QUEST FOR EE

Air pollution can only be controlled by society as a whole. A fundamental cause for environmental degradation and lack of ecological sustainability is institutional failure (Dasgupta et al., 2000). Management measures that are technically viable from a sustainable development perspective cannot circumvent public approval. This must include public debates and public-driven legal decisions as well as the funding of programs of ecological research and ecological monitoring, activities that often provide solutions to environmental issues. Therefore, EE is essential for effective public participation and subsequent political action.

A culture of real sustainable development must be developed in public opinion on the basis of "ecological literacy" (Orr, 1992; Oskamp, 2000), which is still to be achieved across nations and societies. This new literacy must provide the ground for managing boundaries between knowledge and action, by creating efficient knowledge systems capable of mobilizing science and technology for sustainable development (Cash et al., 2003).

From a scientist point of view, ecological literacy and, subsequently, deeper involvement of the public in environmental issues would correspond to what has been called "a new social contract" for science and scientists (Lubchenco, 1998): the contract for sustainable development. In order to achieve these two goals, efficient programs of EE and education for sustainable development must be developed and applied in integrating science and public interests and welfare, and in changing political identities, relationships, and institutions (McMichael et al., 2003; Miller, 2005).

An excellent start and guide in EE is Agenda 21, of the United Nations, which details and integrates all ecological, social, and economic aspects of sustainability (UN Agenda 21, 2005). This is organized in four sections, addressing (1) social and economic dimensions, (2) issues related to the conservation and management of resources for development, (3) the role of major social groups, and (4) means of implementation.

Agenda 21 is a fundamental document framing international action for sustainable development and includes a practical summary of all issues that are essential for sustainable development:

- Integrating environment and development in decision making
- Protecting the atmosphere (Chapter 9)
- Integrating the planning and management of land resources
- Combating deforestation (and promoting reforestation in deforested and degraded areas)
- Priority special management for fragile ecosystems: combating desertification and drought
- Priority special management for fragile ecosystems: sustainable development of mountain areas
- Promoting sustainable patterns of agriculture and rural development
- Conserving biological diversity (i.e., not only "star" species, but with an accent on entire-ecosystem phenomena such as pollution, key species, and community ecology)
- Environmentally sound management of biotechnology
- Protecting the seas, and their living resources (e.g., the capacity of oceans to function as a sink for atmospheric CO_2)
- Protecting the quality and supply of freshwater resources: integrated management

- Environmentally sound management of toxic chemicals, including the international circuit of dangerous products
- Changing consumption patterns
- Demographic long-term dynamics (sustainability)
- Promoting and protecting the conditions required for good human health
- Promoting sustainable human settlement (e.g., preventing human settlements in areas with high flood risk)
- Environmentally sound management of hazardous wastes, solid wastes, sewage-related issues, and radioactive wastes
- International cooperation for speeding up sustainable development in developing countries
- Equitable development
- Fighting poverty
- Increasing the international focus of issues related to women, children, and youth
- Recognizing and strengthening the role of indigenous people and their communities
- Role of nongovernmental organizations (NGOs)
- Role of local authorities in finding and implementing measures toward sustainable development
- Role of scientists and of scientific and technological communities
- Role of business and industry
- Role of workers and their trade unions
- Role of farmers
- Financial resources and mechanisms for implementing sustainable development
- Transferring environmentally sound technology, cooperation, and capacity building
- Developing science in general—for knowledge-based development
- Promoting the science of sustainable development
- Ensuring the continuity between national and international mechanisms, arrangements, and instruments for capacity building and cooperation
- Ensuring information for decision making
- Promoting education, public awareness, and training

To give a general example, a good starting point for educating the public (of all ages) on the continuity between environmental health and human, individual health is point 6.13 of Agenda 21, concerning national health care plans (Box 16.3), and which actually recall many issues that we have chosen to discuss in this chapter earlier.

To give another, but more concrete example, human resources for sustainable development are being developed through a pilot EU project on EE that was started recently under the auspices of the European Commission, within the program *Leonardo da Vinci*, and is concerned with the "development of a European curriculum for methodological training in the field of EE." As particularly in the developing countries of Central-Eastern Europe (the former socialist block) wide ecological

**BOX 16.3.   UN AGENDA 21, CHAPTER 6: PROMOTING
AND PROTECTING HUMAN HEALTH,
PARAGRAPH 6.13 (ACTIVITIES)**

6.13. Each national Government, in accordance with national plans for public health, priorities and objectives, should consider developing a national health action plan with appropriate international assistance and support, including, at a minimum, the following components:

(a) National public health systems:

Programmes to identify environmental hazards in the causation of communicable diseases;

Monitoring systems of epidemiological data to ensure adequate forecasting of the introduction, spread or aggravation of communicable diseases;

Intervention programmes, including measures consistent with the principles of the global AIDS strategy;

Vaccines for the prevention of communicable diseases;

(b) Public information and health education: Provide education and disseminate information on the risks of endemic communicable diseases and build awareness on environmental methods for control of communicable diseases to enable communities to play a role in the control of communicable diseases;

(c) Intersectoral cooperation and coordination:

Second experienced health professionals to relevant sectors, such as planning, housing and agriculture;

Develop guidelines for effective coordination in the areas of professional training, assessment of risks and development of control technology;

(d) Control of environmental factors that influence the spread of communicable diseases: Apply methods for the prevention and control of communicable diseases, including water supply and sanitation control, water pollution control, food quality control, integrated vector control, garbage collection and disposal and environmentally sound irrigation practices;

(e) Primary health care system:

Strengthen prevention programmes, with particular emphasis on adequate and balanced nutrition;

Strengthen early diagnostic programmes and improve capacities for early preventative/treatment action;

Reduce the vulnerability to HIV infection of women and their offspring;

(f) Support for research and methodology development:

Intensify and expand multidisciplinary research, including focused efforts on the mitigation and environmental control of tropical diseases;

> Carry out intervention studies to provide a solid epidemiological basis for control policies and to evaluate the efficiency of alternative approaches;
>
> Undertake studies in the population and among health workers to determine the influence of cultural, behavioural and social factors on control policies;
>
> (g) Development and dissemination of technology:
>
> Develop new technologies for the effective control of communicable diseases;
>
> Promote studies to determine how to optimally disseminate results from research;
>
> Ensure technical assistance, including the sharing of knowledge and know-how.

literacy and public involvement are in the process of being developed society-wide, educators in the field of ecology and sustainable development need to be the subject of particular attention themselves.

The rationale of this project was to bring together representatives of target groups (from university, preuniversity, environmental NGO sectors, environmental agencies, botanical gardens, natural science museums, and parks and natural reserves from Romania and Bulgaria) and EE specialists from various European nations: both developing (Romania and Bulgaria) and developed countries (France, Germany, and Spain). The project builds on the spirit of Agenda 21, and integrates the university expertise of ecologists, biologists, sociologists, psychopedagogists, and economists, and practitioners of formal and nonformal EE from participating countries (Costica et al., 2006).

Regarding the role of atmospheric pollution in sustainable development, most relevant for society at large is Chapter 9 of Agenda 21, on the protection of the atmosphere (Box 16.4).

BOX 16.4.   UN AGENDA 21, CHAPTER 9: PROTECTION OF THE ATMOSPHERE (A SELECTION)

INTRODUCTION

9.1. Protection of the atmosphere is a broad and multidimensional endeavour involving various sectors of economic activity. The options and measures described in the present chapter are recommended for consideration and, as appropriate, implementation by Governments and other bodies in their efforts to protect the atmosphere.

9.2. It is recognized that many of the issues discussed in this chapter are also addressed in such international agreements as the 1985 Vienna Convention for the Protection of the Ozone Layer, the 1987 Montreal Protocol on Substances that Deplete the Ozone Layer as amended, the 1992 United Nations Framework Convention on Climate Change and other international, including regional, instruments. In the case of activities covered by such agreements, it is understood that the recommendations contained in this chapter do not oblige any Government to take measures which exceed the provisions of these legal instruments. However, within the framework of this chapter, Governments are free to carry out additional measures which are consistent with those legal instruments.

9.3. It is also recognized that activities that may be undertaken in pursuit of the objectives of this chapter should be coordinated with social and economic development in an integrated manner with a view to avoiding adverse impacts on the latter, taking into full account the legitimate priority needs of developing countries for the achievement of sustained economic growth and the eradication of poverty.

9.4. In this context particular reference is also made to programme area A of chapter 2 of Agenda 21 (Promoting sustainable development through trade).

9.5. The present chapter includes the following four programme areas:

(a) Addressing the uncertainties: improving the scientific basis for decision-making;
(b) Promoting sustainable development:
    Energy development, efficiency and consumption;
    Transportation;
    Industrial development;
    Terrestrial and marine resource development and land use;
(c) Preventing stratospheric ozone depletion;
(d) Transboundary atmospheric pollution.

   . . .

3.   INDUSTRIAL DEVELOPMENT

Basis for Action

9.16. Industry is essential for the production of goods and services and is a major source of employment and income, and industrial development as such is essential for economic growth. At the same time, industry is a major resource and materials user and consequently industrial activities result in emissions into the atmosphere and the environment as a whole. Protection of the atmosphere can be enhanced, *inter alia*, by increasing resource and materials efficiency in industry, installing or improving pollution abatement technologies and replacing chlorofluorocarbons (CFCs) and other ozone-depleting substances with appropriate substitutes, as well as by reducing wastes and by-products.

Objectives

9.17. The basic objective of this programme area is to encourage industrial development in ways that minimize adverse impacts on the atmosphere by, *inter alia*, increasing efficiency in the production and consumption by industry of all resources and materials, by improving pollution-abatement technologies and by developing new environmentally sound technologies.

Activities

9.18. Governments at the appropriate level, with the cooperation of the relevant United Nations bodies and, as appropriate, intergovernmental and non-governmental organizations, and the private sector, should:

(a) In accordance with national socio-economic development and environment priorities, evaluate and, as appropriate, promote cost-effective policies or programmes, including administrative, social and economic measures, in order to minimize industrial pollution and adverse impacts on the atmosphere;

(b) Encourage industry to increase and strengthen its capacity to develop technologies, products and processes that are safe, less polluting and make more efficient use of all resources and materials, including energy;

(c) Cooperate in the development and transfer of such industrial technologies and in the development of capacities to manage and use such technologies, particularly with respect to developing countries;

(d) Develop, improve and apply environmental impact assessments to foster sustainable industrial development;

(e) Promote efficient use of materials and resources, taking into account the life cycles of products, in order to realize the economic and environmental benefits of using resources more efficiently and producing fewer wastes;

(f) Support the promotion of less polluting and more efficient technologies and processes in industries, taking into account area-specific accessible potentials for energy, particularly safe and renewable sources of energy, with a view to limiting industrial pollution, and adverse impacts on the atmosphere.

. . .

D. TRANSBOUNDARY ATMOSPHERIC POLLUTION

Basis for Action

9.25. Transboundary air pollution has adverse health impacts on humans and other detrimental environmental impacts, such as tree and forest loss and the acidification of water bodies. The geographical distribution of atmospheric pollution monitoring networks is uneven, with the developing countries severely underrepresented. The lack of reliable emissions data outside Europe and North America is a major constraint to measuring transboundary air pollution. There is also insufficient information on the environmental and health effects of air pollution in other regions.

9.26. The 1979 Convention on Long-range Transboundary Air Pollution, and its protocols, have established a regional regime in Europe and North America, based on a review process and cooperative programmes for systematic observation of air pollution, assessment and information exchange. These programmes need to be continued and enhanced, and their experience needs to be shared with other regions of the world.

Objectives

9.27. The objectives of this programme area are:

(a) To develop and apply pollution control and measurement technologies for stationary and mobile sources of air pollution and to develop alternative environmentally sound technologies;

(b) To observe and assess systematically the sources and extent of transboundary air pollution resulting from natural processes and anthropogenic activities;

(c) To strengthen the capabilities, particularly of developing countries, to measure, model and assess the fate and impacts of transboundary air pollution, through, inter alia, exchange of information and training of experts;

(d) To develop capabilities to assess and mitigate transboundary air pollution resulting from industrial and nuclear accidents, natural disasters and the deliberate and/or accidental destruction of natural resources;

(e) To encourage the establishment of new and the implementation of existing regional agreements for limiting transboundary air pollution;

(f) To develop strategies aiming at the reduction of emissions causing transboundary air pollution and their effects.

Activities

9.28. Governments at the appropriate level, with the cooperation of the relevant United Nations bodies and, as appropriate, intergovernmental and non-governmental organizations, the private sector and financial institutions, should:

(a) Establish and/or strengthen regional agreements for transboundary air pollution control and cooperate, particularly with developing countries, in the areas of systematic observation and assessment, modelling and the development and exchange of emission control technologies for mobile and stationary sources of air pollution. In this context, greater emphasis should be put on addressing the extent, causes, health and socio-economic impacts of ultraviolet radiation, acidification of the environment and photo-oxidant damage to forests and other vegetation;

(b) Establish or strengthen early warning systems and response mechanisms for transboundary air pollution resulting from industrial accidents and natural disasters and the deliberate and/or accidental destruction of natural resources;

(c) Facilitate training opportunities and exchange of data, information and national and/or regional experiences;

(d) Cooperate on regional, multilateral and bilateral bases to assess transboundary air pollution, and elaborate and implement programmes identifying specific actions to reduce atmospheric emissions and to address their environmental, economic, social and other effects.

. . .

Environmental education includes the dimension of ecological sustainability, and the ability to develop and operate with sustainability indices, such as ESI, EFI, EVI, the Millennium Development Goal 7 Index—MDG7 (Esty et al., 2005) and the Living Planet Index—LPI (WWF, 2004).

Any program of education for sustainable development must include the following features:

- Interdisciplinary and holistic
- Value- and fairness-driven
- Critical thinking and problem-solving orientation
- Participatory
- Applicable in real-life and local contexts
- Favoring creativity

Education for sustainable development is our best chance in what Edward O. Wilson called the "Century of the Environment" and the "bottleneck" of the immediate future. We subscribe to his view that combining science and technology with foresight and moral courage is the way to secure a long-term perspective of humanity (Wilson, 2002). Above all, this education should aim at developing a reflex of judging immediate economic gains in the context of long-term effects on our ecological–social–economic systems.

Our epoch is characterized by human domination of the Earth (Eldredge, 1995; Vitousek et al., 1997). However, this should be replaced by responsibilization and a return to a more long-lasting relationship, one that is at the origin of any natural ecosystem: dynamic equilibrium.

REFERENCES

Abrahamsen, G. 1983. Effects of lime and artificial acid rain on the enchytraeid (Oligochaeta) fauna in coniferous forest. *Holarctic Ecology* 6(3): 247–254.

Aerts, R. 2006. The freezer defrosting: Global warming and litter decomposition rates in cold biomes. *Journal of Ecology* 94: 713–724.

Aerts, R., Cornelissen, J.H.C., Dorrepaal, E., and Van Logtestijn, R.S.P. 2004. Effects of experimentally imposed climate scenarios on flowering phenology and flower production of subarctic bog species. *Global Change Biology* 10: 1599–1609.

Armstrong, D.P. 2005. Integrating the metapopulation and habitat paradigms for understanding broad-scale declines of species. *Conservation Biology* 19(5): 1402–1410.

Barbier, E.B. 2005. *Natural Resources and Economic Development.* Cambridge University Press, Cambridge, UK, pp. 409.

Barrett, J. and Simmons, C. 2003. An ecological footprint of the UK: Providing a tool to measure the sustainability of local authorities. Report of the Stockholm Environment Institute, Stockholm, Sweden, http://www.york.ac.uk/inst/sei/odpm/ODPM%20Main%20Report. pdf

BCPR. 2004. Reducing disaster risk, a challenge for development, A global report. United Nations Development Programme (UNDP), Bureau for Disaster Risk and Recovery, p. 146, http://www.undp.org/bcpr/disred/documents/publications/rdr/english/rdr_english.pdf

Beckerman, W. 1992. Economic growth and the environment: Whose growth? Whose environment? *World Development* 20: 481–496.

Bergkvist, B., Folkeson, L., and Berggren, D. 1989. Fluxes of Cu, Zn, Pb, Cd, Cr, and Ni in temperate forest ecosystems. A literature review. *Water, Air and Soil Pollution* 47(3–4): 217–286.

Blum, J.D., Klaue, A., Nezat, C.A., et al. 2002. Mycorrhizal weathering of apatite as an important calcium source in base-poor forest ecosystems. *Nature* 417: 729–731.

Bobbink, R. and Roelofs, J.G.M. 1995. Nitrogen critical loads for natural and semi-natural ecosystems: The empirical approach. *Water, Air and Soil Pollution* 85(4): 2413–2418.

Bobbink, R., Homung, M., and Roelofs, J.G.M. 1998. The effects of airborne nitrogen pollutants on species diversity in natural and semi-natural European vegetation. *Journal of Ecology* 86(5): 717–738.

Bolund, P. and Hunhammar, S. 1999. Ecosystem services in urban areas. *Ecological Economics* 29: 293–301.

Boron, S. and Murray, K. 2003. Bridging the unsustainability gap: A framework for sustainable development. *Sustainable Development* (12)2: 65–73.

Breitholtz, M., Ruden, C., Hansson, S.O., and Bengtsson, B.E. 2006. Ten challenges for improved ecotoxicological testing in environmental risk assessment. *Ecotoxicology and Environmental Safety* 63(2): 324–335.

Brenner-Weiss, G. and Obst, U. 2003. Approaches to bioresponse-linked instrumental analysis in water analysis. *Analytical and Bioanalytical Chemistry* 377(3): 408–416.

Brydges, T. 2001. Ecological change and the challenges for monitoring. *Environmental Monitoring and Assessment* 67(1–2): 89–95.

Caldwell, M.M., Bjorn, L.O., Bornmann, J.F., et al. 1998. Effects of increased solar ultraviolet radiation on terrestrial ecosystems. *Journal of Photochemistry and Photobiology B: Biology* 46: 40–52.

Cano-Ruiz, J.A. and McRae, G.J. 1998. Environmentally conscious chemical process design. *Annual Review of Energy and the Environment* 23(1): 499–536.

Cape, J.N. 1993. Direct damage to vegetation caused by acid rain and polluted cloud: Definition of critical levels for forest trees. *Environmental Pollution* 82(2): 167–180.

Cash, D.W., Clark, W.C., Alcock, F., et al. 2003. Knowledge systems for sustainable development. *Proceedings of the National Academy of Sciences* 100(14): 8086–8091.

Chapman, J.L. and Reiss, M.J. 1999. *Ecology: Principles and Applications* (2nd edition). Cambridge University Press, Cambridge, UK.

Cheng, S., Grosse, W., Karrenbrock, F., and Thoennessen, M. 2002. Efficiency of constructed wetlands in a decontamination of water polluted by heavy metals. *Ecological Engineering* 18(3): 317–325.

Cherian, S. and Oliveira, M.M. 2005. Transgenic plants in phytoremediation: Recent advances and new possibilities. *Environmental Science and Technology* 39(24): 9377–9390.

Ciumasu, I.M., Krämer, P.M., Weber, C.M., et al. 2005. A new, versatile single-use field immunosensor for environmental pollutants. Development and proof of principle with TNT, diuron and atrazine. *Biosensors and Bioelectronics* 21, 354–364.

Costanza, R. and Daly, H.E. 1992. Natural capital and sustainable development. *Conservation Biology* 6(1): 37–46.

Cornejo, J.J., Munoz, F.G., Ma, C.Y., and Stewart, A.J. 1999. Studies on the decontamination of air by plants. *Ecotoxicology* 8(4): 311–320.

Costica, N., Ciumasu, I.M., and Costica, M. 2006. Partnership for human resources development in Romania and the European Union: Training the trainers in environmental education. *Proceedings of the 2nd European Fair on Education for Sustainable Development* "Promoting Education for Sustainable Development in Europe," September 13–15, 2006, Hamburg.

Cunningham, S.D., Berti, W.R., and Huang, J.W. 1995. Phytoremediation of contaminated soils. *Trends in Biotechnology* 13(9): 393–397.

Daly, H.E. 1991. *Steady-State Economics* (2nd edition). Freeman & Co., San Francisco; Island Press, Washington, DC.

Dasgupta, P., Levin, S., and Lubchenco, J. 2000. Economic pathways to ecological sustainability. Challenges for the new millennium. *BioScience* 50(4): 339–345.

Dasgupta, S., Laplante, B., Wang, H., and Wheeler, D. 2002. Confronting the environmental Kuznets curve. *Journal of Economic Perspectives* 16(1): 147–168.

DaSilva, E. 1998. Biotechnology: Developing countries and globalization. *World Journal of Microbiology and Biotechnology* 14(4): 463–486.

De Groot, R.S., Wilson, M.A., and Boumans, R.M.J. 2002. A typology for the classification, description and valuation of ecosystem functions, goods and services. *Ecological Economics* 41: 393–408.

De Leo, G.A., Rizzi, L., Caizzi, A., and Gatto, M. 2001. The economic benefits of the Kyoto Protocol. *Nature* 413: 478–479.

Ek, A., Grahn, O., Hultberg, H., and Renberg, I. 1995. Recovery from acidification in Lake Oervattnet, Sweden. *Water, Air and Soil Pollution* 85(3): 1795–1800.

Ekins, P. 2003. Identifying critical natural capital. Conclusions about critical natural capital. *Ecological Economics* 44: 277–292.

Eldredge, N. 1995. *Dominion*. University of California Press, Berkeley, 190pp.

Ehrlich, P.R. and Ehrlich, A.H. 1981. *The Causes and Consequences of the Disappearance of Species*. Random House, New York.

Ehrlich, P.R., Wolff, G., Daily, G.C., et al. 1999. Knowledge and the environment. *Ecological Economics* 30: 260–284.

El-Ali, J., Sorger, P.K., and Jensen, K.F. 2006. Cells on chips. *Nature* 442: 403–411.

Emerton, L. and Bos, E. 2004. Value. *Counting Ecosystems as an Economic Part of Water Infrastructure*. IUCN, Gland, Switzerland and Cambridge, UK, 88pp.

Epstein, P.R. 2005. Climate change and human health. *The New England Journal of Medicine* 353(14): 1433–1436.

Ernst, W.H.O. 1998. Sulfur metabolism in higher plants: Potential for phytoremediation. *Biodegradation* 9(3–4): 311–318.

Esty, D.C., and Levy, M., Srebotnjak, T., and de Sherbinin, A. 2005. *2005 Environmental Sustainability Index: Benchmarking National Environmental Stewardship*. Yale Center for Environmental Law & Policy, New Haven.

Folke, C., Jansson, A., Larsson, J., and Costanza, R. 1997. Ecosystem appropriation by cities. *Ambio* 26(3): 167–172.

Galloway, J.N. 1995. Acid deposition: Perspectives in time and space. *Water, Air and Soil Pollution* 85(1): 15–24.

Galloway, J.N., Likens, G.E., and Hawley, M.E. 1984. Acid precipitation: Natural versus anthropogenic components. *Science* 226: 829–830.

Gavrilescu, M. and Chisti, Y. 2004. Biotechnology—a sustainable alternative for chemical industry. *Biotechnology Advances* 23: 471–499.

Georgescu-Roegen, N. 1971. *The Entropy Law and Economic Process*. Harvard University Press, Cambridge.

Giddings, B., Hopwood, B., and O'Brian, G. 2002. Environment, economy and society: Fitting them together into sustainable development. *Sustainable Development* 10: 187–196.

Gilbert, O.L. 1986. Field evidence for an acid rain effect on lichens. *Environmental Pollution A* EPEBD7 40(3): 227–231.

Gray, R. and Milne, M.J. 2002. Sustainability reporting: Who's kidding whom? *Chartered Accountants Journal of New Zealand* 81(6): 66–70.

Grubb, M.J., Hope, C., and Fouquet, R. 2002. Climatic implications of the Kyoto Protocol: The contribution to the international spillover. *Climatic Change* 54(1–2): 11–28.

Haagvar, S. 1984. Effects of liming and artificial acid rain on Collembola and Protura in coniferous forest. *Pedobiologia* 27(5): 341–354.

Haines, A., McMichael, A.J., and Epstein, P.R. 2000. Environment and health: 2. Global climate change and health. *Canadian Medical Association Journal* 163(6): 728–734.

Hames, R.S., Rosenberg, K.V., Lowe, J.D., Barker, S.E., and Dhondt, A.A. 2002. Adverse effects of acid rain on the distribution of the wood thrush *Hylocichla mustelina* in North America. *Proceedings of the National Academy of Sciences USA* 99(17): 11235–11240.

HDR. 2003. Human Development Report 2003. *Millennium Development Goals: A Compact Among Nations to End Human Poverty*. United Nations Development Programme (UNDP), Oxford University Press, New York. Available at: http://hdr.undp.org.reports/global/2003/pdf/hdr03_complete.pdf

Herman, J.R., Bhartia, P.K., Ziemke, J., Ahmad, Z., and Larko, D. 1996. UV-B increases (1979–1972) from decreases in total ozone. *Geophysical Research Letters* 23: 2117–2120.

Hilton, F.G. 2006. Poverty and pollution abatement: Evidence from lead phase-out. *Ecological Economics* 56(2006): 125–131.

Honney, O., Verheyen, K., Butaye, J., Jacquemyn, H., Bossuyt, B., and Hermy, M. 2002. Possible effects of habitat fragmentation and climate change on the range of forest plant species. *Ecology Letters* 5(4): 525–530.

IPCC (Intergovernmental Panel on Climate Change). 2005. Carbon dioxide capture and storage. Special Report of the Intergovernmental Panel on Climate Change, http://arch.rivm.nl/env/int/ipcc/pages_media/SRCCS-final/ccsspm.pdf

IPCC-WGIII. 2007. IPCC Fourth Assessment Report, Working Group III—Mitigation. International Panel for Climate for Climate Change, http://www.ippc.ch/

IUCN. 2004. A daily report of the 3rd IUCN World Conservation Congress. IUCN Congress Bulletin, Volume 39 (15), published by the International Institute for Sustainable Development, New York.

Jameton, A. and Pierce, J. 2001. Environment and health: 8. Sustainable health care and emerging ethical responsibilities. *Canadian Medical Association Journal* 164(3): 365–369.

Jansson, A.M., Hammer, M., Folke, C., and Costanza, R. (Eds). 1994. *Investing in Natural Capital—the Ecological Economics Approach to Sustainability*. ISEE Press/Island Press, Covelo, CA.

Jha, R. and Bhanu Murthy, K.V. 2003. An inverse global environmental Kuznets cuve. *Journal of Comparative Economics* 31(2): 352–368.

Johnson, D.W. and Taylor, G.E. 1989. Role of air pollution in forest decline in Eastern North America. *Water, Air and Soil Pollution* 48(1–2): 21–43.

Jump, A.S. and Penuelas, J. 2005. Running to stand still: Adaptation and the response of plants to rapid climate change. *Ecology Letters* 8(9): 1010–1020.

Kalkstein, L.S. and Smoyer, K.E. 1993. The impact of climate change on human health: Some international implications. *Cellular and Molecular Life Sciences* 49(11): 969–979.

Kaly, U.L., Pratt, C.R., and Mitchell, J. 2004. The demonstration environmental vulnerability index (EVI) 2004. South Pacific Applied Geoscience Commission (SOPAC), Technical Report 384, http://www.vulnerabilityindex.net/Files/EVI%202004%20Technical%20Report.pdf

Keller, W.B., Gunn, J.M., and Yan, N.D. 1998. Acid rain—perspectives on lake recovery. *Journal of Aquatic Ecosystem Stress and Recovery (Formerly Journal of Aquatic Ecosystem Health)* 6(3): 207–216.

Khemani, L.T., Momin, G.A., Praksa Rao, P.S., Safai, P.D., and Singh, G. 1989. Spread of acid rain over India. *Atmospheric Environment ATENBP* 23(4): 757–762.

Killham, K., Firestone, M.K., and McColl, J.G. 1983. Acid rain and soil microbial activity: Effects and their mechanisms, *Journal of Environmental Quality* 12: 133–137.

Kimbal, K.D. and Levin, S.A. 1985. Limitations of laboratory bioassays: The need for ecosystem-level testing. *BioScience* 35(3): 165–171.

Kosatsky, T. 2005. The 2003 European heat waves. *Eurosurveillance* 10(7–9): 148–149.

Krajick, K. 2001. Acid rain. Long-term data show lingering effects from acid rain. *Science* 292(5515):195–196.

Kutz, S.J., Hoberg, E.P., Polley, L., and Jenkins, E.J. 2005. Global warming is changing the dynamics of Arctic host–parasite systems. *Proceedings of the Royal Society B: Biological Sciences* 272(1581): 2571–2576.

Langelle, O. 2000. Why ecological modernization and sustainable development should not be conflated. *Journal of Environmental Policy & Planning* 2(4): 303–322.

Lawrence, G.B. 2002. Persistent episodic acidification of streams linked to acid rain effects on soil. *Atmospheric Environment* 36(2002): 1589–1598.

Lecocq, F. and Kapoor, K. 2005. *State and Trends of the Carbon Market*. World Bank, Washington, DC. Available at: http://carbonfinance.org/docs/CarbonMarketStudy2005.pdf

Lee, J.J., Neely, G.E., Perrigan, S.C., and Grothaus, L.C. 1981. Effect of simulated sulphuric acid rain on yield, growth and foliar injury of several crops. *Environmental and Experimental Botany* 21(2): 171–185.

Likens, G.E. and Bormann, F.H. 1974. Linkages between terrestrial and aquatic ecosystems. *BioScience* 24(8): 447–456.

Likens, G.E., Driscoll, C.T., and Buso, D.C. 1996. Long-term effects of acid rain: Response and recovery of a forest ecosystem. *Science* 272(5259): 244–246.

Lubchenco, J. 1998. Entering the century of the environment: A new social contract for science. *Science* 279(5350): 491–497.

Manne, A.S. and Richels, R.G. 2001. An alternative approach to establish trade-offs among greenhouse gases. *Nature* 410: 675–677.

McCann, K.S. 2000. The diversity-stability debate. *Nature* 405: 228–233.

McMichael, A.J., Butler, C.D., and Folke, C. 2003. New visions of addressing sustainability. *Science* 302(5652): 1919–1920.

Meijkamp, B., Aerts, R., Van de Staaj, J., Tosserams, M., Ernst, W.H.O., and Rozema, J. 1999. Effects of UV-B on secondary metabolites in plants. In: J. Rozema (Ed), *Stratospheric Ozone Depletion: The Effects of Enhanced UV-B Radiation on Terrestrial Ecosystems*, pp. 71–79. Backhuys Publishers, Leiden, The Netherlands.

Miller, C.A. 2005. New civic epistemologies of quantification: Making sense of indicators of local and global sustainability. *Science, Technology & Human Values* 30(3): 403–432.

Moser, A. 1994. Trends in biotechnology: Sustainable technology development: From high tech to eco tech. *Acta Biotechnologica* 14(4): 315–335.

Newman, J.R. and Schreiber, R.K. 1984. Animals as indicators of ecosystem responses to air emissions. *Environmental Management* 8(4): 309–324.

Niemi, G.J. and McDonald, M.E. 2004. Application of ecological indicators. *Annual Reviews in Ecology and Systematics* 35: 89–111.

Odum, E.P. 1985. Trends expected in stressed ecosystems. *BioScience* 35(7): 419–422.

OECD. 1997. Report of the final ring test of the *Daphnia magna* reproduction test.OCDE/ GD(97)19. Organisation for Economic Co-operation and Development—OECD, Paris. OECD Environmental Health and Safety Publication, Series on Testing and Assessment, No.6, 190pp.

OECD. 2004. *OECD Key Environmental Indicators*. Organisation for Economic Co-operation and Development, Paris.

Oechel, W.C., Hastings, S.J., Vourltis, G., Jenkins, M., Riechers, J., and Grulke, N. 1993. Recent change of Arctic tundra ecosystems from net carbon dioxide sink to a source. *Nature* 361: 520–523.

Orr, D.W. 1992. *Ecological Literacy: Education and the Transition to a Postmodern World*. SUNY Press, Albany, NY, 210pp.

Oskamp, S. 2000. Psychology of promoting environmentalism: Psychological contributions to achieving an ecologically sustainable future for Humanity. *Journal of Social Issues* 56(3): 373–390.

Overrein, L.N., Seip, H.M., and Tollan, A. 1980. Acid precipitation—effects of forest and fish. Norwegian Council for Scientific and Industrial Research, P.O. Box 333, Blindern, Oslo 3, Norway.

Owensby, C.E., Coyne, P.I., and Auen, L.M. 1993. Nitrogen and phosphorus dynamics of a tallgrass prairie ecosystem exposed to elevated carbon dioxide. *Plant, Cell and Environment* 16: 843–850.

Paavola, J. and Adger, W.N. 2005. Fair adaptation to climate change. *Ecological Economics* 56: 594–609.

Panayotou, T. 1993. Empirical tests and policy analysis of environmental degradation at different stages of economic development. International Labour Office, Working paper WP238 Technology and Employment Programme, Geneva.

Parmesan, C. and Yohe, G. 2003. A globally coherent fingerprint of climate change impacts across natural systems. *Nature* 421: 37–42.

Patz, J.A. 2002. A human disease indicator for the effects of recent global climate changes. *Proceedings of the National Academy of Sciences USA* 99(20): 12506–12508.

Patz, J.A., Campbell-Lendrum, D., Holloway, T., and Foley, J.A. 2005. Impact of regional climate change on human health. *Nature* 438: 310–317.

Petrovic, M., Eljarrat, E., Lopez de Alda, M.J., and Barcelo, D. 2004. Endocrine disrupting compounds and other emerging contaminants in the environment: A survey on new monitoring strategies and occurrence data. *Analytical and Bioanalytical Chemistry* 378: 549–562.

Pielke, Jr., R.A., Landsea, C., Mayfield, M., Laver, J., and Pasch, R. 2005. Hurricanes and global warming. *Bulletin of the American Meteorological Society* 86: 1571–1575.

Ramakrishnan, P.S. 1998. Sustainable development, climate change and tropical rainforest landscape. *Climatic Change* 39(2–3): 583–600.

Rapport, D.J., Regier, H.A., and Hutchinson, T.C. 1985. Ecosystem behavior under stress. *The American Naturalist* 125(5): 617–640.

Reuss, J.O., Cosby, B.J., and Wright, R.F. 1987. Chemical processes governing soil and water acidification. *Nature* 329: 27–32.

Reich, P.B. and Amundson, R.G. 1985. Ambient levels of ozone reduce net photosynthesis in tree and crop species. *Science* 230: 566–570.

Reich, J.W. and Schlesinger, W.H. 1992. The global carbon dioxide flux in soil respiration and its relationships to vegetation and climate. *Tellus* 44B: 81–99.

Richard, Y., Rosa, E.A., and Dietz, T. 2004. The ecological footprint intensity of national economies. *Journal of Industrial Ecology* 8(4): 139–154.

Richardson, S.D. 2003. Water analysis: Emerging contaminants and current issues. *Analytical Chemistry* 75(12): 2831–2857.

Rossignol, P.A., Orme-Zavaleta, J., and Rossignol, A.M. 2006. Global climate change and its impact on disease embedded in ecological communities. *Environmental Geosciences* 13(1): 55–63.

Rusek, J. 1993. Air-pollution-mediated changes in Alpine ecosystems and ecotones. *Ecological Applications* 3(3): 409–416.

Salt, D.E., Blaylock, M., Kumar, N.P.B.A., et al. 1995. Phytoremediation: A novel strategy for the removal of toxic metals from the environment using plants. *Bio/Technology* 13: 468–474.

Santiago, S., Becker van Slooten, K., Chèvre, N., et al. 2002. Guide pour l'utilisation des testes ecotoxicologiques avec les daphnies, les bactéries luminescentes et les algues vertes, appliqués aux échantillons de l'environnement. Institut Forel, Université de Genève; Ecole Polytechniques Fédérale de Lausanne, http://www.cipel.org/sp/IMG/pdf/Guide-ECOTOX-Nov2002.pdf

Scheu, S. 2001. Plants and generalist predators as links between the below-ground and above-ground system. *Basic and Applied Ecology* 2(1): 3–13.

Scheu, S., Schlitt, N., Tiunov, A.V., Newington, J.E., and Jones, H.T. 2002. Effects of the presence and community composition of earthworms on microbial community functioning. *Oecologia* 133(2): 254–260.

Schiermeier, Q. 2006. Carbon market survives gas leaks. *Nature* 441: 405.

Schindler, D.W., Curtis, P.J., Parker, B.R., and Stainton, M.P. 1996. Consequences of climate warming and lake acidification for UV-B penetration in North American boreal lakes. *Nature* 379: 705–708.

Scholtens, B. 2005. What drives socially responsible investment? The case of the Netherlands. *Sustainable Development* 13: 129–137.

Schuetz, S., Schoening, M.J., Schroth, P., et al. 2000. An insect-based BioFET as bioelectronic nose. *Sensors and Actuators B* 65: 291–295.

Schwitzguebel, J.-P. 2001. Hype or hope: The potential of phytoremediation as an emerging green technology. *Remediation Journal* 11(4): 63–78.

Shadbegian, R.J. and Gray, W.B. 2005. Pollution abatement expenditures and plant-level productivity: A production function approach. *Ecological Economics* 54: 196–208.

Singh, S.P. 1998. Chronic disturbance, a principal cause of environmental degradation in developing countries. *Environmental Conservation* 25(1): 1–2.

SOLGCS. 2005. Safeguarding the ozone layer and the global climate system: Issues related to hydrofluorocarbons and perfluorocarbons. Special report of the Intergovernmental Panel on Climate Change (IPCC) and the Technology and Economic Assessment Panel (TEAP) and the United Nations Framework Convention on Climate Change and the Montreal Protocol., ISBN 92-9169-118-6, http://arch.rivm.nl/env/int/ipcc/pages_media/SROC-final/srocspmts_en.pdf

SOPAC-UNEP-EVI. 2005. Building resilience in SIDS. The Environmental Vulnerability Index, http://www.vulnerabilityindex.net

Sprinz, D.F. 1992. Why countries supports international environmental agreements: The regulation of acid rain in Europe. PhD Thesis, University of Michigan, 195pp.

SRES. 2000. Special Report on Emissions Scenarios. United Nations Intergovernmental Panel on Climate Change, http://www.grida.no/climate/ipcc/emissions

Stafford, N. 2006. Gas for the greenhouse. *Nature* 422: 499.

Stern, D.I., Common, M.S., and Barbier, E.B. 1996. Economic growth and environmental degradation: The environmental Kuznets curve and sustainable development. *World Development* 24(7): 1151–1160.

SR. 2007. Stern Review: The economics of climate change, http://www.sternreview.org.uk

Straton, A. 2006. A complex systems approach to the value of ecological resources. *Ecological Economics* 56: 402–411.

Sunyer, J. and Grimalt, J. 2006. Global climate change, widening health inequalities, and epidemiology. *International Journal of Epidemiology* 35(2): 213–216.

Sutherland, W.J. 2006. Predicting the ecological consequences of environmental change: A review of the methods. *Journal of Applied Ecology* 43: 599–616.

The Economist. 2005a. Don't despair. Most of the news on the climate change is bad, but not all of it. Issue December 10–16: 11–12; related articles at pages 75–77 and 86–87.

The Economist. 2005b. The Oiloholics. Oil prices could yet go higher—unless the world's guzzlers curb their thirst. Issue August 27 to September 2: 11; related survey at pages 57–58.

The Economist. 2005c. The real trouble with oil. How to avoid the next energy shock. Issue April 30th to May 6th: 11; related survey on oil after page 48.

The Economist. 2005d. Rescuing environmentalism. Market forces could prove the environment's best friend—if only greens could learn to love them. Issue April 23–26; related article at pages 78–80.

The Economist. 2006a. Cleaning up. Power companies and shorting traders have done better than the environment. Issue May 6–12: 79.

The Economist. 2006b. Sins of emission. The idea of offsetting carbon emissions is sound in theory, but not yet in practice. Issue August 5–11: 11.

Travis, C.C. and Hester, S.T. 1991. Global chemical pollution. *Environmental Science and Technology* 25: 814–819.

Travisi, C.M., Nijkamp, P., and Vindigni, G. 2006. Pesticide risk valuation in empirical economics: A comparative approach. *Ecological Economics* 56(4): 455–474.

Tilman, D. and Downing, J.A. 1994. Biodiversity and stability in grasslands. *Nature* 367: 363–365.

Turner, R.E. 2005. On the cusp of restoration: Science and society. *Restoration Ecology* 13(1): 165–173.

UN Agenda 21. 2005. United Nations Agenda 21. Available at: http://www.un.org/esa/sustdev/documents/agenda21/english/agenda21toc.htm

UNDESA-DSD. 2006. United Nations Department of Economic and Social Affairs, Division for Sustainable Development. Trends in sustainable development. United Nations Department of Economic and Social Affairs, Division for Sustainable Development.

UNFCCC. 2005. United Nations Framework Convention on Climate Change. Kyoto protocol to the United Nations Framework Convention on Climate Change, United Nations, http://unfcc.int/resource/docs/convkp/kpeng.pdf

US EPA. 2002. Methods for evaluating wetland condition. Using algae to assess environmental conditions in wetlands. EPA-822-R-02-021. U.S. Environmental Protection Agency, Office of Water.

Van de Staaij, J., Rozema, J., Van Beem, A., and Aerts, R. 2001. Increased solar UV-B radiation may reduce infection by arbuscular mycorrhizal fungi (AMF) in dune grassland plants: Evidence from five years of field exposure. *Plant Ecology* 154: 171–177.

Van Griethuysen, P. 2002. Sustainable development: An evolutionary economic approach. *Sustainable Development* 10: 1–11.

Vincent, K. 2004. Creating an index of social vulnerability for Africa. Working Paper 56, Tyndall Centre for Climate Change Research, University of East Anglia, Norwich, UK.

Vitousek, P.M., Mooney, H., A., Lubchenco, J., and Melillo, J.M. 1997. Human domination of Earth's ecosystems. *Science* 277(5325): 494–499.

Vitousek, P.M., Aber, J.D., Howarth, R.W., et al. 1993. Human alteration of the global nitrogen cycle: Sources and consequences. *Ecological Applications* 7(3): 737–750.

Wackernagel, M. and Rees, W. 1996. *Our Ecological Footprint: Reducing Human Impact on the Earth.* New Society Publishers, Philadelphia.

Wackernagel, M., Schulz, N.B., Deumling, D., et al. 2002. Tracking the ecological overshoot of the human economy. *PNAS* 99(14): 9266–9271.

Waltner-Towes, D. 2004. *Ecosystem Sustainability and Health: A Practical Approach.* Cambridge University Press, Cambridge, UK, 138pp.

Watt, W.D. 1986. A summary of the impact of acid rain on Atlantic salmon (*Salmo salar*) in Canada. *Water, Air and Soil Pollution* 35(1–2): 27–35.

WCCDL. 2006. Workshop climate change and disaster losses: Understanding and attributing trends and projections, May 25–26, 2006, Hohenkammer, Germany. Proceedings available at: http://sciencepolicy.colorado.edu/sparc/research/projects/extreme_events/munich_workshop/workshop_booklet.pdf

WCED. 1987. Our common future. ("The Brundtland Report") Report of the World Commission on Environment and Development (WCED), to the General Assembly of the United Nations (04.08.1987).

WDR. 2004. World Disaster Report 2004. The International Federation of Red Cross and Red Crescent Societies, http://www.ifrc.org/publicat/wdr2004/

Welsh, R., Ervin, D.E. 2006. Precaution as an approach to technology development. *Science, Technology & Human Values* 31(2): 153–172.

Wigley, T.M.L. 1998. The Kyoto Protocol: CO_2, CH_4 and climate implications. *Geophysical Research Letters* 25(13): 2285–2288.

Winner, W.E. 1994. Mechanistic analysis of plant responses to air pollution. *Ecological Applications* 4(4): 651–661.

Wolterbeek, B. 2002. Biomonitoring of trace element air pollution: Principles, possibilities and perspectives. *Environmental Pollution* 120(1): 11–21.

Wright, R.F. 1989. RAIN project: Role of organic acids in moderating pH change following reduction in acid deposition. *Water, Soil and Air Pollution* 46(1–4): 251–259.

Wright, R.F. and Schindler, D.W. 1995. Interaction of acid rain and global changes: Effects on terrestrial and aquatic ecosystems. *Water, Air and Soil Pollution* 85(1): 88–99.

Wurst, S., Langel, R., Reineking, A., Bonkowski, M., and Scheu, S. 2003. Effects of earthworms and organic litter distribution on plant performance and aphid reproduction. *Oecologia* 137(1): 90–96.

WHO. 2002. *The World Health Report 2002. Reducing Risks, Promoting Healthy Lives.* The World Health Organization, Geneva.

WHOE. 2004. *Heat-Waves: Risks and Responses.* Series Health and Global Environmental Change, No. 2. World Health Organization Europe, Copenhagen, http://www.euro.who.int/document/e82629.pdf

Wilson, E.O. 2002. *The Future of Life.* Little, Brown by arrangement with Alfred A. Knopf, Inc. Published again in 2003 by Abacus, 230 pages, ISBN 0-349-115795. Chapter two "The Bottleneck" published in Scientific American, February 24, 2002.

WWF. 2004. The Living Planet Report 2004. World Wildlife Fund in the US and Canada, United Nations Environmental Program—World Conservation Monitoring Program, Global Footprint Network, www.panda.org/downloads/lpr2004.pdf

Yachi, S. and Loreau, M. 1999. Biodiversity and ecosystem productivity in a fluctuating environment: The insurance hypothesis. *Proceedings of the National Academy of Sciences USA* 96: 1463–1468.

Yager, P., Edwards, T., Fu, E., et al. 2006. Microfluidic diagnostic technologies for global public health. *Nature* 442: 412–418.

Yan, N.D., Keller, W., Scully, N.M., Lean, D.R., and Dillon, P.J. 2002. Increased UV-B penetration in a lake owing to drought-induced acidification. *Nature* 381: 141–143.

17 Regional and Global Environmental Issues of Air Pollution

Luisa T. Molina and Bhola R. Gurjar

CONTENTS

17.1 INTRODUCTION

There is growing evidence that human activities in an increasingly globalized, industrialized, and interconnected world are influencing both air quality and climate change ranging from urban and regional to continental and global scales. Rapid population growth and increased energy demand are the primary forces driving unprecedented environmental changes.

In July 2009, the world population has reached 6.8 billion, 313 million more than that in 2005 or a gain of 78 million persons annually (UN, 2009). Out of these, 5.6 billion (or 82% of the world's total) will be living in the less developed regions. Most growth occurs in urban areas; for the first time in history, half of the people in the world were living in urban areas in 2008 (UN, 2008). The world's burgeoning urban areas consume a large fraction of the Earth's current fossil fuel budget to produce electrical energy, propel transportation, power industrial processes, prepare food, and provide heat and ventilation for homes, commercial enterprises, and public buildings. Exhaust emissions from these fossil fuel combustion sources and the processes they power emit large quantities of gases and fine particulate matter (PM) into the atmosphere, resulting in serious health and environmental consequences (Molina and Molina, 2004; Molina et al., 2004).

Substantial reduction of harmful emissions to the atmosphere is achievable through a combination of technology improvement and policy measures. In fact, air pollution has decreased in some cities in different parts of the world where effective emission control strategies and increasing energy efficiency were implemented. However, many areas, especially in Asia where the most polluted cities are now found, still suffer from excessive air pollution. Energy consumption and transport demand increase every year and are responsible for a substantial part of both anthropogenic greenhouse gas (GHG) emissions and adverse health effects from air pollution. This is related to the immense industrial expansion in many developing country cities that are producing goods for the global economy and the relatively high growth in private cars as their citizens become more affluent. Addressing the problem of rapidly growing cities that have severe air pollution, even though their standards of living are improving, has become one of the most important environmental challenges.

17.2 SOURCES AND TRANSPORT OF ATMOSPHERIC POLLUTANTS

Air pollution can be natural or man-made. Gases such as sulfur dioxide (SO_2), hydrogen sulfide (H_2S), and carbon monoxide (CO) are continually released into the atmosphere as by-products of natural events such as volcanic activity, vegetation decay, and forest fires. Fine particles are distributed throughout the atmosphere by winds, forest fires, volcanic eruptions, and other similar natural disturbances. However, air pollution of human origin has become a major, persistent problem in many urban/industrialized areas around the world.

Table 17.1 lists several primary air pollutants, which are emitted *directly* into the atmosphere from specific sources; these may be distinguished from secondary air pollutants that are formed *in situ* in the atmosphere as a result of oxidation or photochemical reactions. The table also lists the major sources of these pollutants, their residence time in the atmosphere and the potential local, regional, and global consequences of the release of these chemicals into the environment. Other harmful pollutants include hazardous air pollutants (HAPs), persistent organic pollutants (POPs), mercury, and other heavy metals.

TABLE 17.1

Major Primary Pollutants and Their Impacts

Pollutant	Anthropogenic Sources	Residence Time[a]	Effects and Consequences
1,3-Butadiene	Vehicle exhaust	2 h	Probable carcinogen; ozone precursor
Formaldehyde[b]	Vehicle exhaust	4 h	Respiratory irritant; probable carcinogen; ozone precursor
Benzene	Vehicle exhaust	10 days	Carcinogen
Alkenes, aromatic hydrocarbons	Vehicle exhaust, solvents	Hours to 2 days	Ozone precursors
Nitrogen oxides (NO_x)	Vehicle exhaust, combustion	1 day	Increased respiratory disease; ozone and acid rain precursors
Polycyclic aromatic hydrocarbons (PAHs)	Incomplete combustion (e.g., gasoline, diesel, biomass)	Hours to days	Some are probable carcinogens
SO_2	Coal and other sulfur-containing fossil fuel burning	Hours to days	Increased respiratory disease; acid rain precursor
Ammonia (NH_3)	Waste treatment, fertilizers, vehicle exhaust, animal waste	1–7 days	Respiratory irritant; neutralizes acids
Respirable PM (PM_{10})[b]	Road dust, windblown dust, incomplete combustion	5–10 days	Increased respiratory disease; reduced visibility
Fine PM ($PM_{2.5}$)[b]	Fuel combustion, diesel exhaust, windblown dust	5–10 days	Increased respiratory and cardiopulmonary disease; reduced visibility
Lead	Leaded gasoline, paint	5–10 days	Kidney and brain damage; learning disabilities
Carbon monoxide (CO)	Incomplete combustion	2 months	Cardiovascular and neurobehavioral disease
Carbon dioxide (CO_2)[b]	Fossil fuel and biomass burning	3–4 years	Global warming
Methane (CH_4)	Rice paddies, livestock, natural gas leaks	8–10 years	Ozone precursor; global warming
Chlorofluorocarbons (CFCs)	Air conditioners, refrigerators, spray cans, foam products	50–100 years	Ozone depletion; global warming

Source: Molina L.T. and Molina M.J. (Eds). 2002. *Air Quality in the Mexico Megacity: An Integrated Assessment.* Kluwer Academic Publishers, Dordrecht, The Netherlands, 384pp.

[a] Arey J. 2000. In *Environmental Medicine*, L. Moller, ed., pp. 52–71. Joint Industrial Safety Council Product 33, Sweden.

[b] Also formed from other pollutants.

The fate of the various compounds emitted at the Earth's surface depends on their stability. The efficiency of their removal processes is affected by direct dispersion and transport, and also by meteorological factors such as temperature, solar intensity, and the presence of clouds and fog. There are two types of processes by which chemicals are removed from the troposphere:

1. *Physical removal:* Removal of a chemical from the atmosphere by rain is called "wet deposition," that is, the chemical dissolves in a raindrop and falls to the earth with the drop. The chemical is no longer an air pollutant, but it may now affect another part of the environment. For example, if the chemical is an acid, it can end up in a lake with possible consequences to the aquatic ecosystem.
2. *Chemical removal:* The second type of removal process, loss by chemical reaction, destroys the original species, converting it into another species. For example, hydrocarbons are eventually converted to water and carbon dioxide (CO_2). When chemical reaction occurs, some of the transformed chemicals (secondary pollutants) are potentially more harmful than their precursors (primary pollutants).

Compounds such as the CFCs are chemically very inert and practically insoluble in water; thus they are not removed by the cleansing mechanisms available in the troposphere. These compounds are able to persist in the atmosphere long enough to diffuse upward to the stratosphere, where they are eventually decomposed by high-energy solar radiation to yield radicals that can destroy stratospheric ozone through a catalytic process (Molina and Rowland, 1974; Molina and Molina, 2005). In contrast, compounds such as NH_3 and hydrogen chloride are rapidly removed by rain, the average time scale for removal being weeks. From a global perspective, hydrocarbons and NO_x are also removed quickly: hydrocarbons are not soluble in water, but are first oxidized by various species such as the hydroxyl radical (OH), which converts them to soluble compounds that are then removed by rain. The predominant fate of NO_x is also removal by rain, after conversion to nitric acid that contributes to acid deposition.

From a local and regional perspective, hydrocarbons and NO_x are responsible for the degradation of air quality by forming ozone and secondary PM. The more photochemically and oxidatively active hydrocarbons may decompose on a time scale of minutes, whereas the less reactive ones may last many hours, contributing to the formation of ozone and particulates downwind from the sources where they are emitted.

17.3   IMPACTS OF AIR POLLUTION

Air pollution used to be considered as a local concern rather than a long-term global change issue. As mentioned in the previous chapters in this book, air pollution can adversely affect human health by direct inhalation and by other exposures such as contamination of drinking water and food and skin transfer. However, the atmosphere is a shared resource that respects no boundaries; air pollutants do not stop when they reach city or country boundaries. Once released into the atmosphere, air pollutants can be carried by winds, mix with other pollutants, undergo chemical transformations, and are eventually deposited on various surfaces. Thus, their impacts can occur far from their sources. The regional and global dispersion of pollutants generated locally has been well established in the case of acid deposition and stratospheric ozone depletion. Recently, long-range transport of tropospheric ozone has been seen to be increasing throughout the Northern Hemisphere. POPs from industrial economies are being transported over long distances to the Arctic. Air pollution and wildfire emissions originating from northern mid-latitudes give rise to the Arctic Haze phenomenon.

Emissions from urban and industrial centers of the developed world, and increasingly from large cities of the developing world, are changing the chemical content of the downwind troposphere in a number of fundamental ways. The degraded atmospheres in large cities often contain high concentrations of PM, O_3, SO_2, CO, NO_x, and volatile organic compounds (VOCs). PM is often reported as mass concentration in the total suspended particulates (TSP), PM_{10}, and $PM_{2.5}$ (particles with aerodynamic diameters of less than ~40, 10, and 2.5 µm, respectively). The major PM chemical components are sulfate ($SO_4^=$), nitrate (NO_3^-), ammonium (NH_4^+), organic carbon (OC), elemental carbon (EC), and soil dust (Molina and Molina, 2004; Molina et al., 2004).

Emissions of CO, NO_x, and VOCs drive the formation of photochemical smog and its associated oxidants, degrading air quality and threatening both human and ecosystem health. On a larger scale, these same emissions drive the production of ozone (a powerful GHG) in the free troposphere, contributing significantly to global warming. Urban and industrial areas are also major sources of the potent GHGs, including CO_2, methane (CH_4), nitrous oxide (N_2O), and halocarbons. NO_x and SO_2 emissions are also processed to strong acids by atmospheric photochemistry on regional to continental scales, driving acid deposition to sensitive ecosystems and damage to materials, including historic buildings and monuments. Direct urban/industrial emissions of carbonaceous aerosol particles are compounded by the emission of high levels of secondary aerosol precursors, including NO_x, VOCs, SO_2, and NH_3. The resulting mix of primary (directly emitted) and secondary (formed in subsequent photochemical and chemical reactions) aerosols is now recognized to play an important role in the Earth's climate (IPCC, 2007). In the following sections, we will discuss some of the regional and global consequences of air pollution.

17.4   VISIBILITY IMPAIRMENT AND REGIONAL HAZE

Cities with poor air quality are easily identifiable by their hazy atmosphere and impaired visibility, which is readily distinguishable from natural fog and clouds. The connection between atmospheric pollutants and visibility is a well-understood phenomenon (Watson, 2002). The effect results predominantly from the presence of small particles in the atmosphere, either directly emitted or formed by chemical transformations of gaseous pollutants. Loss of visibility, in fact, provided early indications of the striking effects on the atmosphere that can result from human activities. For example, Los Angeles recorded its first serious episode of smog in the summer of 1943, when visibility was only three blocks and residents suffered from eye irritation, respiratory discomfort, and nausea (CARB, 1999).

In Beijing, China, the visibility is often low, in part because of the relatively high frequency of foggy days. However, the sky overhead is usually gray even in the absence of fog or clouds. Bergin et al. (2001) conducted a study of the physical and chemical properties of aerosols in Beijing during June 1999 and concluded that, in that period, combustion-related particles were mainly responsible for visibility impairment. However, in the spring months, it is often sand storms and dust that lead to very low visibility (Yang et al., 2002). Dust and sand storms that originate in the dry regions of northern China and Mongolia and blow across parts of China, the Korean peninsula, and Japan occur during the spring months as, during this time, cold air masses from Siberia whip deserts and soils eastward after the dry continental winter. Recently, these storms are growing in intensity and frequency (Molina et al., 2004).

Besides the above-mentioned dust and sand storms, clouds of tiny aerosol particles from anthropogenic emissions hang over a number of regions. These seasonal layers of haze reduce the amount of sunlight that can reach the Earth's surface, which has potential impacts on air quality, climate, and hydrological cycle (Ramanathan et al., 2001). The brown haze, known as atmospheric brown clouds (ABC), will be discussed further in Section 17.8.6.

17.5 REGIONAL ECOSYSTEM IMPACTS

17.5.1 Acid and Fixed Nitrogen Deposition

Acid deposition (acid rain) is the first widely recognized regional-scale ecosystem impact driven by urban and industrial emissions. The process is driven by atmospheric oxidation of NO_x and SO_2 (emitted during combustion) into ambient nitric and sulfuric acid. Problems arise when these acids, and the sulfate and nitrate aerosols they form, are deposited downwind on poorly buffered surface waters or soils, thereby adversely affecting sensitive lakes, streams, forests, and farmlands (NAPAP, 1990). Regional air pollution problems of acidification have been reduced in Europe and North America, but are now a growing policy focus in parts of Asia, where acidic deposition has increased (UNEP, 2007).

A closely related problem involves fertilization effects caused by the deposition of airborne fixed nitrogen species (ammonium and nitrate aerosols and their gas-phase precursors) on buffered soils and fresh or marine surface waters that are not susceptible to acidification. Combined with fixed nitrogen and phosphorus from fertilizer, animal waste, and human sewage sources, atmospheric deposition of fixed nitrogen can help overfertilize soils, lakes, streams, and estuaries, leading to changes in primary productivity (excessive plant growth and decay) and, potentially, to eutrophication (Galloway, 1996). This enhanced plant growth is causing a variety of problems, such as a lack of oxygen in the water needed for fish and shellfish to survive and severe reductions in water quality. More recently, it has been documented that high levels of fixed nitrogen deposition can have significant effects on ecosystem diversity, even when deposition receptor areas are not heavily acidified. For example, Stevens et al. (2004) report that British grasslands subject to long-term chronic levels of nitrogen deposition have significantly lower levels of species diversity than those exposed to lower deposition rates. Nitrogen deposition has been recognized within the Convention on Biological Diversity as a significant driver of species loss; several major global biodiversity hotspots have been identified as being at significant risk (Phoenix et al., 2006).

17.5.2 PHOTOCHEMICAL OXIDANT DAMAGE

As urbanization spreads, it has also been recognized that photochemical oxidant production is increasingly a regional problem (NRC, 1991; Guttikunda et al., 2003). Photochemical oxidants are the products of reactions between NO_x and a wide variety of VOCs. The most well-known oxidants are ozone, peroxyacetyl nitrate (PAN), and hydrogen peroxide (H_2O_2). Photochemically produced oxidants and their precursors flowing out of major cities frequently generated high levels of ozone and other oxidants all the way to neighboring cities, exposing suburbs, forests, and agricultural areas. Regional-scale impacts on crop yields caused by tropospheric ozone have been estimated to cause economic losses for 23 arable crops in 47 countries in Europe in the range US$ 5.7–12 billion/year (Holland et al., 2006).

It has been suggested that photochemical oxidant damage may be having a significant effect on crop yields in important agricultural areas impacted by emissions from major Chinese cities (Chameides et al., 1999a,b) as well as India and Pakistan (Emberson et al., 2003). Gregg et al. (2004) report greater plant growth in New York City compared to a rural environment and attribute the effect to the higher levels of ozone in the rural area investigated. Significant damage to forests surrounding the Mexico City air basin caused by exposure to high levels of photochemical oxidants (mainly ozone) has been documented (Fenn et al., 2002). An assessment of the ozone impact to crops in Asia and Africa is given in Chapter 15.

17.5.3 EFFECTS OF PM ON PHOTOSYNTHETICALLY ACTIVE RADIATION

Aerosol particles are well known to affect the climate system by interacting with solar radiation through direct and indirect processes (Section 17.6). Recently, another consequence of the aerosol direct effect on solar radiation has been investigated, namely, the effect on vegetation carbon gain due to a reduction in total Photosynthetically Active Radiation (PAR) reaching the Earth's surface and the increase of the diffuse fraction of PAR (Cohan et al., 2002). This spectral region corresponds more or less with the range of light visible to the human eye (400–700 nm) that photosynthetic organisms are able to use in the process of photosynthesis.

Recent model analyses demonstrate the large impact of Asian megacity SO_2 emissions on regional SO_2 pollution. High regional SO_2 and other fine PM precursors can create extremely high levels of fine PM with absorption and scattering properties that significantly influence both the direct and diffuse components of PAR (Cohan et al., 2002); the resulting haze over eastern China has measurably decreased solar radiation reaching the surface since 1954 (Kaiser and Qian, 2002). It has been suggested that attenuation of PAR by both atmospheric PM and PM deposited on plant leaves may significantly impact the solar radiation available for photosynthesis in important agricultural regions in China (Chameides et al., 1999b; Bergin et al., 2001).

Yamasoe et al. (2006) showed that high concentrations of aerosol particles in the atmosphere due to biomass burning in the Amazon region decrease the amount of global photosynthetic radiation at varying canopy levels, affecting sensible and latent heat fluxes at the surface.

17.6   REGIONAL CLIMATE CHANGE

High levels of air pollutants emitted from urban/industrialized areas can affect regional climate. Long-lived primary GHG emissions associated with major cities have a direct impact on infrared radiative forcing on a global basis (IPCC, 2007). Furthermore, high regional atmospheric concentrations of the powerful, but shorter-lived tropospheric ozone (Akimoto, 2003) will have a more pronounced regional effect.

A progressive and accelerated long-term warming trend has been reported for Asia for the period 1860–2004. Australia is suffering severe drought in recent years and had its warmest year on record, as well as its hottest April, in 2005. In Europe, the average temperature has increased by about 1.4°C compared to preindustrial levels (UNEP, 2007).

17.6.1   Effects of PM on Sunshine Duration and Temperature

Fine particles can have a direct effect on short-wavelength radiative forcing by scattering and/or by absorbing solar radiation (see Section 17.8). In recent decades, surface temperature records in some urbanized regions of China and India (Menon et al., 2002) show a measurable cooling, despite a general warming observed for most of the globe. Analyses of meteorological data from many monitoring stations in heavily urbanized regions of China demonstrate significant downward trends in both sunshine duration (1–3% per decade) and maximum daily temperatures (0.2–0.6 K per decade) (Qian and Giorgi, 2000; Kaiser and Qian, 2002). The observed cooling trends are consistent with the predicted effects of high measured levels of soot containing fine particles, which heat the air, alter regional atmospheric circulation, and contribute to regional climate change (Menon et al., 2002).

17.6.2   Effects of PM on Precipitation and Clouds

Atmospheric aerosol particles are crucial to the existence of clouds as we know them. High aerosol loadings that increase effective cloud condensation nuclei levels can also influence precipitation levels, lengthening cloud lifetimes, and suppressing rain and snow by nucleating more, but smaller, cloud droplets. Satellite observations show significant rainfall suppression downwind of major cities (Rosenfeld, 2000). High PM loadings with a large fraction of absorbing soot particles are predicted to reduce cloudiness by absorptive heating of air masses (Ackerman et al., 2000), although the actual impact on cloud cover may also be affected by the increased atmospheric circulation that this heating will induce (Menon et al., 2002). Currently, there is considerable difficulty in establishing climatically meaningful relationships among aerosol, clouds, and precipitation. Stevens and Feingold (2009) suggest that clouds react to aerosols in a very complex way and the reaction is strongly dependent on the type and state of the cloud. It is therefore important to study the types of cloud systems in which aerosols have the greatest influence.

17.7 LONG-RANGE TRANSPORT OF AIR POLLUTANTS

With the growth of multicity "megalopolis" regions in North America, Europe, East Asia, and South Asia, the export of air pollutants from urbanized regions to sensitive environments has become a major concern because of wide-ranging consequences for human health and ecosystems, visibility degradation, weather modification, changes in radiative forcing, and tropospheric oxidation capacity. For example, the Arctic is often perceived as a pristine place, yet its atmosphere has served as a receptor for air pollution from the industrial regions of northern mid-latitude continents, as manifested in particular by thick aerosol layers (Arctic haze) and by accumulation of persistent pollutants such as mercury. The haze affects the highly reflective Arctic ice sheet in ways that can increase temperatures both in the atmosphere and on the Earth's surface. A major field study ARCTAS (Arctic Research on the Composition of the Troposphere from Aircraft and Satellites) was launched in 2008 to identify how air pollution contributes to climate change in the region (NASA, 2009).

There is also concern in North America and Europe that efforts to improve air quality locally through domestic emission controls could be partially offset by Asian industrialization and economic growth and the associated transpacific transport of pollution. The Task Force on Hemispheric Transport of Air Pollution was created in 2004 by the United Nations Economic Commission for Europe (UNECE) Convention on Long-range Transboundary Air Pollution with the aim to improve the understanding of intercontinental transport of air pollution and to provide estimates of source–receptor relationships for intercontinental air pollution (UNCEC, 2009).

Observations from the ground, aircraft, and satellites throughout the global atmosphere have provided a wealth of evidence that air pollution can be transported over long distances, for example, from eastern Asia to the western United States, from North America to Europe, and from mid-latitudes to the Arctic (see, e.g., Berntsen et al., 1999; Jacob et al., 1999; Yienger et al., 2000; Edwards et al., 2003; Stohl et al., 2007). Our ability to quantify the magnitude of transport has improved recently from an increasing body of observational evidence, including new information from intensive field campaigns and satellite-borne instruments, improved emission inventories, and global and regional chemical transport models.

Multiple studies have documented the transport of Asian pollution to North America (Jacob et al., 1999; Jaffe et al., 2004; Zhao et al., 2008) and Europe (Lelieveld et al., 2001). This transport is initiated by cold frontal passages over eastern Asia, which lift continental boundary layer air to the free troposphere where it is then carried toward North America by the westerlies. Dramatic evidence for transpacific transport of Asian air to North America is offered by Asian dust plumes, which have been tracked by satellites across the ocean and found to cause large-scale exceedances of air quality standards at sites in the western United States (Husar et al., 2001). Transpacific transport is highly episodic but is most frequent and rapid in spring, when frontal activity is at a maximum and the atmospheric circulation is strong (Yienger et al., 2000). There is, however, evidence to suggest that Asian influences are not limited to the spring period and are present throughout the year (Van Curen et al., 2005).

More recently, results from INTEX-B conducted over the Northeast Pacific and the west coast of North America (Singh et al., 2009) suggested that Asian pollution enhanced surface ozone concentrations by 5–7 ppbv over western North America in Spring 2006. The 2000–2006 rise in Asian anthropogenic emissions increased this influence by 1–2 ppbv (Zhang et al., 2008). Dunlea et al. (2009) studied atmospheric transformation of aerosols as they are transported from Asia across the Pacific and observed that sulfate and organic aerosol in aged Asian pollution layers are consistent with fast formation near the Asian continent, followed by washout during lofting and subsequent transformation during transport across the Pacific.

These field measurements have demonstrated the enormous pollutant potential of major cities and "megalopolis" regions, as well as the fact that significant quantities of longer-lived gaseous pollutants and fine particles can be transported and detected over intercontinental scales. However, relatively few measurements have been carried out on the polluted outflow from the rapidly growing developing world's megacities in tropical and subtropical latitudes. MILAGRO (Megacity Initiative: Local And Global Research Observations) was the first international collaborative project to examine the behavior and export of atmospheric emissions generated in megacities. The Mexico City Metropolitan Area (MCMA)—one of the world's largest megacities and North America's most populous city—was selected as the case study for MILAGRO and took place in March 2006 (Molina et al., 2010). The campaign, together with an earlier intensive field study in 2003 (Molina et al., 2007), have provided a wealth of information on the emissions, dispersion, and transformation processes of the pollutants emitted to the MCMA atmosphere and their urban, regional, and hemispheric impacts (Molina et al., 2008, 2010; Singh et al., 2009). The MCMA motor vehicles produce abundant amounts of primary PM, EC, particle-bound PAHs, CO, and a wide range of air toxics, including formaldehyde, acetaldehyde, benzene, toluene, and xylene. VOC/CO emission ratios are notably higher than those in the United States, and aldehydes emissions are significant. High aerosol concentrations were observed and were composed, in large part, of organics, but black carbon, crustal matter, sulfate, and nitrate were also significant contributors. Biomass burning (agricultural, forest, wood cooking, and trash burning) also contributes to the urban and regional pollution in the Mexico Basin. The pollution plume from Mexico City can be observed several hundreds of kilometers downwind; aircraft-based measurements show ongoing production of secondary organic aerosols and ozone for several days downwind (see Molina et al., 2010 and references therein).

17.8   GLOBAL CLIMATE CHANGE

According to the Intergovernmental Panel on Climate Change (IPCC), there is now visible and unequivocal evidence of climate change impacts, and there is a consensus that GHG emissions (principally CO_2) from human activities are the main drivers of change. The Earth's average temperature has been recorded to increase by approximately 0.74°C over the past century. The visible evidence of this warming includes shrinking mountain glaciers, thawing permafrost, earlier breakup of river and lake ice, changes in precipitation patterns and ocean currents, and increasing frequency and intensity of heat waves, storms, floods, and droughts in some regions around the world (IPCC, 2007). Climate change is now recognized as a major global challenge that will have significant and long-lasting impacts on human well-being and development (UNEP, 2007).

17.8.1   EARTH'S CLIMATE SYSTEM

Climate in a narrow sense is usually defined as the "average weather," or more rigorously, it is described in terms of the mean and variability of temperature, precipitation, and wind over a period of time, ranging from months to millions of years (standard averaging period is 30 years) (IPCC, 2007). Climate in a wider sense is the state of the climate system, which is a complex, interactive system consisting of the atmosphere, land surface, snow and ice, oceans and other bodies of water, and living things (IPCC, 2007).

Climate change refers to the variation in the Earth's global climate or in regional climates over time. It describes changes in the variability or average state of the atmosphere over time scales ranging from decades to millions of years. These changes can be caused by processes internal to the Earth or external factors that affect climate (called "forcing"). External forcings include natural phenomena such as variation in solar intensity and volcanic eruptions, and more recently, human-induced changes in atmospheric composition. Solar radiation powers the climate system. There are three fundamental ways to change the radiation balance of the Earth: (1) by changing the incoming solar radiation; (2) by changing the fraction of solar radiation that is reflected (called "albedo"; e.g., by changes in cloud cover, atmospheric particles, vegetation, or extent of snow and ice cover); and (3) by altering the long-wavelength radiation from the Earth back toward space (e.g., by changing GHG concentrations). Climate, in turn, responds directly to such changes, as well as indirectly, through a variety of feedback mechanisms (IPCC, 2007).

17.8.2   GREENHOUSE EFFECT

GHGs are those gaseous constituents of the atmosphere, both natural and anthropogenic, that absorb and emit radiation at specific wavelengths within the spectrum of infrared radiation emitted by the Earth's surface, the atmosphere, and clouds. Water vapor (H_2O), CO_2, N_2O, CH_4, and O_3 are the primary GHGs in the Earth's atmosphere. Moreover, there are a number of entirely man-made GHGs in the atmosphere, such as halocarbons and other chlorine- and bromine-containing substances, sulfur hexafluoride (SF_6), hydrofluorocarbons (HFCs), and perfluorocarbons (PFCs).

The Earth receives energy from the Sun in the form of radiation, predominantly in the visible or near-visible (e.g., ultraviolet) part of the spectrum, and reflects about one-third of the incoming solar radiation back to space. The remaining two-third is absorbed by the surface and, to a lesser extent, by the atmosphere. To balance the absorbed incoming energy, the Earth must, on average, radiate the same amount of energy back to space. Because the Earth is much colder than the Sun, it radiates at much longer wavelengths, primarily in the infrared part of the spectrum. Much of this thermal radiation emitted by the land and ocean is absorbed by the atmosphere, including clouds, and reradiated back to Earth. This is called the natural greenhouse effect, which maintains the surface of the Earth at a temperature that is suitable for life; otherwise the surface temperature would average less than −15°C and would be too cold for most life as we know it.

There are many feedback mechanisms in the climate system that can either amplify (positive feedback) or diminish (negative feedback) the effects of a change in climate forcing. For example, as rising concentrations of GHGs warm the Earth, snow and ice begin to melt. This melting reveals darker land and water surfaces beneath the snow and ice; these darker surfaces absorb more of the Sun's heat, causing more warming, which causes more melting, and so on, in a self-reinforcing cycle. This feedback loop, known as the "ice-albedo feedback," amplifies the initial warming caused by rising levels of GHGs. Unraveling the complexities of the Earth's climate system requires the integrated and coordinated efforts of scientists from different disciplines (IPCC, 2007).

17.8.3 CLIMATIC IMPACTS OF AIR POLLUTANTS

The global energy budget of the Earth is determined by a balance at the top of the atmosphere between incoming solar radiation, reflected solar radiation, and outgoing terrestrial radiation. The climate is in equilibrium when the absorbed solar radiation (incoming minus reflected) equals the outgoing terrestrial radiation. Increases in the atmospheric emissions of GHGs can change the Earth's radiative balance and the temperature of the atmosphere.

Pollutants that affect the Earth's radiation balance are called climate-active pollutants. One way of describing their impact is through the idea of radiative forcing—a measure of the change caused by anthropogenic emissions of a particular primary pollutant, or of anthropogenic emissions leading to formation of a particular secondary pollutant. Radiative forcing is defined as positive if it results in a gain of energy for the Earth system and negative if it results in a loss. Positive forcing causes warming, and negative forcing causes cooling. GHGs such as tropospheric ozone and CH_4 absorb infrared radiation emitted from the Earth's surface and re-emit it at a lower temperature, thus decreasing the outgoing radiation flux and producing a positive forcing. Fine particles can have a direct effect on short wavelength radiative forcing by scattering (negative forcing) and/or absorbing solar radiation (positive forcing). Decreases of black carbon aerosols mitigate warming, while decreases of sulfate, nitrate, and organic aerosols exacerbate warming. Aerosols also act *indirectly* because they are involved in the formation of clouds, which have a large effect on the balance of radiation in the atmosphere. Figure 17.1 shows the global average radiative forcings of the air pollutants as estimated by IPCC Working Group 1 (IPCC-WG1, 2007). In contrast to the estimate for forcing by the GHGs, there is large uncertainity in estimating the aerosol climate forcing, which influences the

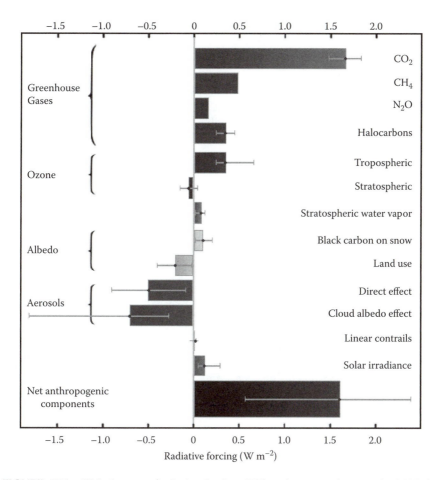

FIGURE 17.1 Global-average radiative forcing (RF) estimates and ranges in 2005 for anthropogenic CO_2, CH_4, N_2O, and other important agents and mechanisms. The net anthropogenic radiative forcing and its range are also shown. (Adapted from IPCC-WG1, 2007.)

estimate of the total anthropogenic forcing and hence the projected temperature increase.

17.8.4 IMPACTS OF CLIMATE CHANGE

Predictions of the future climate are made using computer models. The most advanced are called Global Circulation Models (GCMs). Interactions between the atmosphere, the oceans, the land surface, and ice are included in climate models, as are natural variations in the radiation from the Sun. One element of such calculations is the use of *emissions scenarios*, models of the way in which GHG emissions will change in the future. However, the actual temperature change will depend critically on choices that society makes regarding the reduction in GHG emissions, including human behavior, economic and technical developments, and the effectiveness of international, national, and local initiatives to reduce emissions.

The IPCC Fourth Assessment Report analyzes the results of a number of climate models and estimates that the increase in the global mean temperature by 2090–2099 would be 1.8–4.0°C, relative to 1980–1999 (IPCC, 2007). This is the best estimate, drawing on six emission marker scenarios, while the likely range is 1.1–6.4°C. If CO_2 concentrations in the atmosphere double, the global average surface warming would likely be in the range 2–4.5°C, with the best estimate of about 3°C above preindustrial levels (IPCC, 2007). These figures are for global averages; the temperature increase is widespread over the globe and is greater at higher northern latitudes. Land regions have warmed faster than the oceans (IPCC-SPM, 2007).

It is believed that the twentieth century's anthropogenic GHG emissions have already committed the Earth to an additional 0.1°C of warming per decade owing to the climate system's inertia. Some warming would have occurred even if the concentrations of all GHGs and aerosols in the atmosphere had been kept constant at year 2000 levels, in which case the estimated increase would be 0.3–0.9°C by the end of this century (IPCC, 2007). However, Ramanathan and Feng (2008) suggest that the committed warming will more likely to be higher (2.4°C) as the clean air policy remove the reflecting aerosols from the atmosphere for health reasons.

One of the key issues is that of feedbacks between the changing climate and physical, chemical, and biological processes. The climate system possesses intrinsic positive and negative feedback mechanisms that are generally beyond society's control. The net effect of warming is a strong positive feedback (IPCC, 2001), with several processes within the Earth's complex climate system acting to accelerate warming once it starts. The magnitude of such feedbacks is the subject of intense study. An important positive feedback is the increase in the amount of water vapor in the atmosphere that will result from higher air and ocean temperatures. Higher temperatures generally lead to an increase in natural emissions of pollutants. These include CH_4 from wetlands and tundra, CO_2 from soil and plant respiration, and VOCs from vegetation. These emissions are part of the overall carbon cycle, which is a key component in climate change and includes the exchange of CO_2 with the oceans, soils, and vegetation, in addition to our emissions from burning fossil fuels.

The response of the carbon cycle to changing temperature is a central element in climate models. Large amounts of carbon are released from land and ocean surfaces, for example, from plant respiration and decaying plant matter. At the same time, similar amounts of carbon are absorbed, for example, by the oceans and plants. The rates of these processes will alter in a changing climate. The size and direction (overall emission or overall uptake) of the effects will depend on whether ecosystems are limited by temperature or moisture.

As the global temperature increases, heat waves have become more frequent over most land areas. In the Arctic, average temperatures are rising almost twice as rapidly as in the rest of the world. Widespread melting of glaciers and sea ice and rising permafrost temperatures present further evidence of strong Arctic warming. There is now also evidence of widespread melting of permafrost, both in Alaska and Siberia, which is expected to increase the release of CH_4 from frozen hydrates, giving rise to a significant positive feedback. Changes in snow, ice, and frozen ground have increased the number and size of glacial lakes, have increased ground instability in mountain and other permafrost regions, and have led to changes in some Arctic and Antarctic ecosystems. Some hydrological systems have also been affected through increased runoff and earlier spring peak discharge in many glacier- and snow-fed rivers and through effects on thermal structure and water quality of warming rivers and lakes (IPCC, 2007).

The major health threat of warmer temperatures is the likely increase in more intense and prolonged heat waves that can cause dehydration, heat stroke, and increased mortality. Depending on the location, climate change is expected to increase smog episodes, water- and food-borne contamination, diseases transmitted by insects, and the intensity of extreme weather events (IPCC, 2007; also see Section 16.5 of Chapter 16 on climate impact on ecosystem and health).

17.8.5 AIR QUALITY: CLIMATE CHANGE INTERLINKAGE

Historically air quality and climate change have been treated as two separate and distinct policy issues. Air pollution control strategies have traditionally been focused on reducing emissions of air pollutants that are harmful to health or damage the environment, whereas climate change policy has focused on reducing emissions of GHGs, primarily CO_2. Recently, it has become clear that air pollution and climate change are intimately interrelated—with respect to sources, atmospheric processes, and human and environmental effects. Substantial potential benefits and synergies can be realized from integrated strategies that address both issues together, achieving cobenefits.

Many countries have an excellent record of reducing emissions of air quality pollutants. Improved technology driven by legislation and consequent pollution abatement has played a key role in reducing emissions of air quality pollutants. Examples include fitting of three-way catalysts to gasoline vehicles and low NO_x burners in power stations. Measures to reduce the impact of human activity on climate change and, at the same time, improve air quality are available. However, there are also measures that lead to a reduction in emissions of a climate-active pollutant but to an increase in emissions of air quality pollutants or vice versa. For example, because the cooling effects of sulfate aerosol are believed to have masked some of the warming effects of increased GHG concentrations, regulatory actions to reduce emissions of SO_2 will reduce sulfate aerosol concentrations and may lead to enhanced warming.

Measures that result in benefits for both air quality and climate should be promoted. These include incentives for domestic energy conservation, improved industrial process efficiency, fuel switching, demand management, and measures designed to modify the behavior of individuals so as to reduce the impact of their activities on the atmosphere, particularly the reduction in the use of on-road vehicles, shipping, and aircraft.

Integrated assessment studies can be very helpful in examining mitigation strategies that could benefit both air quality and climate. For example, Fiore et al. (2002, 2008) showed that reductions in CH_4 emissions should lead to global reductions in surface ozone concentrations. The benefits of these reductions to agriculture, forestry, and nonmortality human health were examined by West and Fiore (2005). Ethanol is currently being promoted as a clean and renewable fuel that will reduce air pollution, climate warming, and reliance on imported oil. A recent integrated environmental assessment of the production and use of ethanol as a substitute for gasoline indicates, however, that corn-based ethanol results at best in only small reductions in GHG emissions relative to gasoline (Pimentel and Patzek, 2005; Farrell et al., 2006), and could cause an increase in ozone pollution due to N_2O produced as a by-product of nitrogen fertilizer (Jacobson, 2007). An examination of four megacities (Mexico City, New York City, Santiago, and Sao Paulo) indicated that GHG mitigation would lead to large reductions in ozone and PM concentrations with substantial resulting improvements in public health (Cifuentes et al., 2001). McKinley et al. (2005) found that five proposed control measures in Mexico City that were estimated to reduce annual particle exposure by 1% and maximum daily ozone by 3% would also reduce GHG emissions by 2% for both periods 2003–2010 and 2003–2020. Furthermore, about 4400 Quality Adjusted Life Years (QALYs) would be saved for both time horizons. Assessments of potential cobenefits of GHG mitigation in China have also identified large associated reductions in the emission of air pollutants. When the resulting health improvements are monetized, the emission reductions are found in many cases to be cost-effective and even profitable (Aunan et al., 2004, 2006).

17.8.6 ATMOSPHERIC BROWN CLOUD: LONG-RANGE POLLUTANT TRANSPORT AND CLIMATE CHANGE

Atmospheric Brown Cloud (ABC) is a frequently occurring phenomenon in many regions of the world and is not restricted to urban regions. Recent field studies and satellite observations have revealed that, due to long-range transport, the brown cloud covers vast areas of the world with potentially large impacts on regional and global climate, freshwater budget, agriculture, and health (Ramanathan and Ramana, 2003). This is an example that air pollution and climate change are intricately linked and should be addressed under one common framework.

ABC is composed of numerous submicron-sized aerosols, including BC, sulfates, nitrates, fly ash, and others. ABCs intercept sunlight by absorbing as well as reflecting it, both of which lead to a large surface dimming. The dimming effect is enhanced further because aerosols may nucleate more cloud droplets that makes the clouds reflect more solar radiation. The dimming has a surface cooling effect and decreases evaporation of moisture from the surface, thus slowing down the hydrological cycle. On the other hand, absorption of solar radiation by BC and some organics increases atmospheric heating and tends to amplify greenhouse warming of the atmosphere. In addition to absorbing the reflected solar radiation, black carbon in ABCs absorbs the direct solar radiation and jointly the two processes contribute to a significant enhancement of lower atmosphere solar heating (Ramanathan and Feng, 2009).

ABCs are concentrated in regional and megacity hot spots (e.g., South Asia). Long-range transport from these hot spots causes widespread plumes over the adjacent oceans (e.g., Indian Ocean). Such a pattern of regionally concentrated surface dimming and atmospheric solar heating, accompanied by widespread dimming over the oceans, gives rise to large regional effects. In South Asia and North Africa, for example, the large north–south gradient in the ABC dimming has altered both the north–south gradients in sea surface temperatures and land–ocean contrast in surface temperatures, which in turn slow down the monsoon circulation and decrease the rainfall over the continents. On the other hand, heating by BC warms the atmosphere at elevated levels from 2 to 6 km, where most tropical glaciers are located, thus strengthening the effect of GHGs on retreat of snow packs and glaciers in the Hindu Kush–Himalaya–Tibetan glaciers. This has serious implications on the livelihood of millions of people, which is based on monsoon-fed agriculture such as in the case of India (Ramanathan and Feng, 2009).

Furthermore, in a recent study, Carmichael et al. (2009) present that vast regions in Asia have $PM_{2.5}$ concentrations that exceed, on an annual basis, the WHO guideline of 10 $\mu g/m^3$, often by factors of 2–4. Since the area covered by this study includes >80% of the population, such high aerosol loadings have serious health implications. Besides this, the PM pollution also has important radiative effects, causing a significant dimming at the surface and masking 45% of the warming by GHGs. Moreover, BC concentrations are also high throughout Asia, contributing significantly to atmospheric warming (its warming potential is 55% of that due to CO_2). Simulation studies by Carmichael et al. (2009) indicate that in 2030 the $PM_{2.5}$ levels in large parts of Asia will increase and exacerbate health impacts. However, the aerosols will have a larger masking effect on radiative forcing due to a decrease in BC and an increase in SO_2 emissions. Since the residence time for BC (days to weeks) is significantly less than for the major GHGs, Ramanathan and Carmichael (2008) suggest that focusing most of the particle control efforts on BC would provide additional time to address the longer-term issues of GHGs as well as immediately improve public health.

17.9 STRATOSPHERIC OZONE DEPLETION

Ozone depletion epitomizes the global environmental problems humans face: it is an unintended consequence of human activity. The way it was solved is a success story contributed by many: scientists, technologists, economic and legal experts, environmentalists, and policy makers (see, e.g., Rowland, 2006; WMO and UNEP, 2006; Molina and Molina, 2007).

The CFCs are industrial chemicals that have been used in the past as coolants for refrigerators and air conditioners, propellants for aerosol spray cans, foaming agents for plastics, and cleaning solvents for electronic components, among other uses. These chemicals are also known under trademarks such as Freon (DuPont) and Genetron (Allied Signal). Thomas Midgley (1937) invented the CFCs in the 1930s during a search for nontoxic substances that could be used as coolants in home refrigerators. The two important properties that make the CFCs commercially valuable are their volatility (they can be readily converted from liquid to vapor, and vice versa) and their chemical inertness (they are nontoxic, nonflammable, noncorrosive, and unreactive with most other substances). Because of their chemical inertness and stability, the CFCs were considered to be "miracle" compounds and soon replaced the toxic NH_3 and SO_2 as the standard cooling fluids. Subsequently, the CFCs found uses as propellants for aerosol sprays, blowing agents for plastic foam, and cleansers for electronic components. All this activity doubled the worldwide use of CFCs every

6–7 years and the annual industrial usage eventually reached about 700,000 metric tons by the early 1970s.

In 1973, Lovelock and coworkers (Lovelock et al., 1973) detected measurable levels of CFCs in the atmosphere over the South and North Atlantic. Molina and Rowland (1974) decided to investigate the ultimate atmospheric fate of these wonder compounds and concluded that the only significant sink was solar ultraviolet (UV) photolysis in the middle stratosphere (~25–30 km). The destruction of CFCs by solar radiation leads to the release of chlorine atoms, which participate in ozone destruction cycles. Depletion of stratospheric ozone leads to increases in the level of solar UV radiation reaching the Earth's surface predominantly at wavelengths ~290–320 nm, the so-called UV-B radiation. The mechanism responsible for the UV damage on humans and other biological species are connected with the UV absorption characteristics of DNA, which has a peak around 260 nm, but the spectrum extends beyond 320 nm. UV-B radiation can induce acute skin damage in humans, such as sunburn. Other potential risks of increased UV-B radiation on human health include increases in the morbidity and incidence of skin cancer, eye diseases, and infectious diseases (Molina et al., 2000).

Besides chlorine, bromine also plays an important role in stratospheric chemistry. There are industrial sources of brominated hydrocarbons as well as natural ones; the most important of them are the halons and methyl bromide (CH_3Br). The halons are fully halogenated hydrocarbons, produced industrially as fire extinguishers; examples are CF_3Br and CF_2ClBr. Methyl bromide is both natural and man-made; it is used as an agricultural fumigant. These sources release bromine to the stratosphere at pptv levels, compared with ppbv for chlorine. On the other hand, bromine atoms are about 60 times more efficient than chlorine atoms for ozone destruction on an atom-per-atom basis (WMO and UNEP, 2006). A large fraction of the bromine compounds is present as free radicals, because the temporary reservoirs are less stable and are formed at considerably slower rates than the corresponding chlorine reservoirs.

Various fundamental aspects of the CFC-ozone depletion hypothesis were verified in the late 1970s and early 1980s, following the initial publication of the Molina–Rowland article (Molina and Rowland, 1974). Measurements of the atmospheric concentrations of the CFCs indicated that they accumulate in the lower atmosphere and that they reach the stratosphere in the amounts predicted. On the other hand, a decrease in stratospheric ozone levels was not observable at that time because of the large natural variability of this species. However, the ozone levels in the Antarctic stratosphere dropped dramatically in the spring months starting in the early 1980s, as first reported by Farman and coworkers in 1985 (Farman et al., 1985). Subsequently it became evident that ozone was being depleted in the Northern Hemisphere as well, particularly at high latitudes and in the winter and spring months.

The depletion of ozone over Antarctica—the ozone hole—was not predicted by the atmospheric sciences community. However, the cause of this depletion became very clear in subsequent years: laboratory experiments, field measurements over Antarctica, and model calculations showed unambiguously that the ozone hole can indeed be traced to the industrial CFCs (see, e.g., Molina et al., 1985; Solomon et al., 1986; Molina and Molina, 1987; Molina et al., 1987; Tolbert et al., 1987; Tuck et al., 1989; Anderson et al., 1991; Russell et al., 1996; Rowland, 2006).

Observations and models confirmed that stratospheric sulfate aerosol (SSA) and polar stratospheric clouds (PSCs) play a key role in ozone loss chemistry through heterogeneous reactions that activate halogen species and deactivate nitrogen species. The atmosphere above the Arctic is not as cold as that above the Antarctic, so ozone depletion there is not as severe. Ozone depletion during the Arctic winter and spring is highly variable, due to changes in stratospheric meteorological conditions from one winter to another, as can be seen from the unexpected ozone losses over central Europe in the summer of 2005. A future Arctic ozone hole as severe as that of the Antarctic appears unlikely, but the population at risk from stratospheric ozone depletion in the Arctic is much higher than in the Antarctic (WMO and UNEP, 2006).

17.9.1 Montreal Protocol

Following the discovery in 1974 by Molina and Rowland that stratospheric ozone could be depleted as a consequence of the release of CFCs to the environment, the scientific community worldwide participated in a joint effort to better understand this threat to the ozone layer. Observations of the ozone layer itself showed that depletion was indeed occurring; the most dramatic loss was discovered over Antarctica. In response to the likelihood of increasing ozone depletion, in September 1987, an international agreement limiting the production of CFCs was approved under the auspices of the United Nations Environment Programme (UNEP). At that time, the expeditions to the Antarctic stratosphere were just beginning to gather the data that would eventually firmly link the CFCs to the depletion of polar ozone. This agreement, the Montreal Protocol on Substances that Deplete the Ozone Layer, initially called for a reduction of only 50% in the manufacture of CFCs by the end of the century. In view of the strength of the scientific evidence that emerged in the following years, the initial provisions were strengthened through the London (1990), Copenhagen (1992), Vienna (1995), Montreal (1997), and Beijing (1999) amendments. Figure 17.2 shows the projected abundance of ODS in the stratosphere according to the provisions of the Montreal Protocol and subsequent amendments. It is now widely regarded as one of the most effective multilateral environmental agreements in existence. The production of CFCs in industrialized countries was phased out at the end of 1995, and other compounds such as the halons, methyl bromide, carbon tetrachloride, and methyl chloroform (CH_3CCl_3) were also regulated. A funding mechanism to help developing countries meet the costs of complying with the protocol and with its subsequent amendments was also established (UNEP, 2007).

Although emissions of ODS have decreased over the last 20 years, it is estimated that the ozone layer over the Antarctic will not fully recover to pre-1980 Antarctic ozone levels until between 2060 and 2075, assuming full Montreal Protocol compliance (WMO and UNEP, 2006).

It is clear that continued decreases in ODS production and use, following the Montreal Protocol provisions, are important for ozone layer recovery. It is also important to realize that there are collateral benefits to the implementation of the Montreal Protocol. In addition to destroying the ozone layer, most ODS are potent GHGs. The Global Warming Potential (GWP) of CFCs, halons, and HCFCs are thousands

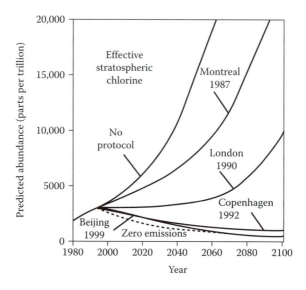

FIGURE 17.2   Effect of the international agreements on the predicted abundance of ozone-depleting substances (ODS) in the stratosphere in 1980–2100. [Adapted from UNEP (United Nations Environment Program), 2007.]

of times more than the most commonly know GHG, CO_2. The ODS phaseout has already provided, and is continuing to provide, significant climate protection benefits. On the other hand, climate change is projected to lead to stratospheric cooling, which, in turn, is predicted to enhance ozone concentrations in the upper stratosphere, but at the same time delay ozone recovery in the lower stratosphere. The extent of the stratospheric ozone–climate interaction is not yet well established; understanding the interconnections between ozone depletion and climate change is crucial for projections of future ozone abundances (WMO-UNEP, 2006; Molina and Molina, 2007; UNEP, 2007).

Although the Montreal Protocol has made considerable progress in the global drive to protect the ozone layer, the goal of achieving a total phaseout of ODS still faces several key challenges. Phasing out the production and use of methyl bromide is one challenge—although alternatives exist, replacement has been slow. Another challenge is the problem of illegal trade in ODS, mostly for servicing air conditioning and refrigeration, which is widespread in the developing world, especially in the Asia Pacific region. Effective control mechanisms for new chemicals threatening the ozone layer are essential; continued monitoring of the ozone layer is needed to ensure that the healing process is taking its expected course (UNEP, 2007).

17.10 FUTURE OUTLOOK: CHALLENGES AND OPPORTUNITIES

The accumulation and dispersion of pollutants, such as tropospheric ozone and airborne PM, not only can affect human health and the ecosystem on a local and regional scale, but also can influence air quality and the Earth's climate on a global scale. In turn, climatic parameters such as temperature and humidity can affect the sources, chemical transformation, transport, and deposition of pollutants (see, e.g., Jacob and Winner, 2008). The net impact of concurrent changes might be difficult to assess. Therefore it is important to integrate air quality and climate stabilization goals in the design of environmental policy to realize potential synergistic benefits and to ensure that actions undertaken to reduce GHG emissions do not result in unintended consequence with regard to air quality or vice versa.

During the last decade, concern for environmental issues has increased greatly. Almost all countries have created environmental institutions and developed new environmental laws and regulations, with some successes in both developed and developing countries, but major problems remain. Millions of people are still being exposed to harmful levels of air pollutants that are caused mainly by emissions from the combustion of fossil fuels in motor vehicles and from industrial processes, heating, and electricity generation. In principle, the problem can be solved to a large extent through the use of clean technologies, and the economic benefits associated with reduced impacts have far outweighed the costs of action. In practice, there are large socioeconomic and political barriers to the transition to new technologies. Furthermore, other tenacious problems associated with unrestrained urban growth, such as traffic congestion, are exacerbating air pollution throughout the world. In addition, there is also geopolitical and socioeconomic dimensions associated with the shift in pollution sources from the developed to developing countries. For example, during the last few decades, important changes have occurred in the nature of industrial production around the world with most of the low-skill-level manufacturing moving from the developed countries to developing economies where lower-cost labor and less stringent emissions standards have provided reduced production costs. Even with the transportation charges added to the production costs, the goods can be sold in Europe and North America for lower prices. Thus, a shift in emissions is occurring from the developed to developing countries. Besides this, new sources of air pollution have emerged as deliberate emissions of toxic, biological, or radioactive material due to war or terrorist acts. Thus, in the future, there are a changing array of possible threats to air quality resulting in new risks for human and environmental health and climate (Hopke, 2009).

Pollution may also be remedied on regional and global scales. At the regional level, examples include the Convention on Long-Range Transboundary Air Pollution (UNECE, 1979–2005), the ASEAN Haze Agreement (ASEAN, 2003), and the Malé Declaration on the Control and Prevention of Air Pollution in South Asia (UNEP/ RRC-AP, 2006).

At the global scale, the Montreal Protocol on Substances that Deplete the Ozone Layer is a model example of international cooperation to protect the environment. Existing mechanisms to tackle ODS are adequate; continued monitoring of the ozone layer is needed to ensure that the healing process is taking its expected course. Unfortunately, recent efforts to imitate it in the area of climate change have not enjoyed comparable success. These represent threats of a magnitude as great as any that have attracted public attention, but their effects take much longer to detect, the control measures are more costly and threatening to special interests, and the causes are multiple and much less easily traced to any jurisdiction. In a situation such as this, one must apply the precautionary principle. If unexpected impacts become more apparent later, it will be more difficult and more costly to reduce the atmospheric concentrations.

A major step forward in controlling GHGs was taken in 1992, when the UN Framework Convention on Climate Change (UNFCCC) was signed. Subsequently Kyoto Protocol was adopted in 1997 and entered into force in 2005, which sets binding targets for 37 industrialized countries and the European community for reducing emissions for the main long-lived GHGs including CO_2, CH_4, and N_2O. This amounts to an average of 5% against 1990 levels over the five-year period 2008–2012. By the end of the first commitment period of the Kyoto Protocol in 2012, a new international framework needs to have been negotiated and ratified that can deliver the stringent emission reductions suggested by IPCC. They were also the focus of intense negotiations during COP-15 in Copenhagen in December 2009. However, the meeting produced only a non-binding agreement, which limits the rise in global temperatures to 2°C. The developed countries also commit to raise US$ 100 billion annually by 2020 to help developing countries combat climate change. However, the Copenhagen Accord does not specify caps on emissions to achieve that objective (UNFCCC, 2009).

In summary, it is now clear that human activities can lead to serious environmental problems not just on a local, but also on a global scale (Fenger, 2008). One of the key steps in any rational approach to addressing regional and global environmental issues is to promote the widespread understanding that all of our human problems are interconnected. Regional and international cooperation will be essential to the solution of environmental problems. Strong involvement of stakeholders at all levels, changes in human behavior, coupled with suitable mechanisms for facilitating technological and financial flows, and the strengthening of human and institutional capacities will be crucial to the future success of efforts to improve air quality and mitigate climate change.

REFERENCES

Ackerman A.S., Toon O.B., Stevens D.E., et al. 2000. Reduction of tropical cloudiness by soot. *Science* 288(5468): 1042–1047.

Akimoto H. 2003. Global air quality and pollution. *Science* 302: 1716–1719.

Anderson J.G., Toohey D.W., and Brune W.H. 1991. Free radicals within the Antarctic Vortex: The role of CFCs in Antarctic ozone loss. *Science* 251: 39–46.

Arey J. 2000. Urban air: Causes and consequences of urban air pollution. In *Environmental Medicine*, L. Moller, ed., pp. 52–71. Joint Industrial Safety Council Product 33, Sweden.

ASEAN. 2003. *ASEAN Haze Agreement*. The Association of South East Asian Nations, Jakarta http://www.aseansec.org/10202.htm (last accessed May 25, 2009)

Aunan K., Fang J.H., Hu T., et al. 2006. Climate change and air quality—measures with co-benefits in China. *Environ. Sci. Technol.* 40: 4822–4829.

Aunan K., Fang J.H., Vennemo, H., et al. 2004, Co-benefits of climate policy—lessons learned from a study in Shanxi, China. *Energy Policy* 32: 567–581.

Bergin M.H., Cass G.R., Xu, J., et al. 2001. Aerosol radiative, physical, and chemical properties in Beijing during June 1999. *J. Geophys. Res.* 106: 17969–17980.

Berntsen T.K., Karlsdottir S., and Jaffe D.A. 1999. Influence of Asian emissions on the composition of air reaching the Northwestern United States. *Geophys. Res. Lett.* 26: 2171–2174.

CARB (California Air Resources Board). 1999. Clean Air—California's Success and Future Challenges; updated August 1999, http://www.arb.ca.gov/omb/50thfinal/50thfinal.html (last accessed May 24, 2009).

Carmichael G.R., Adhikari B., Kulkami S., et al. 2009. Asian Aerosols: Current and Year 2030 Distributions and Implications to Human Health and Regional Climate Change. *Environ. Sci. Technol.* Article ASAP, doi:10.1021/es8036803. Publication Date (Web): July 9.

Chameides W.L., Xingsheng L., Xiaoyan T., et al. 1999a. Is ozone pollution affecting crop yields in China? Geophys. *Res. Lett.* 26: 867–870.

Chameides W.L., Hu H., Liu S.C., et al. 1999b. Case study of the effects of atmospheric aerosols and regional haze on agriculture: An opportunity to enhance crop yields in China through emission controls? *Proc. Natl. Acad. Sci. USA* 26: 13626–13633.

Cifuentes L., Borja-Aburto V.H., Gouveia N., et al. 2001. Climate change: Hidden health benefits of greenhouse gas mitigation. *Science* 293: 1257–1259.

Cohan D.S., Xu J., Greenwald, R., et al. 2002. Impact of atmospheric aerosol light scattering and absorption on terrestrial net primary productivity. *Global Biogeochem. Cycles* 16: doi:10.1029/2001GB001441.

Dunlea E.J., DeCarlo P.F., Aiken A.C., et al. 2008. Evolution of Asian aerosols during transpacific transport in INTEX-B. *Atmos. Chem. Phys.* 9: 7257–7287.

Edwards D.P., Lamarque J.F., Attie J.L., et al., 2003. Tropospheric ozone over the tropical Atlantic: A satellite perspective. *J. Geophys. Res.* 108(D8), 4237, doi:10.1029/2002JD002927.

Emberson L., Ashmore M., and Murray F. (Eds). 2003. *Air Pollution Impacts on Crops and Forests—a Global Assessment*. Imperial College Press, London.

Farman J.C., Gardiner B.G., and Shanklin J.D. 1985. Large losses of total ozone in Antarctica reveal seasonal ClO_x/NO_x interactions. *Nature* 315: 207–210.

Farrell A.E., Plevin R.J., Turner B.T., et al. 2006. Ethanol can contribute to energy and environmental goals. *Science* 311: 506–508.

Fenger J. 2008. Air pollution in the last 50 years—from local to global. *Atmos. Environ.* doi:10.1016/j.atmosenv.2008.09.061.

Fenn M.E., Bauer L.I., and Hernandez-Tejeda T. (Eds). 2002. *Urban Air Pollution and Forests*. Springer, New York.

Fiore A.M., Jacob D.J., Field B.D., et al. 2002. Linking ozone pollution and climate change: The case for controlling methane. *Geophys. Res. Lett.* 29: 1919.

Fiore A.M., West J.J., Horowitz L.W., et al. 2008. Characterizing the tropospheric ozone response to methane emission controls and the benefits to climate and air quality. *J. Geophys. Res.* 113: D08307, doi:10.1029/2007JD009162.

Galloway J.N. 1996. Anthropogenic mobilization of sulphur and nitrogen: Immediate and delayed consequences. *Ann. Rev. Energy Environ.* 21: 261–292.

Gregg J.W., Jones C.G., and Dawson T.E. 2003. Urbanization effects on tree growth in the vicinity of New York City. *Nature* 424: 183–187.

Guttikunda S.K., Carmichael G.R., Calori G., et al. 2003. The contribution of megacities to regional sulfur pollution in Asia. *Atmos. Environ.* 37: 11–22.

Holland M., Kinghorn, S., Emberson, L., et al., 2006. Development of a framework for probabilistic assessment of the economic losses caused by ozone damage to crops in Europe. Part of the UNECE International Cooperative Programme on Vegetation. Contract Report EPG 1/3/205. CEH Project No. CO2309NEW, 49pp.

Hopke P.K. 2009. Contemporary threats and air pollution. *Atmos. Environ.* 43: 87–93, doi:10.1016/j.atmosenv.2008.09.053.

Husar R.B., Tratt D.M., Schichtel B.A., et al. 2001. Asian dust events of April 1998. *J. Geophys. Res.* 106: 18317–18330.

IPCC (Intergovernmental Panel on Climate Change). 2001. *Climate Change 2001: The Scientific Basis. Intergovernmental Panel on Climate Change*. Cambridge University Press, Cambridge.

IPCC (Intergovernmental Panel on Climate Change). 2007. *Climate Change 2007: The Physical Science Basis, Contribution of Working Group I to the Fourth Assessment Report of the Intergovernmental Panel on Climate Change*, S. Solomon, D. Qin,

M. Manning, Z. Chen, M. Marquis, K.B. Avery, M. Tignor, and H.L. Miller (Eds), 996pp. Cambridge University Press, Cambridge, United Kingdom and New York, NY, USA.

IPCC-SPM. 2007. Climate Change 2007: Synthesis Report Summary for the Policy Makers, www.ipcc.ch/pdf/assessment-report/ar4/syr/ar4_syr_spm.pdf (accessed May 25, 2009).

IPCC-WG1. 2007. IPCC Working Group I Fourth Assessment Report Climate Change 2007: The Physical Science Basis Summary for Policymakers, http://ipcc-wg1.ucar.edu/wg1/docs/WG1AR4_SPM_PlenaryApproved.pdf (accessed May 25, 2009).

Jacob D.J., Logan J.A., and Murti P.P. 1999. Effect of rising Asian emissions on surface ozone in the United States. *Geophys. Res. Lett.* 26: 2175–2178.

Jacob D.J. and Winner D.A. 2008. Effect of climate change on air quality. *Atmos. Environ.* 43: 51–63, 2009.

Jacobson M.Z. 2007. Effects of ethanol (E85) versus gasoline vehicles on cancer and mortality in the United States. *Environ. Sci. Technol.* 41: 4150–4157.

Jaffe D., Bertschi I., Jaegle L., et al. 2004. Long-range transport of Siberian biomass burning emissions and impact on surface ozone in western North America. *Geophys. Res. Lett.* 31: L16106, doi: 10.1029/2004GL020093.

Kaiser D.P. and Qian Y. 2002. Decreasing trends in sunshine duration over China for 1954–1998: Indication of increased haze pollution? *Geophys. Res. Lett.* 29: doi:10.1029/2002GL016057.

Lelieveld J., Crutzen P.J., Ramanathan V., et al. 2001. The Indian Ocean Experiment: Widespread air pollution from South and Southeast Asia. *Science* 291: 1031–1036.

Lovelock J.E., Maggs R.J., and Wade R.J. 1973. Halogenated hydrocarbons in and over the Atlantic. *Nature* 241: 194–196.

McKinley G., Zuk M., Hojer M., et al. 2005. Quantification of local and global benefits from air pollution control in Mexico City. *Environ. Sci. Technol.* 39: 1954–1961.

Menon S., Hansen J.E., Nazarenko L., and Luo Y. 2002. Climate effects of black carbon aerosols in China and India. *Science* 297: 2250–2252.

Midgley T. 1937. From the Periodic Table to production. *Ind. Eng. Chem.* 29: 241–244.

Molina L.T., Kolb C.E., de Foy B., et al. 2007. Air quality in North America's most populous city—overview of the MCMA-2003 campaign. *Atmos. Chem. Phys.* 7: 2447–2473.

Molina L.T., Madronich S., Gaffney J.S., and Singh H.B. 2008. Overview of MILAGRO/INTEX-B Campaign. *IGAC Newslett.* 38: 2–15.

Molina L.T. and Molina M.J. 1987. Production of Cl_2O_2 from the self-reaction of the ClO radical. *J. Phys. Chem.* 91: 433–436.

Molina L.T. and Molina M.J. (Eds). 2002. *Air Quality in the Mexico Megacity: An Integrated Assessment*. Kluwer Academic Publishers, Dordrecht, The Netherlands, 384pp.

Molina L.T., Madronich S., Gaffney J., et al. 2010. An overview of the MILAGRO 2006 campaign: Mexico City emissions and their transport and transformation, *Atmos. Chem. Phys. Discussion*, 7819–7983.

Molina L.T., Molina M.J., Stachnick R.A., and Tom R.D. 1985. An upper limit to the rate of the $HCl + ClONO_2$ reaction. *J. Phys. Chem.* 89: 3779–3781.

Molina M.J. and Molina L.T. 2004. 2004 Critical Review: Megacities and atmospheric pollution. *J. Air Waste Manage.* 54: 644–680.

Molina M.J. and Molina L.T. 2007. Chlorofluorocarbons and destruction of the ozone layer. In *Environmental and Occupational Medicine*, Fourth edition, W.N. Rom (Ed.), pp. 1605–1615. Lippincott, Williams and Wilkins, Philadelphia.

Molina M.J. and Rowland F.S. 1974. Stratospheric sink for chlorofluoromethanes: Chlorine-atom catalyzed destruction of ozone. *Nature* 249: 810–812.

Molina M.J., Tso T.-L., Molina L.T. and Wang F.C.-Y. 1987. Antarctic stratospheric chemistry of chlorine nitrate, hydrogen chloride and ice. Release of active chlorine. *Science* 238: 1253–1260.

Molina M.J., Molina L.T., Fitzpatrick T.B., and Nghiem P.T. 2000. Ozone depletion and human health effects. In *Environmental Medicine*, L. Moller (Ed.). Joint Industrial Safety Council Product 33, Sweden.

Molina M.J., Molina L.T., Slott R., et al. 2004. 2004 Critical Review Supplement: Air Quality in Selected Megacities, http://www.awma.org

NASA. 2009. Arctic Research of the Composition of the Troposphere from Aircraft and Satellites (ARCTAS), http://www.espo.nasa.gov/arctas/ (accessed May 25, 2009).

NAPAP. 1990. Integrated Assessment Report, National Acid Precipitation Assessment Program, Washington, DC.

NRC. 1991. *Rethinking the Ozone Problem in Urban and Regional Air Pollution.* National Academy Press, Washington, DC.

Phoenix G.K., Hicks W.K., Cinderby S., et al. 2006. Atmospheric nitrogen deposition in world biodiversity hotspots: The need for a greater global perspective in assessing N deposition impacts. *Global Change Biol.* 12: 470–476.

Pimentel D. and Patzek T.W. 2005. Ethanol production using corn, switchgrass, and wood: Biodiesel production using soybean and sunflower, *Nat. Resour. Res.* 14: 67–76.

Qian Y. and Giorgi F. 2000. Regional climatic effects of anthropogenic aerosols? The case of southwestern China. *Geophys. Res. Lett.* 27: 3521–3524.

Ramanathan V. and Carmichael G. 2008. Global and regional climate changes due to black carbon. *Nat.Geosci.* 1: 221–227.

Ramanathan V., Crutzen P.J., Kiehl J.T., and Rosenfeld D. 2001. Aerosols, climate, and the hydrological cycle. *Science* 294: 2119–2124, doi:10.1126/science.1064034.

Ramanathan V. and Feng Y. 2008. On avoiding dangerous anthropogenic interference with the climate system: Formidable challenges ahead. *Proc. Natl. Acad. Sci.*, 105, 14245–14258. www.pnas.org_cgi_doi_10.1073_pnas.0803838105

Ramanathan V. and Ramana, M.V. 2003. Atmospheric Brown Clouds. *Environ. Manage.* December 2003, 28–33.

Rosenfeld D. 2000. Suppression of rain and snow by urban and industrial air pollution. *Science* 287: 1793–1796.

Rowland F.S., 2006. Stratospheric ozone depletion, *Trans. R. Soc. B.* 361: 769–790, doi:10.1098/rstb.2005.1783.

Russell J.M. III, Luo M., Cicerone R.J., and Deaver L.E. 1996. Satellite confirmation of the dominance of chlorofluorocarbons in the global stratospheric chlorine budget. *Nature* 379: 526–529.

Singh H.B., Brune W.H., Crawford, J.H., et al. 2009. Chemistry and transport of pollution over the Gulf of Mexico and the Pacific: Spring 2006 INTEX-B campaign overview and first results. *Atmos. Chem. Phys.* 9: 2301–2318.

Solomon S., Garcia R.R., Rowland F.S., and Wuebbles D.J. 1986. On the depletion of Antarctic ozone. *Nature* 321: 755–758.

Stevens C.J., Dise N.B., Mountford J.O., and Gowing D.J. 2004. Impact of nitrogen deposition on the species richness of grasslands. *Science* 303: 1876–1879.

Stohl A., Forster C., Huntrieser H., et al., 2007. Aircraft measurements over Europe of an air pollution plume from Southeast Asia—aerosol and chemical characterization. *Atmos. Chem. Phys.* 7: 913–937.

Tolbert M.A., Rossi M.J., Malhotra R., and Golden D.M. 1987. Reaction of chlorine nitrate with hydrogen chloride and water at Antarctic stratospheric temperatures. *Science* 238: 1258–1260.

Tuck A.F., Watson R.T., Condon E.P., Margitan J.J., and Toon O.B. 1989. The planning and execution of ER-2 and DC-8 aircraft flights over Antarctica, August and September 1987. *J. Geophys. Res.* 94: 11, 181–222.

UN (Department of Economic and Social Affairs, Population Division). 2008. United Nations expert group meeting on population distribution, urbanization, internal migration and development. Working Paper No. UN/POP/EGM-URB/2008/01, United Nations, New York.

UN (Department of Economic and Social Affairs, Population Division). 2009. World Population Prospects: The 2008 Revision Highlights, Working Paper No. ESA/P/WP.210. United Nations, New York.

UNECE. 1979–2005. The Convention on Long-range Transboundary Air pollution website. United Nations Economic Commission for Europe, Geneva, http://unece.org/env/lrtap/lrtap_h1.htm (last accessed July 5, 2009).

UNEP (United Nations Environment Program). 2007. *Global Environmental Outlook (GEO-4): Environment for Development*. Progress Press Ltd., Malta.

UNECE (United Nations Economic Commision for Europe) website: http://www.unece.org/env/lrtap/welcome.html (accessed 5 July 2009).

UNEP/RRC-AP. 2006. Malé Declaration on the Control and Prevention of Air Pollution in South Asia and its Likely Transboundary Effects. UNEP Regional Resource Centre for Asia and the Pacific, Bangkok, http://www.rrcap.unep.org/issues/air/maledec/baseline/indexpak.html (last accessed July 5, 2009).

VanCuren R.A., Cliff S.S., Perry K.D., and Jimenez-Cruz M. 2005. Asian continental aerosol persistence above the marine boundary layer over the eastern North Pacific: Continuous aerosol measurements from Intercontinental Transport and Chemical Transformation 2002 (ITCT 2K2), *J. Geophys. Res.*, 110, D09S90, doi:10.1029/2004JD004973.

Watson J.G. 2002. Visibility: Science and regulation. *J. Air Waste Manage.* 52: 628–713.

West J.J. and Fiore A.M. 2005. Management of tropospheric ozone by reducing methane emissions. *Environ. Sci. Technol.* 39: 4685–4691.

WMO (World Meteorological Organization). 1995. Scientific assessment of ozone depletion: 1994. WMO Global Ozone Research and Monitoring Project, Report No. 37. WMO, Geneva.

WMO (World Meteorological Organization). 1999. Scientific assessment of ozone depletion: 1998. WMO Global Ozone Research and Monitoring Project, Report No. 44. WMO, Geneva.

WMO (World Meteorological Organization). 2003. Scientific assessment of ozone depletion: 2002, Global research and monitoring project, Report No. 47, WMO, Geneva.

WMO and UNEP. 2006. Executive Summary of the Scientific Assessment of Ozone Depletion: 2006. Scientific Assessment Panel of the Montreal Protocol on Substances that Deplete the Ozone Layer, Geneva and Nairobi, http://ozone.unep.org/Publications/Assessment_Reports/2006/Scientific_ Assessment_2006_Exec_Summary.pdf (last accessed May 25, 2009).

Yamasoe M.A., von Randow C., Manzi A.O., et al. 2006. Effect of smoke and clouds on the transmissivity of photosynthetically active radiation inside the canopy. *Atmos. Chem. Phys.* 6: 1645–1656.

Yang D., Yan P., and Xu X. 2002. Characteristics of aerosols under dust and sand weather in Beijing. *J. Appl. Meteorol.* 13: 185–194.

Yienger J.J., Galanter M., Holloway T.A., et al. 2000. The episodic nature of air pollution transport from Asia to North America. *J. Geophys. Res.* 105: 26931–26945.

Zhang L., Jacob D.J., Boersma K.F., et al. 2008. Transpacific transport of ozone pollution and the effect of recent Asian emission increases on air quality in North America: An integrated analysis using satellite, aircraft, ozonesonde, and surface observations. *Atmos. Chem. Phys.* 8: 6117–6136.

Zhao T.L., Gong S.L., Zhang X.Y., and Jaffe D.A. 2008. Asian dust storm influence on North American ambient PM levels: Observational evidence and controlling factors, *Atmos. Chem. Phys.*, 8, 2717–2728.

Index

A